# Aflatoxigenic Fungi

Aflatoxigenic fungi, primarily belonging to the *Aspergillus* genus (*A. flavus*, *A. parasiticus*), are responsible for producing aflatoxins—highly toxic and carcinogenic secondary metabolites. These fungi pose significant risks to plant health, animal health, and human well-being due to their widespread contamination of food and feed. Aflatoxigenic fungi impact plants by causing diseases, and yield loss leads to economic loss and contamination. In connection with animals, aflatoxigenic fungi create toxicity in livestock, contamination of animal products, mortality, and economic loss. Finally, concerning humans, aflatoxigenic fungi lead to acute toxicity, carcinogenic effects, immune suppression and nutritional risk.

**Features:**

- Provides a systemic review of the recent developments in aflatoxigenic fungi on various biotic organisms.
- Includes an overview of the impact of aflatoxins on various plants and on the agricultural economy.
- Describes the role of aflatoxins in human health, including infants, and in cancer.
- Contains special emphasis on the control of aflatoxigenic fungi by nanomaterials and plant bioactive components.
- Discusses the detection of aflatoxigenic fungi by surface-enhanced Raman spectroscopy (SERS).

# Aflatoxigenic Fungi

## Its Impact on Plant, Animal and Human Health

Edited by
A. Sankara Narayanan, S. Gomathi,
V. Ambikapathy, and A. Panneerselvam

CRC Press
Taylor & Francis Group
Boca Raton London New York

CRC Press is an imprint of the
Taylor & Francis Group, an **informa** business

Designed cover image: Shutterstock images

First edition published 2026
by CRC Press
2385 NW Executive Center Drive, Suite 320, Boca Raton FL 33431



ISBN: 978-1-032-78262-1 (hbk)
ISBN: 978-1-032-78264-5 (pbk)
ISBN: 978-1-003-48704-3 (ebk)

DOI: 10.1201/9781003487043

Typeset in Times
by KnowledgeWorks Global Ltd.

# Contents

## SECTION 4 Advancement in the Detection of Aflatoxigenic Fungi

# Preface

*Aflatoxigenic Fungi: Its Impact on Plant, Animal and Human Health* is an edited book that focuses mainly on the current research trends on aflatoxigenic fungi and its impact on plant, animal, and human health.

Aflatoxigenic fungi produce aflatoxins, a class of mycotoxins (toxic compounds produced by fungi) that can have significant detrimental effects on living organisms, including humans, animals, and plants. These fungi are primarily members of the genus *Aspergillus*, especially *A. flavus* and *A. parasiticus*. They are commonly found in agricultural products such as corn, peanuts, and tree nuts, where they grow under favorable conditions (such as high humidity and temperature). Aflatoxins are known to be potent carcinogens, mutagens, and immunosuppressive agents.

These fungi cause various consequences in the living system, especially concerning humans, plants, and animals. This includes carcinogenic effects, acute aflatoxicosis, and immunosuppression in human beings. It creates various health hazards in animals, including liver damage, reproductive issues, and death, from various sources, including animal feed and fish feed. The agricultural impacts include crop losses, soil and plant health, and persistence in the environment, a notable point in this context. This book also addresses the recent advancement in the detection of aflatoxigenic fungi by various high-end techniques and its control by various methods, including nanomaterials and plant bioactive components.

This book includes four sections: Section 1: Overview of Aflatoxigenic Fungi in Plant, Animal, and Human health; Section 2: Aflatoxigenic Fungi: Impact on Soil, Plant, and Human Health, and the Agricultural Environment; Section 3: Control of Aflatoxigenic Fungi; and Section 4: Advancement in the Detection of Aflatoxigenic Fungi. Thirty-three chapters have been contributed by authors with expertise from different parts of the world, including Mexico, China, Turkey, Bulgaria, Cameroon, and India.

The consequences of aflatoxigenic fungi on living systems are significant, with major impacts on human and animal health, agricultural productivity, and the environment. Effective prevention and control measures are critical for minimizing these impacts and protecting public health. Further, the editors hope the book may be a good resource for researchers and post-graduate students working in the area of aflatoxigenic fungi and a reference source for beginners in the field.

**Dr. A. Sankara Narayanan**
**Dr. S. Gomathi**
**Dr. V. Ambikapathi**
**Dr. A. Panneerselvam**
*- Editors*
*Aflatoxigenic Fungi: Its Impact on Plant, Animal and Human Health, CRC Press, Taylor and Francis Publications*

# Editor Biographies

**Dr. A. Sankara Narayanan** has been an Associate Professor in the Department of Life Sciences, School of Science, GSFC University, Vadodara of Gujarat state, India, since July 2024. His current research focus is on microbiome studies, fermented food products, and environmental microbiology. He has published 20 books with different international publishers, 70 chapters, and 80 research articles in international and national journals of repute, guided 5 PhDs, 16 MPhil scholars, and operated 5 funded projects in microbiology. From 2002 to 2015, he worked as an Assistant Professor and Head, Department of Microbiology, K.S.R. College of Arts and Science, Tiruchengode, Tamil Nadu, and from August 2015 to May 2021, he was associated with Uka Tarsadia University, Surat, Gujarat, India. From June 2021 to June 2024, he worked as an Associate Professor and Head in the Department of Life Sciences, Sri Sathya Sai University for Human Excellence (SSSUHE), Kalaburagi, India. He has been awarded the Indian Academy of Sciences (IASc), National Academy of Sciences (NAS), and the National Academy of Sciences (TNAS) sponsored summer research fellowship for young teachers consecutively for three years, and his name is included as a Mentor in the DST-Mentors/Resource persons for summer/winter camps and other INSPIRE initiatives, Department of Science and Technology, Government of India, New Delhi. He is a grant reviewer for the British Society for Antimicrobial Chemotherapy (BSAC), UK. He is an active member of the editorial boards of various national and international journals. He is a Doctoral Committee member, Board of Study member in Microbiology in various colleges and universities in India, and a reviewer for many reputed international journals. One of his edited books received the Outstanding Academic Title (OAT) Award from an international publisher in 2020. This OAT Award represents the highest caliber of scholarly titles and overall excellence in presentation and importance relative to other literature in the field. His name was included in the World Scientists and University Rankings for 2023; this ranking list is released by AD Scientific Index.

**Dr. S. Gomathi** is an Assistant Professor in the Department of Botany and Microbiology, A.V.V.M. Sri Pushpam College (Autonomous), Poondi, Thanjavur District, Tamil Nadu, India. She has approximately 15 years of experience in teaching and research. Her area of specialization is mycology, medicinal plant diversity, and antifungal activity studies.

She has published more than 25 papers in internationally reputed journals and 15 chapters in books from internationally reputed publishers. She has guided 10 MPhil and 2 PhD scholars, and PhD scholars are working under her guidance now at Bharathidasan University, Tiruchirappalli, India. She is involved in the organization of various seminars and workshops, and acts as a resource person and participant in various workshops to enhance her research skills. She has filed and published two Indian patents. She is a life member of various professional and scientific bodies. She has participated in various orientation programs and shared her expertise with young students as a resource person.

**Dr. V. Ambikapathy** is an Associate Professor and Head, Department of Botany and Microbiology, A.V.V.M. Sri Pushpam College (Autonomous), Poondi, Thanjavur District, Tamil Nadu, India. He has 29 years of experience in teaching and research. He has specialized in the fields of mycology and plant pathology. He has visited the Udayana University, Bali, Indonesia. He received the INSA visiting scientist fellow award and has filed two patents.

He has deposited 92 gene sequences in GenBank, conducted 4 metagenome bioprojects, and maintains more than 500 microbial cultures in his laboratory. He is a reviewer for reputed journals and a doctoral committee member for more than 50 research scholars working at Bharathidasan University. He has published 149 research and review articles, 8 books, and 27 book chapters. He has guided 21 PhD scholars, 32 MPhil scholars, and organized several symposia, seminars, workshops, DST-INSPIRE programs for "plus one" students, YSSP programs, and so on. Dr Ambikapathy has conducted more than 50 training programs on edible mushroom cultivation technology for the welfare of women under a self-help group scheme. He has received several grants from various Indian funding agencies such as UGC-SERO, DST-INSPIRE, NMPB, UGC, and Tamil Nadu State Council for Science and Technology (TNSCST), India.

**Prof. A. Panneerselvam**, D.Sc., is an Associate Professor and Head (Retd), Department of Botany and Microbiology, A.V.V.M. Sri Pushpam College (Autonomous), Poondi, Thanjavur District, Tamil Nadu, India. He has 40 years of experience in teaching and research. His areas of specialization are mycology, plant pathology, and actinobacteriology. He has visited the National University of Singapore, Nanyang Technical University, TEMSEK Laboratory Singapore, Malaysian Agricultural University, Kuala Lumpur, Malaysia, Bangkok, Udayana University, Bali, Indonesia, the Institute of Fundamental and Applied Research, National Research University (TIIAME), Tashkent, Uzbekistan, and national-level institutes in India.

He has received 15 awards, including an INSA visiting fellowship, New Delhi; Member of Research Board of Advisors; Excellent Faculty award in Botany; Man of Year Commemorative Model; Life Time Achievement Award 2022; Senior Scientist Award 2023; and the MSI Fellow Award 2024. He has deposited 191 nucleotide sequences in GenBank, conducted 4 metagenome bioprojects, and maintains more than 500 microbial cultures in his laboratory. He is a reviewer for reputed journals and a doctoral committee member for 127 research scholars working at Bharathidasan University. He has published 476 research papers and review articles, 23 popular articles, and chapters in more than 12 books. He has guided 43 PhD scholars, 67 MPhil scholars, and organized several symposia, seminars, workshops, DST-INSPIRE programs for "plus one" students, the YSSP program, and so on. He has conducted more than 100 training programs on edible mushroom cultivation technology for the welfare of women under a self-help group scheme, Dissemination of Innovative Technology (DIT). He has received grants from various funding agencies such as the Department of Science and Technology (DST), the University Grants Commission (UGC), the Council of Scientific and Industrial Research (CSIR), the Tamil Nadu State Council for Science and Technology (TNSCST), the National Medicinal Plants Board (NMPB), and the District Rural Development Agency (DRDA). He is a life member of various academic bodies and an editorial board member of reputed journals. He trained a women's self-help group on edible mushroom cultivation technology under various schemes. He has also been awarded a D.Sc. by the Periyar University, Salem, Tamil Nadu of India.

# Contributors

**Juan Salvador Acevedo-Gil**
Plant Protection Laboratory
Department of Microbiology, Escuela Nacional de Ciencias Biológicas
Instituto Politécnico Nacional
Mexico City, Mexico

**Anandhi Anandaraj**
Centre for Drug Discovery and Development
Sathyabama Institute of Science and Technology
Chennai, Tamil Nadu, India

**K. Ananthi**
PG and Research Department of Microbiology
Dhanalakshmi Srinivasan College of Arts and Science for Women (Autonomous)
Perambalur, Tamil Nadu, India

**K. Anuradha**
Department of Microbiology
Bhavan's Vivekananda College of Science, Humanities & Commerce
Secunderabad, Telangana, India

**K. Aravindhan**
Department of Biotechnology
Microbiology & Bioinformatics, National College (Autonomous)
Tiruchirappalli, Tamil Nadu, India

**Gabriel Betanzos-Cabrera**
Universidad Autónoma del Estado de Hidalgo
Mexico

**Juan Carlos Cancino-Diaz**
Department of Microbiology
Escuela Nacional de Ciencias Biológicas
Instituto Politécnico Nacional
Mexico

**Mario E. Cancino-Diaz**
Escuela Nacional de Ciencias Biológicas
Instituto Politécnico Nacional
Mexico

**Peng Chen**
School of Pharmacy
Lanzhou University
Gansu, PR China

**Sanchita Choubey**
Dr. D. Y. Patil Arts Commerce and Science College
Pimpri-Chinchwad, Maharashtra, India

**Blessy Cleatus**
Department of Medical Laboratory Technology
Sathyabama Institute of Science and Technology
Chennai, Tamil Nadu, India

**Juan Antonio Cruz-Maya**
Unidad Profesional Interdisciplinaria en Ingeniería y Tecnologías Avanzadas
Instituto Politécnico Nacional
Mexico City, Mexico

**Priyanka Dahiya**
Plant-Microbe Interaction Laboratory, Department of Microbiology
Maharshi Dayanand University
Rohtak, Haryana, India

**Amita Suneja Dang**
Centre for Medical Biotechnology
Maharshi Dayanand University
Rohtak, Haryana, India

**Miguel A. De la Rosa-Ramos**
Universidad Nacional Autónoma de México
Mexico

**S. Manju Devi**
Department of Microbiology
Bhavan's Vivekananda College of Science, Humanities & Commerce
Secunderabad, Telangana, India

**S. Shalini Devi**
Department of Microbiology
Bhavan's Vivekananda College of Science, Humanities & Commerce
Secunderabad, Telangana, India

**S. Uma Devi**
Department of Botany
Sree Ayyappa College for Women
Nagercoil, Tamil Nadu, India

**Liqian Du**
School of Pharmacy
Lanzhou University
Gansu, PR China

**Kajal Gaba**
Molecular Diagnostics Laboratory, Centre for Medical Biotechnology
Maharshi Dayanand University
Rohtak, Haryana, India

**Marieta Petrova Georgieva**
Faculty of Pharmacy, Department of Pharmacology, Toxicology and Pharmacotherapy
Medical University – Varna Prof. Dr. Paraskev Stoyanov
Bulgaria, Varna

**Difo Voukang Harouna**
Department of Biological Sciences, Faculty of Science
University of Maroua
Cameroon

**Janet Jan-Roblero**
Department of Microbiology, Escuela Nacional de Ciencias Biológicas
Instituto Politécnico Nacional
Mexico City, Mexico

**K.R. Jeya**
Department of Biotechnology
Government Arts and Science College (Women)
Perambalur, Tamil Nadu State, India

**Mahesh Kedharashetty**
Department of Microbiology
Bhavan's Vivekananda College of Science, Humanities & Commerce
Secunderabad, Telangana, India

**Shruti Kharod**
Molecular diagnostics laboratory, Centre for Medical Biotechnology
Maharshi Dayanand University
Rohtak, Haryana, India

**K. Ajith Kumar**
Department Life Sciences, Kristu Jayanti College (Autonomous)
Bengaluru, India

**S. Kumaran**
PG Research Department of Microbiology
Sri Sankara Arts and Science College
Kanchipuram, Tamil Nadu, India

**Anil Kumar**
Molecular diagnostics laboratory, Centre for Medical Biotechnology
Maharshi Dayanand University
Rohtak, Haryana, India

**S. Chaitanya Kumari**
Department of Microbiology
Bhavan's Vivekananda College of Science, Humanities & Commerce
Secunderabad, Telangana, India

**G.R. Shree Kumari**
School of Biosciences and Technology
Vellore Institute of Technology
Vellore, Tamil Nadu, India

**Archana Lakshmaiah**
Sri Sathya Sai University for Human Excellence
Kalaburagi, Karnataka, India

**I. Lavkor**
Biological Control Research Institute
Kisla Str, Yuregir, Adana, Turkey

**Yunhao Ma**
School of Pharmacy
Lanzhou University
Gansu, PR China

**Hamadou Mamoudou**
Department of Biological Sciences, Faculty of Science
University of Maroua
Cameroon

**S. Meenakshi**
PG & Research Department of Microbiology
Sadakathullah Appa College
Tirunelveli, Tamil Nadu, India

**José Silvestre Mendoza-Figueroa**
Department of Microbiology, Escuela Nacional de Ciencias Biológicas
Instituto Politécnico Nacional
Mexico City, Mexico

**Pritty Mishra**
Bhavan's Vivekananda College of Science, Humanities & Commerce
Secunderabad, Telangana, India

**Vajiha Banu Habeeb Mohammed**
PG and Research Department of Microbiology
Jamal Mohamed College (Autonomous)
Tiruchirappalli, Tamil Nadu, India

**V. Mohanasrinivasan**
School of Biosciences and Technology
Vellore Institute of Technology
Vellore, Tamil Nadu, India

**Martin Alain Mune Mune**
Department of Biological Sciences, Faculty of Science
University of Maroua
Cameroon

**Jairus Olumasai Nandwa**
Department of Genetics
Hebrew University of Jerusalem
Israel

**Nagashri Nanjundeshwara**
Sri Sathya Sai University for Human Excellence
Kalaburagi, Karnataka, India

**A. Sankara Narayanan**
Department of Life Sciences, School of Science (SoS)
GSFC University
Vadodara, Gujarat, India

**Yutika Narzary**
Department of Botany
Bodoland University
Rangalikhata, Assam, India

**Harita Parmar**
Department of Life Sciences, School of Science
GSFC University
Vadodara, Gujarat, India

**Roshni Patel**
Sun Pharma Advanced Research Co. Ltd.
Vadodara, Gujarat, India

**Sai Prasad**
Department of Life Sciences
Sri Sathya Sai University for Human Excellence
Kalaburagi, Karnataka, India

**Samridhi Pushkarna**
Molecular Diagnostics Laboratory, Centre for Medical Biotechnology
Maharshi Dayanand University
Rohtak, Haryana, India

**Ramya Raghavan**
Sri Sathya Sai University for Human Excellence
Kalaburagi, Karnataka, India

**K. Deepak Raj**
Department of Microbiology
Bhavan's Vivekananda College of Science, Humanities & Commerce
Secunderabad, Telangana, India

**Simran Rani**
Plant-microbe interaction laboratory, Department of Microbiology
Maharshi Dayanand University
Rohtak, Haryana, India

**Chetna Rathi**
Plant-microbe interaction laboratory, Department of Microbiology
Maharshi Dayanand University
Rohtak, Haryana, India

**Lokesh Ravi**
Department of Biomedical Sciences, School of Health Sciences
The Apollo University
Chittoor, Andhra Pradesh, India
and
Centre for Digital Health and Precision Medicine
The Apollo University
Chittoor, Andhra Pradesh, India

**R. Saravanan**
Department of Zoology
Sri Kumara Gurupara Swamigal Arts College
Srivaikuntam, Tamil Nadu, India

**G. Sathvik**
Department of Botany, School of Life Sciences
St Joseph's University
Bengaluru, India

**Maheswaran Solayappan**
Faculty of Applied Sciences
AIMST University
Bedong, Kedah, Malaysia

**Stanila Seryozheva Stoeva**
Faculty of Pharmacy, Department of Pharmacology, Toxicology and Pharmacotherapy
Medical University – Varna Prof. Dr. Paraskev Stoyanov
Varna, Bulgaria

**Pooja Suneja**
Plant-Microbe Interaction Laboratory, Department of Microbiology
Maharshi Dayanand University
Rohtak, Haryana, India

**Vaishnavi Thirukumaran**
Centre for Drug Discovery and Development
Sathyabama Institute of Science and Technology
Chennai, Tamil Nadu, India

**Rajasekar Thirunavukkarasu**
Centre for Drug Discovery and Development
Sathyabama Institute of Science and Technology
Chennai, Tamil Nadu, India

**S. Umayaparvathi**
Department of Microbiology
New Prince Shri Bhavani Arts and Science College
Chennai, Tamil Nadu, India

**M. Veerapagu**
Department of Biotechnology, Microbiology & Bioinformatics
National College (Autonomous)
Tiruchirappalli, Tamil Nadu State, India

**D.M. Yashaswini**
Department of Botany, School of Life Sciences
St Joseph's University
Bengaluru, India

**Vandi Yonas**
Department of Biological Sciences, Faculty of Science
University of Maroua
Cameroon

**Baizhuo Zhang**
School of Pharmacy
Lanzhou University
Gansu, PR China

**Zhongkun Zhou**
School of Pharmacy
Lanzhou University
Gansu, PR China

# Section 1

## Overview of Aflatoxigenic Fungi in Plant, Animal, and Human Health

# 1.1 Overview and Sources of Aflatoxigenic Fungal Diversity in Plants

*Priyanka Dahiya, Simran Rani, Chetna Rathi, Sai Prasad, Sankara Narayanan, Amita Suneja Dang, and Pooja Suneja*

## INTRODUCTION

Mycotoxins are a group of toxic secondary metabolites synthesized by soil-borne fungi such as *Aspergillus, Fusarium*, and *Penicillium*. They present a major challenge to global health, agriculture, and international trade by adversely affecting human and animal well-being and agricultural productivity. Among these toxins, aflatoxins pose a particular concern as they contaminate a wide range of staple foods and animal products globally, including cereals, spices, nuts, milk, and milk products [1]. Approximately 20 different types of aflatoxins exist, but B1, B2, G1, and G2 are the most significant due to their frequent occurrence in food. The designations B1 and G2 indicate their fluorescence under ultraviolet light (blue or green) and relative abundance (1 for the major type, 2 for the minor type). These toxins are known carcinogens and mutagens [2]. Aflatoxin B1 is particularly concerning because it is considered a Group 1 human carcinogen by the International Agency for Research on Cancer (IARC). It leads to genotoxicity by binding to DNA and causing structural changes. Various health problems, such as acute and chronic toxicity, teratogenicity, and immunotoxicity have been linked to B1 [3, 4]. Crop contamination by these toxins leads to significant economic losses due to the rejection of produce in markets [5]. For instance, the annual grain loss due to aflatoxins in the United States was assessed to be USD 500 million [6]. India, as the world's second-largest groundnut producer, can only export 800,000 tons annually due to aflatoxin contamination [7]. The export of groundnut products in India reduced from 550 metric tons (USD 42.5 million) to 265 metric tons (USD 32.5 million) over ten years. In China, aflatoxins cause crop yield losses of approximately 21 million tons, equivalent to 4.2% of the total annual crop production [8].

## GENERAL FEATURES OF AFLATOXIN-PRODUCING *ASPERGILLUS* SPECIES

The two primary species responsible for most aflatoxin contamination events in agricultural products globally are *Aspergillus flavus* and *Aspergillus parasiticus*. However, other species from the *Aspergillus* and *Emericella* genera have also been reported to produce aflatoxins [9]. It is essential to know the population biology and biodiversity of these fungi to control and prevent aflatoxin contamination. *A. flavus* is a common soil resident that can thrive in various organic materials such as plant residues, animal feed, compost, stored grains, decaying wood, and dead insects [10]. This adaptable species exhibits variations in its appearance and aflatoxin production. Depending on the specific strain, it produces large (L-type) or small (S-type) sclerotia. The L-strains mainly produce aflatoxin B1 and B2, whereas S-strains additionally produce G1 and G2 [11]. The L-strains tend to produce more conidia, fewer large sclerotia, and low aflatoxins whereas S-strains produce many small sclerotia, fewer conidia, and high aflatoxins [10]. In terms of its life strategy, abundant

DOI: 10.1201/9781003487043-2

sporulation allows the exploitation of new nutrient environments, and ample sclerotia production might provide a survival advantage in soil with high microbial competition. On the other hand, using more resources, mycelial synthesis may offer an edge in colonizing host tissues and nutrient uptake. *A. flavus* generally thrives at temperatures above 25°C, with higher temperatures favoring its growth, sporulation, and aflatoxin production [12]. *A. parasiticus* is another significant pathogen causing aflatoxin contamination in crops. It majorly produces L-type sclerotia, and the colony color and spore ornamentation serve as distinguishing features from *A. flavus* [4]. It consistently synthesizes high levels of both B and G aflatoxins (B1, B2, G1, and G2) and colonizes crops at lower temperatures with optimal growth between 22 and 25°C. Its growth slows down compared to *A. flavus* at higher temperatures [12].

## FACTORS INFLUENCING *ASPERGILLUS* GROWTH AND AFLATOXIN PRODUCTION

The growth of *Aspergillus* and aflatoxin production are affected by different **extrinsic and intrinsic factors along with** interactions between coexisting fungal species in the same environment [13].

## EXTRINSIC FACTORS

### Water Activity (Aw) and Temperature

These are important factors that significantly affect fungal growth and aflatoxin production. These factors also play crucial roles in the transcription of two important regulatory genes (*afl*R and *afl*S) in the aflatoxin biosynthesis pathway [14]. Aflatoxin production primarily occurs at temperatures ranging from 29 to 30°C, and it is inhibited at temperatures above 37°C or below 25°C. The optimal water activity (aw) for aflatoxin production is approximately 0.99, and levels below 0.85 aw significantly hinder fungal growth and toxin production. Complete inhibition occurs at 0.70–0.75 aw [15]. The influence of these two significant factors on *Aspergillus* growth and aflatoxin production has been documented in numerous studies. For instance, brown Korean rice is more susceptible to *A. flavus* infection and aflatoxin production compared to rough or white Korean rice. The highest levels of *A. flavus* and aflatoxins were found in brown rice stored at 97% relative humidity at 21°C [16]. In sorghum grains, ideal conditions for fungal growth and AFB1 production were 0.99 aw and a temperature of 37°C, whereas the minimum water activity required for fungal growth was 0.91 at moderate temperatures [17]. Gallo et al. reported maximum fungal growth and AFB1 production at 28°C and 0.96 aw, whereas AFB1 production was inhibited at 20°C (with aw of 0.90 and 0.93) and 37°C. Similarly, the regulatory genes for aflatoxin production showed the highest expression at 28°C, and at 20°C and 37°C, low expression was observed, suggesting temperature's key role in fungal growth or AFB1 production [18]. In maize, the optimum condition for fungal growth was 0.99 aw and 30°C, whereas AFB1 production peaked at 0.98 aw and 30°C. *A. flavus* growth was completely inhibited at 0.90 aw and 20°C [19]. Mannaa and Kim discovered that temperature and relative humidity had significant effects on *A. flavus* growth in stored rice grains. A one-unit increase in temperature had a greater impact on fungal growth than a one-unit increase in relative humidity, indicating that temperature is a more critical factor for fungal growth [20]. *A. flavus* could grow in the range 0.92–0.96 aw and 28–37°C, but optimal conditions for AFB1 production were 33°C and 0.96 aw in polished rice. The fungal growth and AFB1 production in paddy rice was highest at 37°C and 0.94 aw [21]. Krulj et al. observed ideal conditions for B1 production in several cereals at 30°C across a broad temperature range (15–37°C) and moisture content (0.85–0.99 aw). Grain hulls played a protective role by concentrating B1 10–170 times more than dehulled grain [22].

## Abiotic Stress

Heat and drought stress weaken plant defenses and make crops like peanuts, maize, and pistachios more susceptible to fungal infection and aflatoxin contamination. Stressed plants produce fewer phytoalexins, antifungal compounds that hinder fungal growth and aflatoxin production, and have weaker physical barriers (cracked hulls, damaged kernels), facilitating fungal colonization and seed infection [23]. Drought conditions lower aw levels below 0.95 in peanuts, leading to reduced phytoalexin production and increased susceptibility to aflatoxin contamination [24]. A positive correlation between elevated aflatoxin levels and climatic conditions such as drought and hot summer temperatures was noted after an examination of maize samples collected in Serbia during 2012–2016. Leaf and soil moisture also affect the extent of aflatoxin contamination. Dry soil and plant surfaces, often related to high air temperatures, can promote aflatoxin outbreaks [25].

## Soil Conditions

The proportions of sand, silt, and clay particles in the soil influence aflatoxin persistence and leaching. Silty clay loam soils and soils rich in clay and organic matter have a high adsorption capacity, preventing aflatoxins from leaking into groundwater. However, these soil types also create reservoirs for long-term contamination because aflatoxins can persist in them for up to 120 days [26, 27]. Densely packed soils, like those often found in peanut fields, have been linked to higher populations of *Aspergillus* species compared to looser soils [28]. Achaglinkame et al. reported light, sandy soils promote *A. flavus* growth, leading to increased aflatoxin contamination [29].

## $CO_2$ and $O_2$ Levels

A high $CO_2$ level and a low $O_2$ level inhibit aflatoxin production and fungal growth [30]. In paddy, $CO_2$ levels influenced *A. flavus* growth and aflatoxin production. The highest $CO_2$ level (80%) caused an 88% inhibition in growth and a 97% reduction in aflatoxin synthesis. The 20% $CO_2$ caused 59% and 47% reductions in growth and aflatoxin production, respectively, whereas 80% $CO_2$ caused 88% and 97% reductions in growth and aflatoxin production, respectively [31]. Giorni et al. reported a 25% $CO_2$ level effectively reduced *A. flavus* development in maize, but to decrease aflatoxin synthesis substantially, a minimum of 50% $CO_2$ was essential [32]. The presence of light also impacts fungal proliferation and aflatoxin production. Light inhibits aflatoxin synthesis, whereas it is promoted by darkness [33].

## Mechanical Injury

Damage to crops by insects such as lesser cornstalk borer, nitidulid beetles, rice weevils and rodents create entry points for fungal infection. Aphids and root-knot nematodes can also increase infection rates [11].

## Intrinsic Factors

Substrate composition: Carbohydrate-rich substrates, such as glucose, sucrose, xylose and ribose, are more favorable for aflatoxin production because they readily provide energy for fungal growth. However, lactose and sorbose were unable to promote aflatoxin production. In addition to carbohydrates, lipids also contribute to aflatoxin production by acting as signaling molecules in biosynthesis. The addition of corn oil to defatted wheat increased aflatoxin production compared to oil-free media [14, 34].

### Amino Acids

Glycine, glutamate, and alanine, when present with metal ions such as zinc and magnesium, promote aflatoxin production. Amino acids like tyrosine promote aflatoxin production, whereas tryptophan inhibits it [35]. A significant increase in aflatoxin production was observed with increasing zinc concentration [14]. Metals like cadmium and iron slow down fungal growth and aflatoxin production [36].

### pH

*A. flavus* growth was higher at higher pH values within the 3.5–6.5 range [37]. Aflatoxin production was more strain-specific, with some strains producing maximum aflatoxin at a pH of 5.0, whereas others showed no significant difference between pH 5.5 and 6.0 [38].

Agricultural practices: Monoculture and late planting lead to increased fungal inoculum and pest damage, resulting in high fungal infection and aflatoxin production [39]. Crop rotation and intercropping help minimize toxin contamination. For instance, rotating legumes such as cowpea and soybean with maize helps disrupt the infection cycle and improve soil nutrients [28]. Agricultural practices such as a combination of supplementary irrigation and tied ridges have been reported to decrease *Aspergillus* infection levels in the field by 68% [40].

## INFECTION PROCESS OF *ASPERGILLUS*

*Aspergillus* thrives in warm environments and moist soils, forming hyphae and sclerotia for survival [41]. It has a biphasic life cycle, which involves colonizing plant debris in soil and invading seeds or grains of actively growing crops. During its growth cycle, the fungus exists as a rapidly growing mycelium. When conditions are unfavorable, the mycelium condenses into resilient sclerotia for survival. Favorable conditions induce these sclerotia to germinate, producing new hyphae or conidia, with soil acting as a reservoir for conidia. The conidia are the primary source of infection for susceptible crops and are dispersed via different biotic and abiotic factors, such as wind, nitidulid beetles, and other insects [10, 12, 42]. Air-dispersed conidia are a major threat to aboveground crops like maize and tree nuts. In maize, the fungus targets the silk and kernels, with young kernels at the late milk–early dough stage being especially vulnerable. A higher colonization incidence on the silk of mature maize ears compared to younger ones has been observed [11]. Corn silks and insect-damaged kernels are ideal targets for spore colonization [43]. Overwintering corncobs further contribute to the airborne spore load, and nitidulid beetles facilitate infection by making additional entry points. Peanuts are directly infected due to their subterranean fruiting habit, making them highly susceptible to soil-borne spores. Infected peanut surfaces release conidia, which act as secondary inoculum, resulting in multiple infection series within the same season [41]. Again, in cotton, wind plays a crucial role in spreading conidia, which infects cotton bolls via floral structures and insect wounds [43]. Peanut and cotton seeds can also become contaminated through rain splash dispersal of spores [11]. Insects and mites serve as carriers of fungal spores on their bodies, with mites contributing to increasing *A. flavus* growth by transporting fungal spores and introducing them into fresh grains [44]. In addition, the respiration process of insects raises the temperature and moisture content within stored grains, creating ideal conditions for fungal growth [45]. *A. flavus* produces sclerotia and conidia in crop waste and in the soil, which act as the primary source of infection for plants in the subsequent growing season. The conidia produced on crop debris or infected plants throughout the growing season serve as secondary inoculum when environmental conditions suitable for infection arrive [46] (Figure 1.1.1).

Aflatoxin contamination, caused by *Aspergillus* species *flavus* and *parasiticus*, presents a global challenge. The fungal growth and production of aflatoxins are influenced by both extrinsic and intrinsic factors. Optimal conditions for aflatoxin production require warm temperatures, high

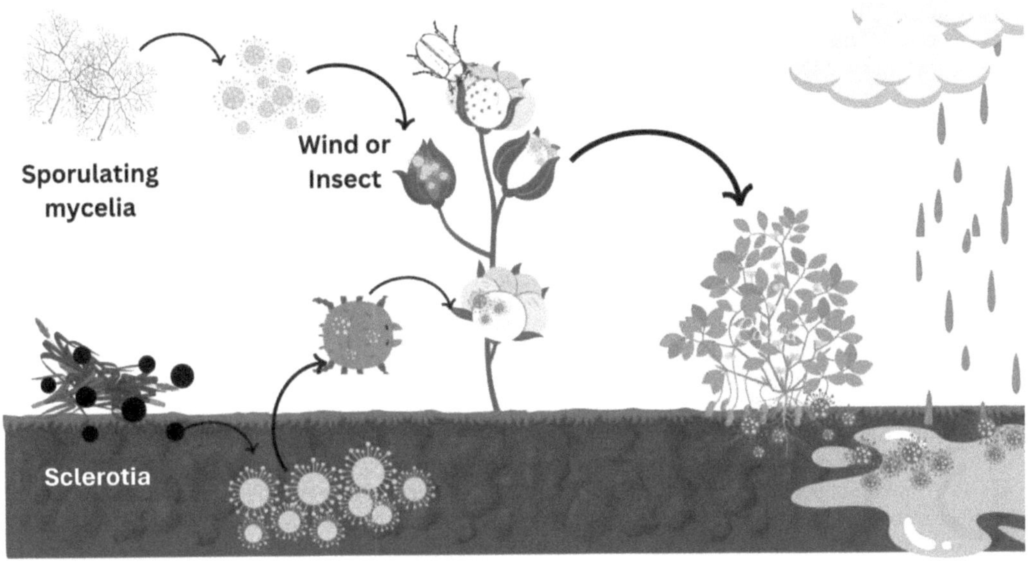


**FIGURE 1.1.1** Infection process of *Aspergillus* in the agricultural ecosystem.

humidity, and host susceptibility. It is crucial to understand these interrelated factors to develop strategies for minimizing aflatoxin contamination in crops along with implementing good agricultural practices, proper storage, and breeding resistant varieties. Continuous research is urgently needed to enhance the detection, decontamination, and mitigation of aflatoxin, to safeguard food safety and reduce economic losses.

## ACKNOWLEDGMENT

The authors are grateful for the infrastructural support provided by the Department of Microbiology and the Centre for Medical Biotechnology (CMBT) at Maharshi Dayanand University (MDU), Rohtak, India. PD acknowledges the University Grants Commission (UGC), New Delhi, India, for the UGC fellowship. SR and CR are thankful to the Council of Scientific & Industrial Research (CSIR), New Delhi, India, for their CSIR fellowships.

## REFERENCES

1. Shabeer, Saba, Shahzad Asad, Atif Jamal, and Akhtar Ali. "Aflatoxin contamination, its impact and management strategies: An updated review." *Toxins* 14, no. 5 (2022): 307. https://doi.org/10.3390/toxins14050307
2. Pickova, Darina, Vladimir Ostry, and Frantisek Malir. "A recent overview of producers and important dietary sources of aflatoxins." *Toxins* 13, no. 3 (2021): 186. https://doi.org/10.3390/toxins13030186
3. Benkerroum, Noreddine. "Chronic and acute toxicities of aflatoxins: Mechanisms of action." *International Journal of Environmental Research and Public Health* 17, no. 2 (2020): 423. https://doi.org/10.3390/ijerph17020423
4. Frisvad, Jens Christian, V. Hubka, C. N. Ezekiel, S. B. Hong, A. Nováková, A. J. Chen, M. Arzanlou et al. "Taxonomy of *Aspergillus* section *Flavi* and their production of aflatoxins, ochratoxins and other mycotoxins." *Studies in Mycology* 93, no. 1 (2019): 1–63. https://doi.org/10.1016/j.simyco.2018.06.001
5. Soni, Pooja, Sunil S. Gangurde, Alejandro Ortega-Beltran, Rakesh Kumar, Sejal Parmar, Hari K. Sudini, Yong Lei et al. "Functional biology and molecular mechanisms of host-pathogen interactions for aflatoxin contamination in groundnut (*Arachis hypogaea* L.) and maize (*Zea mays* L.)." *Frontiers in Microbiology* 11 (2020): 227. https://doi.org/10.3389/fmicb.2020.00227

6. Wu, Fang. "Global impacts of aflatoxin in maize: Trade and human health." *World Mycotoxin Journal* 8, no. 2 (2015): 137–142. 10.3920/WMJ2014.1737
7. Suneja, K. "India ropes in global institution to curb contamination in spice shipments." Geneva: World Trade Organization, 2019.
8. Gong, Andong, Mengge Song, and Jingbo Zhang. "Current strategies in controlling *Aspergillus flavus* and aflatoxins in grains during storage: A review." *Sustainability* 16, no. 8 (2024): 3171. https://doi.org/10.3390/su16083171
9. Ahmad, Malik Mobeen, Mahboob Ahmad, Athar Ali, Rifat Hamid, Saleem Javed, and Malik Zainul Abdin. "Detection of *Aspergillus flavus* and *Aspergillus parasiticus* from aflatoxin-contaminated peanuts and their differentiation using PCR-RFLP." *Annals of Microbiology* 64 (2014): 1597–1605. https://doi.org/10.1007/s13213-014-0803-5
10. Fora, Fatuma Sharamo. "Efficacy of atoxigenic *Aspergillus flavus* (Aflasafe ke01) on the population of aflatoxigenic fungi and aflatoxin contamination of maize (*Zea mays* l.) in alternate seasons in lower eastern Kenya." PhD diss., Egerton University, 2016.
11. Amaike, Saori, and Nancy P. Keller. "*Aspergillus flavus*." *Annual Review of Phytopathology* 49, no. 1 (2011): 107–133. https://doi.org/10.1146/annurev-phyto-072910-095221
12. Connel, Ching'anda. "Impacts of competition between *Aspergillus* section Flavi species on aflatoxin contamination of crops." PhD diss., The University of Arizona, 2021.
13. Tai, Bowen, Jinghua Chang, Yang Liu, and Fuguo Xing. "Recent progress of the effect of environmental factors on *Aspergillus flavus* growth and aflatoxins production on foods." *Food Quality and Safety* 4, no. 1 (2020): 21–28. https://doi.org/10.1093/fqsafe/fyz040
14. Liu, Xiao, Xuanli Guan, Fuguo Xing, Cong Lv, Xiaofeng Dai, and Yang Liu. "Effect of water activity and temperature on the growth of *Aspergillus flavus*, the expression of aflatoxin biosynthetic genes and aflatoxin production in shelled peanuts." *Food Control* 82 (2017): 325–332. https://doi.org/10.1016/j.foodcont.2017.07.012
15. Bhatnagar, Deepak, Jeffrey W. Cary, Kenneth Ehrlich, Jiujiang Yu, and Thomas E. Cleveland. "Understanding the genetics of regulation of aflatoxin production and *Aspergillus flavus* development." *Mycopathologia* 162, no. 3 (2006): 155–166. https://doi.org/10.1007/s11046-006-0050-9
16. Choi, Seonyeong, Hyejung Jun, Jihyun Bang, Soo-Hyun Chung, Yoonsook Kim, Byeong-Sam Kim, Hoikyung Kim, Larry R. Beuchat, and Jee-Hoon Ryu. "Behaviour of *Aspergillus flavus* and *Fusarium graminearum* on rice as affected by degree of milling, temperature, and relative humidity during storage." *Food Microbiology* 46 (2015): 307–313. https://doi.org/10.1016/j.fm.2014.08.019
17. Lahouar, Amani, Sonia Marin, Ana Crespo-Sempere, Salem Saïd, and Vicente Sanchis. "Effects of temperature, water activity and incubation time on fungal growth and aflatoxin B1 production by toxicogenic *Aspergillus flavus* isolates on sorghum seeds." *Revista Argentina De Microbiologia* 48, no. 1 (2016): 78–85. https://doi.org/10.1016/j.ram.2015.10.001
18. Gallo, Antonia, Michele Solfrizzo, Filomena Epifani, Giuseppe Panzarini, and Giancarlo Perrone. "Effect of temperature and water activity on gene expression and aflatoxin biosynthesis in *Aspergillus flavus* on almond medium." *International Journal of Food Microbiology* 217 (2016): 162–169. https://doi.org/10.1016/j.ijfoodmicro.2015.10.026
19. Bernaldez, Victoria, Juan J. Cordoba, Naresh Magan, Ana Belen Peromingo Arévalo, and Alicia Rodriguez. "The influence of ecophysiological factors on growth, *afl*R gene expression and aflatoxin B1 production by a type strain of *Aspergillus flavus*." *LWT-Food Science and Technology* 83 (2017): 283–291. https://doi.org/10.1016/j.lwt.2017.05.030
20. Mannaa, Mohamed, and Ki Deok Kim. "Effect of temperature and relative humidity on growth of *Aspergillus* and *Penicillium* spp. and biocontrol activity of *Pseudomonas* protegens AS15 against aflatoxigenic *Aspergillus flavus* in stored rice grains." *Mycobiology* 46, no. 3 (2018): 287–295. https://doi.org/10.1080/12298093.2018.1505247
21. Lv, Cong, Jing Jin, Ping Wang, Xiaofeng Dai, Yang Liu, Mumin Zheng, and Fuguo Xing. "Interaction of water activity and temperature on the growth, gene expression and aflatoxin production by *Aspergillus flavus* on paddy and polished rice." *Food Chemistry* 293 (2019): 472–478. https://doi.org/10.1016/j.foodchem.2019.05.009
22. Krulj, Jelena, Siniša Markov, Aleksandra Bočarov-Stančić, Lato Pezo, Jovana Kojić, Nataša Ćurčić, Elizabet Janić-Hajnal, and Marija Bodroža-Solarov. "The effect of storage temperature and water activity on aflatoxin B1 accumulation in hull-less and hulled spelt grains." *Journal of the Science of Food and Agriculture* 99, no. 7 (2019): 3703–3710. https://doi.org/10.1002/jsfa.9601
23. Cotty, Peter J., and Ramon Jaime-Garcia. "Influences of climate on aflatoxin producing fungi and aflatoxin contamination." *International Journal of Food Microbiology* 119, no. 1–2 (2007): 109–115. https://doi.org/10.1016/j.ijfoodmicro.2007.07.060

24. Lavkor, Isilay, and Isil Var. "The control of aflatoxin contamination at harvest, drying, pre-storage and storage periods in peanut: The new approach." In *Aflatoxin-Control, Analysis, Detection and Health Risks*, pp. 45–64. Intechopen, 2017. https://doi.org/10.5772/intechopen.68675
25. Dövényi-Nagy, Tamás, Csaba Rácz, Krisztina Molnár, Károly Bakó, Zsombor Szláma, Ákos Jóźwiak, Zsuzsa Farkas, István Pócsi, and Attila Csaba Dobos. "Pre-harvest modelling and mitigation of aflatoxins in maize in a changing climatic environment—A review." *Toxins* 12, no. 12 (2020): 768. https://doi.org/10.3390/toxins12120768
26. Madden, Uford A., and Henry M. Stahr. "Preliminary determination of mycotoxin binding to soil when leaching through soil with water." *International Biodeterioration & Biodegradation* 31, no. 4 (1993): 265–275. https://doi.org/10.1016/0964-8305(93)90022-T
27. Accinelli, Cesare, H. K. Abbas, R. M. Zablotowicz, and J. R. Wilkinson. "*Aspergillus flavus* aflatoxin occurrence and expression of aflatoxin biosynthesis genes in soil." *Canadian Journal of Microbiology* 54, no. 5 (2008): 371–379.
28. Nji, Queenta Ngum, Olubukola Oluranti Babalola, and Mulunda Mwanza. "Soil *Aspergillus* species, pathogenicity and control perspectives." *Journal of Fungi* 9, no. 7 (2023): 766. https://doi.org/10.3390/jof9070766
29. Achaglinkame, Matthew Atongbiik, Nelson Opoku, and Francis Kweku Amagloh. "Aflatoxin contamination in cereals and legumes to reconsider usage as complementary food ingredients for Ghanaian infants: A review." *Journal of Nutrition & Intermediary Metabolism* 10 (2017): 1–7. https://doi.org/10.1016/j.jnim.2017.09.001
30. Mahbobinejhad, Zohre, Heshmatolah Aminian, Leila Ebrahimi, and Kourosh Vahdati. "Reduction of aflatoxin production by exposing *Aspergillus flavus* to CO2." *Journal of Crop Protection* 8, no. 4 (2019): 441–448.
31. Mousa, Wael, Farinazleen Mohd Ghazali, S. Jinap, Hasanah Mohd Ghazali, Son Radu, and Abd El-Raziq Salama. "Temperature, water activity and gas composition effects on the growth and aflatoxin production by *Aspergillus flavus* on paddy." *Journal of Stored Products Research* 67 (2016): 49–55. https://doi.org/10.1016/j.jspr.2016.01.003
32. Giorni, Paola, Paola Battilani, Amedeo Pietri, and Naresh Magan. "Effect of aw and CO2 level on *Aspergillus flavus* growth and aflatoxin production in high moisture maize post-harvest." *International Journal of Food Microbiology* 122, no. 1–2 (2008): 109–113. https://doi.org/10.1016/j.ijfoodmicro.2007.11.051
33. Rushing, Blake R., and Mustafa I. Selim. "Aflatoxin B1: A review on metabolism, toxicity, occurrence in food, occupational exposure, and detoxification methods." *Food and Chemical Toxicology* 124 (2019): 81–100. https://doi.org/10.1016/j.fct.2018.11.047
34. Scarpari, Marzia, Marta Punelli, Valeria Scala, Marco Zaccaria, Chiara Nobili, Matteo Ludovici, Emanuela Camera, Anna A. Fabbri, Massimo Reverberi, and Corrado Fanelli. "Lipids in *Aspergillus flavus*-maize interaction." *Frontiers in Microbiology* 5 (2014): 74. https://doi.org/10.3389/fmicb.2014.00074
35. Chang, Perng-Kuang, Sui Sheng T. Hua, Siov Bouy L. Sarreal, and Robert W. Li. "Suppression of aflatoxin biosynthesis in *Aspergillus flavus* by 2-phenylethanol is associated with stimulated growth and decreased degradation of branched-chain amino acids." *Toxins* 7, no. 10 (2015): 3887–3902. https://doi.org/10.3390/toxins7103887
36. Khan, Rahim, Farinazleen Mohamad Ghazali, Nor Ainy Mahyudin, and Nik Iskandar Putra Samsudin. "Biocontrol of aflatoxins using non-aflatoxigenic *Aspergillus flavus*: A literature review." *Journal of Fungi* 7, no. 5 (2021): 381. https://doi.org/10.3390/jof7050381
37. Kosegarten, Carlos E., Nelly Ramírez-Corona, Emma Mani-López, Enrique Palou, and Aurelio López-Malo. "Description of *Aspergillus flavus* growth under the influence of different factors (water activity, incubation temperature, protein and fat concentration, pH, and cinnamon essential oil concentration) by kinetic, probability of growth, and time-to-detection models." *International Journal of Food Microbiology* 240 (2017): 115–123. https://doi.org/10.1016/j.ijfoodmicro.2016.04.024
38. Casquete, Rocío, María José Benito, María de Guía Córdoba, Santiago Ruiz-Moyano, and Alberto Martín. "The growth and aflatoxin production of *Aspergillus flavus* strains on a cheese model system are influenced by physicochemical factors." *Journal of Dairy Science* 100, no. 9 (2017): 6987–6996. https://doi.org/10.3168/jds.2017-12865
39. Meissle, Michael, Patrik Mouron, Tomke Musa, Franz Bigler, Xavier Pons, Vasileios P. Vasileiadis, Stefan Otto et al. "Pests, pesticide use and alternative options in European maize production: Current status and future prospects." *Journal of Applied Entomology* 134, no. 5 (2010): 357–375. https://doi.org/10.1111/j.1439-0418.2009.01491.x
40. Gebreselassie R., A. Dereje, and H. Solomon. "On Farm Pre Harvest Agronomic Management Practices of Aspergillus Infection on Groundnut in Abergelle, Tigray." *Journal of Plant Pathology and Microbiology* 5 (2014): 228. https://doi.org/10.4172/2157-7471.1000228

41. Payne, Gary A. "Process of contamination by aflatoxin-producing fungi and their impact on crops." *Mycotoxins in Agriculture and Food Safety* 9 (1998): 279–306.
42. Jaime-Garcia, Ramon, and Peter J. Cotty. "*Aspergillus flavus* in soils and corncobs in south Texas: Implications for management of aflatoxins in corn-cotton rotations." *Plant Disease* 88, no. 12 (2004): 1366–1371. https://doi.org/10.1094/PDIS.2004.88.12.1366
43. Dowd, Patrick F., Eric T. Johnson, and W. Paul Williams. "Strategies for insect management targeted toward mycotoxin management." In *Aflatoxin and Food Safety*, pp. 542–567. CRC Press, 2005.
44. Magan, Naresh "Ecophysiology of biocontrol agents for improved competence in the phyllosphere." In *Microbial Ecology of Aerial Plant Surfaces*, pp. 149–164. Wallingford UK: CABI, 2006.
45. Reddy, K. R. N., C. S. Reddy, and K. Muralidharan. "Potential of botanicals and biocontrol agents on growth and aflatoxin production by *Aspergillus flavus* infecting rice grains." *Food Control* 20, no. 2 (2009): 173–178. https://doi.org/10.1016/j.foodcont.2008.03.009
46. Kimberly A. Scheidegger, and G. A. Payne. "Unlocking the secrets behind secondary metabolism: A review of *Aspergillus flavus* from pathogenicity to functional genomics." *Journal of Toxicology: Toxin Reviews* 22, no. 2–3 (2003): 423–459. https://doi.org/10.1081/TXR-120024100

# 1.2 Overview and Sources of Aflatoxigenic Fungal Diversity in Animals

*Sanchita Choubey*

## INTRODUCTION

Several living things contribute to the complex tapestry of life on Earth; some are essential to the ecosystem's health, while others are subtle threats. The latter includes a wide range of fungi that are classified as aflatoxigenic fungi. These minuscule enemies skulk in the animal food chain, secretly releasing strong toxins called aflatoxins that have a disastrous effect on animal health [1]. This introduction explores the diverse range of aflatoxigenic fungus in animals, their complex nature, the harmful consequences of aflatoxin and several sources of contamination [2].

Although aflatoxins are a class of chemical compounds with similar structures, their effects on animal health vary widely. The International Agency for Research on Cancer (IARC) has classified these chemicals as carcinogenic, meaning that animals exposed to them may develop cancer. Still, the negative impacts go well beyond the danger of cancer. Aflatoxin-contaminated food or feed causes a host of detrimental health effects in animals who consume it. Aflatoxin-induced disruption of essential liver functions, including decreased protein synthesis, compromised detoxification pathways, and eventually liver failure, is the main cause for concern [3].

Aflatoxin pollution affects not just the health of individual animals but also entire livestock populations and the agriculture industry as a whole. Animals exposed to aflatoxins have been shown in studies to have stunted growth, decreased weight gain and decreased immunity [4]. This results in lower milk and egg production, lower cattle output and higher vulnerability to infectious illnesses. Significant economic repercussions result from this, affect farmers' livelihoods, and possibly create a scarcity of food items obtained from animals [5].

Developing successful mitigation techniques requires an understanding of the causes of aflatoxin contamination in animals. Aflatoxigenic fungal infections before harvest are a serious concern. Crop vulnerability to fungal colonisation and subsequent aflatoxin generation can be considerably increased by environmental conditions such as high temperatures and drought stress during crop development. These elements emphasise the intricate interactions that exist between the environment, farming methods and fungal development [6].

Improper storage practices after harvest further exacerbate the problem. Grains and other feedstuffs stored in humid and poorly ventilated conditions create an ideal environment for aflatoxigenic fungi to thrive, leading to a surge in aflatoxin production after harvest. These storage practices become breeding grounds for the very toxins that threaten animal health [7]. The complex nature of aflatoxigenic fungus, as well as the many sources of aflatoxin contamination, require a multidisciplinary strategy to protect animal health. Crop monitoring programmes play an important part in this endeavour. By frequently testing crops for aflatoxin levels before harvest, tainted fruit can be diverted to non-food applications, successfully preventing the poison from entering the animal food chain. Furthermore, optimising grain and feed storage conditions is critical [8]. Proper drying and ventilation measures during storage can considerably lower the danger of aflatoxin development by inhibiting fungal growth.

DOI: 10.1201/9781003487043-3 

Beyond preventative measures, researchers are actively investigating alternative mitigation mechanisms that account for fungal variety. For example, the discovery of selective mould inhibitors that are effective against specific aflatoxigenic strains is promising [9]. Furthermore, introducing particular binding agents into animal feed shows promise in decreasing aflatoxin bioavailability dependent on the fungal toxins present. Finally, nutritional therapies are being investigated as a potential method for mitigating the negative consequences of aflatoxin exposure. According to research, supplementing diets with specific antioxidants and other detoxifying agents may assist in supporting animal health in the face of aflatoxin contamination, with the supplements potentially customised to the dominant aflatoxin types [10].

## A SPECTRUM OF FUNGAL THREATS: AFLATOXIGENIC DIVERSITY [11]

Aflatoxigenic fungi are a varied group within the genus *Aspergillus*, specifically *Aspergillus flavus* and *Aspergillus parasiticus*. This diversity poses a complex challenge, as different strains show variances in the following:

- **Growth patterns:** Some strains develop in hot, humid environments, whereas others may establish themselves in cooler climates.
- **Toxin production capabilities:** The quantity and types of aflatoxins produced might vary greatly between strains.
- **Environmental tolerances:** Susceptibility to variables such as fungicides and drought stress varies between strains.

## A TOXIC ARSENAL: THE DEVASTATING EFFECTS OF AFLATOXINS

Aflatoxins are a class of structurally similar chemical substances that have a chilling effect on animal health. The International Agency for Research on Cancer (IARC) classifies these chemicals as carcinogens, meaning they have the potential to cause cancer in exposed animals [12]. However, the negative impacts extend far beyond cancer risk:

- **Liver damage:** Aflatoxins alter critical liver functions, resulting in decreased protein synthesis, poor detoxification processes, and, eventually, liver failure [13]. This can appear as slower development rates, weight loss and an increased susceptibility to various illnesses.
- **Stunted growth and development:** Animals exposed to aflatoxins frequently show stunted growth, decreased weight gain, and altered immunological function. This results in decreased cattle productivity, lower milk and egg production, and greater susceptibility to infectious illnesses [14].
- **Economic ramifications:** Aflatoxin poisoning has serious economic consequences for farmers. Contaminated crops and livestock products may be rejected or decontaminated, resulting in financial losses. Furthermore, aflatoxin exposure leads to decreased productivity, poorer overall output, and probable shortages of animal-derived food products [15].

The varied nature of aflatoxins, with variable toxicities and potencies, exacerbates the problem.

## SOURCES OF AFLATOXIGENIC FUNGAL DIVERSITY IN ANIMALS

Aflatoxigenic fungal diversity in animals does not refer to the origin of the fungus, but rather the many methods in which these fungi can penetrate the animal environment and contaminate it [16]. Here are the main causes of aflatoxin contamination in animals:

- **Pre-harvest infection:** This is the primary cause of contamination. Aflatoxigenic fungus, specifically *A. flavus* and *A. parasiticus*, can infect crops such as corn, peanuts, cottonseed and rice prior to harvest. When these diseased crops are fed to animals, they become a source of aflatoxins [2].
- **Environmental factors:** Environmental variables during crop growth can have a substantial impact on the risk of aflatoxin contamination. High temperatures, drought stress, and insect damage to crops can all contribute to fungal colonisation and subsequent aflatoxin formation [17].
- **Improper storage:** Even if crops are free of aflatoxins upon harvest, poor storage conditions can promote fungal growth and aflatoxin generation. Storing grains and feedstuffs in humid, poorly ventilated circumstances permits aflatoxigenic fungi to thrive and contaminate the feed [18].
- **Contaminated bedding:** Aflatoxigenic fungus can thrive in bedding materials such as straw and sawdust. Aflatoxin exposure can occur when animals come into contact with contaminated bedding.

## REDUCING AFLATOXIN CONTAMINATION IN ANIMALS

Aflatoxins are a serious hazard to animal health, but there are numerous measures that can be used to reduce contamination.

### Pre-Harvest Strategies [19]

- **Crop selection:** Using crop cultivars that are less vulnerable to aflatoxin contamination can be an effective first step.
- **Good agricultural practices:** Implementing techniques that promote healthy plant growth and reduce stress, such as regular irrigation and fertilisation, can limit fungal growth and aflatoxin formation.
- **Monitoring crops:** Regularly testing crops for aflatoxin levels prior to harvest enables the rejection or diversion of affected produce to non-food applications.
- **Biological control:** Introducing non-aflatoxin generating *Aspergillus* fungi can compete with aflatoxigenic strains, potentially lowering total aflatoxin production.

### Post-Harvest Strategies [19]

- **Proper storage:** Maintaining ideal storage conditions for grains and feedstuffs is critical. This includes thorough drying to reduce moisture content, storing in cold, well-ventilated areas, and utilising appropriate storage containers to avoid moisture penetration.
- **Mould inhibitors:** Mould inhibitors in animal feed can help limit the growth of aflatoxigenic fungi and prevent increased aflatoxin synthesis. However, rules and specific applications for these inhibitors must be considered.
- **Physical separation:** Some approaches, such as density separation, can be used to distinguish infected grain fractions from those with lower aflatoxin concentrations.

## ANIMAL MANAGEMENT STRATEGIES [20]

- **Feeding practices:** Avoiding feeding obviously mouldy or damaged grains to animals might help prevent exposure. Diluting contaminated feed with aflatoxin-free feed may also be a possibility, depending on the extent of contamination and legislation.
- **Binding agents:** Adding certain binding agents to animal feed can assist in reducing aflatoxin bioavailability, which means that the animal's digestive tract absorbs less of the poison. The efficiency of these agents may vary according to the type of aflatoxin present.

**TABLE 1.2.1**
**Aflatoxigenic Fungal Diversity in Animals**

| Category | Information |
|---|---|
| **Animals Susceptible** | Various animals including livestock (cattle, pigs, poultry), companion animals (dogs, cats) and wildlife |
| **Symptoms of Aflatoxin Poisoning** | Reduced appetite, weight loss, lethargy, diarrhoea, vomiting, jaundice, liver damage death |
| **Common Aflatoxigenic Fungi** | *A. flavus, A. parasiticus, A. nomius* |
| **Sources of Contamination** | Contaminated feed (grains, peanuts, cottonseed meal) |
| **Risk Factors** | Warm, humid climates |
| **Prevention Methods** | Cool, dry feed storage, high-quality feed ingredients, crop monitoring for fungal growth, feed testing for aflatoxin contamination |

- **Nutritional interventions:** Supplementing animal diets with antioxidants and other detoxifying agents may assist in improving animal health and mitigating the detrimental effects of aflatoxin exposure. Supplements can be adjusted to the main types of aflatoxin.

## ADDITIONAL CONSIDERATIONS

- **Regulations:** Aflatoxin levels in animal feed are frequently regulated by national and international guidelines. These regulations must be addressed while implementing mitigating solutions.
- **Economic feasibility:** Some techniques, such as some mould inhibitors or binding agents, may incur expenses that must be evaluated against the possible economic losses from aflatoxin exposure.
- **Research and development:** Ongoing studies are looking at new ways to control aflatoxigenic fungus and mitigate the effects of aflatoxin contamination (Table 1.2.1).

## CONCLUSION

Aflatoxigenic fungal variety is a substantial hazard to animal health due to the many sources of aflatoxin contamination. This chapter has given an overview of the various natures of these fungi, the devastation caused by aflatoxins and the several sources of contamination. Understanding these complexities is the first step towards establishing effective mitigation techniques and ensuring the welfare of animals in the food chain.

## ACKNOWLEDGEMENT

We wish to express our sincere thanks and gratitude to Dr. D. Y. Patil, Arts Commerce & Science College, Pimpri, Pune, for providing the necessary facilities to carry out this study.

## REFERENCES

1. Rahmati-Joneidabad, M., & Alizadeh Behbahani, B. (2021, August 1). Boswellia sacra essential oil: Antioxidant activity and antifungal effect on some spoilage fungi causing strawberry rot. *Food Science and Technology*, *18*(114), 25–34. https://doi.org/10.52547/fsct.18.114.25.
2. Ehrlich, K. C. (2014). Non-aflatoxigenic *Aspergillus flavus* to prevent aflatoxin contamination in crops: Advantages and limitations. *Frontiers in Microbiology*, *5*. https://doi.org/10.3389/fmicb.2014.00050.

3. Fritz, W. (1985, January). WHO-IARC monographs on the evaluation of the carcinogenic risk of chemicals to humans. Vol. 33: Polynuclear aromatic compounds, part 2, carbon blacks, mineral oils and some nitroarenes. 245 seiten. International Agency for Research on Cancer, Lyon 1984. Preis: 50,00 Sw. fr.; 25.00 $. *Food/Nahrung*, *29*(4), 390–390. https://doi.org/10.1002/food.19850290423.
4. Chen, X., He, K., Wei, C., Yang, W., & Geng, Z. (2020, February 25). Green tea powder decreased egg weight through increased liver lipoprotein lipase and decreased plasma total cholesterol in an indigenous chicken breed. *Animals*, *10*(3), 370. https://doi.org/10.3390/ani10030370.
5. Baqwa, M., Swanepoel, E., & Venter, M. (2022, September 21). Challenges affecting Mzongwana cattle farmers to commercialise livestock production in Eastern Cape, from a local economic development perspective. *Africa's Public Service Delivery & Performance Review*, *10*(1). https://doi.org/10.4102/apsdpr.v10i1.576.
6. Ortega-Beltran, A., Moral, J., Puckett, R. D., Morgan, D. P., Cotty, P. J., & Michailides, T. J. (2018, June 20). Fungal communities associated with almond throughout crop development: Implications for aflatoxin biocontrol management in California. *PLoS One*, *13*(6), e0199127. https://doi.org/10.1371/journal.pone.0199127.
7. Cwalina-Ambrozik, B. (2013). Fungi colonization of potato tubers (*Solanum tuberosum L.*) after harvest and after storage. *Acta Agrobotanica*, *55*(2), 133–140. https://doi.org/10.5586/aa.2002.051.
8. Sharafi, K., Kiani, A., Khalid Omer, A., Parnoon, K., & Massahi, T. (2024, March 13). Importance of rigorous and ongoing monitoring of animal feed quality and storage conditions to mitigate aflatoxin M1 contamination in dairy products in Iran: A health policy brief. *Journal of Health Reports and Technology*, *10*(2). https://doi.org/10.5812/jhrt-146654.
9. Foubister, V. (2003, April). Anticancer drug shows promise against lupus. *Drug Discovery Today*, *8*(8), 333–334. https://doi.org/10.1016/s1359-6446(03)02664-3.
10. Zahra, N., Tanveer, R., Zaheer, M., Moosa, H., Saeeed, M. K., Shahzad, K., Saleem, Y., & Nawaz, S. (2021, November 3). Aflatoxin types, permissible level, factors responsible for aflatoxin contamination, determination and detoxification methods in animal feed. *Chemical Science and Engineering Research*, *3*(8). https://doi.org/10.36686/ariviyal.cser.2021.03.08.040.
11. Moubasher, H., Taleb, A. A., & Senousy, H. H. (2013, September). Molecular differentiation between aflatoxigenic and non-aflatoxigenic strains of *Aspergillus flavus* and *Aspergillus parasiticus*. *Microbiology*, *82*(5), 642–646. https://doi.org/10.1134/s0026261713050160.
12. Kligman, A. M., & Kligman, L. H. (1994, December). Carcinogens show comedogenic activity: A potential animal screen for tumorigenic substances. *Cancer Letters*, *87*(2), 171–178. https://doi.org/10.1016/0304-3835(94)90219-4.
13. Baker, K., & Nasraway, S. A. (2000, July). Multiple organ failure during critical illness: How organ failure influences outcome in liver disease and liver transplantation. *Liver Transplantation*, *6*(S1), S5–S9. https://doi.org/10.1002/lt.500060503.
14. Lakhanpal, J., & Gupta, J. (2017). Decreased antibiotic susceptibility and enhanced probiotic production potential of goat milk fermented curd in comparison with cow and buffalo milk. *BioTechnologia*, *2*, 121–129. https://doi.org/10.5114/bta.2017.68311.
15. Young, A. G. (2017, June). Risky resources: Household production, food contamination, and perceptions of aflatoxin exposure among Zambian female farmers. *Economic Anthropology*, *4*(2), 200–212. https://doi.org/10.1002/sea2.12088.
16. Barakat, S., & Swaileh, K. M. (2022, June 6). Fungal contamination, aflatoxigenic fungi and levels of aflatoxin B1 in spices marketed in the West Bank of Palestine. *Food Additives & Contaminants: Part B*, *15*(4), 245–253. https://doi.org/10.1080/19393210.2022.2085330.
17. Hamidou, F., Rathore, A., Waliyar, F., & Vadez, V. (2014, February). Although drought intensity increases aflatoxin contamination, drought tolerance does not lead to less aflatoxin contamination. *Field Crops Research*, *156*, 103–110. https://doi.org/10.1016/j.fcr.2013.10.019.
18. Kodape, A., Lama, A., & Vivek Babu, C. (2024, April). Metagenomic insights of fungal diversity of peanuts under storage conditions and mitigation of aflatoxigenic fungi through competitive exclusion and phytochemicals. *Food Bioscience*, *58*, 103711. https://doi.org/10.1016/j.fbio.2024.103711.
19. Yan, L., Song, W., Chen, Y., Kang, Y., Lei, Y., Huai, D., Wang, Z., Wang, X., & Liao, B. (2021, May). Effect of non-aflatoxigenic strains of *Aspergillus flavus* on aflatoxin contamination of pre-harvest peanuts in fields in China. *Oil Crop Science*, *6*(2), 81–86. https://doi.org/10.1016/j.ocsci.2021.04.004.
20. Moschini, M., Gallo, A., Piva, G., & Masoero, F. (2008, December). The effects of rumen fluid on the in vitro aflatoxin binding capacity of different sequestering agents and in vivo release of the sequestered toxin. *Animal Feed Science and Technology*, *147*(4), 292–309. https://doi.org/10.1016/j.anifeedsci.2008.01.010.

# 1.3 Overview and Sources of Aflatoxigenic Fungal Diversity in Humans

*Vajiha Banu Habeeb Mohamed, A. Sankara Narayanan, and K. Anandhi*

## INTRODUCTION

Mycotoxins are harmful substances that certain moulds produce as by-products; they are mainly associated with the *Aspergillus*, *Penicillium*, and *Fusarium* species. Several agricultural products, such as cereals, nuts, fruits and animal feed, are susceptible to contamination by microorganisms. They can infiltrate the food chain and induce toxicity when ingested by humans or animals, posing a serious hazard to both human and animal health. Of the approximately 300 distinct mycotoxins, aflatoxins are the most well-known group. Toxic compounds known as aflatoxins have the potential to stimulate both acute and chronic poisoning in humans, as well as mutagenic, teratogenic, and cancerous effects. Long-term human exposure to aflatoxins has been associated with several health problems, such as immunosuppression and organ impairment; aflatoxin M1 (AFM1) is hepatotoxic and can harm hepatic cells [1].

A wide variety of climatic conditions, from dry to tropical wet to temperate, allow *Aspergillus* fungi to flourish and multiply virtually anywhere in the world. To develop actively, *A. flavus* needs water with an activity of at least 0.98 and temperatures between 25 and 33°C. Temperatures between 30 and 40°C are still sufficient for it to grow. For this reason, such fungi develop in all the main hot weather growing regions of cereal, peanuts, tree nuts and cotton seeds [2]. Aflatoxin-producing fungi can produce hazardous compounds in the preharvest stages of maize, rice, and pupuru when the environmental conditions are right. Long-term storage of these items in non-hermetic containers may also promote the development of aflatoxin. Prolonged exposure to aflatoxins causes harmful health effects in humans, including malignancies, child stunting, and even death [3]. As one of the most dangerous pollutants that can have a major impact on the food chain, aflatoxins are currently a topic of great concern. As apex predators, humans frequently eat tainted food originating from both plants and animals. In addition to harming people's health, aflatoxin contamination-related food insecurity can have negative social, political and economic effects on humanity [4] (Figure 1.3.1).

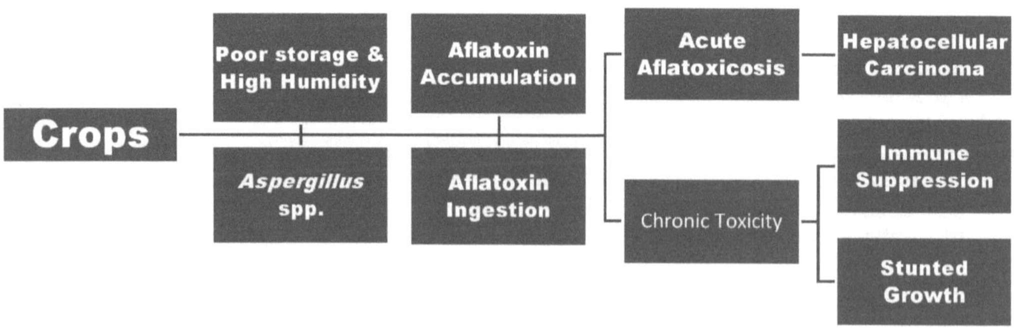


**FIGURE 1.3.1** Effects of aflatoxin on human health.


DOI: 10.1201/9781003487043-4

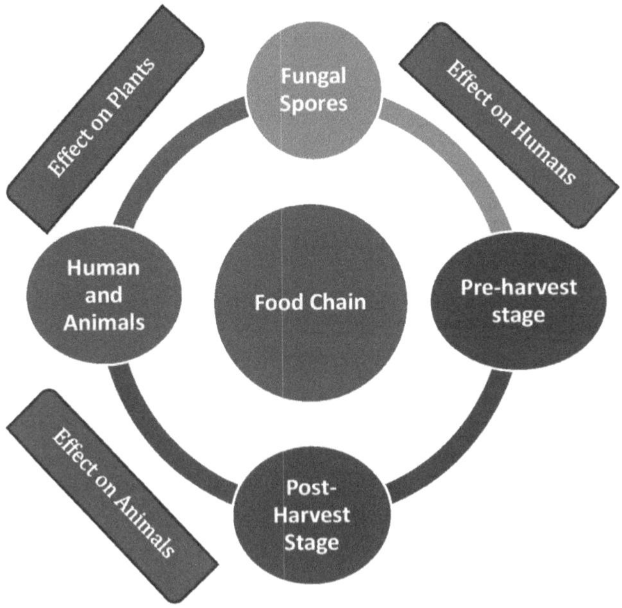


**FIGURE 1.3.2** *Aspergillus*-derived mycotoxins in food.

Mycotoxin contamination of food, feed, and agricultural products has been a significant concern recently due to the potential for these poisonous compounds to cause various health problems in humans and animals, including poisoning. Worldwide, mycotoxins are present in a wide range of food and agricultural goods. Any fungus that generates one or more mycotoxins is considered toxic. Toxigenic fungus creates secondary metabolites called mycotoxins. The three most significant genera of fungi that create mycotoxins are *Aspergillus*, *Fusarium*, and *Penicillium* [5]. Among the highly dangerous secondary metabolites produced by fungi like *A. flavus*, *A. parasiticus*, and *A. nomius* are aflatoxins, which are polyketides. Cereal crops such as wheat, walnut, corn, cotton, peanuts, and tree nuts are usually impacted by these fungus [6]. They can cause a range of problems, including teratogenicity, immunotoxicity and hepatotoxicity, which can cause serious health hazards for both humans and animals [7]. The biodiversity of aflatoxigenic strains is significantly impacted by changes in global temperature [8]. Temperature, the water activity of the commodity, and the amount of carbon dioxide in the air are the main elements that affect biodiversity [9] (Figure 1.3.2).

Fungicidal preservatives such as propionic acid and phosphine are artificial substances used in grain storage. Additionally, compounds such as fullerol, fullerene, tiadiazines, and coumarins can prevent *A. flavus* from proliferating and from generating aflatoxins. By stopping the growth of mould, chemical preservatives can ensure the safety of food, but their usage may be limited because of potential health risks and environmental concerns. Low-toxicity natural antifungal medications have garnered a lot of attention lately. Various bioactive substances, such as essential oils obtained from plants and marine animals, including phenols, aldehydes and alcohols, possess potent antifungal properties [10].

## MAJOR SOURCES OF HUMAN EXPOSURE TO AFLATOXIN

### Aflatoxin from Contaminated Animal Products

Because animal products constitute a major pathway for human exposure to aflatoxins, our attention should be directed toward aflatoxins that are known to be transferred from feed in terms of mycotoxins and food safety [11]. There will be aflatoxins in eggs, meat and milk if animals consume contaminated grains. When large quantities of aflatoxin are present, milk can also contain the aflatoxin B1 product AFM1, which is created during aflatoxin metabolism. Humans can absorb aflatoxin by consuming milk or dairy products, such as yoghurt and breast milk, primarily in places where livestock and other animals are fed substandard grains [12].

## Animal Feed and Dairy Products

Aflatoxin-producing peanut-based diets that originated in South America were the cause of the "turkey X" sickness, which led to the discovery of aflatoxins in the early 1960s [13]. The long-term exposure of animals to sub-acutely toxic levels of aflatoxins has since been linked to several negative effects on animal health, including feed refusal, increased susceptibility to disease, weight loss, poor eggshell quality, decreased milk yield, defects in carcass quality, reproduction defects and other serious effects like liver lesions/tumours, hepatotoxicity, nephrotoxicity, gastroenteritis and teratogenicity [14, 15].

The presence of microscopic fungi has an impact on the nutritional value, organoleptic characteristics, and overall quality of feeds. Moulds can synthesise various mycotoxins, which can harm nutritional and organoleptic qualities. Any sort of feed might be conducive to aflatoxin contamination and fungal growth, just like any other food commodity, if it is not properly harvested, dried, processed and stored [16]. Furthermore, supplying tainted feed to animals causes aflatoxins to be transferred into the milk as AFMs. As an illustration, 1%–3% of the AFB1 found in feedstuffs manifests as AFM1 in milk [17]. When AFMs are present in milk, dairy products become contaminated because standard food processing methods like heating do not remove these aflatoxins from food. Because cow's milk is a vital nourishment for both adults and children, contamination with aflatoxin puts human health at risk. Consequently, it is deemed undesirable for AFM1 to be present in milk and milk products [18].

## Aflatoxin from Contaminated Plant: Maize, Sorghum, Peanut and Cashew Nut

Mycotoxin contamination of maize often lowers crop value and increases health hazards for household animals and people. Aflatoxin B1, which is classified as carcinogenic to humans by the International Agency for Research on Cancer (IARC), is one of the very dangerous forms of aflatoxin [19].

The potential for sorghum to contain mycotoxin contamination and the grain's increasing significance as food and feed in many nations [20] is of concern. Due to their ability to help people in rural areas meet their nutritional needs, peanuts are highly valued and regarded as a "woman's crop". They are cultivated second only to maize. *A. flavus* and *A. parasiticus* are the main producers of aflatoxins, which are toxic substances that suppress the immune system, cause cancer, and retard growth. Aflatoxin levels in oilseeds have been steadily rising worldwide, and mitigating concerns are becoming increasingly pressing. As a result, efficient aflatoxin control strategies are essential for public health, food safety and economic growth [21, 22].

The primary source of inoculum for *A. flavus* that causes infection in peanut seeds is the peanut pods that grow beneath the soil. Consequently, peanuts have a greater aflatoxin level than other food commodities, making them a major source of human carcinogen exposure [23]. Owing to their high protein, mineral, and lipid content, cashew nuts are susceptible to fungal infestation. When infections can arise at other phases, such as preharvest or during harvest, they typically result from improper handling and storage of the harvested nuts. A moisture level of less than 8% is thought to be safe for raw cashew nuts, and when the seeds are not adequately dried before being stored, they promote the growth and infestation of fungi. Furthermore, high temperatures and humidity during storage promote the growth of fungi in cashew nuts and the potential contamination by mycotoxin [24].

Regrettably, the agriculture sector is greatly concerned about the susceptibility of many cereal crops, particularly rice, to AF contamination. These harmful compounds, which are created as secondary metabolites by *Aspergillus section flavi* fungus, can taint grains prior to harvest, frequently by contact with contaminated soil. On the other hand, incorrect post-harvest grain storage methods and pest infestation can also foster an atmosphere that encourages the growth of fungi that produce AF, increasing the likelihood of AF generation in the future [25]. Among the major chemical forms of AFs, AFB1 has received the most attention due to its toxicity profile and occurrence in various agricultural crops consumed by humans, such as corn, rice, wheat, peanuts, dried fruits, sorghum, and others [26].

### Aflatoxin from Contaminated Grains: Rice Wheat

Most rice is grown in subtropical climates, which are defined by warmth and humidity. Following harvest, rice is often dried, and when stored improperly, it can serve as a suitable substrate for the growth of fungi. Fifteen percent of the rice crop is lost annually, according to the Food and Agriculture Organization (FAO) [27], due to improper storage conditions that encourage the growth of fungi and other harmful agents. Aflatoxin-containing mycotoxins are frequently found in rice. Aflatoxin production is stimulated by storage temperatures and moisture levels, which leads to annual losses of valuable food bioresources like rice and impacts the economies of rice-producing nations. The four fungal genera that produce higher quantities of mycotoxins in food products are *Fusarium, Penicillium, Aspergillus* and *Alternaria* [27].

According to the World Health Organization (WHO), aflatoxin is a global food security concern. In considering toxicity and carcinogenicity, the presence of aflatoxin in rice is a serious public health concern, especially in developing countries where people are at risk of aflatoxin exposure. Rice and other staple foods are susceptible to aflatoxin contamination. The presence of aflatoxins in rice attracts worldwide attention because of the significant economic losses associated with their negative impacts on animal and human health and trade [28].

## EFFECTS ON IMMUNE SYSTEM

It has been observed that eating foods infected with aflatoxin regularly can weaken the immune systems in both humans and animals globally [29]. Aflatoxin exposure harms the immune system, especially reactions involving cell-mediated immunity. The high prevalence of human immunodeficiency virus (HIV) has been linked to immunological suppression and nutritional consequences resulting from chronic aflatoxin exposure. HIV infection and compromised CD4 proteins brought on by aflatoxin exposure are related to this. Moreover, there is a connection between elevated aflatoxin levels and a higher risk of tuberculosis in HIV-positive people. Aflatoxin exposure increases the risk of acquiring bacterial, viral and parasitic diseases [30].

## MECHANISM OF ACTION ON HUMANS

All aflatoxin species are lipolytic in nature; they enter the circulation and travel to different tissues, including the liver, after being absorbed via cell membranes at the site of exposure, which includes the respiratory and gastrointestinal tracts. AFB1 needs to be bioactivated to be harmful, which mostly happens in hepatocytes. The first absorption of AFB1 occurs in the small intestine, specifically in the duodenum. They are converted by the liver into less harmful aflatoxin M1 or reactive epoxide intermediates. The cytochrome P450 enzyme changes AFB1 in humans and vulnerable animals into an epoxide, which attaches to blood albumin and DNA and damages DNA. Hepatocarcinogenesis occurs when the epoxide attaches itself to DNA in the mitochondria. The genetic code of enzymes that regulate cell proliferation is impacted by AFB1's binding to DNA at the guanine site in liver cells. This mechanism results in the formation of tumours. Aflatoxin-containing molecules bind to and disrupt enzymes and substrates required for the initiation, transcription and translation phases of protein synthesis by creating adducts with DNA, RNA, and proteins [31].

## ACKNOWLEDGEMENT

The authors acknowledge the PG and Research Department of Microbiology, Jamal Mohamed College (Autonomous), Tiruchirappalli.


## REFERENCES


1. Massahi, T., Omer, A. K., Habibollahi, M. H., et al. (2024). Human health risk assessment of aflatoxin M1 in various dairy products in Iran: A literature review. *Journal of Food Composition and Analysis*, 129,106124.
2. Mamo, F. T., Abate, B. A., Zheng, Y., Nie, C., He, M., & Liu, Y. (2021). Distribution of *Aspergillus* fungi and recent aflatoxin reports, health risks, and advances in developments of biological mitigation strategies in China. *Toxins*, *13*(10), 678.
3. Ekpakpale, D. O., Kraak, B., Meijer, M., Ayeni, K. I., Houbraken, J., & Ezekiel, C. N. (2021). Fungal diversity and aflatoxins in maize and rice grains and cassava-based flour (Pupuru) from Ondo state, Nigeria. *Journal of Fungi*, *7*(8), 635.
4. Pickova, D., Ostry, V., & Malir, F. (2021). A recent overview of producers and important dietary sources of aflatoxins. *Toxins*, *13*(3), 186.
5. Milki, S., & Abdeta, D. (2023). Public health impact of aflatoxin. *Journal of Bacteriology & Mycology: Open Access*, *11*(1), 34–39.
6. Payne, G. A., & Brown, M. P. (1998). Genetics and physiology of aflatoxin biosynthesis. *Annual Review of Phytopathology*, *36*(1), 329–362.
7. Severns, D. E., Clements, M. J., Lambert, R. J., & White, D. G. (2003). Comparison of *Aspergillus* ear rot and aflatoxin contamination in grain of high-oil and normal-oil corn hybrids. *Journal of Food Protection*, *66*(4), 637–643.
8. Magan, N., Medina, A., & Aldred, D. (2011). Possible climate-change effects on mycotoxin contamination of food crops pre-and postharvest. *Plant Pathology*, *60*(1), 150–163.
9. AboDalam, T. H., Amra, H., Sultan, Y., Magan, N., Carlobos-Lopez, A. L., Cumagun, C. J. R., & Yli-Mattila, T. (2020). New genotypes of aflatoxigenic fungi from Egypt and the Philippines. *Current Research in Environmental & Applied Mycology (Journal of Fungal Biology*, *10*(1), 142–155.
10. Jerković, I., Kranjac, M., Marijanović, Z., Šarkanj, B., Cikoš, A. M., Aladić, K., Pedisić, S., & Jokić, S. (2019). Chemical diversity of *Codium bursa* (Olivi) C. Agardh headspace compounds, volatiles, fatty acids and insight into its antifungal activity. *Molecules*, *24*, 842.
11. FAO, I. (2010). Good practices of the feed industry implementing the Codex Alimentarius code of practice in good animal feeding. *FAO Animal production and Health / Manual Rome*, *9*, 103.
12. Piermarini, S., Micheli, L., Ammida, N. H., et al. (2007). Electrochemical immunosensor array using a 96-well screen-printed microplate for aflatoxin B1 detection. *Biosensors and Bioelectronics*, *22*(7), 1434–1440.
13. Kensler, T. W., Roebuck, B. D., Wogan, G. N., & Groopman, J. D. (2011). Aflatoxin: A 50-year Odyssey of mechanistic and translational toxicology. *Toxicological Sciences*, *120*, 28–48.
14. Bhat, R., Rai, R. V., & Karim, A. A. (2010). Mycotoxins in food and feed: Present status and future concerns. *Comprehensive Reviews in Food Science and Food Safety*, *9*, 57–81.
15. Zain, M. E. (2011). Impact of mycotoxins on humans and animals. *Journal of Saudi Chemical Society*, *15*, 129–144.
16. Pahlow, G., Muck, R. E., & Driehuis, F. (2003). Microbiology of ensiling. In *Silage Science and Technology*; Buxton, D.R., Muck, R.E., Harrison, J.H., Eds.; American Society of Agronomy; pp. 31–93.
17. Diaz, D. E., Hagler, W. M. Jr., Blackwelder, J. T., Eve, J. A., Brinton, A., Anderson, K. L., Jones, F. T., & Whitlow, L. W. (2004). Aflatoxin binders II: Reduction of aflatoxin M1 in milk by sequestering agents of cows consuming aflatoxin in feed. *Mycopathologia*, *157*, 233–241.
18. Giovati, L., Magliani, W., Ciociola, T., Santinoli, C., Conti, S., & Polonelli, L. (2015). AFM1 in milk: Physical, biological, and prophylactic methods to mitigate contamination. *Toxins*, *7*, 4330–4349.
19. Probst, C., Bandyopadhyay, R., & Cotty, P. J. (2014). Diversity of aflatoxin-producing fungi and their impact on food safety in sub-Saharan Africa. *International Journal of Food Microbiology*, *174*, 113–122.
20. Lahouar, A., Crespo-Sempere, A., Marín, S., Saïd, S., & Sanchis, V. (2015). Toxigenic molds in Tunisian and Egyptian sorghum for human consumption. *Journal of Stored Products Research*, *63*, 57–62.
21. Kodape, A., Lama, A., & Babu, C. V. (2024). Metagenomic insights of fungal diversity of peanuts under storage conditions and mitigation of aflatoxigenic fungi through competitive exclusion and phytochemicals. *Food Bioscience*, 58,103711.

22. Akello, J., Alamu, E. O, Mwila, C. M., et al. (2024). Impact of conservation farming practices on *Aspergillus* population density, peanut aflatoxin level, and exposure risk in Zambia. *Frontiers in Sustainable Food Systems*, *8*, 1331188.
23. Sultana, T., Malik, K., Raja, N. I., et al. (2024). Aflatoxins in peanut (Arachis hypogaea): Prevalence, global health concern, and management from an innovative nanotechnology approach: A mechanistic repertoire and future direction. *ACS Omega*, *9*(24), 25555–25574.
24. Musangi, C. R., Juma, B. S., Mukhebi, D. W., Isoe, E. M., Kibiti, C. M., & Mbinda, W. M. (2024). *Aspergillus* population diversity and its role in aflatoxin contamination of cashew nuts from coastal Kenya. *PLoS One*, *19*(1), e0292519.
25. Ouma, F., Luthra, K., Oduola, A., & Atungulu, G. G. (2024). Investigating safe storage conditions to mitigate aflatoxin contamination in rice. *Food Control*, *163*, 110529.
26. Oduola, A. A., Callewaert, P., Devlieghere, F., Bluhm, B. H., & Atungulu, G. G. (2022). Growth and aflatoxin B1 biosynthesis rate of model *Aspergillus flavus* NRRL 3357 exposed to selected infrared wavelengths. *Food Control*, *141*, 109204.
27. Al-Zoreky, N. S., & Saleh, F. A. (2019). Limited survey on aflatoxin contamination in rice. *Saudi Journal of Biological Sciences*, *26*(2), 225–231.
28. Ali, N. (2019). Aflatoxins in rice: Worldwide occurrence and public health perspectives. *Toxicology Reports*, *6*, 1188–1197.
29. Ammann, H. M. (2003). Is indoor mold contamination a threat to health? Part two. *Journal of Environmental Health*, *66*(2), 43–44.
30. Williams, J. H., Timothy, D. P., Pauline, E. J., et al. (2004). Human aflatoxicosis in developing countries: A review of toxicology, exposure, potential health consequences, and interventions. *The American Journal of Clinical Nutrition*, *80*(5), 1106–1022.
31. Usha, P. S., Preetida, J. B., Prameela, D., et al. (2017). Aflatoxins: Implications on health. *Indian Journal of Clinical Biochemistry*, *32*(2), 124–133.

# 1.4 Climate Change and Its Impact on Aflatoxigenic Fungi

*Shruti Kharod, Samridhi Pushkarna, Kajal Gaba, Anil Kumar, M Veerapagu, K. R. Jeya, Pooja Suneja, and Amita Suneja Dang*

## INTRODUCTION

Climate change is one of the biggest risks to human health in the 21st century. Growing population and climate change are creating new problems for the safety of food and feed (1). The sustainability of human societies, the health of Earth's ecosystems, the key agricultural systems, the availability of food, and the productivity of plants and animals are all impacted by climate change (2). This shift in climate will also lead to an increase in microbial hazards such as enhanced mycotoxin production, which might harm crops and raise food safety concerns. Aflatoxin-producing *Aspergillus* species are the primary source of the most researched mycotoxins (3). Aflatoxins are genotoxic, mutagenic, and carcinogenic, providing a significant risk to both animal and human health (4,5). They are fungal metabolites that can harm infant development, cause immune suppression, gastroenteritis, cancer, and lead to death in critical cases of acute exposure (6). In the early 1960s, nearly 100,000 turkeys died from acute liver necrosis caused by contaminated peanut powder in animal diets, known as Turkey-X sickness (7). Acute aflatoxicoses and fatalities may result from changes in weather patterns in certain regions (8). Aflatoxins are mostly produced by *A. parasiticus* and *A. flavus*, which can be found in cereals, cotton, peanuts, tree nuts, pepper, spices, and other agricultural goods (9). Despite their native habitat, species are found in soil and decaying vegetation and can infiltrate various organic substrates with suitable conditions for growth (10). Aflatoxin-contaminated crops cause significant economic losses for farmers, handlers, processors, and marketers (11). A concerning contamination in maize was found in several reports from countries in Southern and Eastern Europe. Drought and semi-arid to arid conditions are associated with contamination in tropical regions, and the impoverished often survive on contaminated staple foods (2,12,13). It is a matter of concern that climate change could disrupt the incidence patterns of aflatoxin in food and feed highlighting the importance of establishing preventive measures (Figure 1.4.1).

## ORIGIN OF AFLATOXINS

Aflatoxins are the harmful secondary metabolites of fungal origin(14). Their production has been studied most among the mycotoxins, and are dangerous to both human and animal health because they are highly toxic and carcinogenic (15,16). Approximately 20 aflatoxins are commonly known and reported to be structural derivatives of difuranocoumarins (6). Among them, four aflatoxins, that is, B1, B2, G1 and G2, are mostly reported as contaminating food and feed. The chromatographic mobility of aflatoxins B1, B2, G1 and G2 via thin-layer chromatography and their blue or green fluorescence under UV light were used to identify them (17) and many structural analogues of aflatoxins, such as aflatoxins P1, Q1, B2a and G2a, have been discovered. The most prevalent naturally occurring carcinogen, aflatoxin B1, is widely present in animal tissues and fluids as a


DOI: 10.1201/9781003487043-5

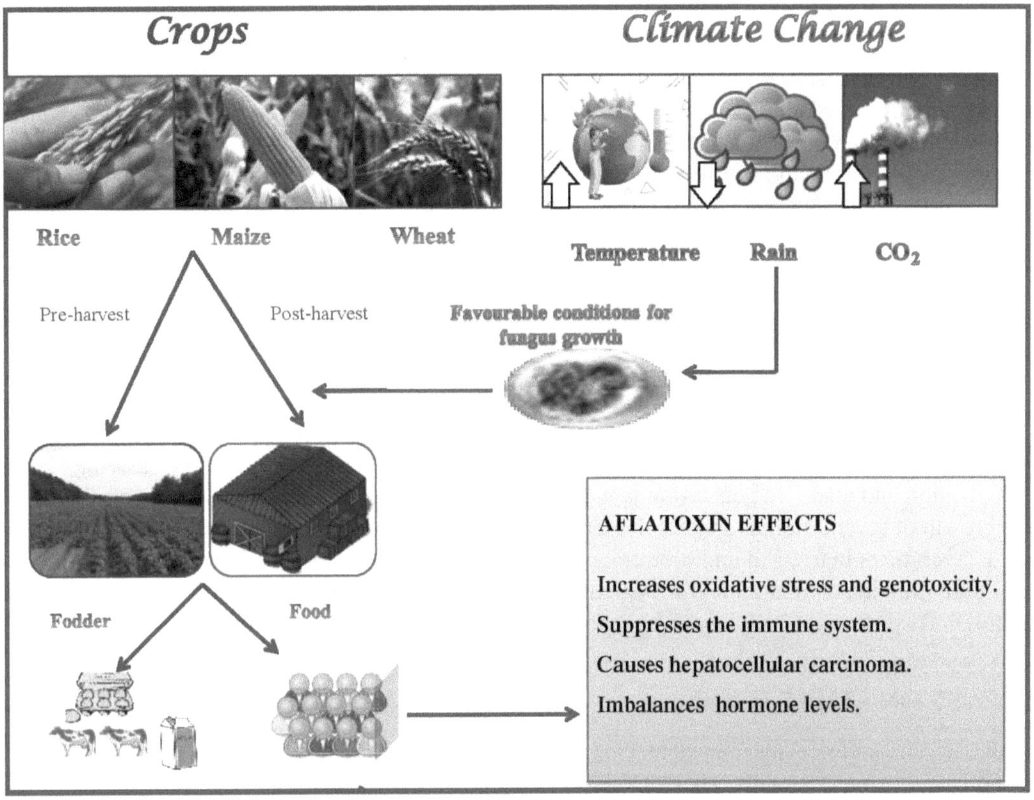


**FIGURE 1.4.1** A schematic representation of climate change effect on aflatoxin contamination.

result of contaminated food or other sources of nourishment, causing acute and chronic aflatoxicosis (18).

## AFLATOXINS PRODUCTION

A series of complex redox mechanisms are involved in the formation of aflatoxins (19). An almost 75-kilobase(kb) DNA cluster shared by *A. flavus* and *A. parasiticus* contains two different transcriptional regulators (*aflR* and *aflS*) and at least 30 co-regulated downstream metabolic genes that are involved in the aflatoxin production pathway(20). Aflatoxins are formed through the conversion of a hexanoyl and seven malonyl-CoA to a condensed polyketide noranthrone by the products of two fatty acid synthase genes, *aflA* (fas-2) and *aflB*(as-1), and a polyketide synthase gene, *aflC* (pksA)(21). The subsequent enzymatic reactions including reduction and oxidation, transform the polyketide intermediate into aflatoxin.

## IMPACT OF CLIMATE CHANGE ON AFLATOXINS CONTAMINATION

Modern techniques like genome sequencing projects, microarray technologies, quantitative proteomics, and RNA-sequencing technologies facilitate understanding the relationship between climate change and molecular reactions. Growth, toxin production, host interaction, and development are all altered by many factors associated with climate change. Aflatoxin synthesis at lower temperatures is strongly correlated with increased water activity (22). Under stressed conditions, the regulatory gene *aflS*, which interacts with *aflR*, is shown to be strongly upregulated, and a high *aflS*:*aflR* ratio is associated with high aflatoxin production (23). In relation to the availability of water, *aflR* and *aflD* are extremely responsive to temperature. Numerous studies indicate that rising $CO_2$, temperature, and water activity all interact to increase stress in plants, which in turn leads to the formation of aflatoxin (24). There are numerous variables involved in the incredibly intricate relationship between aflatoxins and climate change. When addressing the consequences of climate change, it has been observed that there has been a rise in average worldwide air temperatures and changes in the distribution of rainfall; yet, there has also been an increase in weather variability with more extreme occurrences like heat waves, droughts, and excessive precipitation (25). It is challenging to forecast how climate change will affect the distribution and activity of fungal species because they are impacted by various factors, including fungal traits, host availability features, and competitive interactions among microbiota (26,27). Furthermore, the interplay between environmental factors like temperature, water availability, and atmospheric $CO_2$ makes it challenging. Climate, plant, and storage-related issues, as well as non-infectivity factors like insect damage and the bioavailability of (micro)nutrients, all have a significant impact on aflatoxin levels (28). Aflatoxin contamination can occur in two different stages: first, during the crop's growing stage in the field, and second during storage. Consequently, the key component of the agro-ecosystem driving fungal colonization and aflatoxin generation is the climate. It is well known that rising temperatures favor the growth of several fungus species that have the potential to be toxic (29). Climate change is causing temperatures in tropical and subtropical regions to rise, impacting all continents and increasing the contamination of aflatoxin. Warm-weather crops are more susceptible to *A. flavus* infection, and aflatoxin B1 production will be enhanced by the anticipated temperature increase (30).

## IMPACT ON STAPLE FOOD

Maize, wheat, and rice are important food crops in agriculture commodities. Corn crops are sensitive during bloom, especially when their silks are browning (31). Airborne or post-borne spores are most potent during this time. It has been observed that high temperatures and droughts, which frequently coincide during the growing season, probably lead to poor kernel development, the growth of fungi, and maximum production of toxins. Poorly managed storing conditions increase aflatoxin contamination by *A. flavus*. The risk of insect damage is higher in areas exposed to aflatoxins (32). Fungal presence in the field is one of the major factors contributing to post-harvest infection; during the storage stage, *A. flavus* colonized kernels serve as the inoculum for infection (30). The growth of fungus is directly proportional to the moisture content in grains (33). Depending on the method used, rice can be harvested with a moisture content of 16% to 28% before being dried to a level of 13.5% (34). Rice can develop aflatoxins when it is subjected to the right temperature and high humidity (35). As a result, it is critical to assess the likelihood of aflatoxin contamination in rice and byproducts. The health of people and cattle is at risk due to aflatoxin contamination in the food supply. Aflatoxins cause public health risks, food safety issues, and economic challenges due to crop contamination in both developed and developing countries. Billions of tons of agricultural products are produced each year, comprising oilseeds, groundnuts, wheat, rice, barley, corn, sorghum, and soybeans. Every year, more than 25% of the world's harvest is contaminated, resulting in losses of more than 1 million metric tons of food and edibles. The United States (US) maize industry loses an estimated $160 million annually, with poor countries such as Saharan Africa suffering even greater losses (36). The European Union (EU)'s control of aflatoxins is predicted to result in a 64% decrease in African exports of nuts, cereals, oil seeds, and dried fruits, at an estimated annual cost of US$670 million (37). Gambia suffered an overall economic loss in both internal and foreign commerce between 2000 and 2014, accounting for almost US$23 million annually (38). The annual loss resulting from aflatoxin contamination is estimated by the International Institute of Tropical Agriculture (IITA) to be US$1.2 billion worldwide (39). Aflatoxins contamination in Thailand, the Philippines, and Indonesia was determined to cause yearly economic losses exceeding $900 million. In 2018 the United States faced economic damage of $655 million due to contaminated crops. Africans also suffer financial damages of up to millions of dollars due to aflatoxin contamination in peanuts (40). Adequate control measures and excellent agricultural practices must be put in place to lower aflatoxin contamination, safeguard national economies, and preserve farmers' livelihoods (41). Aflatoxin poisoning of food crops and goods derived from them has become a health issue for humans and animals (42,43). The problem is particularly serious in poor nations where, in addition to posing a risk to public health, the products tainted with toxins are leading to huge economic losses in the world food market.

## IMPACT ON HUMAN HEALTH

Rainfall and temperature patterns have been altered due to climate change, leading to extended fungal growth and aflatoxin production in areas that were not favorable to it earlier. Aflatoxin exposure through diet carries many serious health hazards to humans (44). It is recognized that prolonged exposure to aflatoxins can cause cancer, development problems, and damage DNA. Acute poisoning of aflatoxin can cause liver damage and be fatal. Aflatoxin B1, a genotoxic hepatocarcinogen, is a major risk factor for hepatocellular carcinoma (45). Aflatoxins are also responsible for hormonal disturbances in males and females. A notable drop in weight and cholesterol levels has also been observed on exposure to aflatoxins. Interleukin (IL)-1b, tumor necrosis factor alpha (TNF-a), and IL-6 and are among the pro-inflammatory cytokines that are released in response to AFB1 (46).

## STRATEGIES FOR MANAGING AFLATOXIN CONTAMINATION

Aflatoxin contamination can be prevented using various methods, including adequate food storage to avoid mold development, agricultural monitoring for aflatoxin contamination, and the implementation of regulatory measures to limit aflatoxin levels in food items. Furthermore, dietary variety and education on food safety procedures might help limit aflatoxin exposure in affected communities. To protect public health, numerous nations have imposed strict laws regarding the use of aflatoxin-contaminated feed and food for animal and human consumption. Various techniques have been developed to reduce the level of aflatoxin contamination in crops, such as chemical, biological, and physical methods. The introduction of nontoxic molds, bacteria, and yeasts lowers the amount of aflatoxin contamination in food products and agriculture. The detrimental impacts of these microbes, such as space and nutritional conflict, or biological interactions, such as antibiosis, are employed to regulate aflatoxins. Use of non-toxigenic strains of *A. flavus* based on the competition between non-toxicogenic and toxicogenic strains and the employment of a variety of bacteria to lower aflatoxin levels including *Stenotrophomonas sp.,Bacillus sp., Enterococcus sp.,* has helped in the management of aflatoxin production (47). A biological control product called Aflasafe is being used to control aflatoxin contamination in crops, especially in grains like sorghum, maize, and groundnuts. Aflasafe also removes native strains of *A. flavus* from their natural habitat. These strains serve as the foundation for producing Aflasafe formulations that are unique to the area from which they are gathered(48). Aflatoxin control strategies often use chemical agents to limit the growth and production of aflatoxin. The first chemicals employed to tackle aflatoxins were insecticides and fumigants. Several chemicals, including NaHSO3, $Ca(OH)_2$, formaldehyde, NaClO, $Na_2[B_4O_5(OH)_4]{\cdot}8H_2O$, and sorbents, have been shown to considerably reduce the amounts of aflatoxins in various foods. Citric acid and ozone gas can decrease AFB1 levels in walnuts and poultry feed, respectively; however, the high expense of ozonation may limit its use in post-harvest activities. There are many fungicides available in the market, such as Voriconazole, Amphotericin B, caspofungin and posaconazole, that reduce aflatoxin contamination in crops. Many chemicals, acids, and oxidizing agents like NaHSO3, calcium hydroxide, formaldehyde, and NaOCl are also being used (49). Physical treatments, such as steam under pressure, roasting, and other cooking processes, successfully regulate or decrease aflatoxin contamination in some crops. High fluid pressure, thermal processing, and electric field treatments can lower aflatoxin levels in crops. Ultraviolet-C (UVC) and gamma irradiation have significantly lowered AFB1 levels in crops. Preventive efforts, such as the establishment of excellent practices in manufacturing and agriculture and suitable storage conditions, have been established to reduce aflatoxin contamination; nevertheless, these methods are not always successful (50,51).

## CONCLUSION

*Aspergillus* species produce aflatoxins, which are among the most well-known naturally occurring compounds that cause cancer. *A. parasiticus, A. flavus*, and *A. nomius* are the three primary species that create aflatoxins. Millions of hectares of agricultural land are affected by aflatoxin pollution worldwide, with some regions losing production of crops. Aflatoxins are a persistent problem that threatens the viability of agricultural communities. Climate change is bound to have an impact on mold colonization patterns, agriculture, and the excretion of several secondary metabolites, including aflatoxins. Aflatoxin pollution may significantly reduce the area that can be used for commercial crop cultivation as the temperature warms and weather patterns become more erratic. An integrated strategy that incorporates viewpoints from several disciplines, like economics, food technology, and social studies, is recommended for successful management of the risks that climate change poses to food safety. Methods for managing aflatoxins, detoxifying, and altering cropping patterns are all viable possibilities. The research findings and reliable methods may help prioritize approaches to dealing with the diverse effects of climate change–related hazards around the world. Further research should concentrate on finding the regulatory mechanisms that are influenced by climate change–related factors, particularly those that are known to affect aflatoxin production.

## ACKNOWLEDGMENT

The authors are grateful to the Centre for Medical Biotechnology and the Department of Microbiology, Maharshi Dayanand University, Rohtak, for the infrastructure to carry out this work. SK and KG acknowledge the University Research Scholarship, Maharshi Dayanand University for financial support and SP is thankful to the Department of Science & Technology, Ministry of Science and Technology, Government of India, New Delhi for providing the INSPIRE Fellowship.

## REFERENCES

1. Lee, H., Calvin, K., Dasgupta, D., et al., 2023. Climate change 2023: Synthesis report. Contribution of working groups I, II and III to the sixth assessment report of the intergovernmental panel on climate change. The Australian National University.
2. Assunção, R., Martins, C., Viegas, S., et al., 2018. Climate change and the health impact of aflatoxins exposure in Portugal–an overview. *Food Additives & Contaminants: Part A 35*(8): 1610–1621.
3. Van der Fels-Klerx, H.J., Liu, C. and Battilani, P., 2016. Modelling climate change impacts on mycotoxin contamination. *World Mycotoxin Journal 9*(5):717–726.
4. Milani, J.M., 2013. Ecological conditions affecting mycotoxin production in cereals: A review. *VeterinarniMedicina 58*(8): 405–411.
5. Valencia-Quintana, R., Milić, M., Jakšić, D., et al., 2020. Environment changes, aflatoxins, and health issues, a review. *International Journal of Environmental Research and Public Health 17*(21): 7850.
6. Gholami-Shabani, M., Shams-Ghahfarokhi, M. and Razzaghi-Abyaneh, M., 2017. Aflatoxins and aflatoxigenic fungi in Iran: A systematic review of the past, present, and future. *MycologiaIranica 4*(2): 65–84.
7. Blount, W.P., 1961. Turkey" X" disease. *The Journal of the British Turkey Federation 9*: 52–54.
8. Navale, V., Vamkudoth, K.R., Ajmera, S. and Dhuri, V., 2021. *Aspergillus* derived mycotoxins in food and the environment: Prevalence, detection, and toxicity. *Toxicology Reports 8*: 1008–1030.
9. Mahato, D.K., Lee, K.E., Kamle, M., et al., 2019. Aflatoxins in food and feed: An overview on prevalence, detection and control strategies. *Frontiers in Microbiology 10*: 483502.
10. Paterson, R.R.M. and Lima, N., 2010. How will climate change affect mycotoxins in food? *Food Research International 43*(7): 1902–1914.
11. Kos, J., Radić, B., Radović, R., et al., 2024. Aflatoxins in maize, milk and dairy products from Serbia. *Food Additives & Contaminants: Part B 17(4)*: 296-307.

12. Battilani, P., Toscano, P., Van der Fels-Klerx, H.J., et al., 2016. Aflatoxin B1 contamination in maize in Europe increases due to climate change. *Scientific Reports 6*(1): 24328.
13. Iqbal, S.Z., Selamat, J. and Arino, A., 2016. Mycotoxins in food and food products: Current status. In *Food Safety: Basic Concepts, Recent Issues, and Future Challenges* (113–123). Springer Germany.
14. Kobayashi, T., Abe, K., Asai, K., et al., 2007. Genomics of *Aspergillus oryzae*. *Bioscience, Biotechnology, and Biochemistry 71*(3): 646–670.
15. Rokas, A., Payne, G., Fedorova, N.D., et al., 2007. What can comparative genomics tell us about species concepts in the genus *Aspergillus*? *Studies in Mycology 59*(1) 11–17.
16. Morales-Moo, T., Hernández-Camarillo, E., Carvajal-Moreno, M., et al., 2020. Human health risk associated with the consumption of aflatoxins in popcorn. In *Risk Management and Healthcare Policy* 13 (2583–2591). Dove Medical Press.
17. Kumar, A., Pathak, H., Bhadauria, S. and Sudan, J., 2021. Aflatoxin contamination in food crops: Causes, detection, and management: A review. *Food Production, Processing and Nutrition 3*(1): 9.
18. Wang, B., Mahoney, N.E., Pan, Z., Khir, R., Wu, B., Ma, H. and Zhao, L., 2016. Effectiveness of pulsed light treatment for degradation and detoxification of aflatoxin B1 and B2 in rough rice and rice bran. *Food Control 59*: 461–467.
19. Georgianna, D.R. and Payne, G.A., 2009. Genetic regulation of aflatoxin biosynthesis: From gene to genome. *Fungal Genetics and Biology 46*(2): 113–125.
20. Ehrlich, K.C., Mack, B.M., Wei, Q., et al., 2012. Association with AflR in endosomes reveals new functions for AflJ in aflatoxin biosynthesis. *Toxins 4*(12) :1582–1600.
21. Liao, J., He, Z., Xia, Y., Lei, Y. and Liao, B., 2020. A review on biosynthesis and genetic regulation of aflatoxin production by major *Aspergillus* fungi. *Oil Crop Science 5*(4): 166–173.
22. OBrian, G.R., Georgianna, D.R., Wilkinson, J.R., et al., 2007. The effect of elevated temperature on gene transcription and aflatoxin biosynthesis. *Mycologia 99*(2): 232–239.
23. Schmidt-Heydt, M., Rüfer, C.E., Abdel-Hadi, A., Magan, N. and Geisen, R., 2010. The production of aflatoxin B1 or G1 by *Aspergillus parasiticus* at various combinations of temperature and water activity is related to the ratio of aflS to aflR expression. *Mycotoxin Research 26*: 241–246.
24. Schmidt-Heydt, M., Abdel-Hadi, A., Magan, N. and Geisen, R., 2009. Complex regulation of the aflatoxin biosynthesis gene cluster of *Aspergillus flavus* in relation to various combinations of water activity and temperature. *International Journal of Food Microbiology 135*(3): 231–237.
25. Paterson, R.R.M. and Lima, N., 2011. Further mycotoxin effects from climate change. *Food Research International 44*(9): 2555–2566.
26. Frisvad, J.C., 2015. Taxonomy, chemodiversity, and chemoconsistency of *Aspergillus*, *Penicillium*, and *Talaromyces* species. *Frontiers in Microbiology 5*: 126088.
27. Madgwick, J.W., West, J.S., White, R.P., et al., 2011. Impacts of climate change on wheat anthesis and fusarium ear blight in the UK. *European Journal of Plant Pathology 130*: 117–131.
28. Medina, A., Rodríguez, A., Sultan, Y. and Magan, N., 2015. Climate change factors and *Aspergillus flavus*: Effects on gene expression, growth and aflatoxin production. *World Mycotoxin Journal 8*(2): 171–179.
29. Gregory, P.J., Johnson, S.N., Newton, A.C. and Ingram, J.S., 2009. Integrating pests and pathogens into the climate change/food security debate. *Journal of Experimental Botany 60*(10): 2827–2838.
30. Shekhar, M., Singh, N., Bisht, S., Singh, V. and Kumar, A., 2018. Effects of climate change on occurrence of aflatoxin and its impacts on maize in India. *International Journal of Current Microbiology and Applied Sciences 7*(6): 109–116.
31. Focker, M., van Eupen, M., Verweij, P., et al., 2023. Effects of climate change on areas suitable for maize cultivation and aflatoxin contamination in Europe. *Toxins 15*(10): 599.
32. Dorner, J.W., Cole, R.J., Connick, W.J., et al., 2003. Evaluation of biological control formulations to reduce aflatoxin contamination in peanuts. *Biological Control 26*(3): 318–324.
33. Reed, C., Doyungan, S., Ioerger, B. and Getchell, A., 2007. Response of storage molds to different initial moisture contents of maize (corn) stored at 25 C, and effect on respiration rate and nutrient composition. *Journal of Stored Products Research 43*(4): 443–458.
34. Saad-Hussein, A., Taha, M.M., Beshir, S., et al., 2014. Carcinogenic effects of aflatoxin B1 among wheat handlers. *International Journal of Occupational and Environmental Health 20*(3): 215–219.
35. Reddy, K.R.N., Reddy, C.S. and Muralidharan, K., 2009. Potential of botanicals and biocontrol agents on growth and aflatoxin production by *Aspergillus flavus* infecting rice grains. *Food Control 20*(2): 173–178.
36. Jallow, A., Xie, H., Tang, X., Qi, Z. and Li, P., 2021. Worldwide aflatoxin contamination of agricultural products and foods: From occurrence to control. *Comprehensive Reviews in Food Science and Food Safety 20*(3): 2332–2381.

37. Omari, R., Tetteh, E.K., Baah-Tuahene, S., Karbo, R., Adams, A. and Asante, I.K., 2020. Aflatoxins and their management in Ghana: A situational analysis. *FARA Research Report 5*(20) :80.
38. Turner, P.C., Collinson, A.C., Cheung, Y.B., et al.,2007. Aflatoxin exposure in utero causes growth faltering in Gambian infants. *International Journal of Epidemiology36*(5):1119–1125.
39. Soler-Labajos, N. and Jiménez-Zarco, A.I., 2016. Country brand management: Assessing the role of social media in creating the image of Marca Espana (Spain brand). In *Managing public relations and brand image through social media* (89–115): IGI Global.
40. Schmale, D.G. and Munkvold, G.P., 2009. Mycotoxins in crops: A threat to human and domestic animal health. *The Plant Health Instructor 3*(3): 340–353.
41. Fox, E.M. and Howlett, B.J., 2008. Secondary metabolism: Regulation and role in fungal biology. *Current Opinion in Microbiology 11*(6): 481–487.
42. Adeyeye, S.A., 2016. Fungal mycotoxins in foods: A review. *Cogent Food & Agriculture 2*(1): 1213127.
43. Jelinek, C.F., Pohland, A.E. and Wood, G.E., 1989. Worldwide occurrence of mycotoxins in foods and feeds an update. *Journal of the Association of Official Analytical Chemists 72*(2): 223–230.
44. Peles, F., Sipos, P., Kovács, S., Győri, Z., Pócsi, I. and Pusztahelyi, T., 2021. Biological control and mitigation of aflatoxin contamination in commodities. *Toxins 13*(2): 104.
45. Hatipoglu, D. and Keskin, E., 2022. The effect of curcumin on some cytokines, antioxidants and liver function tests in rats induced by aflatoxin B1. *Heliyon 8*(7).
46. Islam, F., Das Trisha, A., Hafsa, J.M., Hasan, A., Degen, G.H. and Ali, N., 2021. Occurrence of aflatoxin M1 in human breast milk in Bangladesh. *Mycotoxin Research 37*(3): 241–248.
47. Ali, S., Hassan, M., Essam, T., Ibrahim, M.A. and Al-Amry, K., 2021. Biodegradation of aflatoxin by bacterial species isolated from poultry farms. *Toxicon 195*: 7–16.
48. Bandyopadhyay, R., Ortega-Beltran, A., Akande, A., et al., 2016. Biological control of aflatoxins in Africa: Current status and potential challenges in the face of climate change. *World Mycotoxin Journal 9*(5): 771–789.
49. Carvajal, M. and Castillo, P., 2009. Effects of aflatoxins contaminating food on human health. *Tropical Biology and Conservation Management 7*: 60–84.
50. Adejumo, T.O. and Adejoro, D.O., 2014. Incidence of aflatoxins, fumonisins, trichothecenes and ochratoxins in Nigerian foods and possible intervention strategies. *Food Science and Quality Management 31*: 127–147.
51. Fandohan, P., Zoumenou, D., Hounhouigan, D.J., et al., 2005. Fate of aflatoxins and fumonisins during the processing of maize into food products in Benin. *International Journal of Food Microbiology 98*(3): 249–259.

# Section 2

## Aflatoxigenic Fungi: Impact on Soil, Plant, and Human Health, and the Agricultural Environment

# 2.1 Aflatoxigenic Fungi and Their Ecology in Soil

*Janet Jan-Roblero, Juan Antonio Cruz-Maya,*
*José Silvestre Mendoza-Figueroa,*
*Juan Carlos Cancino-Díaz*

## INTRODUCTION

Toxins are protein compounds produced by organisms such as bacteria and fungi that specifically affect the function of target cells, tissues, or organs, sometimes leading to cell death and even organism death. Fungi are ubiquitous and present in significant quantities in soil, with high fungal diversity. Soil saprophytic fungi perform several functions, including the degradation of organic matter. Although they contribute to soil equilibrium, they can produce compounds, such as toxins, that affect humans and animals. These mycotoxins can contaminate agricultural products during harvest or post-harvest. Aflatoxin is one such mycotoxin of major food concern; among the aflatoxin-producing fungi, the genus *Aspergillus* is the major producer of aflatoxin. The International Agency for Research on Cancer (IARC) has classified this mycotoxin as carcinogenic. Aflatoxin ingested by humans or animals through contaminated agricultural foods can have toxic effects on liver tissue. The most important damage is the induction of liver cancer, but it has also been seen to affect other organs such as the pancreas, kidney and bladder. Other effects caused by aflatoxin include DNA mutations, immunosuppression, and malnutrition (1).

Aflatoxin produced by the fungus can contaminate agricultural products such as grains and oilseeds, animal feeds, nuts, herbs, and other crops, but the ingestion of aflatoxin by animals means that this mycotoxin can also be present in products derived from animals such as meat, eggs, milk, and other products. Aflatoxin contamination of agricultural and livestock products causes major economic losses to the food industry and poses a high risk to consumer health. The food industry worldwide has significant problems with aflatoxin contamination, and the global import and export market for agricultural products has adopted control measures to reduce aflatoxin contamination of food. According to reports, 25% of global crops contain molds and their toxins, and some agricultural commodities (2%) contain very high levels of mycotoxins (2). The death of 100,000 turkeys in the United Kingdom in the 1960s due to aflatoxin contamination was the most emblematic case of the impact of aflatoxin. In addition, economic losses have been reported in the peanut crop in the state of Georgia in the United States, due to aflatoxin contamination (3).

It is estimated that approximately 500 million people worldwide are exposed to aflatoxin-contaminated food (4). This has led to the introduction of regulatory measures to control the contamination of food by this mycotoxin. The European Union has established permissible limits for fungal toxins in food and feed (5). China has also implemented a strict program for aflatoxin B1 (AFB1) in food; the maximum allowed in cereals is 5–20 μg/kg, whereas in dietetic foods it is 0.5 μg/kg for AFB1 and aflatoxin M1 (AFM1) and total absence in infant formula (6).

The source of crops, the safety of agricultural products and the study of aflatoxin-producing fungi in soil have been addressed with great interest. This chapter discusses important points about aflatoxin-producing fungi from the ecological perspective to the factors involved in the production of this mycotoxin, with the aim of proposing some alternatives for pre- and post-harvest control.

DOI: 10.1201/9781003487043-7

## GENERAL CHARACTERISTICS OF SOIL FUNGI

Fungi belonging to the *Aspergillus* group are classified as imperfect filamentous fungi. This group of fungi includes 250 species known to date. In general, this group of fungi are producers of secondary metabolites that are beneficial to humans (7); for example, *A. terreus* is the producer of the compound lovastatin, which is used as a drug for the regulation of cholesterol levels. In contrast, within this group of *Aspergillus* fungi there are others that produce secondary metabolites that are not beneficial to humans. This group of fungi produces toxins that can cause cancer when ingested by humans or animals, and are therefore called mycotoxins.

*A. flavus* is an important producer of mycotoxins. One of them is known as aflatoxin and causes great economic losses in the food industry. In soil, *A. flavus* is one of the most common fungi because soil is its natural environment. In the soil, it lives as a saprophytic fungus and can therefore survive on many sources of organic nutrients present in the soil, but it can also be found in stored grains (8). *A. flavus* grows at a temperature of 12–48°C, but 28–37°C is the optimum temperature for its growth.

Morphologically, the fungus has the important characteristic of cell differentiation depending on environmental conditions. In the reproductive form, the fungus adopts the mycelial form, or the asexual spore form known as conidia. When nutritional conditions are unfavorable, there is a process of cell differentiation from the mycelial form to resistant structures called sclerotia, which allows the fungus to survive in unfavorable environmental conditions. When conditions improve, sclerotia germinate to produce new mycelium and conidiophores, which store conidia (9) that eventually form colonies.

*A. flavus* does not have the ability to infect plants, requires the plant to be in unfavorable conditions to do it, and is considered an opportunistic pathogen. Once the fungus becomes established on plants, the effect spreads to other surrounding plants, causing agricultural losses: corn, cotton, peanuts, and nut trees such as Brazil nuts, pecans, pistachios, and walnuts. However, the presence of the *A. flavus* fungus does not indicate food contamination because there are *A. flavus* strains that do not produce aflatoxins (non-aflatoxigenic). The aflatoxin-producing fungi are the ones of concern because they can contaminate pre-harvest crops with aflatoxins. Another ability of *A. flavus* is to colonize grains after harvest and persist for long periods during storage, ultimately causing grain deterioration and aflatoxin contamination. Because *A. flavus* lacks a specific host (10), it can be established in seeds of both monocots and dicots, and seeds produced both above ground (maize) and below ground (peanuts) (11).

## THE ECOLOGY OF *ASPERGILLUS* IN SOIL

Basically, *A. flavus* is a saprophytic member of the soil, which can survive and colonize the soil with ease. As a nutritional source, this fungus assimilates the waste organic matter contained in the soil. The life cycle of *A. flavus* can be divided into two phases: one corresponding to the colonization of the fungus on the soil organic matter, and the other corresponding to the infection of the plant. The seasons of the year influence the life cycle of *A. flavus*. Spring and late summer are suitable for the beginning of fungal growth; in this situation the morphological structures of sclerotia or conidia present in plant residues or on the soil surface germinate to produce mycelia and subsequently produce numerous conidiophores, increasing the number of conidia (12). These new conidia are potential inoculants that infect crop plants. During these seasons, there is an abundant amount of *A. flavus* dispersed in the soil and in the organic matter, which represents the initial source of primary inoculation of the underground parts of the plants, especially in peanuts. Another means of plant infestation is that conidia can be airborne or carried by vectors such as conidia-bearing insects (13). Subsequently, growing plants, whose tissues are infected with conidia can serve as sources of secondary conidial inoculation, in which conidia are dispersed to new, uninfected plant tissues and thus spread (Figure 2.1.1).

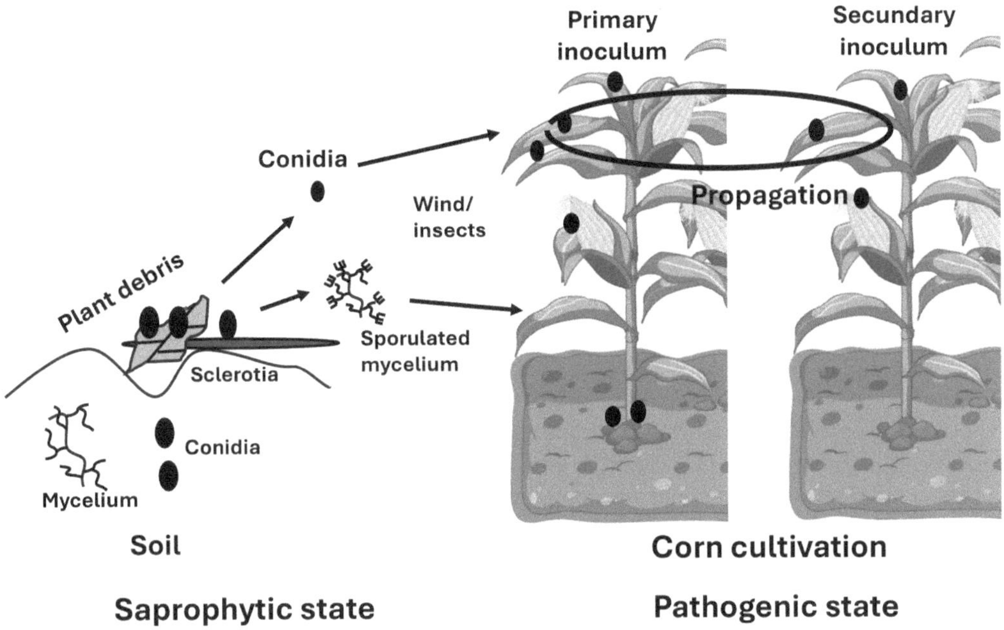


**FIGURE 2.1.1** Ecology of *Aspergillus flavus*. In the saprophytic stage, the fungus is found in soil and plant debris (organic matter) In the form of mycelia, sclerotia, and conidia (more abundant). In the pathogenic stage, conidia and sporulated mycelium are transported to crops by air or insects for primary inoculation of the plant. Conidia in the soil can infect the plant root. Secondary inoculation occurs from plants infected by the conidia of the fungus to another new plant by air or insects that spread the fungus to the crop.

Colonization of *A. flavus* in the plant can be successful depending on the health status of the plant. Because this fungus is an opportunistic pathogen, healthy plants are not affected by *A. flavus*, whereas plants suffering from stress, such as heat stress or drought, are highly susceptible to *A. flavus* (14). This shows that plant health is important for the fungus to take hold. Therefore, the inoculation potential of the fungus is environmental and host dependent.

Despite our understanding of how the conidia of the fungus inoculate plants, there is little information about the saprophytic activities of this fungus in soil, although it has been shown that under these saprophytic conditions, there is aflatoxin synthesis by *A. flavus* in soils. On the other hand, the colonization of plant structures by *A. flavus* does not necessarily mean that aflatoxin is produced in large quantities. This indicates that aflatoxin production by the fungus is regulated and depends on environmental conditions for its production.

Although *A. flavus* is a soil inhabitant, it does not mean that it is the only mycotoxin producing fungus. There are other species such as *A. parasiticus* and *A. nomius*, although the ecology of these in soil is being studied.

## *ASPERGILLUS* IN AGRICULTURAL SOILS

As mentioned above, *Aspergillus* is a soil inhabitant and is therefore likely to be found in agricultural soils. Although *A. flavus* is one aflatoxin-producing fungus, this does not mean that all *A. flavus* isolates present in agricultural soils are aflatoxin producers because there are isolates that do not produce aflatoxin and are referred to as non-aflatoxigenic fungi. Overall, aflatoxigenic fungi have been shown to be common in any setting, with their frequency ranging from 50% to 80% in soil, the remainder being non-aflatoxigenic fungi, and so the relative distribution is modulated by many factors. These factors include cropping history, crop species, crop management, soil composition and environmental conditions (15). Similarly, the drought is another factor that influences the population of the fungus; soils affected by severe drought significantly affect the population of *A. flavus* with a significant reduction; however, *A. flavus* conidia can remain dormant in the soil during drought and when good soil conditions, nutrient sources and adequate humidity are available, they can germinate and restore the population of *A. flavus*.

Corn soils have *A. flavus* population densities ranging from 200 to more than 300,000 colony-forming units (CFUs) per gram of soil. This represents less than 0.2% to less than 8% of the total fungal population of the arable soil, indicating a significant abundance of this fungal species. Soil organic matter is the predominant factor associated with the maintenance of the *A. flavus* population. Regarding the presence of the fungus on the plant, it has been observed that the amount of fungus in a maize crop, *A. flavus* multiplies in a wide range of log 3 to log 6 CFUs per gram of plant tissue (16). Similar levels have been observed in vegetative tissues of soybean and cotton residues.

The composition of the fungal rhizosphere community associated with the tea plant *Camellia sinensis* has been studied, and it was found that *Camellia* disease is strongly associated with the regulation of soil properties (17). Another study concerning *Rehmannia* cultivars showed that its root system depends on the amount and composition of fungal species in the soil rhizosphere (18).

Due to technical limitations in the isolation and detection of fungi in soil, only a few species have been isolated from experimental environments. Thus, they do not fully reflect what occurs in agricultural or cultivated soils. Another limitation in isolating cultivable fungi from soil is the culture media, which has allowed the recovery of only 17% of the known fungal species, which also limits the general information on soil fungi (19).

Despite technological limitations, it is now known that a single fungus can produce multiple mycotoxins, and that a single mycotoxin can be produced by multiple fungi, complicating the study of mycotoxins. Approximately 300 mycotoxins produced by 150 fungal species are currently known. Soil geography is associated with the presence of some specific fungi, suggesting that the predominant mycotoxins may vary depending on the region. In the case of aflatoxins and ochratoxins that are predominant in tropical and subtropical regions, and regarding *A. flavus*, maize, rice, wheat, cottonseed, peanuts, and nuts are the most affected crops (20). Within the same species of aflatoxin-producing *A. flavus*, it has been shown that there is diversity among isolates obtained from different geographical locations. As mentioned above, *A. flavus* is widely distributed in soils. It contaminates grains worldwide (3). *A. flavus* grows at 12–34°C, but toxin production occurs at 20–30°C (21, 22). In the soil of the Yangtze River Basin, *A. flavus* is widely distributed in the soil, so this soil is at the highest risk of contamination (23).

Studies on the distribution of soil microbiome and contaminant toxins have been conducted worldwide. These studies have shown a correlation between mycotoxin contamination and soil type. The soil type and its physicochemical characteristics should be taken into consideration because it is dependent on the distribution of fungal microbiota and the degree of mycotoxin contamination. This indicates that there is an important relationship between soil type and fungal community. Based on this principle, the distribution of strains was analyzed to determine the aflatoxin contamination in different agricultural production areas in China. The results showed that aflatoxin is not present in all peanut growing areas because *A. flavus* non-toxigenic is prevalent in some areas and aflatoxin-producing isolates are present in other areas (24). A procedure for searching for non-toxigenic strains of the fungus in peanut soil was proposed by Zhang et al. (25).

Aflatoxin contamination in peanut crop soils in China was predicted by the number of aflatoxin-producing fungal colonies, and these prediction data were used to regulate and prevent the presence of aflatoxins in peanut crops (26). Similar results were obtained by Wang et al. (23). These studies showed that the Yangtze River basin has the highest distribution of aflatoxins and the highest risk of aflatoxin contamination due to the unique climatic and environmental conditions in this region.

On the other hand, in the peanut producing areas of Xiangyang, Hubei Province, the microbial load of *A. flavus* and the virulence of the fungus are higher in clay loam soils than in sandy loam soils. In addition, the distribution of the fungus is associated with irrigation conditions (25).

## CONTAMINATION WITH AFLATOXIN

Aflatoxin is colorless, odorless, and extremely toxic to humans. Coumarin and difuran rings are the main components of the chemical structure, and there may be derivatives and isomers (27). As mentioned above, there are different types of mycotoxins, and aflatoxins are no exception. Based on their blue (B) or green (G) fluorescence under UV light and their relative mobility by thin-layer chromatography on silica gel, aflatoxins can be classified into types B1, B2, G1 and G2. Types B1 and B2 are produced by *A. flavus*, whereas types B1, B2, G1 and G2 are produced by *A. parasiticus*. In addition, *A. flavus* can also produce other mycotoxins such as cyclopiazonic acid, kojic acid, beta-nitropropionic acid, aspertoxin, aflatrem and aspergillic acid (28), but these mycotoxins have been little studied.

To date, the detection of mycotoxins in soil is low, in the microgram range, and there are no reports of higher levels, suggesting that aflatoxin contamination has not increased recently. The soil type contributes to the mycotoxin retention capacity. Clay soil has high retention capacity for toxic compounds, but sandy soil has the potential to leach compounds (29), indicating that the physico-chemical composition of the soil is an important factor in the retention of mycotoxins. On the other hand, soil can degrade mycotoxins. In 1980, the rate of aflatoxin degradation in soil was determined using C14-labeled aflatoxin. The result showed that degradation by soil microorganisms is a contributing factor in the degradation of mycotoxins (30). Thus, the risk of aflatoxin contamination in soil was low, and the main risk of contamination is during the storage period of agricultural products after harvest. Another study relating the presence of aflatoxin in soil and maize crops showed that AFB1 can be rapidly degraded in soil at a temperature of 28.8°C (31).

Regarding the presence of aflatoxin in water, *Aspergillus oryzae* was detected in a tank of stored water in 1997, showing that there is a potential risk of mycotoxins (32). However, there are no reports of their detection in aquatic sediments, which is an indication that these environments are of low risk.

Aflatoxin-producing fungi such as *A. flavus, A. parasiticus,* and *A. nomius* are present in soil and are possible sources of aflatoxin contamination of crops and agricultural products, particularly cereals, oilseeds, and nuts. Several studies have shown that soil is the major source of aflatoxin contamination in crops, and that the fungus *A. flavus* is very common in crop soils (33, 34). Horn et al. (35) suggested that climate and crop composition are strongly associated with aflatoxin colonization and toxicity. Peanut is a crop highly susceptible to infection by aflatoxin-producing fungi because peanut shells are in direct contact with the soil; therefore, aflatoxin contamination in peanuts is mainly due to aflatoxin-producing *Aspergillus* isolates present in the soil.

On the other hand, the direct contact between soil and plant roots and nutrient exchange are the main routes of aflatoxin entry into agricultural products, causing aflatoxin contamination of crops (36). Although it is impossible to eliminate fungi from the soil to prevent aflatoxin contamination of crops, there are cultivation techniques that control or regulate this contamination process. It has been reported that crop rotation can change the population structure of fungi in crop soils and help control mycotoxin contamination. Crop rotation has been used on soils historically cultivated with cotton, and switching to groundnut, which is highly affected by mycotoxins, has reduced the population of *A. flavus* (37). In addition, temperature is another factor that can control contamination because aflatoxin production is affected by temperature.

Regarding the different derivatives of aflatoxins, it has been documented that not all *A. flavus* isolates are aflatoxin producers; 50% of *A. flavus* strains produce aflatoxin and the aflatoxin B variant. Regarding other fungal species, most strains of *A. parasiticus* have been found to produce both aflatoxins B and G (38). In the case of crops, the main aflatoxins present are AFB1, AFG1, AFB2 and AFG2, and their toxicity levels are AFB1>AFG1>AFB2>AFG2>AFG2 (27). In this regard, AFB1 has been regulated, and the International European Standards has reported that AFB1 limits are less than or equal to 2 μg/kg and should not exceed 4 μg/kg of total aflatoxin (AFT) (39). The standard in China indicates that the maximum limit for AFB1 is 20 μg/kg in peanuts, maize and their products, 10 μg/kg in rice and oils, and 5 μg/kg in cereals, beans, fermented foods and spices.

In terms of microscopic morphology, *A. flavus* can be classified as either L-type or S-type, and this morphology is associated with virulence and the size of the sclerotia. S-type strains of *A. flavus* produce a few sclerotia <400 μm in size, whereas L-type strains are characterized by fewer but longer sclerotia (40). It has been observed that most L-type strains are not aflatoxigenic, whereas most S-type strains are highly toxigenic. These morphological types can be controlled by crop rotation and soil temperature, with an increase in soil temperature favoring S-type strains (41). Therefore, *A. flavus* contamination can be controlled by changes in soil temperature and crop rotation.

## FACTORS INFLUENCING AFLATOXIN PRODUCTION

As mentioned above, not all strains of *A. flavus* are aflatoxin producing, and those that do produce the mycotoxin are regulated in their production or expression by many factors, including carbon and nitrogen source, water activity and temperature, pH, and reactive agents. These factors, which regulate aflatoxin production in *A. flavus*, have been evaluated at the experimental level and represent a promising control to reduce aflatoxin production. These factors that control the expression of genes involved in aflatoxin synthesis are discussed below (Figure 2.1.2).

## CARBON SOURCE

The synthesis of aflatoxin in the fungi can be controlled by the source of carbon, nitrogen, amino acids, lipids, trace elements and other nutritional factors. Among them, the biosynthesis of aflatoxin is influenced by carbon and nitrogen sources. Simple sugars such as glucose, sucrose, fructose, and maltose induce aflatoxin formation, whereas peptone, sorbose, or lactose do not induce aflatoxin formation in *A. flavus* (42). It has been observed that enzymes involved in the degradation of sugars are associated with aflatoxin production, as in the case of the high activity of the enzyme alpha-amylase and the induction of aflatoxin production in *A. flavus* (43). At present, the genes involved in aflatoxin synthesis are known, and it has been observed that the genes involved in sugar utilization are in *loci* genetically close to those of aflatoxin, indicating that there is a genetic relationship between sugar metabolism and aflatoxin synthesis (44). On the other hand, lipid substrates are good carbon sources, and their utilization is related to aflatoxin production because it has been shown

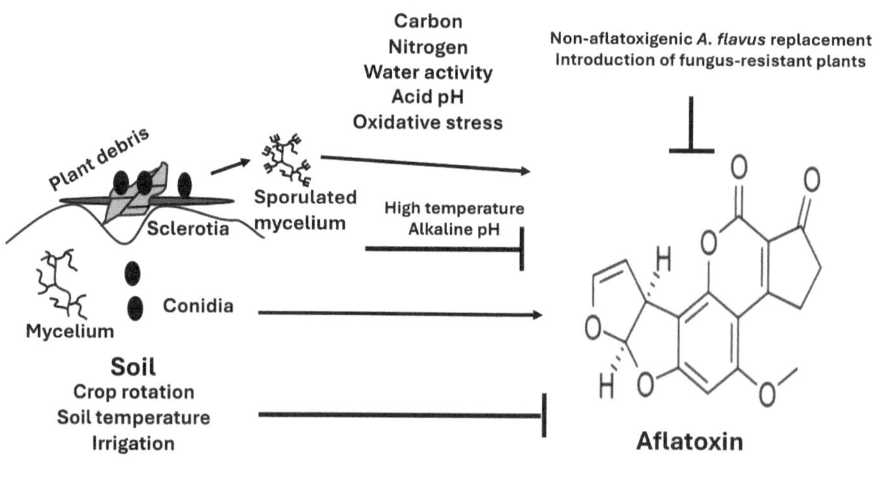


**FIGURE 2.1.2** Control strategies for aflatoxin production. Sporulating mycelium can induce aflatoxin production by carbon and nitrogen sources, water activity, acid pH and oxidative stress. High temperature and alkaline pH inhibit aflatoxin production. Sporulation induces aflatoxin formation. On the other hand, other strategies are soil management for cultivation such as crop rotation, soil temperature and irrigation. Non-aflatoxigenic fungi or fungus-resistant plants can be introduced into the soil to reduce aflatoxin production.

that lipid substrates induce the expression of the *lipA* gene (a gene involved in lipid metabolism) and subsequent aflatoxin production (45).

## NITROGEN

Nitrogen is closely associated with aflatoxin production. Culture media containing some nitrogen compounds such as asparagine, aspartate, alanine, ammonium nitrate, ammonium nitrite, ammonium sulfate, glutamate, glutamine and proline can induce aflatoxin production, whereas culture media containing sodium nitrate and sodium nitrite as nitrogen sources do not induce aflatoxin. Nitrate suppresses the formation of averufin and aflatoxin in the fungus because nitrate inhibits the expression of genes involved in the aflatoxin synthesis (46). In *A. flavus*, tryptophan and tyrosine are inhibitors of aflatoxin synthesis (47).

## TEMPERATURE

Aflatoxin synthesis is directly affected by temperature (14); when the temperature was increased to more than 36°C, aflatoxin production was completely inhibited. Thus, a microarray assay showed that the fungus grown at a high temperature decreased the expression of genes involved in aflatoxin synthesis.

## CROP pH

The biosynthesis of aflatoxin in *A. flavus* takes place in an acidic medium but is inhibited in an alkaline medium. The mechanism by which pH is related to aflatoxin synthesis appears to be at the molecular level, involving the transcriptional regulatory protein PacC (48). pH is capable of regulating PacC protein expression, and PacC is a transcriptional regulator that controls the expression of genes involved in aflatoxin biosynthesis; in alkaline conditions, this factor is suppressed, and the absence of PacC prevents the transcription of these genes.

## DEVELOPMENTAL STAGE OF THE FUNGUS

Spore formation is associated with aflatoxin synthesis because mutant strains deficient in the sporulation process are unable to produce aflatoxin. On the other hand, compounds that inhibit sporulation in *A. parasiticus* also inhibit aflatoxin production. Aflatoxin production may gradually decrease in subcultures that cause morphological changes in the fungi. In addition, sporulation and aflatoxin synthesis are negatively affected when *A. parasiticus* and *A. nidulans* are treated with chemicals that inhibit polyamine biosynthesis (49).

## OXIDATIVE STRESS

In *A. parasiticus*, the oxidative stress induces aflatoxin synthesis. Treatment of *A. flavus* and *A. parasiticus* with tert-butyl hydroperoxide induces an increase in aflatoxin production (50), whereas hydrolysable tannins, such as gallic acid and caffeic acid, inhibit aflatoxin synthesis. The addition of ascorbic acid to the culture of *A. flavus* reduces aflatoxin synthesis, with no effect on fungal growth (51).

## PLANT METABOLITES

One of the plant metabolites, n-decyl aldehyde, reduced the synthesis of aflatoxin by 95% (52). In contrast, octanal has a 500% increase in aflatoxin production and hexanal causes a 50% reduction in fungal growth but has no effect on aflatoxin production. The 13(S)-hydroperoxide derivative of linoleic acid reduces production of aflatoxins (53).

## CONTROL OF AFLATOXIN CONTAMINATION

Aflatoxin contamination in food causes major economic losses, and its potential impact on human health has been clearly demonstrated. Therefore, the need to eliminate it from food is imperative. Efforts include monitoring, management, and control of aflatoxin levels in agricultural products from pre-harvest to post-harvest. Experimental studies of aflatoxin synthesis in the fungus may be another alternative to control food contamination by this mycotoxin. However, no studies have been conducted to date.

## PRE-HARVEST CONTROL

The presence of aflatoxins can be regulated by irrigation, pest control and soil temperature. A comprehensive knowledge of the environmental and physiological factors of the fungi associated with aflatoxin synthesis is required for good aflatoxin control. Therefore, understanding the biology and ecology of the fungus and how it interacts with plants will help in proposing control strategies.

## POST-HARVEST CONTROL

Aflatoxin contamination during post-harvest storage is common in most tropical countries due to heat, humidity and moisture. The growth of the fungus and the presence of aflatoxin in food is often due to poor harvesting, handling and storage practices. There has been a great deal of focus on the storage conditions that are most conducive to reducing the growth of the fungus. Detoxifying aflatoxin-contaminated grains to reduce contamination is another strategy used in maize and peanuts (54).

## BIOLOGICAL CONTROL

Non-aflatoxigenic fungal isolates have been used as biocontrol agents. Thus, the introduction of non-aflatoxigenic isolates that adequately colonize, compete, and outcompete toxigenic fungi is an alternative to reduce contamination levels. In fact, non-aflatoxigenic native *A. flavus* strains have been used to displace toxigenic *A. flavus* isolates in the field, and there has been a significant reduction in aflatoxin contamination (40). Another strategy is to use plants with resistance to fungal infections, such as *A. flavus*. Naturally, fungus-resistant germplasm is available, but the identification of specific biochemical factors associated with resistance to *A. flavus* could be important for the genetic manipulation of germplasm to increase the level of resistance to the fungi. Aflatoxin contamination is a complex problem and a combination of approaches to both fungal control and aflatoxin-resistant crops would be appropriate to achieve the greatest success in pre-harvest contamination.

## FUNGUS-RESISTANT COMMERCIAL CROPS

The most economical and effective strategy to reduce aflatoxin contamination is the use of commercial crop varieties that are resistant to fungal infection or aflatoxin production. There is great interest in introducing aflatoxin-resistant or -tolerant plants into agricultural crops. Currently, plant varieties or germplasm lines have been identified for crops such as maize, cotton and peanut, but the resistance of these plants to the fungi is low and therefore not yet adequate. Some lines with low to moderate resistance in corn are under testing and development.

## CONCLUSIONS

Aflatoxin produced by the fungus *A. flavus* is a worldwide problem in the food industry, and large losses are caused by contamination with this mycotoxin. Knowledge of the ecology of *A. flavus* in agricultural soil can be an important tool to control its environmental contamination, and technological application can help in this problem. Currently, the ecology of *A. flavus* in soil as well as the infective form in plant is not known in detail; therefore, a thorough investigation can generate more knowledge. On the other hand, the introduction of non-aflatoxigenic *A. flavus* strains into agricultural crops could be a microbial advantage to reduce the aflatoxin concentration in plants. Knowledge of the factors affecting aflatoxin synthesis in the fungi is also important for its application in the field. In addition, crop management and other environmental factors may contribute to control. Although the elimination of *A. flavus* from the environment is not easy, the possibility of managing new strains of this fungus with adverse effects on the aflatoxigenic ones could reduce the population of the undesirable fungus and avoid an ecological imbalance in the soil and in the agricultural sector.

## ACKNOWLEDGMENTS

The authors would like to acknowledge the financial support provided by Grants SIP20241031 and SIP20253699 from the Instituto Politécnico Nacional (IPN) México, as well as COFAA and EDI, IPN fellowships, and support from SNII-CONAHCyT.

## REFERENCES

1. Ajmal, M., W. Bedale, A. Akram, and J.H. Yu. "Comprehensive review of aflatoxin contamination, impact on health and food security, and management strategies in Pakistan." *Toxins* 14 (2022): 845. https://doi.org/10.3390/toxins14120845
2. Wild, C.P., and Y.Y. Gong. "Mycotoxins and human disease: A largely ignored global health issue." *Carcinogenesis* 31 (2010): 71–82. https://doi.org/10.1093/carcin/bgp264
3. Klich, M.A. "*Aspergillus flavus*: The major producer of aflatoxin." *Molecular Plant Pathology* 8 (2007): 713–722. https://doi.org/10.1111/j.1364-3703.2007.00436.x
4. Smith, L.E., A.J. Prendergast, P.C. Turner, J.H. Humphrey, and R.J. Stoltzfus. "Aflatoxin exposure during pregnancy, maternal, anemia, and adverse birth outcomes." *American Journal of Tropical Medicine and Hygiene* 96 (2017): 770–776. https://doi.org/10.4269/ajtmh.16-0730
5. Alshannaq, A., and J.H. Yu. "Analysis of E.U. Rapid alert system (RASFF) notifications for aflatoxins in exported U.S. Food and feed products for 2010-2019." *Toxins* 13 (2021): 90. https://doi.org/10.3390/toxins13020090
6. GB 2761-2017. Food safety national standard-limit of mycotoxins in food. national health and family planning commission of the People's Republic of China. State Administration of Food and Drug Administration: Beijing, China, 2017. https://www.codeofchina.com/standard/GB2761-2017.html
7. Brakhage, A.A., J. Schuemann, S. Bergmann, K. Scherlach, V. Schroeckh, and C. Hertweck. "Activation of fungal silent gene clusters: A new avenue to drug discovery." *Progress in Drug Research* 66 (2008): 3–12. https://doi.org/10.1007/978-3-7643-8595-8_1
8. Sweeney, M.J., and A.D Dobson. "Mycotoxin production by *Aspergillus, fusarium* and *Penicillium* species." *International Journal of Food Microbiology* 43 (1998): 141–158. https://doi.org/10.1016/s0168-1605(98)00112-3
9. Chang, P.K., J.W. Bennett, and P.J. Cotty. "Association of aflatoxin biosynthesis and sclerotial development in *Aspergillus parasiticus*." *Mycopathologia* 153 (2002): 41–48. https://doi.org/10.1023/a:1015211915310
10. St Leger, R.J., S.E. Screen, and B. Shams-Pirzadeh. "Lack of host specialization in *Aspergillus flavus*." *Applied Environmental of Microbiology* 66 (2000): 320–324. https://doi.org/10.1128/AEM.66.1.320-324.2000
11. Robens, J.F., and K. Cardwell. "The costs of mycotoxin management to the USA: Management of aflatoxins in the United States." *Journal of Toxicology* 22 (2003): 139–152. https://doi.org/10.1081/TXR-120024089
12. Abbas, H.K., C. Accinelli, R.M. Zablotowicz, C.A. Abel, H.A. Bruns, Y. Dong, and W.T Shier. "Dynamics of mycotoxin and *Aspergillus flavus* levels in aging Bt and non-Bt corn residues under Mississippi no-till conditions." *Journal of Agricultural and Food Chemistry* 56 (2008): 7578–7585. https://doi.org/10.1021/jf801771a
13. Horn, BW. "Biodiversity of *Aspergillus* section *flavi* in the United States: A review." *Food Additives and Contaminants* 24 (2007): 1088–1101. https://doi.org/10.1080/02652030701510012

14. O'Brian, G.R., D.R. Georgianna, J.R. Wilkinson, H.K. Abbas, J. Wu, D. Bhatnagar, T.E. Cleveland, W. Nierman, and G.A. Payne. "The effect of elevated temperature on gene expression and aflatoxin biosynthesis." *Mycologia* 90 (2007): 232–239. https://doi.org/10.3852/mycologia.99.2.232
15. Abbas, H.K., R.M. Zablotowicz, and M.A. Locke. "Spatial variability of *Aspergillus flavus* soil populations under different crops and corn grain colonization and aflatoxins." *Canadian Journal of Botany* 82 (2004): 1768–1775. https://doi.org/10.1139/b04-131
16. Hawksworth, D.L. "The fungal dimension of biodiversity: Magnitude, significance, and conservation." *Mycological Research* 95 (1991): 641–655. https://doi.org/10.1016/S0953-7562(09)80810-1
17. Liu, R.C., Z.Y. Xiao, A. Hashem, E.F. Abd-Allah, and Q.S. Wu. "Mycorrhizal fungal diversity and its relationship with soil properties in *Camellia oleifera*." *Agriculture* 11(2021): 47. https://doi.org/10.3390/agriculture11060470
18. Wu, L., J. Chen, M.U. Khan, J. Wang, H. Wu, Z. Xiao, Z. Zhang, and W. Lin. "Rhizosphere fungal community dynamics associated with *Rehmannia glutinosa* replant disease in a consecutive monoculture regime." *Phytopathology* 108 (2018): 1493–1500. https://doi.org/10.1094/PHYTO-02-18-0038-R
19. Bridge, P., and B. Spooner. "Soil fungi: Diversity and detection." *Plant and Soil* 232 (2001): 147–154. https://doi.org/10.1023/A:1010346305799
20. Amaike, S., and N.P. Keller. "*Aspergillus flavus*." *Annual Review of Phytopathology* 49 (2011): 107–133. https://doi.org/10.1146/annurev-phyto-072910-095221
21. Daou R., K. Joubrane, R.G. Maroun, L. R. Khabbaz, A. Ismail, and A. El Khoury. "Mycotoxins: Factors influencing production and control strategies." *AIMS Agriculture and Food* 6(2021): 416–447. https://doi.org/10.3934/agrfood.2021025
22. Cleveland T.E., J. Yu, N. Fedorova, D. Bhatnagar, G.A. Payne, W.C. Nierman, and J.W. Bennett. "Potential of *Aspergillus flavus* genomics for applications in biotechnology." *Trends in Biotechnology* 27(2009): 151–157. https://doi.org/10.1016/j.tibtech.2008.11.008
23. Wang, X., D. Wang, S. Zhang, M. Zhu, Q. Yang, J. Dong, Q. Zhang, and P. Feng. "Research progress related to aflatoxin contamination and prevention and control of soils." *Toxins* 15 (2023): 475. https://doi.org/10.3390/toxins15080475
24. Wei, D.D., L. Zhou, C.S. Zhang, and Y. Liu. "Atoxigenic *Aspergillus flavus* competitively reduce aflatoxin production by the toxigenic *Aspergillus flavus*." *Modern Food Science and Technology* 30 (2014): 92–97.
25. Zhang, S., X. Wang, D. Wang, Q. Chu, Q. Zhang, X. Yue, M. Zhu, J. Dong, L. Li, X. Jiang, et al. 2022. Discovery of the relationship between distribution and aflatoxin production capacity of *Aspergillus* species and soil types in peanut planting areas. *Toxins* 14: 425. https://doi.org/10.3390/toxins14070425
26. Yang, B., J. Shan, F. Xing, X. Dai, G. Wang, J. Ma, T. V. Adegoke, X. Zhang, Q. Yu, and X. Yu. "Distribution, accumulation, migration and risk assessment of trace elements in peanut-soil system." *Environmental Pollution* 304 (2022): 119193. https://doi.org/10.1016/j.envpol.2022.119193.
27. Abdel-Haq, H., M. Palmery, M.G. Leone, L. Saso, and B. Silvestrini. "Relaxant effects of aflatoxins on isolated Guinea pig trachea." *Toxicological Science* 55 (2000): 162–170. https://doi.org/10.1093/toxsci/55.1.162
28. Goto, T., D.T. Wicklow, and Y. Ito. "Aflatoxin and cyclopiazonic acid production by a sclerotium-producing *Aspergillus tamarii* strain." *Applied and Environmental Microbiology* 62 (1996): 4036–4038. https://doi.org/10.1128/aem.62.11.4036-4038.1996
29. Juraschek, L.M., A. Kappenberg, and W. Amelung. "Mycotoxins in soil and environment." *The Science of the Total Environment* 814 (2022): 152425. https://doi.org/10.1016/j.scitotenv.2021.152425
30. Angle, J.S., and G.H. Wagner. "Decomposition of aflatoxin in soil." *Soil Science Society of America Journal* 44 (1980): 1237–1240. https://doi.org/10.2136/sssaj1980.03615995004400060021x
31. Accinelli, C., H.K. Abbas, R.M. Zablotowicz, and J.R. Wilkinson. "*Aspergillus flavus* aflatoxin occurrence and expression of aflatoxin biosynthesis genes in soil." *Canadian Journal of Microbiology* 54 (2008): 371–379. https://doi.org/10.1139/w08-018
32. Paterson, R.R., J. Kelley, and M. Gallagher. "Natural occurrence of aflatoxins and *Aspergillus flavus* (link) in water." *Letters in Applied Microbiology* 25 (1997): 435–436. https://doi.org/10.1111/j.1472-765x.1997.tb00012.x
33. Khan, R., F.M. Ghazali, N.A. Mahyudin, and N.I.P. Samsudin. "Biocontrol of aflatoxins using non-aflatoxigenic *Aspergillus flavus*: A literature review." *Journal of Fungi* 7 (2021): 381. https://doi.org/10.3390/jof7050381
34. Tran-Dinh, N., I. Kennedy, T. Bui, and D. Carter. "Survey of vietnamese peanuts, corn and soil for the presence of *Aspergillus flavus* and *Aspergillus parasiticus*." *Mycopathologia* 168 (2009): 257–268. https://doi.org/10.1007/s11046-009-9221-9

35. Horn, B.W. "Ecology and population biology of aflatoxigenic fungi in soil." *Journal of Toxicology: Toxin Review* 22 (2003): 351–379. https://doi.org/10.1081/TXR-120024098
36. Sakudo, A., Y. Toyokawa, T. Misawa, and Y. Imanishi. "Degradation and detoxification of aflatoxin B1 using nitrogen gas plasma generated by a static induction thyristor as a pulsed power supply." *Food Control* 73 (2017): 619–626. https://doi.org/10.1016/j.foodcont.2016.09.014
37. Horn, B.W., and J.W. Dorner. "Soil populations of *Aspergillus* species from section *flavi* along a transect through peanut-growing regions of the United States." *Mycologia* 90 (1998): 767–776. https://doi.org/10.1080/00275514.1998.12026969
38. Xu, J., and X.Y. Luo. "Molecular biology of aflatoxin biosynthesis." *Journal of Hygiene Research* 32 (2003): 628–631.
39. Wang, S.X., S.S. Lu, Y. Zhang, G. Xie, and H.O. Wang. "Current status and advances in the standards of mycotoxin detection." *Science Technology of Food Industry* 3 (2011): 408–412.
40. Cotty, P.J. "Aflatoxin-producing potential of communities of *Aspergillus* section *flavi* from cotton producing areas in the United States." *Mycological Research* 101 (1997): 698–704. https://doi.org/10.1017/S0953756296003139
41. Jaime-Garcia, R., and P.J. Cotty. "Crop rotation and soil temperature influence the community structure of *Aspergillus flavus* in soil." *Soil Biology and Biochemistry* 42 (2010): 1842–1847. https://doi.org/10.1016/j.soilbio.2010.06.025
42. Payne, G.A., and M.P. Brown. "Genetics and physiology of aflatoxin biosynthesis." *Annual Review of Phytopathology* 36 (1998): 329–362. https://doi.org/doi:10.1146/annurev.phyto.36.1.329
43. Woloshuk, C.P., J.R. Cavaletto, and T.E. Cleveland. "Inducers of aflatoxin biosynthesis from colonized maize kernels are generated by an amylase activity from *Aspergillus flavus*." *Phytopathology* 87 (1997): 164–169. https://doi.org/10.1094/PHYTO.1997.87.2.164
44. Yu, 44., J. Chang, P.K. Bhatnagar, and T.E. Cleveland. "Cloning of a sugar utilization gene cluster in *Aspergillus parasiticus*." *Biochimica Biophysica Acta* 1493 (2000): 211–214. https://doi.org/10.1016/s0167-4781(00)00148-2
45. Yu, 45., J. Mohawed, S.M. Bhatnagar, and T.E. Cleveland. "Substrate-induced lipase gene expression and aflatoxin production in *Aspergillus parasiticus* and *Aspergillus flavus*." *Journal Applied Microbiology* 95 (2003): 1334–1342. https://doi.org/10.1046/j.1365-2672.2003.02096.x
46. Kachholz, T., and A.L. Demain. "Nitrate repression of averufin and aflatoxin biosynthesis *Aspergillus parasiticus*." *Journal of Natural Products* 46 (1983): 499–506. https://doi.org/10.1021/np50028a013
47. Wilkinson, J.R., J. Yu, J.M. Bland, W.C. Nierman, D. Bhatnagar, and T.E. Cleveland. "Amino acid supplementation reveals differential regulation of aflatoxin biosynthesis in *Aspergillus flavus* NRRL 3357 and *Aspergillus parasiticus* SRRC 143." *Applied Microbiology Biotechnology* 74 (2007): 1308–1319. https://doi.org/10.1007/s00253-006-0768-9
48. Keller, N.P., C. Nesbitt, B. Sarr, T.D. Phillips, and G.B. Burow. "pH regulation of sterigmatocystin and aflatoxin biosynthesis in *Aspergillus* spp." *Phytopathology* 87 (1977): 643–648. https://doi.org/10.1094/PHYTO.1997.87.6.643
49. Guzman-de-Peña, D., J. Aguirre, and J. Ruiz-Herrera. "Correlation between the regulation of sterigmatocystin biosynthesis and asexual and sexual sporulation in *Emericella nidulans*." *Antonie Van Leeuwenhoek* 73 (1998): 199–205. https://doi.org/10.1023/a:1000820221945
50. Reverberi, M., S. Zjalic, A. Racelli, A.A. Fabbri, and C. Fanelli. "Oxidant/antioxidant balance in *Aspergillus parasiticus* affects aflatoxin biosynthesis." *Mycotoxin Research* 22 (2006): 39–47. https://doi.org/10.1007/BF02954556
51. Kim, J.H., B.C. Campbell, R. Molyneux, N. Mahoney, K.L. Chan, J. Yu, J.R. Wilkinson, J. Cary, D. Bhatnagar, and T.E. Cleveland. "Gene Targets for fungal and mycotoxin control." *Mycotoxin Research* 22 (2006): 3–8. https://doi.org/10.1007/BF02954550
52. Wright, M.S., D.M. Greene-McDowelle, H.J. Zeringue, D. Bhatnagar, and T.E. Cleveland. "Effects of volatile aldehydes from *Aspergillus*-resistant varieties of corn on *Aspergillus parasiticus* growth and aflatoxin biosynthesis." *Toxicon* 38 (2000): 1215–1223. https://doi.org/10.1016/s0041-0101(99)00221-4
53. Criseo, G., A. Bagnara, and G. Bisignano. "Differentiation of aflatoxin-producing and non-producing strains of *Aspergillus flavus* group." *Letters in Applied Microbiology* 33 (2001): 291–295. https://doi.org/10.1046/j.1472-765x.2001.00998.x
54. Sipos, P., F. Peles, D. L. Brassó, B. Béri, T. Pusztahelyi, I. Pócsi, and Z. Győri. "Physical and chemical methods for reduction in aflatoxin content of feed and food." *Toxins* 13 (2021): 204. https://doi.org/10.3390/toxins13030204

# 2.2 Aflatoxins
## *Metabolic Interactions with Plants and Soil Biota*

*Chetna Rathi, Simran Rani, Priyanka Dahiya, Sankara Narayanan, Amita Suneja Dang, and Pooja Suneja*

## INTRODUCTION

*Aspergillus* species, particularly *A. flavus* and *A. parasiticus*, are well known for producing aflatoxins, which pose major threats to food safety and public health. Aflatoxins mainly include AFB1, AFB2, AFG and AFG2; are known for their cytotoxic, mutagenic, and carcinogenic properties; and represent a major threat to agricultural and public health [1, 2]. It is important to study the metabolic interactions between soil biota, plants and aflatoxigenic fungus to understand the process of aflatoxin contamination (Figure 2.2.1). *Aspergillus* species associated with plants exhibit various lifestyles, ranging from saprophytes to weak phytopathogens. These lifestyle shifts are primarily induced by changes in the global transcriptome, which influences secondary metabolite synthesis. Aflatoxin and other secondary metabolites are present in many agricultural commodities, but our understanding of the regulatory mechanisms that control their formation remains limited, making it difficult to effectively minimize aflatoxin contamination. Mycotoxins like

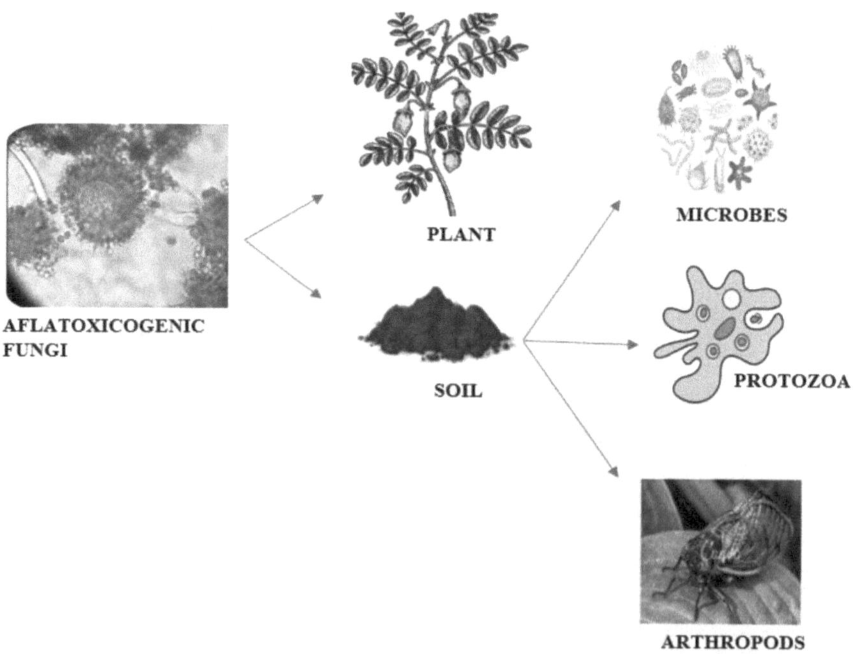


**FIGURE 2.2.1** Interactions between aflatoxigenic fungi, plants and soil biota.


DOI: 10.1201/9781003487043-8

aflatoxins were formerly assumed to be fungal defense mechanisms against competing species [3]. They are now acknowledged to act as chemical messengers, mediating interactions between various kingdoms. They act as defense mechanisms against pathogen virulence factors, quorum sensing inhibitors and fungal structure protectors. Despite extensive research on aflatoxin toxicity and control measures, the ecological consequences of its contamination in soil ecosystems is still unclear. This is also true in terms of aflatoxin interactions with soil biota and the subsequent effects on soil health and productivity [4]. Aflatoxin-contaminated crops are currently disposed in various ways, including incorporation into soil for microbial breakdown, which may unintentionally aggravate environmental pollution and disrupt natural equilibrium. Aflatoxin poisoning is anticipated to become more common in the future due to changes in temperature and precipitation patterns caused by climate change, raising major concerns for soil ecosystems, food quality and human health [5].

## CEREAL–ASPERGILLUS INTERACTIONS

Cereals such as wheat, barley, rice and maize are the primary staple food for both humans and animals. Cereals grains, although widely consumed worldwide due to their nutritional value, have not been spared aflatoxin contamination. Aflatoxin growth in grains has a number of detrimental effects including heating, discoloration and loss of biomass. Aflatoxins can also degrade proteins and lipids, alter their digestibility, and produce volatile compounds that give food an unpleasant odor. As a result, it reduces the grains' ability to germinate, limits their utilization as seeds, and lowers the quality of baking and malting [6]. *Aspergillus* species, particularly *A. flavus*, prefers maize as a host, raising concern about fungal infection and aflatoxin contamination. Some of the factors that influence *Aspergillus* pathogenesis in maize include environmental circumstances, the metabolic status of the kernels, the fungus physiological state and the period of time after infection. When *A. flavus* mycelia infect the scutellum, they form structures similar to biofilms at the endosperm–scutellum interface. These formations are comparable to those found in *A. fumigatus*-induced lung infections in people. Expression analysis of maize sucrose synthase-encoding genes and pathogenesis-related protein genes demonstrated an increase in response to *A. flavus* infection, indicating the activation of defense mechanisms [7]. Co-infections with other fungal taxa, such as *Fusarium*, are common in maize fields despite a lack of extensive investigation into these interactions. There is little evidence to support the concept that *Fusarium* inhibits *A. flavus's* production of AFB1 and that *A. flavus* inhibits *Fusarium* species in turn [8]. To effectively reduce aflatoxin contamination and ensure food safety, it is critical to understand the complicated interactions between *Aspergillus* fungus and maize. Because maize is prone to fungal infection, much effort has been expended to identify the quantitative trait loci (QTL) associated with aflatoxin resistance and generate resistant maize genotypes. These studies have found that resistance is polygenic, which means that multiple factors contribute to the resistant phenotype [9]. The aflatoxin infection in commercially farmed maize was reduced by 90% compared with untreated maize upon application of the commercial biocontrol product F-X1, which contains the non-aflatoxigenic *A. flavus* strain MUCL54911 [10]. Rice is also contaminated by aflatoxins in pre- and post-harvest conditions due to alterations in agricultural practices [11]. It is more susceptible to aflatoxin contamination than other cereals due to inappropriate drying of rice grains and high moisture content. This leads to grain discoloration along with deterioration of its quality. Increased aflatoxin contamination has been observed in corn plants facing various abiotic stresses during their reproductive stage. Abiotic stresses, such as drought and heat, lead to reduction in photosynthetic pigments, leaf water potential and cell membrane thermostability. Based on their physiological parameters, the stressed corn genotypes had maximum aflatoxin contamination, whereas the least stressed plants had low aflatoxin contamination. There exists a complex relationship between aflatoxin contamination and physiological responses of corn kernels, which needs to be explored to understand the differences among the responses of corn genotypes to aflatoxins [12].

## LEGUME–ASPERGILLUS INTERACTIONS

Legumes are important agriculture crops that are consumed primarily by humans, serve as forage and silage for livestock, and enhance soil fertility. Commonly grown legumes include peanuts *(Arachis hypogaea)*, chickpeas (*Cicer arietinum*), lentils (*Lens culinars*), lupins (*Lupinus albus),* tamarind (*Tamarindus indica*), alfalfa (*Medicago sativa*), beans (*Phaseolus vulgaris*) and clover (*Trifolium*). However, aflatoxin contamination severely affects these legumes and in-turn has a negative impact on human health, specifically infants and children [13]. The interaction between *A. flavus* and legumes has implications for both food safety and agricultural output. Peanut (*Arachis hypogaea*) is an economically important legume, a significant source of protein, and non-drying oil. Peanuts have evolved complex defense systems to combat infections, including the ability to avoid invasion and initiate various defense reactions. Research has identified 842 possible genes connected to *A. flavus* resistance in post-harvest seeds, and genes involved in defensive responses found to be upregulated [14]. A cell wall, composed of polysaccharides such as cellulose, hemicellulose and pectin, is an important component of plant defense. Pathogenic fungi commonly release enzymes that break down polysaccharides to help them eliminate these barriers efficiently [15]. Important mechanisms include pod infection, seed invasion, and aflatoxin production in the cotyledon. The pathogen resistance is determined by the structure of the pod shell, which is the first point of contact between plant and mold. After *A. flavus* successfully invades, it colonizes the seed cotyledon, resulting in the production of aflatoxin [16]. Under drought stress, aflatoxin production was consistent in peanut pods if the roots of those plants were well-watered, whereas in well-watered peanut pods, aflatoxin was not produced while roots were under drought stress. This highlights the significance of drought stress in the interaction between plants and fungi [17]. As a result, in addition to improving plant tolerance through genetic modification or selection, proper field watering is essential. Understanding the complicated connection between *A. flavus* and legumes is critical for developing effective measures to reduce aflatoxin contamination, ensuring agricultural efficiency and food safety. Soybean (*Glycine max*), a widely known legume, is a rich source of vegetarian protein with all the essential amino acids and dietary fiber. *A. flavus* and aflatoxin contamination of soybean is very low due to less moisture at the time of soybean maturity, closed pods and the presence of a fungal growth inhibitor preventing within its seed. Nevertheless, fungal growth increases due to the heat labile nature of these inhibitors after cooking of soyabean-based food, leading to enhanced aflatoxin production. *A. flavus* colonizes the surface of soybean seed and its exposure to high temperature, even for a short period, promotes high production of aflatoxins [18].

## FIBER–ASPERGILLUS INTERACTIONS

Fiber crops provide raw materials for the textile industry derived from their bast, leaves, grasses, and so on, and are crucial to the economy. One of the most serious concerns about aflatoxin contamination in fiber crops is food and feed safety. Cotton production confronts significant challenges due to *Aspergilli*'s tendency to contaminate cottonseed before and after harvesting [12]. A comparative transcriptome analysis of cotton, corn and peanut, was done to investigate changes in gene expression in these crops under aflatoxigenic *A. flavus* infection. The analysis revealed 26 common genes as plausible candidates for resistance to *A. flavus*, six of which encode 2-oxoglutarate and belong to the Fe(II)-dependent oxygenase superfamily, indicating a common defense mechanism. Aflatoxigenic and non-aflatoxigenic strains induced the activation of genes encoding helix-loop-helix proteins, uridine diphosphate (UDP) glycosylation transferase and alcohol dehydrogenase [19]. The primary metabolism was activated, altering the signal transduction pathways involved in plant defense mechanisms. Infection with non-aflatoxigenic *A. flavus* increased sucrose, starch and glycerolipid metabolism in the pericarp. Unlike aflatoxigenic strains, non-aflatoxigenic *A. flavus* stimulates metabolic pathways, indicating potential target genes for cotton resistance and tolerance to fungal stress [20]. A greater knowledge of the molecular mechanisms underlying *A. flavus* resistance in cotton will help in developing techniques to limit fungal contamination and ensuring the quality and safety of cottonseed. Flax, an important source of bast fibers and oil, has also become an increasingly popular food ingredient because of its nutrient-rich seeds having potential health benefits. The leftover seed cakes are used as animal feed. Late-maturing varieties of flax like Szaphir and Symphonia face a rainy period at the end of the maturation stage, making it susceptible to fungal infection. The flax of cultivar Lu-5 was checked for internal contamination during seed storage in which propagules of *Alternaria* and *Penicillium* were found to increase on seeds and inside the seed coat, whereas those of *Fusarium* decreased compared with their contents at harvesting. Aflatoxin production might occur subsequently if the ground seeds are stored in an environment supporting fungal growth [21].

## ASPERGILLUS–SOIL MICROBIOTA INTERACTION

Soil microbiota shares a variety of intricate ways to interact with fungi. These interactions are usually mediated by direct physical contact or indirect biochemical and enzymatic interactions along with the release of volatile organic compounds (VOCs). Several studies have been conducted to screen microbes as well as their metabolites from endophytic and rhizospheric pool with biocontrol ability to inhibit the growth of aflatoxigenic fungi. Aflatoxins present in soil are prone to detoxification, degradation or microbial biotransformation [22]. There are two detoxification mechanisms: complete degradation and conversion to less dangerous forms. Studies have been conducted with AFB1 because of its toxic and carcinogenic properties. Numerous fungi and bacteria, including non-aflatoxigenic *A. flavus* isolates, *A. niger, A. glaucus, Rhizopus* sp., *Corynebacterium rubrum* and *Hypomyces rosellus*, and *Flavobacterium aurantiacum*, reduce C-3 keto on the cyclopentanone ring and convert AFB1 to aflatoxicol. Few bacteria can also convert AFB1 into less toxic AFB2, whereas some fungi can transform it into a fluorescent product [23]. Several other *Rhizopus* species also have been known to transform AFG1. Large numbers of fungi have been found to reduce the AFB1 concentration to 70%–95%. It has been observed that interactions between bacteria and fungi can activate the silent biosynthesis-related genes in *Aspergillus*, resulting in the generation of a diverse spectrum of secondary metabolites. This close relationship may have an impact on the metabolism of both organisms, potentially leading to antifungal and antibacterial defence [24]. Biodegradation techniques can be developed based on the latest bioinformatics tools and generating 'omics' data, identifying microorganisms capable of degrading aflatoxins, confirming non-toxicity of degradation compounds, and improving their toxin tolerance, degradation abilities, and validating modes of application [25].

Volatile organic compounds (VOCs) released by pathogenic fungi play a significant role in pathogenesis. VOCs include terpenoids, phenylpropanoids and fatty acid derivatives and can damage plants even without direct physical contact. A diverse range of fungal-bacterial interactions selectively activate fungal secondary metabolite (SM) genes [26]. The virulence factor of *Ralstonia solanacearum*, an important soil pathogen of peanut, was reduced in response to VOCs of *A. flavus*, whereas *A. flavus* responded to the bacterial VOCs by reducing conidiospore production and enhancing the production of aflatoxins. Arbuscular mycorrhizae are also affected by *Aspergilli. Glomus mosseae* was found to negatively impact an *A. niger* population through the plant. *A. niger*, in turn, inhibited *G. mosseae* in its extramatrical stage through VOCs. A few yeasts have been reported to directly reduce filamentous fungi's ability to create aflatoxin. *Wickerhamomyces anomalus* produces 2-phenyl ethanol (2-PE), a volatile molecule that suppresses the expression of chromatin-modifying and aflatoxin biosynthesis genes, reducing *A. flavus* growth and biosynthesis [27]. Similarly, ethyl acetate aids in the biocontrol actions of several yeasts against certain molds, including *A. carbonarius*. However, the efficiency of VOC-mediated biocontrol varies depending on the species. Streptomyces isolates, and several yeast species, including *Hanseniaspora uvarum* and *Pichia kluyveri*, create VOCs such as dimethyl trisulfide and benzenamine, which prohibit *A. flavus* and *A. ochraceus* from making aflatoxins. Thus, interaction-specific VOCs reveal the complexities of fungal connections and chemical communication and serve ecologically important functions of antagonistic processes via their antifungal activity [28].

## ASPERGILLUS–SOIL MACROBIOTA INTERACTION

In addition to microorganisms, animals such as centipedes, millipedes (arthropods), earthworms (annelids) and macroalgae, form an important component of the soil. These organisms, widely referred to as soil macrobiota, interact with *Aspergilli* in myriad ways. Aflatoxin production might have exerted selective pressure on arthropoda to evolve a mechanism that can detoxify the same. Different arthropod species, such as soldier fly larvae (*Hermetia illucens*) [29], the maize weevil (*Sitophilus zeamais*) [30], yellow and lesser mealworms (*Tenebrio molitor* and *Alphitobius diaperinus*) [31], and the cabbage looper (*Trichoplusia ni*) [32], possess varying susceptibility towards aflatoxins and a variety of other mycotoxins. It is also hypothesized that insect species that frequently feed on aflatoxin-contaminated food develop a high degree of tolerance against aflatoxins. Maize weevils survived while fed on food sources containing up to 30,000 μg kg-1 of AFB1. Non-aflatoxigenic knockout and *Aspergillus* strains that produce lower amounts of toxins are less antagonistic towards insect populations [33]. Molds have evolved a variety of defense systems against predators in their natural habitats, including aflatoxigenic fungi, owing to their ability to release mycotoxins and evade phagocytosis [34]. One well-known example is *A. fumigatus*, which interacts with free-living amoeba species such as *Acanthamoeba castellanii* [35]. *A. fumigatus* conidia can tolerate amoeba ingestion. They can escape from phagocytic vacuoles and germinate in an intracytoplasmic manner, resulting in amoeba death. This fungal interaction with amoeba has also been compared with mammalian and avian macrophages, demonstrating the ability of the fungus to kill and escape macrophages is a pre-adoptive trait developed in their original niche [36]. DHN (1,8-dihydroxynaphthalene) melanin produced by *A. fumigatus* and related ascomycetes help the spores to survive phagocytosis and environmental stresses. *A. terreus*, which lacks DHN melanin, produces tyrosinase (TyrP) and a unique non-ribosomal peptide synthetase (NRPS)-like enzyme (MelA) that produces aspulvinone E polymerized by TyrP. This newly formed pigment, Asp-melanin, in addition to its usual function conferring resistance against ultraviolet (UV) radiation, hinders phagocytosis by soil amoeba [37]. In addition, *A. fumigatus* also produces fumagillin, an antibiotic molecule that is toxic to mammals yet effective against a wide range of amoebas and microsporidia [38]. Several recent studies have reported the complexity of the ecological network of soil microbiota, the macrobiota that influence the ecological niche and the secondary metabolite production by *Aspergilli* [27]. Therefore, it is important to explore soil-inhabiting microorganisms such as actinomycetes, fungi that exhibit antagonistic activity against *Aspergillus*, arthropods, protists, and so on, that interact with them and help reduce their growth and convert aflatoxins into less toxic forms.

## CONCLUSION

Cereals, legumes, and fiber crops are the lifeline of every country's economy owing to their utilization as staple food and fibers. The prevalence of aflatoxigenic fungi has become a significant menace for economically important crops all over the world. The main reason for this is the inability to control the factors promoting their growth. Proper management practices like effective pesticides, timely irrigation, and utmost care during the packaging and transport of crops are required to minimize and eradicate aflatoxins. Research on pre- and post-harvest contamination of food and feed commodities has received major attention, but knowledge of aflatoxin contamination and its toxicological effects in soil ecosystems is scarce. To thoroughly understand the ecology of aflatoxigenic fungi and the evolutionary pressures leading to the production of aflatoxins, its metabolic interactions with plants and soil biota must be explored and investigated.

## ACKNOWLEDGMENT

The authors are thankful for the infrastructural support provided by the Department of Microbiology and the Centre for Medical Biotechnology at Maharshi Dayanand University (MDU), Rohtak, India. CR and SR are grateful to the Council of Scientific & Industrial Research (CSIR) in New Delhi, India, for the CSIR fellowship. PD acknowledges the fellowship support from the University Grants Commission (UGC), New Delhi, India. The authors are thankful to Department of Science and Technology, Govt. of Haryana, India (Grant No. HSCSIT/R&D/2022/2968) for providing infrastructural and financial support to undertake this research work.

## REFERENCES

1. Reverberi, M., Punelli, M., Scala, V. et al., 2013. Genotypic and phenotypic versatility of *Aspergillus flavus* during maize exploitation. *PLoS One*, *8*(7):e68735.
2. Keller, N.P., Turner, G. and Bennett, J.W., 2005. Fungal secondary metabolism—From biochemistry to genomics. *Nature Reviews Microbiology*, *3*(12):937–947.
3. Malysheva, S.V., Arroyo-Manzanares, N., Cary, J.W. et al., 2014. Identification of novel metabolites from *Aspergillus flavus* by high resolution and multiple stage mass spectrometry. *Food Additives & Contaminants: Part A*, *31*(1):111–120.
4. Janzen, D.H., 1977. Why fruits rot, seeds mold, and meat spoils. *The American Naturalist*, *111*(980): 691–713.
5. Rohlfs, M., Albert, M., Keller, N.P. et al., 2007. Secondary chemicals protect mould from fungivory. *Biology Letters*, *3*(5):523–525.
6. Kumar, P., Gupta, A., Mahato, D.K. et al., 2022. Aflatoxins in cereals and cereal-based products: Occurrence, toxicity, impact on human health, and their detoxification and management strategies. *Toxins*, *14*(10):687.
7. O'Brien, J. and Wright, G.D., 2011. An ecological perspective of microbial secondary metabolism. *Current Opinion in Biotechnology*, *22*(4):552–558.
8. Ouadhene, M.A., Ortega-Beltran, A., Sanna, M. et al., 2023. Multiple year influences of the aflatoxin biocontrol product af-x1 on the A. flavus communities associated with maize production in Italy. *Toxins*, *15*(3):184.
9. Fountain, J.C., Khera, P., Yang, L. et al., 2015. Resistance to *Aspergillus flavus* in maize and peanut: Molecular biology, breeding, environmental stress, and future perspectives. *The Crop Journal*, *3*(3):229–237.
10. Mahato, D.K., Lee, K.E., Kamle, M. et al., 2019. Aflatoxins in food and feed: An overview on prevalence, detection and control strategies. *Frontiers in Microbiology*, *10*:483502.
11. Oyebamiji, Y.O., Shamsudin, N.A., Adebayo, I.A. et al., 2023. Aflatoxins in cereals: Prevalence, health implications and management strategies: A review. *Egyptian Journal of Botany*, *64*(4):1–15.
12. Achaglinkame, M.A., Opoku, N. and Amagloh, F.K., 2017. Aflatoxin contamination in cereals and legumes to reconsider usage as complementary food ingredients for Ghanaian infants: A review. *Journal of Nutrition & Intermediary Metabolism*, *10*:1–7.
13. Chalivendra, S.C., DeRobertis, C., Chang, P.K. et al., 2017. Cyclopiazonic acid is a pathogenicity factor for *Aspergillus flavus* and a promising target for screening germplasm for ear rot resistance. *Molecular Plant-Microbe Interactions*, *30*(5):361–373.
14. Boughalleb-M'Hamdi, N., Salem, I.B. and M'Hamdi, M., 2018. Evaluation of the efficiency of *Trichoderma*, *Penicillium*, and *Aspergillus* species as biological control agents against four soil-borne fungi of melon and watermelon. *Egyptian Journal of Biological Pest Control*, *28*(1):12.
15. Wang, H., Lei, Y., Wan, L. et al., 2016. Comparative transcript profiling of resistant and susceptible peanut post-harvest seeds in response to aflatoxin production by *Aspergillus flavus*. *BMC Plant Biology*, *16*:54.
16. Nigam, S.N., Waliyar, F., Aruna, R. et al., 2009. Breeding peanut for resistance to aflatoxin contamination at ICRISAT. *Peanut Science*, *36*(1):42–49.
17. Nnamani, G.C., Gana, A.S., Shaahu, A. et al., 2021. Effects of aflatoxin on soybean. *Journal of Plant Development*, 28:161–168.
18. Reddy, K.N., Abbas, H.K., Zablotowicz, R.M. et al., 2007. Mycotoxin occurrence and *Aspergillus flavus* soil propagules in a corn and cotton glyphosate-resistant cropping systems. *Food Additives and Contaminants*, *24*(12),1367–1373.

19. Holbrook, C.C. and Stalker, H.T., 2003. Peanut breeding and genetic resources. *Plant Breeding Reviews*, *22*:297–356.
20. Diener, U.L. 1989. Preharvest Aflatoxin Contamination of Peanuts, Corn and Cottonseed: A Review. In O'Rear, C.E., and Llewellyn, G.C. (eds) *Biodeterioration Research* 2 (217–244). Springer, Boston, MA. https://doi.org/10.1007/978-1-4684-5670-7_20
21. Gruzdeviene, E., Mankeviciene, A., Lugauskas, A. et al., 2006. The effect of environmental conditions on the variation of fungi and mycotoxin contents in oil flax seed. *Ekologija*, *3*:64–70.
22. Sanders, T.H., Cole, R.J., Blankenship, P.D. et al., 1993. Aflatoxin contamination of peanuts from plants drought stressed in pod or root zones. *Peanut Science*, *20*(1):5–8.
23. Bedre, R., Rajasekaran, K., Mangu, V.R. et al., 2015. Genome-wide transcriptome analysis of cotton (*Gossypium hirsutum L.*) identifies candidate gene signatures in response to aflatoxin producing fungus *Aspergillus flavus*. *PLoS One*, *10*(9):e0138025.
24. Jin, L., Yang, L., Li, W. et al., 2021. Diversity and biocontrol potential of culturable endophytic fungi in cotton. *Frontiers in Microbiology*, *12*:698930.
25. El Ariebi, N., Hiscox, J., Scriven, S.A. et al., 2016. Production and effects of volatile organic compounds during interspecific interactions. *Fungal Ecology*, *20*:144–154.
26. Savchenko, T., Walley, J.W., Chehab, E.W. et al., 2010. Arachidonic acid: An evolutionarily conserved signaling molecule modulates plant stress signaling networks. *The Plant Cell*, *22*(10):3193–3205.
27. Pfliegler, W.P., Pócsi, I., Győri, Z. et al., 2020. The *Aspergilli* and their mycotoxins: Metabolic interactions with plants and the soil biota. *Frontiers in Microbiology*, *10*:488850.
28. Veras, F.F., Correa, A.P.F., Welke, J.E. et al., 2016. Inhibition of mycotoxin-producing fungi by *Bacillus* strains isolated from fish intestines. *International Journal of Food Microbiology*, *238*:23–32.
29. Drott, M.T., Lazzaro, B.P., Brown, D.L. et al., 2017. Balancing selection for aflatoxin in *Aspergillus flavus* is maintained through interference competition with, and fungivory by insects. *Proceedings of the Royal Society B: Biological Sciences*, *284*(1869):20172408.
30. Camenzuli, L., Van Dam, R., De Rijk, T. et al., 2018. Tolerance and excretion of the mycotoxins aflatoxin B1, zearalenone, deoxynivalenol, and ochratoxin A by *Alphitobius diaperinus* and *Hermetia illucens* from contaminated substrates. *Toxins*, *10*(2):91.
31. El Khoury, R., Mathieu, F., Atoui, A. et al., 2017. Ability of soil isolated actinobacterial strains to prevent, bind and biodegrade ochratoxin A. *Toxins*, *9*(7):222.
32. Ianiri, G., Idnurm, A., Wright, S.A. et al., 2013. Searching for genes responsible for patulin degradation in a biocontrol yeast provides insight into the basis for resistance to this mycotoxin. *Applied and Environmental Microbiology*, *79*(9):3101–3115.
33. Zeringue, H.J. Jr, Bhatnagar, D. and Cleveland, T.E., 1993. C15H24 volatile compounds unique to aflatoxigenic strains of *Aspergillus flavus*. *Applied and Environmental Microbiology*, *59*(7):2264–2270.
34. Hua, S.S.T., Beck, J.J., Sarreal, S.B.L. et al., 2014. The major volatile compound 2-phenylethanol from the biocontrol yeast, Pichia anomala, inhibits growth and expression of aflatoxin biosynthetic genes of *Aspergillus flavus*. *Mycotoxin Research*, *30*:71–78.
35. Verheecke, C., Liboz, T., Anson, P. et al., 2015. *Streptomyces–Aspergillus flavus* interactions: Impact on aflatoxin B accumulation. *Food Additives & Contaminants: Part A*, *32*(4):572–576.
36. Hillmann, F., Novohradská, S., Mattern, D.J. et al., 2015. Virulence determinants of the human pathogenic fungus *Aspergillus* fumigatus protect against soil amoeba predation. *Environmental Microbiology*, *17*(8):2858–2869.
37. Van Waeyenberghe, L., Baré, J., Pasmans, F. et al., 2013. Interaction of *Aspergillus fumigatus* conidia with Acanthamoeba castellanii parallels macrophage–fungus interactions. *Environmental Microbiology Reports*, *5*(6):819–824.
38. Stevanovic, J., Stanimirovic, Z., Radakovic, M. et al., 2008. In vitro evaluation of the clastogenicity of fumagillin. *Environmental and Molecular Mutagenesis*, *49*(8):594–601.

# 2.3 Aflatoxigenic Fungal Impact on the Agricultural Environment

## *Sugarcane Field Soil*

*G. Sathvik, D.M. Yashaswini, and Lokesh Ravi*

## INTRODUCTION

Fungi are eukaryotic, unicellular, multicellular, non-phototropic and mostly saprophytic organisms that are classified under the kingdom Fungi or Eumycota, which includes organisms like mushrooms, bread moulds, yeasts and more. Although we usually associate fungi with being beneficial to man, such as mushrooms, several of them are pathogenic, parasitic and toxic to plants and animals [1]. The toxic origin of these fungi is because of secondary metabolites produced by these fungi called mycotoxins, which are produced by three main genera, namely *Aspergillus, Fusarium* and *Penicillium*. According to [2] there are more than 400 types of mycotoxins that have been discovered, however, this chapter addresses aflatoxins (a type of mycotoxin) and their impact on agriculture and sugarcane field soil.

## AFLATOXINS

Some of the most important and extensively studied mycotoxins are aflatoxins. These are acutely toxic, hepatotoxic, immunosuppressive, mutagenic and carcinogenic compounds. They are chemically difuranocoumarin derivatives with a bifuran group in association with a coumarin nucleus and either a pentone ring (aflatoxin AFBs, B1 and B2) or lactone ring (aflatoxin AFGs, G1 and G2). These four major types of aflatoxins are named according to how they fluoresce under UV light. The B group fluoresces blue in far ultraviolet (UV) light, and the G group fluoresces green [3–5] (Figures 2.3.1–2.3.4).

As of 2020, more than 20 derivatives of aflatoxins have been identified from four sections of *Aspergillus*, namely *Flavi, Ochraceorosei, Nidulantes* and *Usti* [10]. The known extrolites produced by various *Aspergillus* sp. are summarised in Table 2.3.1.

1. ***Aspergillus* section *Flavi*** is known to be the most abundant producer of AFBs and some amount of AFGs. The most common species from this section are *A. flavus* and *A. parasiticus*. Some isolates of *A. flavus* and *A. parasiticus* have also been known to produce a derivative of AFGs called AFM1 and AFM2. A few other lesser-known species that produce aflatoxins from *A. Flavi* section are *A. nomius, A. pseudotamarii, A. bombycis, A. toxicarius, A. parvisclerotigenus, A. zhaoqingensis, A. flavus var. columnaris, A. minisclerotigenes and A. arachidicola* [10].
2. ***Aspergillus* section *Ochraceorosei*** deals with only 2 species: *A. ochraceoroseus* and *A. rambellii*. These are the only species to accumulate aflatoxin B1 and sterigmatocystin together. The metabolite 3-O-methylsterigmatocystin was also produced in these cultures. In 2005, Frisvad et al. concluded that *A. ochraceoroseus* accumulates AFB1 more than *A. rambellii* [10].
3. ***Aspergillus* section *Nidulantes*** has a major genus *Emericella*. There are more than 28 species as of 2005 in this genus. Major species like *Emericella astellata* and *E. venezuelensis*

DOI: 10.1201/9781003487043-9

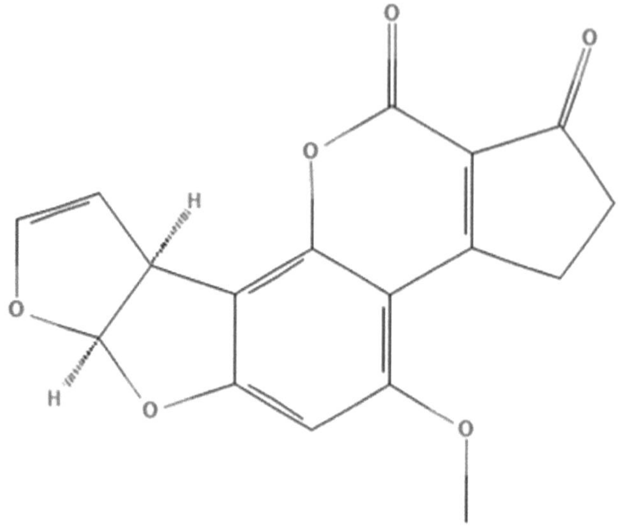


**FIGURE 2.3.1** Aflatoxin B1; CID 186907 [6].

produce AFB1, arugosins, asperthecin, shamixanthone, sterigmatocystin, terrein, varieco-xanthones and more [10].

4. ***Aspergillus* section *pseudotamarii*** lists species like *A. parasiticus, A. nomius, A. bombycis, A. parvisclerotigenus, A. minisclerotigenes, A. arachidicola and A. toxicarius*, which are known for their B1, B2, G1, G2, aspergillic acid, cyclopiazonic acid, kojic acid and parasiticolides production. These are isolated from peanuts (*Arachis hypogea)* [10].
5. ***Aspergillus* section *Usti.*** Kfir et al. [11] concluded that *A. ustus* and *A. puniceus* to produce the precursor autocystin and compounds related to sterigmatocystin, which are later converted to 3-methoxysterigmatocystin and eventually aflatoxin [10, 11].

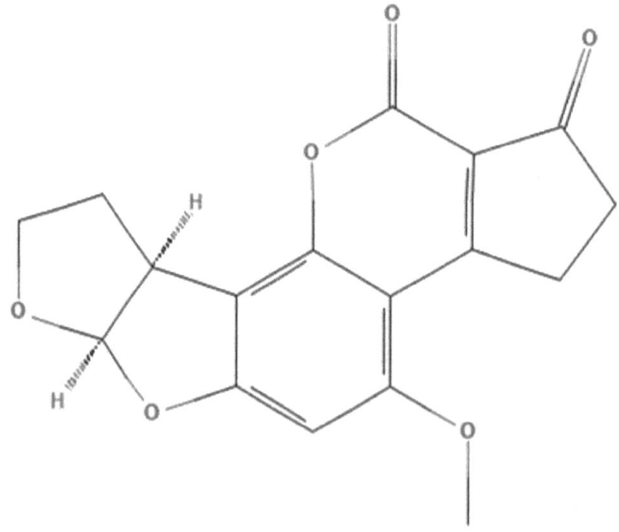


**FIGURE 2.3.2** Aflatoxin B2; CID 2724360 [7].

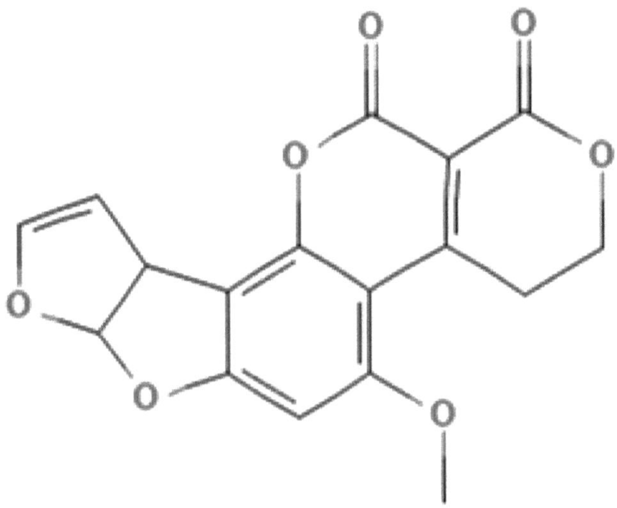


**FIGURE 2.3.3** Aflatoxin G1; CID 14421 [8].

## EFFECTS OF AFLATOXINS IN SOIL

We already know that the major class of aflatoxins are produced by *Aspergillus* section *Flavus*. Most aflatoxins occur commonly in soil, decaying vegetation such as composts and in unplanned food storage systems. Aflatoxins are not only carcinogenic and cause autoimmune disorders in humans and animals; these foodborne toxins can contaminate groundwater and in turn affect the sustainability and fertility of the soil. According to the World Health Organization (WHO)'s Food Safety Digest report of 2018, 25% of maize and peanut crops are destroyed by aflatoxins alone [19]. According to a study done by [20], plants grown in aflatoxin-rich environments absorb them into the leaves, shoots and roots, disrupting the lifecycle and fertility rates of organisms that feed on these plants. Aflatoxins are the silent killers of an ecosystem. In addition to the leaching of aflatoxins into the soil, human interference with the environment causes changes in climatic conditions, leading to

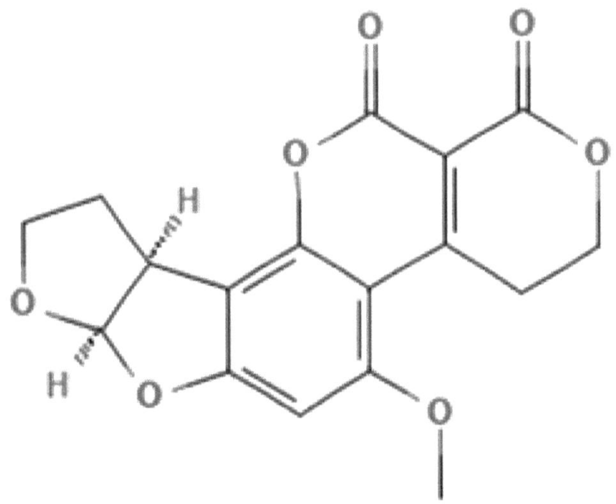


**FIGURE 2.3.4** Aflatoxin G2: CID 2724362 [9].

**TABLE 2.3.1**
**Summary of Known Extrolites Produced by Various *Aspergillus* Sp., Reported to Date**

| Section | Species | Extrolites Produced | Reference |
|---|---|---|---|
| Flavi | *A. bombycis* | Aflatoxins B & G, kojic acid, aspergillic acid | [12] |
| | *A. flavus* | Aflatoxins B1 & B2, kojic acid, cyclopiazonic acid, aspergillic acid, asperfuran, paspalinine, paspaline | [13] |
| | *A. nomius* | Aflatoxins B & G, kojic acid, aspergillic acid, tenuazonic acid, nominine | [14] |
| | *A. parasiticus* | Aflatoxins B & G, kojic acid, aspergillic acid, paspalinine, paspaline | [10] |
| | *A. parvisclerotigenus* | Aflatoxins B & G, parasiticol, cyclopiazonic acid, kojic acid, 3-O-methylsterigmatocystin, versicolorins, a and b aflatrem, paspalinine, paspaline, aflavarin, aspirochlorin | [15] |
| *Pseudotamarii l* | *A. minisclerotigenes* | Aflatoxins B & G, kojic acid, aspergillic acid, paspalinine, paspaline, cyclopiazonic acid, aflavarins, paspalinine, paspaline, aflatrems, aflavinines | [16] |
| | *A. arachidicola* | Aflatoxins B & G, kojic acid, aspergillic acid, parasiticolides, chrysogine | [16] |
| | *A. pseudotamarii* | Aflatoxin B1, kojic acid, cyclopiazonic acid | [12] |
| *Ochraceorosei* | *A. ochraceoroseus* | Aflatoxins B1 & B2, sterigmatocystin, 3-O-methylsterigmatocystin, kotanin-like metabolites, wortmannin-like metabolites, indole alkaloids | [15] |
| | *A. rambellii* | Aflatoxins B1 & B2, sterigmatocystin, 3-O-methylsterigmatocystin, versicolorins, averufin, norsolorinic acid, kotanin/desertorin- wortmannin-, emerin/xanthocillin- and indole-like extrolites | [15] |
| *Nidulantes* | *E. astellata* | Aflatoxins B1, arugosin A & B, asperthecin, shamixanthone, sterigmatocystin, terrein, variecoxanthone A, B & C | [17] |
| | *E. olivicola* | Aflatoxins B1, arugosin E, siderin, shamixanthone, sterigmatocystin, terrein, varitriols | [18] |
| | *E. venezuelensis* | Aflatoxins B1, sterigmatocystin, terrein, compounds with chromophores of the shamixanthone, emerin and desertorin type | [17] |
| *Usti* | *A. ustus* *A. puniceus* | Produces austocystins, which are related to sterigmatocystin, which in turn converts to 3-methoxysterigmatocystin, which is a precursor for various aflatoxins | [10] |

loss of soil moisture and increases in soil pH. Additionally, increased temperatures can cause everlasting and irreversible changes in the soil ecosystem with the potential to disrupt food production quality for human and animal consumption [19].

*A. flavus* is the most abundant aflatoxigenic fungus species present in the soil, and most of the species belonging to the *A. flavi* section are also infectious. AFB1 is considered the most toxic aflatoxin, but in contrast to its danger as described in the previous sections of this chapter, it surprisingly degrades quickly due to the presence of soil microbes, or is adsorbed to soil binding sites [19, 21]. However, in a study done by [22], contaminated plants that are untreated, left behind, or disposed of by burial methods on agricultural soil after uprooting, have been known to increase soil contamination by aflatoxins [15].

## EXPERIMENTS ON LEACHING OF AFLATOXINS INTO SOIL

Leaching experiments of aflatoxins into soil done by [23] showed that different types of soils retain different concentrations of aflatoxins. Over a study period of 120 days, it was determined that silt, clay loamy soils (fit for sugarcane growth), in our case. the sugarcane field soil retained (adsorbed), the highest percentage of aflatoxin AFB1 and formed conjugates with clay or organic matter present in the soil [19, 24] another study concluded that, due to the high adsorption efficiency of silt and clay loam soil, 80 to 90% of AFB1 would bound to aflatoxins degraded at a much slower pace solely because of them being bound to the soil binding sites in clay silt and loamy soils [19]. This however does not eliminate the possibility of 20% aflatoxins leaching out to groundwater or getting absorbed by vegetation [19] which in our case is sugarcane.

## STUDY OF AFLATOXINS IN SUGARCANE FIELDS

*Saccharum officinarum* L., family Poaceae, commonly known as sugarcane, is one of the most commercially and agriculturally important crops. India is the second-largest producer of sugarcane, and rightly so due to the popularity of its direct consumption, such as chewing raw sugarcane, sugarcane juice, or its derivatives like sugar and jaggery in Indian households. This chapter has thoroughly discussed the challenges and contaminants faced by agricultural crops when grown in soil. Sugarcane also inevitably faces challenges posed by the aflatoxigenic fungi *A. flavus* and *A. parasiticus* [20]. As stated previously, AFB1 and various other aflatoxins can leach into the leaves, roots and other parts of the plants when grown in aflatoxin-infested soil. In the case of sugarcane, [20] conducted an extensive study on two major parameters: (1) Natural occurrence of aflatoxigenic fungi and aflatoxin in soil and (2) the aflatoxin uptake ability of sugarcane and its persistence in jaggery [20].

## RESULTS FROM STUDY OF ABOVE PARAMETERS

1. The natural occurrence of aflatoxigenic fungi and aflatoxin in soil: [20] analysed approximately 25 fields in the region of Mysore and Mandya in the state of Karnataka. Eighty percent of the soil samples in these 25 fields were contaminated with aflatoxins ranging from 0.5–22 ppb, with a variable incidence of 7%–26%. Most of the soil samples contained the AFB1 type and AFB2 was rarely found. AFGs were not found in any of the samples [20].
2. The aflatoxin uptake ability of sugarcane and its persistence in jaggery: 40% of the sugarcane juice samples from these 25 fields were contaminated with aflatoxins ranging from 1.0 to 9.5 ppb, and in the jaggery collected from these same batches, almost all contained aflatoxin contamination of up to 1.0 ppb. In 57 sugarcane juice samples collected from local vendors from Mysore and Mandya, 22.2% and 19% had aflatoxin contamination ranging from 0.5 to 6.5ppb, and the 71 jaggery samples from local vendors across Mysore and Mandya exhibited 4.8% and 6.6% aflatoxin contamination of up to 1.0 ppb, respectively [20].

## DISCUSSION

Soil is the natural habitat for microorganisms such as *A. flavus*, which is the single most abundant producer of aflatoxins. These carcinogenic, mutagenic and hepatoxic chemicals are particularly abundant in agricultural soils and their presence is different in different types of soils. In this chapter, we understood that silty clay loam soil, suitable for a sugarcane plantation, retained the highest percentage of aflatoxins over a 120-day incubation period because AFB1 can form conjugates with clay and organic content in these types of soils [19]. These soils also have the highest ability to prevent leaching of AFs into groundwater. However, sandy-type soils, which are also suitable for a sugarcane plantation, can influence the proliferation and occurrence of *A. flavus*. Due to fewer adsorption capabilities, this type of soil can lead to contamination of groundwater and the microbial ecosystem of the soil [19]. This chapter also discussed that plants grown in aflatoxin-rich toxic environments often take up the toxic chemicals and store them in their leaves, sap, shoots and roots [20]. Their study offers accurate results of sugarcane field soil contamination and the presence of these aflatoxins in the direct and indirect products obtained, namely sugarcane juice and jaggery. The study of the sandy loam to silty clay soils of Mandya and Mysore from Karnataka in the southern state of India show how 80% of the 25 (field) soil samples from around these sugarcane growing areas had aflatoxin contamination. Varying results obtained from sugarcane juice and jaggery contamination with aflatoxins were discussed. These results show that the highest percentage of aflatoxin contamination was in the juice obtained from the 25 field samples, with 40% contamination. The lowest contamination was 19% in the sugarcane juice samples sold by vendors across Mandya. The highest aflatoxin contamination of jaggery was also observed in the same 25 field samples obtained, wherein 100% were contaminated with aflatoxins. The lowest concentration of aflatoxins was found in the jaggery samples obtained from vendors in Mysore, with 4.8% contamination [20].

## REFERENCES

1. Ayofemi Olalekan Adeyeye S. Aflatoxigenic fungi and mycotoxins in food: A review. Crit Rev Food Sci Nutr 2020;60:709–21. https://doi.org/10.1080/10408398.2018.1548429.
2. Bhat R, Rai RV, Karim AA. Mycotoxins in food and feed: Present status and future concerns. Compr Rev Food Sci Food Saf 2010;9:57–81. https://doi.org/10.1111/j.1541-4337.2009.00094.x.
3. Adejumo T, Adejoro DO. Incidence of aflatoxins, fumonisins, trichothecenes and ochratoxins in Nigerian foods and possible intervention strategies. n.d.
4. Gourama H, Bullerman LB. *Aspergillus flavus* and *Aspergillus parasiticus*: Aflatoxigenic fungi of concern in foods and feeds : A review. J. Food Prot 1995; 58(12):1395–1404. https://doi.org/10.4315/0362-028X-58.12.1395
5. Pickova D, Ostry V, Malir F. A recent overview of producers and important dietary sources of aflatoxins. Toxins (Basel) 2021;13. https://doi.org/10.3390/TOXINS13030186.
6. National Center for Biotechnology Information. (2024). *PubChem Compound Summary for CID 186907, Aflatoxin B1.* Retrieved April 14, 2024, from https://pubchem.ncbi.nlm.nih.gov/compound/Aflatoxin-B1
7. National Center for Biotechnology Information. (2024). *PubChem Compound Summary for CID 2724360, Aflatoxin B2.* Retrieved April 14, 2024, from https://pubchem.ncbi.nlm.nih.gov/compound/Aflatoxin-B2
8. National Center for Biotechnology Information (2025). PubChem Compound Summary for CID 2724361, *Aflatoxin G1.* Retrieved June 15, 2025 from https://pubchem.ncbi.nlm.nih.gov/compound/Aflatoxin-G1.
9. National Center for Biotechnology Information. (2024). *PubChem Compound Summary for CID 2724362, Aflatoxin G2.* Retrieved April 14, 2024, from https://pubchem.ncbi.nlm.nih.gov/compound/Aflatoxin-G2
10. Varga J, Frisvad JC, Samson RA. A reappraisal of fungi producing aflatoxins. World Mycotoxin J 2009;2:263–77. https://doi.org/10.3920/WMJ2008.1094.
11. Kfir R, Johannsen E, Vleggaar R. Mutagenic activity of austocystins-secondary metabolites of *Aspergillus ustus.* Bull. Environ. Contam. Toxicol. 1986;37:643–650. https://doi.org/10.1007/BF01607817
12. Peterson SW, Ito Y, Horn BW, Goto T. *Aspergillus bombycis,* A new aflatoxigenic species and genetic variation in its sibling species, A. nomius. Mycologia 2001;93:689–703. https://doi.org/10.1080/00275514.2001.12063200.
13. Sargeaxt K, Yeo R. Production of aflatoxin by the culture of strains of *Aspergillus* flaw us-oryzae on sterilized peanuts. Biotechnol Bioeng 1963;5:185–192. https://doi.org/10.1002/bit.260050303.
14. Kurtzman CP, Horn BW, Hesseltine CW. *Aspergillus nomius*, a new aflatoxin-producing species related to *Aspergillus flavus* and *Aspergillus tamarii.* Antonie van Leeuwenhoek 1987;53:147–158. https://doi.org/10.1007/BF00393843.
15. Frisvad JC, Skouboe P, Samson RA. Taxonomic comparison of three different groups of aflatoxin producers and a new efficient producer of aflatoxin B1, sterigmatocystin and 3-O-methylsterigmatocystin, *Aspergillus rambellii* sp. nov. Syst Appl Microbiol 2005;28:442–53. https://doi.org/10.1016/j.syapm.2005.02.012.
16. Pildain MB, Frisvad JC, Vaamonde G, Cabral D, Varga J, Samson RA. Two novel aflatoxin-producing *Aspergillus* species from Argentinean peanuts. Int J Syst Evol Microbiol 2008;58:725–35. https://doi.org/10.1099/ijs.0.65123-0.
17. Frisvad JC, Samson RA, Smedsgaard J. *Emericella astellata*, a new producer of aflatoxin B1, B2 and sterigmatocystin. Lett Appl Microbiol 2004;38:440–5. https://doi.org/10.1111/j.1472-765X.2004.01520.x.
18. Zalar P, Frisvad JC, Gunde-Cimerman N, Varga J, Samson RA. Four new species of *Emericella* from the Mediterranean region of Europe. Mycologia 2008;100:779–95. https://doi.org/10.3852/08-078.

19. Fouché T, Claassens S, Maboeta M. Aflatoxins in the soil ecosystem: An overview of its occurrence, fate, effects and future perspectives. Mycotoxin Res 2020;36:303–9. https://doi.org/10.1007/s12550-020-00393-w.
20. Hariprasad P, Vipin AV, Karuna S, Raksha RK, Venkateswaran G. Natural aflatoxin uptake by sugarcane (*Saccharum officinaurum* L.) and its persistence in jaggery. Environ Sci Pollut Res 2015;22:6246–53. https://doi.org/10.1007/s11356-014-3851-2.
21. Angle JS. Aflatoxin decomposition in various soils. J Environ Sci Health B 1986;21:277–88. https://doi.org/10.1080/03601238609372524.
22. Accinelli C, Abbas HK, Zablotowicz RM, Wilkinson JR. *Aspergillus flavus* aflatoxin occurrence and expression of aflatoxin biosynthesis genes in soil. Can J Microbiol 2008;54:371–9. https://doi.org/10.1139/W08-018.
23. Angle JS, Wagner GH. Aflatoxin B, effects on soil microorganisms. Soil Biol Biochem 1981;13:381–384. https://doi.org/10.1016/0038-0717(81)90081-X
24. Goldberg BS, Angle JS. Aflatoxin movement in soil. J. Environ. Qual 1985;14:224–228. https://doi.org/10.2134/jeq1985.00472425001400020014x

# 2.4 Aflatoxigenic Fungal Impact on the Agricultural Environment

## *Rice Field Soil*

*S. Chaitanya Kumari, K. Deepak Raj, Mahesh Kedharashetty, and K. Anuradha*

## INTRODUCTION

Aflatoxins, a category of toxic compounds known as mycotoxins, are synthesized by fungi, notably *Aspergillus flavus, Aspergillus parasiticus*, and *Aspergillus nomius*. Aflatoxigenic fungi are a group of imperfect, filamentous fungi originating from the soil. They primarily function as saprophytes for most of their life, thriving on a diverse array of substrates, including decomposing plant and animal matter (Figure 2.4.1). The population dynamics of these fungi in the soil are primarily influenced by soil moisture and temperature. These fungi are adaptable to a wide temperature range of 12–48°C, with their growth being optimal between 25°C and 42°C. They can also withstand low water potentials, down to -35 MPa. In conditions of high soil temperatures combined with low moisture levels, often due to drought, aflatoxigenic fungi become more competitive and predominant. In such environments, they generate significant amounts of spores and outcompete other microbes, particularly in crops like rice, corn, cotton, and peanuts. These fungi are found everywhere in nature and play crucial roles in both natural ecosystems and the agricultural economy. Species of *Aspergillus*, a common aflatoxigenic fungus, are particularly efficient at breaking down complex organic materials such as starches, hemicelluloses, celluloses, pectins, and other polysaccharides. They can also break down tougher substances like fats, oils, chitin, and keratin. The decomposition

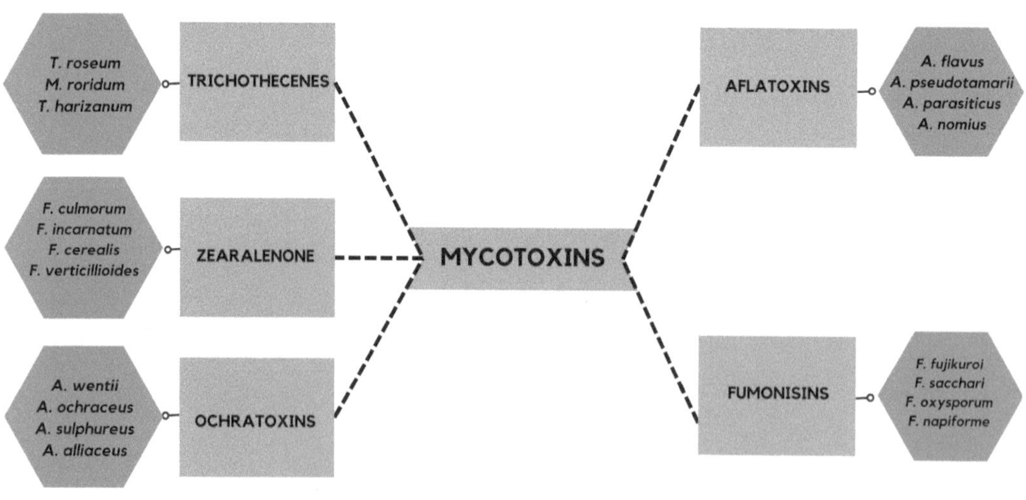


**FIGURE 2.4.1** Major types of mycotoxins.



DOI: 10.1201/9781003487043-10

process is most effective when there is an adequate supply of nitrogen, phosphorus, and other vital inorganic nutrients. Additionally, foods consumed by humans and livestock provide excellent nutrients for *Aspergillus* species, enhancing their growth and proliferation [1].

It has been estimated that up to 25% of global food crops are impacted by mycotoxins, leading to significant economic losses worldwide. Further studies, such as those conducted by Njobeh et al. [2], suggest that the prevalence of aflatoxin contamination could be even more widespread. Exposure to elevated levels of these toxins is linked to an increased risk of developing cancers, including hepato-cellular carcinoma. Certain commodities, particularly grains like rice, cereals like corn, various nuts including peanuts, pistachios, and Brazil nuts, as well as oil seeds like cottonseed and copra (dried coconut meat), are more susceptible to aflatoxin contamination, as highlighted in studies by Eskola et al. [3].

Aflatoxigenic fungi in the soil can infect rice plants at various growth stages, leading to the accumulation of aflatoxins in the grains by the time of harvest. The extent of contamination in the soil directly correlates with the potential for toxin production in the grains [4]. Aflatoxin-producing fungi can thrive on rice under conditions that are often seen with floods and significant rainfall during harvest and storage periods. When rice is not dried sufficiently under the sun and stored improperly, it becomes susceptible to fungal infestations. These conditions create an ideal environment for these fungi to grow and potentially produce toxins [5].

Rice grains are typically harvested with their husks intact, which helps protect them from mycotoxin contamination in the field. However, the presence of mycotoxins in rice is often linked to fungal growth driven by poor storage conditions [6]. Despite these protective layers, paddy rice remains vulnerable to contamination post-harvest. This is particularly problematic because most global rice production occurs in subtropical climates, which are warm and humid—conditions favorable for fungal growth. Additionally, rice is frequently stored for extended periods before being consumed, further increasing the risk of fungal contamination. According to a report by the Food and Agriculture Organization (FAO), fungal growth and mycotoxin contamination contribute to approximately 15% of annual rice crop losses [7].

The occurrence of mycotoxin contamination in food is linked to the presence of fungi, improper agricultural practices, and the conditions under which crops are harvested and stored. Mycotoxins are notably thermostable, meaning they can withstand the temperatures typically used in food processing and cooking. Therefore, the most effective strategy to mitigate the presence of mycotoxins is to prevent their development from the outset. Mycotoxins can contaminate food at various points throughout its lifecycle, including during pre-harvest, post-harvest, processing, packaging, distribution, and storage phases. Rice, a major global staple second only to wheat, forms a vital part of the diet for half of the world's population. It accounts for 27% of the global dietary consumption and 20% of the dietary protein intake in low- and middle-income countries. Typically grown in flooded conditions that maintain high moisture levels, rice fields are environments conducive to mold growth, thereby increasing the risk of mycotoxin contamination. These conditions underline the importance of adopting rigorous and effective agricultural and storage practices to safeguard this essential food source from fungal contamination and mycotoxin risks [8].

The growth and development of rice crops are significantly influenced by temperature, which directly affects plant photosynthesis. When exposed to temperature stress, rice plants experience a reduction in physiological activities, making temperature a critical factor for rice production. As the global climate continues to change, it is likely to have profound effects, not only on rice grain production, but also on the contamination levels of mycotoxins. Predictions suggest that Earth's temperature could rise by 1.5–4.5°C by the end of the 21st century. This increase in temperature is expected to accelerate water evaporation from the surface, leading to higher moisture levels in the atmosphere. Such conditions are conducive to the growth of fungal populations, thereby increasing the likelihood of mycotoxin contamination. Both temperature and humidity are pivotal factors for the growth of fungi, and as these environmental conditions become more prevalent due to climate change, the challenge of controlling mycotoxin contamination in crops like rice will likely become more severe [9].

## AFLATOXIN CONTAMINATION PATHWAYS IN RICE CULTIVATION

Aflatoxins, primarily produced by *A. flavus* and *A. parasiticus*, are potent carcinogens and pose significant health risks. These toxins can contaminate rice through various pathways during the cultivation cycle. Aflatoxigenic fungi are commonly present in soil and can survive as spores or in a dormant state. These fungi can colonize rice plants directly from the soil or through wind and water transmission, especially in fields previously contaminated with aflatoxins [10]. The fungi can infect rice plants through the roots or during flowering. High humidity and temperatures, typical in rice-growing areas, facilitate the growth and spread of these fungi. The colonization is more pronounced during periods of plant stress, such as drought or excessive moisture. Water used for irrigation can also be a significant source of aflatoxins if it contains runoff from contaminated fields or storage facilities. This water can transport fungal spores across large areas, spreading contamination to previously uninfected fields [**11,** 12]. The use of contaminated crop residues as mulch or compost, as well as contaminated planting material (seeds), can introduce aflatoxins into the soil and the new crop cycle [**13**, 14]. The growth of aflatoxigenic fungi in rice paddies is significantly influenced by environmental factors. These conditions not only affect the presence and proliferation of fungi but also the production of aflatoxins. High humidity and warm temperatures are known to favor the growth of *Aspergillus* species. Rice paddies, often flooded during part of the growing season, provide ideal conditions for fungal growth [15]. Fluctuations in moisture levels, particularly during the drying phase post-harvest, can further exacerbate fungal growth and toxin production. The presence of certain nutrients and the pH of the soil can either inhibit or promote the growth of aflatoxigenic fungi. For example, soils with high clay content and neutral pH are less conducive to the growth of these fungi compared to sandy soils with higher acidity. The use of certain agricultural practices such as crop rotation, proper drainage, and organic amendments can mitigate the risk of aflatoxin contamination. Conversely, poor crop residue management and delayed harvest can increase the susceptibility of crops to fungal invasion and aflatoxin contamination [16] (Figure 2.4.2).

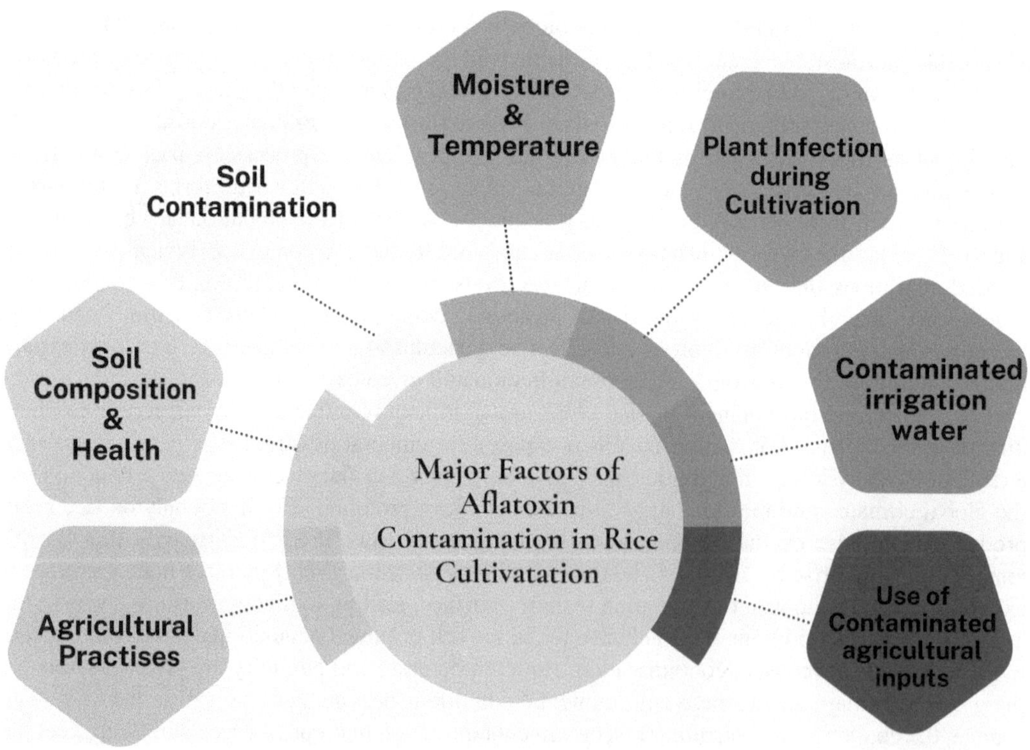


**FIGURE 2.4.2** Major factors of aflatoxin contamination in rice cultivation.

## IMPACT OF AFLATOXIGENIC FUNGI ON RICE FIELD SOIL

Aflatoxigenic fungi, primarily *A. flavus* and *A. parasiticus*, can adversely affect soil health and fertility in rice fields. These fungi produce aflatoxins that are not only toxic to humans and animals, but can also influence the microbial diversity and biochemical processes in the soil. Research indicates that aflatoxins can inhibit the growth of beneficial soil microbes, which are crucial for nutrient cycling and organic matter decomposition. This inhibition can lead to reduced soil fertility by affecting nitrogen fixation and the availability of essential nutrients (Figure 2.4.3). Studies have shown that continuous contamination with aflatoxigenic fungi leads to an imbalance in soil microbial communities, potentially increasing the dominance of pathogenic or non-beneficial organisms. Aflatoxins have a notable persistence in soil environments, influenced by soil type, pH, and organic matter content. This persistence can lead to long-term soil contamination, impacting crop health and productivity in successive seasons [17].

## CONTAMINATION MITIGATION

Mycotoxin-producing fungi can impact rice at various stages of its lifecycle, leading to the development of multiple strategies aimed at mitigating these effects, from prevention to decontamination. Among these, chemical control has been employed to reduce the presence of mycotoxigenic fungi in fields. Although effective for protecting crops, the chemicals can acidify the soil, potentially harming plant growth and reducing beneficial organisms. This has spurred a movement towards reducing the use of chemicals like insecticides, fungicides, and herbicides to promote more sustainable agricultural practices [18].

Post-harvest strategies focus on ensuring optimal storage conditions, which are critical because most mycotoxin contamination in rice is linked to poor storage. Key practices include the use of ultra-hermetic containers, and maintaining appropriate levels of temperature, humidity, and ventilation to prevent fungal growth and subsequent toxin accumulation. However, the higher nutritional value of brown rice drives the need for additional detoxification methods [6].

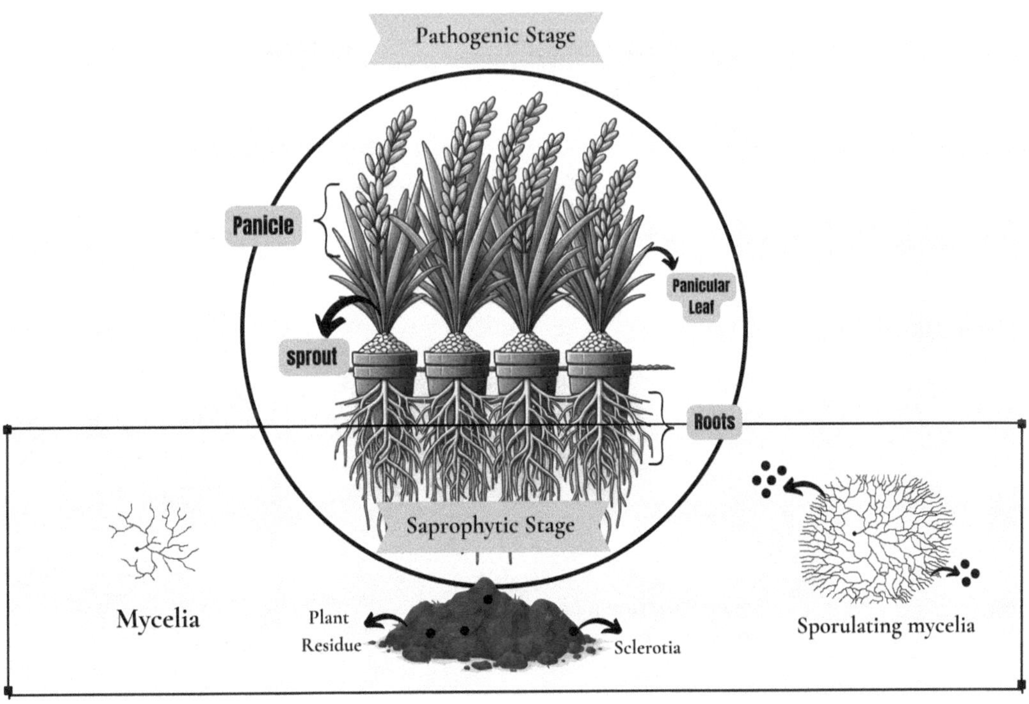


**FIGURE 2.4.3** Lifecycle of aflatoxigenic fungi.

The distribution, concentration, and properties of mycotoxins can change during processing, affecting their toxicity. This underscores the importance of understanding these variations and their implications. One study indicates that aflatoxin B1 and aflatoxin B2 are more prevalent in brown rice and bran, whereas levels in white rice are typically lower, highlighting bran removal as a crucial step in reducing mycotoxin levels. Despite these efforts, some mycotoxin presence is inevitable, prompting the development of safe, effective, and environmentally friendly decontamination methods that also offer a favorable cost-benefit ratio. These methods should inactivate, remove, or destroy the mycotoxins while preserving the food's nutritional and technological properties without producing harmful byproducts [19].

Several decontamination strategies have been proposed, including physical methods such as separation and thermal inactivation, chemical methods like irradiation and solvent extraction, and biological approaches involving microbial inactivation and fermentation. Each method aims to eradicate or neutralize toxins while maintaining the integrity and nutritional value of the food [20]. However, no single technique has proven universally effective against the variety of mycotoxins that can contaminate food, emphasizing the need for strategies that can address multiple toxins simultaneously.

Physical techniques involve segregating impaired or polluted crops from unaffected ones through methods like sorting, sieve cleaning, density segregation, washing, dehulling, and steeping. These approaches aid in reducing the level of mycotoxins, contributing to improved crop quality. The methods for eliminating mycotoxins include heat treatment and irradiation. A study on Aflatoxin Content Reduction in Feed and Food indicates that removing external grain parts, such as dehulling and polishing, effectively reduced aflatoxins by 88%–92%. Furthermore, high moisture thermal treatments like roasting, extrusion, cooking, high-pressure cooking, and instant catapult steam explosion showed reductions of 25%–88% in aflatoxins in rice. Processes like ultraviolet (UV)-light exposure and near-infrared radiation also play a significant role in reducing mycotoxin content during processing, although they are unable to completely eliminate these contaminants [21] (Figure 2.4.4). Essential oils have been used in rice to control the formation of mycotoxins and fungal growth, proving to be an effective method. Taheri et al. [22], conducted a study to assess the impact of thyme, lemongrass,

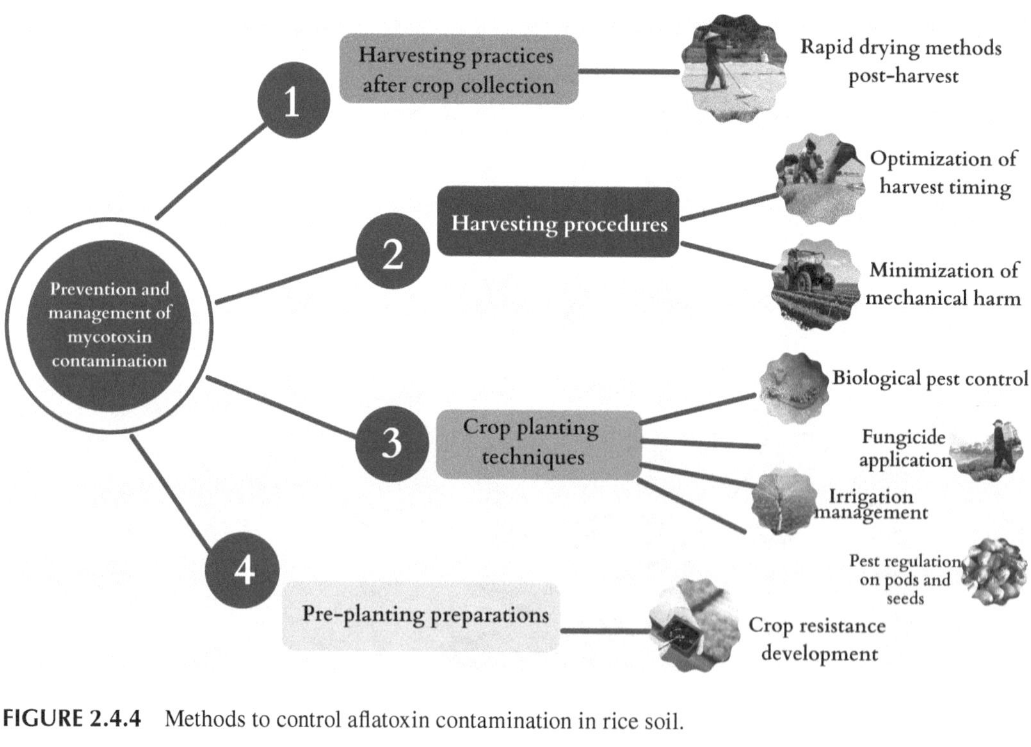


**FIGURE 2.4.4** Methods to control aflatoxin contamination in rice soil.

cinnamon, peppermint, and clove oils on deoxynivalenol production in contaminated rice. The contaminated rice samples were exposed to these essential oils for 5 days, resulting in significant reductions in mycotoxin production. Another research study highlights the chemically identified essential oil from *Myristica fragrans* as a natural food preservative, effectively combating fungal and aflatoxin B1 contamination in scented rice varieties. *Myristica fragrans Houtt.* (family: Myristicaceae) is an aromatic plant indigenous to Indonesia, and also cultivated on a large scale in India. In this paper, the authors see the efficacy of *Myristica fragrans* essential oil against isolated fungal species and aflatoxin B1 secretion by AF LHP R14 cells, an aflatoxin strain, the antioxidant activity of *Myristica fragrans* essential oil, and the phytotoxicity assay of *Myristica fragrans* essential oil. Additionally, this shows that due to its pronounced antifungal, anti-aflatoxigenic, antioxidant activity, and nonphytotoxic nature, the *Myristica fragrans* essential oil can be recommended as a plant-based food preservative for the protection of scented rice varieties and other agri-food commodities from fungal and mycotoxin contamination as well as oxidative biodeterioration. In addition, another study shows that *Apium graveolens* essential oil and their major components linalyl acetate and geranyl acetate can inhibit the growth of a wide range of toxigenic food-borne molds as well as aflatoxin B1 secretion, and recommends its possible deployment for development of a novel plant-based safe food preservative [23].

## PHENOLIC RESPONSES TO FUNGAL STRESS IN RICE DURING KEY GROWTH STAGES

Giorni et al. [24] devised a unique methodology aimed at comprehensively evaluating the relationship between phenolic compounds present in rice grains and fungal contamination throughout the cultivation cycle. Their study included the examination of samples from four different rice varieties across two successive years. The research specifically concentrated on quantifying both the total phenolic content (TPC) and the levels of eight distinct phenolic acids during three key stages of plant growth: medium milk, early dough, and ripening. This approach facilitated a thorough investigation of phenolic alterations in response to fungal pressures under varying environmental conditions, as illustrated in Figure 2.4.5.

Fungal presence and mycotoxin levels were measured using established mycological procedures. The results revealed notable yearly fluctuations, indicating the impact of environmental variables and rice types on fungal growth and mycotoxin synthesis. Specifically, sterigmatocystin was found nearly everywhere, whereas deoxynivalenol levels were higher during the ripening phase. Accurate quantification of mycotoxins was achieved through rigorous analytical methods such as gas chromatography-mass spectrometry (GC-MS) for deoxynivalenol and liquid chromatography-tandem

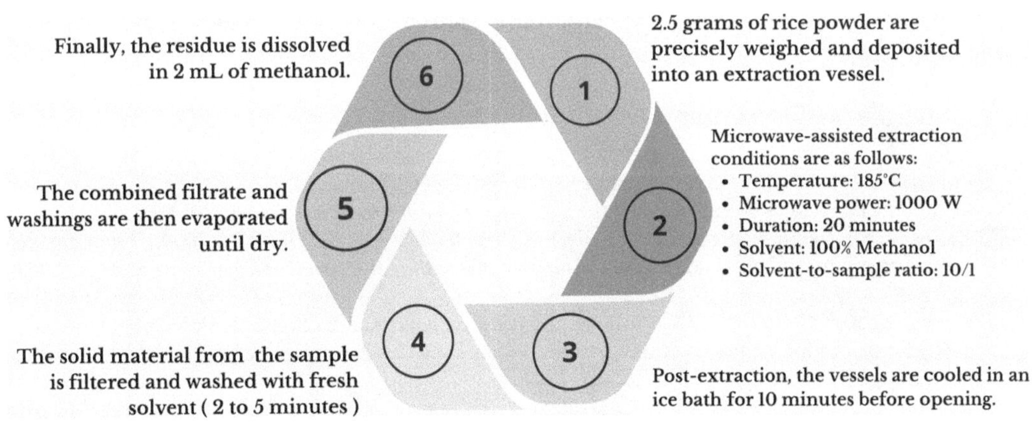


**FIGURE 2.4.5** Extraction of phenolic compounds from rice.

mass spectrometry (LC-MS/MS) for sterigmatocystin. These advanced techniques highlight the sophisticated chemical analysis required for such investigations [25].

Statistical methods such as analysis of variance (ANOVA) and Spearman's correlation were utilized to assess the connections between phenolic profiles, fungal contamination, and mycotoxin levels. The findings revealed significant relationships, particularly indicating that ferulic acid and protocatechuic acid have protective effects against increased fungal presence and mycotoxin accumulation. Specifically, ferulic acid consistently showed higher concentrations during the ripening stage across all varieties, implying its role in bolstering plant defense because fungal threats rise towards the end of the growing season. The study's conclusions underscored the potential of phenolic acids in enhancing crop resilience against fungal pathogens and reducing mycotoxin risks. The observed correlation between elevated levels of certain phenolic acids and reduced mycotoxin content suggests promising directions for developing agricultural practices and breeding strategies focused on enhancing these phenolic compounds in rice grains [26].

A comprehensive methodological framework that includes detailed sampling, precise chemical analysis, and rigorous statistical evaluation along with specific data points, makes a substantial contribution to understanding how plant phenolic compounds interact with fungal contamination in agricultural crops. This approach not only helps unravel the intricate dynamics of plant–pathogen interactions but also offers valuable insights that can guide future crop management and breeding strategies.

## CONCLUSION

Aflatoxins are a major concern due to the significant health risks they present, especially within the context of rice farming. Various environmental conditions, such as high humidity, temperature changes, and farming methods, contribute to the growth of aflatoxin-producing fungi. This amplifies the vulnerability of rice crops to these harmful toxins, posing threats to global food safety and economic stability. To effectively address aflatoxin contamination, a holistic approach is necessary. This includes adopting improved agricultural practices, creating resistant rice strains using advanced breeding techniques, and implementing innovative post-harvest strategies. Additionally, exploring new decontamination methods like essential oils and plant-based preservatives shows promise in improving food safety and promoting sustainable agriculture.

Further exploration is needed to delve into the ecological and biological intricacies of aflatoxin production in the context of climate change. It is important to prioritize research that investigates how environmental shifts may intensify these issues. Developing comprehensive global strategies that incorporate technological advancements, community outreach, and regulatory frameworks is imperative to mitigate aflatoxin contamination and guarantee the safety of rice, a staple food. These holistic approaches are critical for protecting crops throughout their lifecycle, from cultivation to storage, and are essential for ensuring food safety and nutritional quality on a global scale amidst changing climatic conditions.

## REFERENCES

1. Pfliegler, W. P., Pócsi, I., Győri, Z., & Pusztahelyi, T. (2020). The *Aspergilli* and their mycotoxins: Metabolic interactions with plants and the soil biota. *Frontiers in Microbiology, 10*, 488850.
2. Njobeh, P. B., Dutton, M. F., Koch, S. H., Chuturgoon, A., Stoev, S., & Seifert, K. (2009). Contamination with storage fungi of human food from Cameroon. *International Journal of Food Microbiology, 135*(3), 193–198.
3. Eskola, M., Kos, G., Elliott, C. T., Hajšlová, J., Mayar, S., & Krska, R. (2020). Worldwide contamination of food-crops with mycotoxins: Validity of the widely cited 'FAO estimate' of 25%. *Critical Reviews in Food Science and Nutrition, 60*(16), 2773–2789.
4. Bediako, K. A., Ofori, K., Offei, S. K., Dzidzienyo, D., Asibuo, J. Y., & Amoah, R. A. (2019). Aflatoxin contamination of groundnut (*Arachis hypogaea* L.): Predisposing factors and management interventions. *Food Control, 98*, 61–67.

5. Xu, F., Baker, R. C., Whitaker, T. B., Luo, H., Zhao, Y., Stevenson, A., … Zhang, G. (2022). Review of good agricultural practices for smallholder maize farmers to minimise aflatoxin contamination. *World Mycotoxin Journal*, *15*(2), 171–186.
6. Santos, A. R., Carreiró, F., Freitas, A., Barros, S., Brites, C., Ramos, F., & Sanches Silva, A. (2022). Mycotoxins contamination in rice: Analytical methods, occurrence and detoxification strategies. *Toxins*, *14*(9), 647.
7. Thakur, A. K., & Uphoff, N. T. (2017). How the system of rice intensification can contribute to climate-smart agriculture. *Agronomy Journal*, *109*(4), 1163–1182.
8. Daou, R., Joubrane, K., Maroun, R. G., Khabbaz, L. R., Ismail, A., & El Khoury, A. (2021). Mycotoxins: Factors influencing production and control strategies. *AIMS Agriculture and Food*, *6*(1), 416–447.
9. Ren, H., Bao, J., Gao, Z., Sun, D., Zheng, S., & Bai, J. (2023). How rice adapts to high temperatures. *Frontiers in Plant Science*, *14*, 1137923.
10. Fouché, T., Claassens, S., & Maboeta, M. (2020). Aflatoxins in the soil ecosystem: An overview of its occurrence, fate, effects and future perspectives. *Mycotoxin research*, *36*(3), 303–309.
11. Nazareth, T. D. M., Soriano Pérez, E., Luz, C., Meca, G., & Quiles, J. M. (2024). Comprehensive review of aflatoxin and ochratoxin a dynamics: Emergence, toxicological impact, and advanced control strategies. *Foods*, *13*(12), 1920.
12. Peles, F., Sipos, P., Kovács, S., Győri, Z., Pócsi, I., & Pusztahelyi, T. (2021). Biological control and mitigation of aflatoxin contamination in commodities. *Toxins*, *13*(2), 104.
13. Liu, Y. (2022). Approaches to reduce Aspergillus flavus and aflatoxin contamination through utilization of agricultural by-products.
14. Moral, J., Garcia-Lopez, M. T., Camiletti, B. X., Jaime, R., Michailides, T. J., Bandyopadhyay, R., & Ortega-Beltran, A. (2020). Present status and perspective on the future use of aflatoxin biocontrol products. *Agronomy*, *10*(4), 491.
15. Tischner, Z., Páldy, A., Kocsubé, S., Kredics, L., Dobolyi, C., Sebők, R., … Magyar, D. (2022). Survival and growth of microscopic fungi derived from tropical regions under future heat waves in the Pannonian Biogeographical Region. *Fungal Biology*, *126*(8), 511–520.
16. Awuchi, C. G., Ondari, E. N., Eseoghene, I. J., Twinomuhwezi, H., Amagwula, I. O., & Morya, S. (2021). Fungal growth and mycotoxins production: Types, toxicities, control strategies, and detoxification. In *Fungal Reproduction and Growth*, 100207. InTech Open Access Publisher.
17. Shabeer, S., Asad, S., Jamal, A., & Ali, A. (2022). Aflatoxin contamination, its impact and management strategies: An updated review. *Toxins*, *14*(5), 307.
18. Pandey, A. K., Samota, M. K., Kumar, A., Silva, A. S., & Dubey, N. K. (2023). Fungal mycotoxins in food commodities: Present status and future concerns. *Frontiers in Sustainable Food Systems*, *7*, 1162595.
19. Trucksess, M. W., Abbas, H. K., Weaver, C. M., & Shier, W. T. (2011). Distribution of aflatoxins in shelling and milling fractions of naturally contaminated rice. *Food Additives & Contaminants: Part A*, *28*(8), 1076–1082.
20. Oliveira, C. A. F., Bovo, F., Corassin, C. H., Jager, A. V., & Reddy, K. R. (2013). Recent trends in microbiological decontamination of aflatoxins in foodstuffs. *Aflatoxins-Recent Advances and Future Prospects*, *1*, 59–62.
21. Karlovsky, P., Suman, M., Berthiller, F., De Meester, J., Eisenbrand, G., Perrin, I., … Dussort, P. (2016). Impact of food processing and detoxification treatments on mycotoxin contamination. *Mycotoxin Research*, *32*, 179–205.
22. Taheri, P., Soweizy, M., & Tarighi, S. (2023). Application of essential oils to control some important fungi and bacteria pathogenic on cereals. *Journal of Natural Pesticide Research*, *6*, 100052.
23. Das, S., Kumar Singh, V., Kumar Dwivedy, A., Kumar Chaudhari, A., Upadhyay, N., Singh, A., … Dubey, N. K. (2020). Assessment of chemically characterised Myristica fragrans essential oil against fungi contaminating stored scented rice and its mode of action as novel aflatoxin inhibitor. *Natural Product Research*, *34*(11), 1611–1615.
24. Giorni, P., Rastelli, S., Fregonara, S., & Bertuzzi, T. (2020). Monitoring phenolic compounds in rice during the growing season in relation to fungal and mycotoxin contamination. *Toxins*, *12*(5), 341.
25. Awuchi, C. G., Ondari, E. N., Ogbonna, C. U., Upadhyay, A. K., Baran, K., Okpala, C. O. R., … Guiné, R. P. (2021). Mycotoxins affecting animals, foods, humans, and plants: Types, occurrence, toxicities, action mechanisms, prevention, and detoxification strategies—A revisit. *Foods*, *10*(6), 1279.
26. Kazimierczak, R., Średnicka-Tober, D., Leszczyńska, D., Nowacka, A., Hallmann, E., Barański, M., … Gnusowski, B. (2020). Evaluation of phenolic compounds and carotenoids content and mycotoxins occurrence in grains of seventeen barley and eight oat cultivars grown under organic management. *Applied Sciences*, *10*(18), 6369.

# 2.5 Aflatoxigenic Fungal Impact on the Agricultural Environment

## *Peanut Field Soil*

*S. Shalini Devi, Manju Devi, and Prity Mishra*

## INTRODUCTION TO PEANUT SOIL

Peanuts rank among the leading cash crops worldwide[1]. According to the United States Department of Agriculture (USDA), global peanut production reached 49.8 million metric tons in 2022–2023. The cultivation of peanuts benefits from sandy soils, which offer high porosity that facilitates root and pod development. Clayey soils with abundant organic matter are also favored because they form aggregates that further enhance soil porosity[2].

Soils for peanuts: The ideal soil conditions for growing peanuts include well-drained, light-colored, loose, sandy loams that are rich in calcium and contain a moderate amount of organic matter. Although heavier soils can yield higher crops, they can also cause pods to become coated with soil. Unlike many legumes, peanuts are less sensitive to soil pH; however, excessively high pH levels can lead to iron deficiency. Peanuts are more adaptable to acidic soils compared to cotton, which can experience reduced yields in soils with a pH as low as 5.0[3]. A study conducted by Arunyanark et al.[4] indicated that aflatoxin production in peanuts is favored by elevated soil temperatures and low soil moisture levels. The higher temperatures lead to increased moisture loss from the peanut kernels, subsequently lowering the water activity (aw) level. This decrease in aw diminishes the presence of phytoalexins, which play a crucial role in defending the plant against pathogens.

Studies have shown that crop rotation is beneficial for peanut growth. Continuous monoculture negatively impacts the peanut yield[5]. The interaction of plants with the microbial community in the soil is essential for crop health. Continuous monoculture leads to multiplication of more fungi than bacteria in the soil, which negatively impacts the soil health[6].

The peanut yield has been affected by various biotic and abiotic stress factors. The biotic stress factors include insects, fungi, virus and bacteria[7]. Among the biotic factors, the pathogenic infestation of *Aspergillus flavus* and *A. parasiticus* on peanuts is well documented[8]. The decrease in the crop yield and economic loss to farmers as a result of this fungal pathogen has been reported in several studies[9,10]. Because *Aspergillus* species infect peanuts both pre- and post-harvest, the impact of this fungal pathogen poses a serious threat to peanut farmers[11].

## *ASPERGILLUS* IN SOIL

Although *Aspergillus* species have been found in various environments, their primary reservoir is soil. Despite this, most of these species have not been extensively studied in terms of their role in soil fungal ecology. Soil serves as the natural habitat for toxin-producing fungi, providing them with an advantage for mycotoxin production. The diversity of *Aspergillus* species populations in soil and on plants, as well as their stability, remains unclear. The impact of climate change on the emergence of new *Aspergillus* species and mycotoxin production is uncertain. If climate change continues at its current rate, the risk of mycotoxin contamination in soil ecosystems may increase or decrease

 DOI: 10.1201/9781003487043-11

depending on the region. *Aspergillus* species are most commonly found in regions with tropical and subtropical climates, typically between latitudes 25° and 35°. Studies have shown that high-altitude regions have the highest levels of mycotoxin contamination in crops, whereas crops grown at middle and low altitudes have lower levels of contamination. Furthermore, rare and new species of *Aspergillus* are frequently reported in tropical and subtropical soils. The high adaptability of *Aspergillus* species to different environments allows them to survive at various temperatures, low water levels and different soil pH and oxygen concentrations. The interactions of *Aspergillus* species with biotic and abiotic components determine the types and roles of the secondary metabolites, such as aflatoxins, gliotoxin, patulin, cyclopiazonic acid and ochratoxin, that are produced[12].

## AFLATOXIN: TOXIC EFFECTS

Under conducive conditions, *A. flavus* and *A. parasiticus* produce aflatoxin as secondary metabolites. Aflatoxins are potent toxic, carcinogenic, mutagenic and immunosuppressive agents. Among the 18 different types of aflatoxins identified, major members are aflatoxins $B_1$, $B_2$, $G_1$ and $G_2$. Aflatoxin $B_1$ ($AFB_1$) is normally predominant in cultures and food products. Pure $AFB_1$ is a pale-white to yellow crystalline, odourless solid. Aflatoxins are soluble in methanol, chloroform, acetone and acetonitrile. *A. flavus* typically produces $AFB_1$ and $AFB_2$, whereas *A. parasiticus* produces $AFG_1$, $AFG_2$, $AFB_1$ and $AFB_2$. Four other aflatoxins, $M_1$, $M_2$, $B_2A$ and $G_2A$, which may be produced in minor amounts, were subsequently isolated from cultures of *A. flavus* and *A. parasiticus*. Several closely related compounds, namely aflatoxin $GM_1$, parasiticol and aflatoxicol, are also produced by *A. flavus*. Aflatoxins $M_1$ and $M_2$ are major metabolites of aflatoxin $B_1$ and $B_2$, respectively, and are found in the milk of animals that have consumed feed contaminated with aflatoxins[13].

The toxic effects of aflatoxins were initially discovered in England, where 100,000 turkeys died from a new disease called Turkey-X disease. Ducklings and young pheasants were also affected by the disease. It was found that the Brazilian groundnut meal consumed by the turkeys was contaminated with *A. flavus* fungi, and the fungi produced a toxin that led to the death of turkeys. Because the toxin was produced from *A. flavus,* it was termed *aflatoxin* and was a breakthrough in the study of mycotoxins. Aflatoxin contamination has been reported in various countries. An outbreak of trichothecene mycotoxicosis associated with consumption of mold-damaged wheat products occurred in Kashmir Valley, India, in 1989. In the United States, aflatoxin contamination led to rejection of approximately 50% of the maize crop in 1998. In 2004, in rural eastern Kenya, 125 people died due to acute aflatoxicosis[14]. Between 2004 and 2013, there were 18 recalls in the United States, of food and feed due to aflatoxin contamination[15]. Among all the mycotoxins, aflatoxin continues to receive attention due to its toxic effects.

The adverse effects of aflatoxin include liver toxicity, immune suppression, growth retardation and mutagenic effect[13]. At high doses in humans, $AFB_1$ is oxidized by cytochrome P450 in the liver into $AFB_1$- 8,9- epoxide, which exerts hepatotoxic effects. The epoxide binds to the N7 guanine of DNA, forming adducts. This $AFB_1$- DNA adduct opens the imidazole ring to form more stable $AFB_1$ formamidopyrimidine ($AFB_1$-FAPY) adducts, which are mainly involved in the genotoxicity and carcinogenicity of aflatoxins[16]. Aflatoxin induces the expression of various cytochrome P450s (CYPs), which may aggravate cellular and genetic toxicity through bioactivation. Aflatoxin also causes damage to the mitochondria of hepatocytes and thereby reduces oxidative phosphorylation and adenosine triphosphate (ATP) synthesis. Aflatoxin also induces oxidative stress and reactive oxygen species production in the mitochondria of hepatocytes, leading to collapse of the antioxidative system. Because hepatocytes are damaged due to the effect of aflatoxins, several inflammatory cytokines are produced, creating a local inflammatory microenvironment in the hepatocytes, which further exacerbates the hepatotoxicity[17].

$AFB_1$ formamidopyrimidine ($AFB_1$-FAPY) has been reported to be the most mutagenic form because it induces persistent DNA damage. $AFB_1$-FAPY lesions are usually repaired by the base excision repair mechanism, but the exposure to aflatoxins leads to various epigenetic changes in the repair genes, which hinders the repair mechanism. Genetic polymorphism of repair genes further exacerbates the risk. Aflatoxin also induces mutation in the p53 gene. Usually, cells exposed to genotoxic stimuli undergo either p53-dependant or -independent apoptosis. However, in the case of p53 mutation by aflatoxin, cell division occurs with damaged DNA, leading to tumors. Additionally, p53 mutation deregulates many genes that are involved in various cellular and metabolic functions, including tumor suppression[16].

Aflatoxin also induces oxidative stress damage to DNA and lipid peroxidation of membrane lipids. $AFB_1$ metabolism in the liver by the CYP 450 enzyme induces oxidative stress and production of reactive oxygen species, which damage DNA by producing DNA adducts. One of the main studied DNA adducts, 7,8-dihydro-8-oxo-20-deoxyguanosine, is produced by the hydroxyl radicals in $AFB_1$ injected rats in a dose-dependent and time-dependent manner[18].

Aflatoxin toxicity also involves immunosuppression and immunotoxicity. $AFB_1$ epoxide interferes with immune cell proliferation and immune response mediated by the cells. Both cell-mediated and humoral immunity are affected by aflatoxins. The structural and functional integrity of the intestinal epithelial cells are altered by aflatoxin toxicity. The intestinal mucosa, which provides protection as the mechanical, chemical and immunological barrier, is affected by $AFB_1$ exposure. The key cells of innate immunity, such as macrophages, natural killer cells and dendritic cells, have also been affected by aflatoxins. The viability function and the expression of cytokines by these cells have been reduced. The phagocytic and cytotoxic activities of neutrophils have also been hindered by aflatoxins. The complement system, which plays a significant role in innate immunity, has also been shown to be hampered by aflatoxins. Adaptive immune response is also affected by aflatoxins, including failure of the vaccination effect[16].

Aflatoxin also causes teratogenic effects. Studies shown that exposure to high doses of aflatoxin in pregnant women, resulted in the transmission of toxic AF metabolites to fetuses, leading to various complications such as fetus loss and premature birth[19]. In addition to the above-described toxic effects, aflatoxins also exert various other adverse conditions like malnutrition, retarded growth and nervous system diseases[20].

## PEANUT SOIL: IMPACT ON *ASPERGILLUS* GROWTH

The soil type in which groundnut seeds are cultivated can impact the growth of *A. flavus* and the resulting aflatoxin contamination in the area. Light sandy soil can facilitate the rapid spread of *A. flavus*, particularly in dry conditions during the later stages of growth. Conversely, heavier soils with higher water-holding capacity may reduce the level of aflatoxin contamination in groundnut seeds. Studies by Zhang et al.[21] showed that there was a significant difference in the *Aspergillus* species isolates in the clay and sandy soils. In sandy loam samples, the percentage of isolated atoxigenic *Aspergillus* strains was higher at 58.5% compared to clay loam samples. Conversely, the proportion of highly toxin-producing strains was much higher in clay loam samples (39.7%) than in sandy loam samples (7.3%)[21]. The density of *Aspergillus* species in soil may not reliably predict the frequency of peanut seed infection, as indicated by Griffin and Garren[22] and Horn et al.[23]. The susceptibility of plants to fungal invasion under drought stress and elevated soil temperatures appears to be more influential. Furthermore, factors such as the species composition of populations, including the presence of *A. parasiticus*, and the genetic capacity of isolates to produce aflatoxins may play crucial roles in peanut contamination[24].

## SOIL CHEMISTRY AND AFLATOXIN

Soil chemistry is a field dedicated to examining the chemical properties and processes within soil. This discipline covers a broad spectrum of factors impacting soil fertility, plant development, environmental integrity and ecosystem dynamics. Soil chemistry is known to influence the growth and yield of crops. It also affects the growth of beneficial and pathogenic microbial communities. Secretion of microbial secondary metabolites and their further fate is also impacted by the soil chemistry in a particular field. Aflatoxin can enter soil directly by *Aspergillus* residing in soil, or during natural degradation of aflatoxin-contaminated crops. These toxins are capable of causing alterations in physicochemical properties and biotic factors in that soil ecosystem. They have negative implications for soil and groundwater quality. Deterioration of soil quality can potentially pose a risk to sustainable food production[25]. Aflatoxins in soil are, in turn, absorbed by the root system and reach various parts of the plant[26].

Different types of soils are known to exhibit differences in aflatoxin retention capacities. AFB1 forms a conjugate with the clay or organic component in the soil, hence silty clay loam soil allows the highest percentage of aflatoxin persistence[27]. On the other hand, due to the high adsorption coefficient, aflatoxin degradation is slower and a lower environmental risk in silty clay loam soils[28]. Variations in aflatoxin contamination levels at several locations can be attributed to differences in soil chemistry. For instance, lower potassium levels at one location correlated with higher aflatoxin contamination, indicating the potential role of specific soil nutrients in mitigating aflatoxin production.

Certain strategies can aid in degradation or adsorption of aflatoxin in soil, so that the risk of peanuts contaminated by these toxins can be minimized. Application of potassium-based soaps, oyster shell homogenate or a fertilizer containing calcium in peanut growing fields are known to reduce aflatoxin levels in harvested peanuts[29]. Nanoparticles can be employed to impact the peanut yield; conversely, nanoparticles were proved to be effective in fungal inhibition, mycotoxin adsorption and toxicity reduction. Aflatoxin B1 is adsorbed by nanodiamond aggregates of approximately 40 nm or nanobody (Nb) containers composed of polyacrylic acid[30].

## SOIL TYPES IN PEANUT PLANTING AREAS: EFFECT ON AFLATOXIN-PRODUCING STRAINS

Different soil types in peanut planting areas can indeed have varying effects on aflatoxin-producing strains as follows:

1. **Moisture content:** The level of moisture in soil is a critical factor influencing aflatoxin contamination. Soils rich in clay retain moisture more effectively than sandy soils. Excessive moisture can create favorable conditions for the growth of *Aspergillus* fungi, which produce aflatoxins. Conversely, sandy soils drain water more rapidly, reducing the risk of prolonged moisture that encourages fungal growth. However, sandy soils may experience rapid drying during drought periods, causing stress in peanut plants and increasing their vulnerability to aflatoxin contamination.
2. **Soil pH:** The pH level of soil can impact the growth of fungi that produce aflatoxins. Although *A. flavus*, a primary aflatoxin-producing fungus, can thrive across various pH levels, it typically prefers slightly acidic to neutral conditions, ranging from pH 5.0 to 7.0. Soils within this pH range may create a more conducive environment for aflatoxin production.
3. **Organic matter:** Soil organic matter significantly impacts microbial activity, which can influence aflatoxin production. Soils rich in organic matter tend to harbor diverse microbial communities. These communities compete with aflatoxin-producing fungi for resources, potentially limiting their dominance and reducing aflatoxin contamination risk. Applying compost or organic amendments can be a strategy to enhance soil organic matter content and foster microbial diversity, ultimately mitigating aflatoxin concerns in crops.
4. **Texture and structure:** Soil texture, which refers to the relative proportions of sand, silt and clay particles, can impact aflatoxin contamination. Clay soils tend to have finer particles and higher water-holding capacity, creating conditions conducive to fungal growth. In contrast, sandy soils have larger particles and lower water-holding capacity, which may limit fungal proliferation but can lead to increased susceptibility to drought stress in peanut plants.

5. **Soil microbial communities:** Soil microbial communities play a crucial role in shaping soil ecology and can affect aflatoxin contamination. Beneficial microbes, such as certain bacteria and fungi, can compete with aflatoxin-producing fungi for nutrients and space, thereby reducing their population and mitigating aflatoxin risk. Soil types that support diverse and abundant microbial communities may exhibit lower aflatoxin contamination levels.
6. **Soil drainage:** Proper soil drainage is essential for reducing aflatoxin contamination risk. Poorly drained soils can become waterlogged, creating anaerobic conditions that favor the growth of aflatoxin-producing fungi. Sandy soils with good drainage may experience less fungal growth due to their ability to quickly remove excess water, but they may also be prone to nutrient leaching, which can impact plant health.
7. **Soil temperature:** Soil temperature can influence aflatoxin production, with optimal temperatures for fungal growth typically ranging from 25 to 35°C. Soil types that retain heat and maintain warmer temperatures may provide a more favorable environment for aflatoxin-producing fungi. Conversely, cooler soil temperatures may inhibit fungal growth and aflatoxin production, especially during periods of low moisture availability.
8. **Soil management practices**: Soil management practices, such as tillage, crop rotation and cover cropping, can influence soil properties and microbial communities, thereby indirectly affecting aflatoxin contamination. For example, reduced tillage systems may enhance soil organic matter content and microbial diversity, which can contribute to reduced aflatoxin risk. Additionally, crop rotation with non-host plants and cover cropping can disrupt the lifecycle of aflatoxin-producing fungi and reduce their prevalence in the soil.

In conclusion, the diverse array of soil types found in peanut planting areas profoundly influences aflatoxin-producing strains through their multifaceted effects on moisture levels, pH balance, organic matter content, nutrient availability, soil texture, microbial communities, drainage patterns, temperature regimes and agricultural management practices. A detailed understanding of these complex interactions is imperative for devising effective strategies to mitigate aflatoxin contamination risks and uphold the safety and quality of peanut crops. Although soil type alone may not directly dictate aflatoxin contamination levels, its impact on critical factors such as moisture retention, pH equilibrium, organic matter richness and nutrient accessibility significantly shapes the soil environment conducive to aflatoxin-producing strains. Implementing meticulous soil management techniques, encompassing precise irrigation protocols, judicious organic matter supplementation and meticulous nutrient maintenance, stands as a pivotal approach for curtailing aflatoxin contamination hazards in peanut cultivation areas.

## SOIL AMENDMENTS FOR MITIGATION OF AFLATOXIN CONTAMINATION

The organic content and texture of soil are important because mycotoxins strongly bind to soil organic carbon and clay minerals. Soil texture is determined by the relative proportions of sand, silt and clay particles. Silty clay loam soil, for example, has a 50% adsorption capacity, which helps prevent aflatoxins and their metabolites from leaching into groundwater. Soils with higher organic matter and clay content can hold more moisture, which increases the survival chances of *Aspergillus* species' propagules[31,32]. Aflatoxins produced by these species bind to soil particles, resist microbial degradation and lead to prolonged contamination that can last up to 120 days. Soil density also influences the prevalence of *Aspergillus* species; soils with high densities, such as those used for peanut cultivation, tend to have higher numbers of *Aspergillus* species compared to lower-density soils. Light, sandy soils have been shown to affect the occurrence and growth of *A. flavus*, increasing aflatoxin contamination. Therefore, soil type can provide insights into fungal biodiversity expectations. Researchers have noted that *Aspergillus* species are closely associated with highly productive soils[33].

Aflatoxin contamination presents a significant risk to food safety, especially in crops such as peanuts (*Arachis hypogaea* L.)[34]. In regions like Zambia, where peanut farming is crucial for both food security and economic stability, tackling aflatoxin contamination is of utmost importance[34]. The key challenge is to find sustainable solutions that enhance crop productivity while reducing the health hazards linked to aflatoxin exposure.

One promising avenue for tackling aflatoxin contamination is through soil amendments, with composted cattle manure emerging as a potential solution. Compost offers a holistic approach to soil health improvement, providing essential nutrients, enhancing soil structure, and promoting beneficial microbial activity. These factors play a crucial role in mitigating aflatoxin contamination throughout the peanut production cycle.

A study conducted at the Kasisi Agricultural Training Centre in Zambia shed light on the efficacy of compost application in reducing aflatoxin levels in peanuts. The research revealed significant enhancements in soil properties post-compost application, such as increased plant-available water and improved soil microbial respiration. These improvements translated into tangible benefits for peanut cultivation, with observed increases in pod and kernel yield[34]. The study demonstrated a clear correlation between compost application and reduced aflatoxin contamination in harvested peanuts. By enhancing soil health and creating a more favorable growth environment, compost helps mitigate the conditions conducive to aflatoxin production by *A. flavus* and *A. parasiticus.* The findings underscore the potential of soil amendments, particularly composted cattle manure, as a sustainable approach to aflatoxin contamination mitigation in peanuts. By addressing both soil health and crop productivity, compost offers a multifaceted solution to the complex challenge of aflatoxin contamination, contributing to safer and more sustainable peanut production practices in Zambia and beyond[34].

Farmyard manure (FYM) is an organic material derived from decomposed animal waste, bedding materials and other agricultural residues. It serves as a valuable source of nutrients and organic matter for soil, thereby enhancing soil fertility and overall soil health. Studies have shown that FYM can reduce aflatoxin contamination levels in groundnuts when assessed using enzyme-linked immunosorbent assay (ELISA). Particularly effective is the combination of FYM and gypsum in decreasing aflatoxin contamination. Notably, these soil amendments not only reduce aflatoxin contamination but also improve crop yield compared to untreated control conditions. Timing of soil amendment application is crucial, with applications made 50 days after sowing proving more effective than earlier interventions, highlighting the importance of pre-harvest management strategies[35].

The reduction in aflatoxin contamination by application of compost can be attributed to several factors:

- Improved soil health: Composted cattle manure boosts soil microbial activity and overall soil health, creating less favorable conditions for aflatoxin-producing fungi like *A. flavus* and *A. parasiticus.*
- Increased moisture retention: Compost application enhances soil moisture retention, reducing the likelihood of soil moisture deficits that encourage aflatoxin production by fungi.
- Nutrient availability: Compost provides nutrients to plants and soil microbes, potentially promoting robust plant growth and aiding beneficial microorganisms in outcompeting aflatoxin-producing fungi, thus reducing aflatoxin contamination.
- Microbial competition: The introduction of diverse microbial communities through compost, including beneficial fungi and bacteria, may compete with aflatoxin-producing fungi for resources, thereby diminishing their population and activity.
- Soil structure and moisture: Compost can improve soil structure, increasing water retention and aeration. These changes can influence the environmental conditions within the soil, making it less favorable for aflatoxin-producing fungi.

- pH and chemical composition: Compost can alter soil pH and chemical composition, which may affect the growth and activity of soil microorganisms, including aflatoxin producers.
- Detoxification: Some studies suggest that certain compost amendments can directly bind aflatoxins in the soil, reducing their bioavailability to plants and potential for uptake into food crops.

The use of composted cattle manure as a means of reducing aflatoxin contamination offers a sustainable solution that aligns with environmentally friendly agricultural practices. Composting utilizes organic waste materials, reduces the need for chemical fertilizers, and promotes soil fertility and long-term agricultural sustainability.

The use of composted cattle manure is a practical and effective strategy for reducing aflatoxin contamination in peanut soil. This approach not only improves soil health and crop productivity but also contributes to food safety by minimizing the presence of harmful aflatoxins in harvested peanuts.

## ANTI-AFLATOXIGENIC EFFECTS OF NANOPARTICLES ON PEANUTS

In the contemporary research, nanotechnology has gained momentum due to its wide range of applications. Metallic nanoparticles are extensively used in the realm of science. Metallic nanoparticles exhibit extensive chemical, electronic, thermal, catalytic and biological properties[36]. Studies have been conducted to analyze the inhibition of *Aspergillus* growth using nanoparticles. Silver nanoparticles were found to be effective antifungal and anti-aflatoxigenic agents[37,38]. The antifungal effect of nanoparticles on peanuts was also investigated. Ag-loaded titanium dioxide composites were highly effective, with over 90% inhibition of *A. flavus* when exposed to visible light for 15 minutes, and also significantly reduced the contamination level of *A. flavus*, thereby preventing the production of aflatoxins in peanuts. The concentrations of aflatoxins B1, B2 and G2 decreased by 96.02% ± 0.19%, 92.50% ± 0.45% and 89.81% ± 0.52%, respectively. Furthermore, the treatment did not have any noticeable effect on peanut quality[39]. $Fe_2O_3$ nanorods under sunlight irradiation were able to inhibit *A. flavus* growth and aflatoxin production[40].

## ROLE OF NONTOXIGENIC *A. FLAVUS* IN REDUCTION OF AFLATOXINS IN PEANUT SOIL

Nontoxigenic *A. flavus* strains play a crucial role in reducing aflatoxin contamination in peanut soil. Aflatoxin contamination in peanuts is a significant concern due to its well-documented carcinogenic and toxic properties. Research has shown that nontoxigenic *A. flavus* strains can be effective agents for reducing aflatoxin contamination in various crops, including peanuts.

Studies have shown that applying nontoxigenic strains of *A. flavus* to soil can be an effective strategy to reduce aflatoxin contamination in crops before harvest. The underlying principle is that these nontoxigenic strains outcompete toxigenic strains for colonization sites and vital nutrients, thereby limiting aflatoxin production. Research has demonstrated significant reductions in aflatoxin levels, ranging from 88% to 90%, in peanuts grown in various regions when treated with nontoxigenic *A. flavus* strains.

According to a field study in China, two native nontoxigenic *A. flavus* strains, NAFFHB396 and 707, were evaluated for their efficacy in reducing aflatoxin production in peanuts. The results indicated that both strains significantly reduced the density of the *A. flavus* population in the soil compared to plots treated solely with AF2202, a high aflatoxin-producing strain. The study also reported the extent of aflatoxin reduction achieved by the nontoxigenic strains[41].

Peanuts from plots treated with nontoxigenic strains showed significantly lower percentages of *A. flavus* infection compared to those from plots treated with AF2202 alone, especially under drought stress conditions. Aflatoxin analysis showed significant reductions in total aflatoxin content in peanuts treated with nontoxigenic strains, particularly under drought stress conditions.

## RECOMMENDED PRACTICE FOR PREVENTION AND REDUCTION OF AFLATOXIN IN PEANUTS

Continued peanut cultivation on the same land can lead to a buildup of high populations of *A. flavus and A. parasiticus* in the soil, increasing the risk of infection and aflatoxin contamination[42]. The FAO has recommended various agricultural practices for reducing fungal growth aflatoxin contamination and in peanuts.

Studies on the impact of crop rotation on aflatoxin contamination have been conducted; however, in semi-arid environments, where *Aspergillus* populations can be very high, crop rotations may have limited effectiveness in reducing fungal activity. Different cropping systems, including cultivation and fertilizer practices, may affect the survival and population growth of toxigenic fungi. Peanuts grown in different soil types may show varying levels of infection by molds, with light sandy soils favoring rapid fungal proliferation, especially in dry conditions. Heavier soils with higher water-holding capacity are less likely to experience drought stress, which could contribute to lower levels of aflatoxin contamination in peanuts grown in such soils.

In erosion-prone areas, no-till practices may be necessary for soil conservation. Farmers should use soil test results to determine if fertilizer or soil conditioners are needed to maintain adequate soil pH and plant nutrition, reducing the risk of fungal infestation, especially during seed development. Choosing the right peanut variety is crucial, and farmers should consult plant breeding authorities or agricultural extension services to select cultivars adapted to their region and resistant to factors such as insect and fungal attacks. It is important to select a cultivar that matures at the end of the rainy season to allow for post-harvest field drying under favorable conditions.

Irrigation is recommended to combat heat and drought stress, especially during the last 4–6 weeks of crop growth, to minimize pre-harvest aflatoxin contamination. Water used for irrigation and other purposes should be of suitable quality. Proper spacing of plants and weed control are essential to avoid soil moisture depletion and minimize insect damage and fungal infection. Cultivation and crop protection practices that reduce soil insects, mites and nematodes can help lower aflatoxin contamination. Although no fungicides have been widely adopted for pre-harvest control of *A. flavus or A. parasiticus*, some studies on post-harvest treatments have produced mixed results (FAO).

## REFERENCES


1. Archer, P. (2016). Chapter 9 - Overview of the peanut industry supply chain. In H. T. Stalker & R. F. Wilson (Eds.), *Peanuts* (pp. 253–266): AOCS Press.
2. Zhao, C. X., Jia, L. H., Wang, Y. F., Wang, M. L., & McGiffen, M. E. Jr (2015). Effects of different soil texture on peanut growth and development. *Commun Soil Sci Plant Anal*, *46*(18), 2249–2257. doi:10.1080/00103624.2015.1059845
3. Nji, Q. N., Babalola, O. O., & Mwanza, M. (2023). Soil *Aspergillus* species, pathogenicity and control perspectives. *J Fungi (Basel), 9*(7). doi:10.3390/jof9070766
4. Arunyanark, A., Jogloy, S., Wongkaew, S., Akkasaeng, C., Vorasoot, N., Wright, G. C., & Patanothai, A. (2009). Association between aflatoxin contamination and drought tolerance traits in peanut. *Field Crops Res*, *114*(1), 14–22. doi:10.1016/j.fcr.2009.06.018
5. Li, H., Li, C., Song, X., Liu, Y., Gao, Q., Zheng, R., Li, J., Zhang, P., & Liu, X. (2022). Impacts of continuous and rotational cropping practices on soil chemical properties and microbial communities during peanut cultivation. *Sci Rep*, *12*, 2758. doi:10.1038/s41598-022-06789-1
6. Yu, T., Hou, X., Fang, X., Razavi, B., Zang, H., Zeng, Z., & Yang, Y. (2024). Short-term continuous monocropping reduces peanut yield mainly via altering soil enzyme activity and fungal community. *Environ Res*, *245*, 117977. doi:10.1016/j.envres.2023.117977
7. Huang, R., Li, H., Gao, C., Yu, W., & Zhang, S. (2023). Advances in omics research on peanut response to biotic stresses. *Front Plant Sci*, *14*, 1101994. doi:10.3389/fpls.2023.1101994.

8. Asare Bediako, K., Ofori, K., Offei, S. K., Dzidzienyo, D., Asibuo, J. Y., & Adu Amoah, R. (2019). Aflatoxin contamination of groundnut (*Arachis hypogaea* L.): Predisposing factors and management interventions. *Food Control, 98*, 61–67. doi:10.1016/j.foodcont.2018.11.020
9. N'dede, C., Jolly, C., Vodouhe, S., & Jolly, P. (2012). Economic risks of aflatoxin contamination in marketing of peanut in Benin. *Econ Res Int*, 2012. doi:10.1155/2012/230638.
10. El tawila, M. M., Neamatallah, A., & Serdar, S. A. (2013). Incidence of aflatoxins in commercial nuts in the holy city of Mekkah. *Food Control, 29*(1), 121–124. doi:10.1016/j.foodcont.2012.06.004
11. Perrone, G., Susca, A., Cozzi, G., Ehrlich, K., Varga, J., Frisvad, J. C., & Samson, R. A. (2007). Biodiversity of *Aspergillus* species in some important agricultural products. *Stud Mycol, 59*, 53–66. doi:10.3114/sim.2007.59.07
12. Fapohunda, S. O., & Adewunmi, A. A. (2019). Climate change and mycotoxins—The African experience. *Croat J Food Sci Technol, 11*(2), 283–290.
13. Kumar, P., Mahato, D. K., Kamle, M., Mohanta, T. K., & Kang, S. G. (2016). Aflatoxins: A global concern for food safety, human health and their management. *Front Microbiol, 7*, 2170. doi:10.3389/fmicb.2016.02170
14. Lewis, L., Onsongo, M., Njapau, H., Schurz-Rogers, H., Luber, G., Kieszak, S., & Rubin, C. (2005). Aflatoxin contamination of commercial maize products during an outbreak of acute aflatoxicosis in eastern and central Kenya. *Environ Health Perspect, 113*(12), 1763–1767. doi:10.1289/ehp.7998
15. Mitchell, N. J., Bowers, E., Hurburgh, C., & Wu, F. (2016). Potential economic losses to the US corn industry from aflatoxin contamination. *Food Addit Contam Part A Chem Anal Control Expo Risk Assess, 33*(3), 540–550. doi:10.1080/19440049.2016.1138545
16. Benkerroum, N. (2020). Chronic and acute toxicities of aflatoxins: Mechanisms of action. *Int J Environ Res Public Health, 17*(2). doi:10.3390/ijerph17020423
17. Hua, Z., Liu, R., Chen, Y., Liu, G., Li, C., Song, Y., & Liu, Y. (2020). Contamination of aflatoxins induces severe hepatotoxicity through multiple mechanisms. *Front Pharmacol, 11*, 605823. doi:10.3389/fphar.2020.605823
18. Altyar, A. E., Kensara, O. A., Sayed, A. A., Aleya, L., Almutairi, M. H., Zaazouee, M. S., & Abdel-Daim, M. M. (2023). Acute aflatoxin B1-induced hepatic and cardiac oxidative damage in rats: Ameliorative effects of morin. *Heliyon, 9*(11), e21837. doi:10.1016/j.heliyon.2023.e21837
19. Lamplugh, S. M., Hendrickse, R. G., Apeagyei, F., & Mwanmut, D. D. (1988). Aflatoxins in breast milk, neonatal cord blood, and serum of pregnant women. *Br Med J (Clin Res Ed), 296*(6627), 968. doi:10.1136/bmj.296.6627.968
20. Williams, J. H., Phillips, T. D., Jolly, P. E., Stiles, J. K., Jolly, C. M., & Aggarwal, D. (2004). Human aflatoxicosis in developing countries: A review of toxicology, exposure, potential health consequences, and interventions. *Am J Clin Nutr, 80*(5), 1106–1122. doi:10.1093/ajcn/80.5.1106
21. Zhang, S., Wang, X., Wang, D., Chu, Q., Zhang, Q., Yue, X., & Zhang, Q. (2022). Discovery of the relationship between distribution and aflatoxin production capacity of *Aspergillus* species and soil types in peanut planting areas. *Toxins (Basel), 14*(7). doi:10.3390/toxins14070425
22. Griffin, G. J. (1974). Population levels of *Aspergillus flavus* and the A. Niger Group in Virginia peanut field soils. *Phytopathology, 64*, 322.
23. Horn, R., & Lebert, M. (1994). Soil compactability and compressibility. In *Developments in agricultural engineering* (Vol. 11, pp. 45–69). Elsevier.
24. Horn, B. W., & Dorner, J. W. (1998). Soil populations of *Aspergillus* species from section Flavi along a transect through peanut-growing regions of the United States. *Mycologia, 90*(5), 767–776. doi:10.1080/00275514.1998.12026969
25. Fouché, T., Claassens, S., & Maboeta, M. (2020). Aflatoxins in the soil ecosystem: An overview of its occurrence, fate, effects and future perspectives. *Mycotoxin Res, 36*, 303–309. doi:10.1007/s12550-020-00393-w
26. Hariprasad, P., Vipin, A. V., Karuna, S., Raksha, R. K., & Venkateswaran, G. (2015). "Natural aflatoxin uptake by sugarcane (*Saccharum officinarum L.*) and its persistence in jaggery. *Environ Sci Pollut Res, 22*, 6246–6253.
27. Angle, J. S., & Wagner, G. H. (1981). "Aflatoxin B1 effects on soil microorganisms. *Soil Biol Biochem, 13*(5), 381–384.
28. Goldberg, B. S., & Angle, J. S. (1985). *Aflatoxin movement in soil* (Vol. 14, No. 2, pp. 224–228). American Society of Agronomy, Crop Science Society of America, and Soil Science Society of America.
29. Abudulai, M., Mahama, G., Dzomeku, I., Seidu, A., Sugri, I., Nboyine, J., Opoku, N., Alhassan, M., Appaw, W., Ellis, W., Akromah, R., Mochiah, M., Dankyi, A., Jordan, D., Brandenburg, R., Bravo-Ureta, B., Jelliffe, J., Boote, K., MacDonald, G., Chen, J., Phillips, R., Mallikarjunan, K., Balota, M., Hoisington, D., & Rhoads, J. (2020). Evaluation of agricultural practices to increase yield and financial return and minimize aflatoxin contamination in peanut in Northern Ghana. *Peanut Sci.*

30. Horky, P., Skalickova, S., Baholet, D., & Skladanka, J. (2018). "Nanoparticles as a solution for eliminating the risk of mycotoxins. *Nanomaterials*, *8*(9), 727.
31. Madden, U. A., & Stahr, H. M. (1993). Preliminary determination of mycotoxin binding to soil when leaching through soil with water. *Int Biodeterior Biodegradation*, *31*(4), 265–275.
32. Van Rensburg, C. J., Van Rensburg, C., Van Ryssen, J., Casey, N., & Rottinghaus, G. (2006). In vitro and in vivo assessment of humic acid as an aflatoxin binder in broiler chickens. *Poultry Sci*, *85*(9), 1576–1583.
33. Achaglinkame, M. A., Opoku, N., & Amagloh, F. K. (2017). Aflatoxin contamination in cereals and legumes to reconsider usage as complementary food ingredients for Ghanaian infants: A review. *J Nutr Intermed Metab*, *10*, 1–7.
34. Chalwe, H. M., Lungu, O. I., Mweetwa, A. M., Phiri, E., Njoroge, S. M. C., Brandenburg, R. L., & Jordan, D. L. (2019). "Effects of compost manure on soil microbial respiration, plant-available-water, peanut (*Arachis hypogaea* L.) yield and pre-harvest aflatoxin contamination. *Peanut Sci*, *46*(1), 42–49.
35. Ijaz, M., Afzal, A., Ali, S. R., Nawaz, H. H., Altaf, R., Jannat, F. T., & Syed, S. (2022). "Aflatoxin: Can be managed by soil amendments in groundnut? *Pakistan J Phytopathol*, *34*(1), 65–70.
36. Mody, V. V., Siwale, R., Singh, A., & Mody, H. R. (2010). Introduction to metallic nanoparticles. *J Pharm Bioallied Sci*, *2*(4), 282–289. doi:10.4103/0975-7406.72127
37. Mousavi, S. A. P., & Pourtalebi, S. M. (2015). Inhibitory effects of silver nanoparticles on growth and aflatoxin B1 production by *Aspergillus parasiticus*. *Iran J Med Sci*, *40*(6), 501–506.
38. Sultana, T., Javed, B., Raja, N. I., & Mashwani, Z.-U.-R. (2021). Silver nanoparticles elicited physiological, biochemical, and antioxidant modifications in rice plants to control *Aspergillus flavus*. *Green Process Synth*, *10*(1), 314–324. doi:10.1515/gps-2021-0034
39. Yang, D., Wei, H., Yang, X., Cheng, L., Zhang, Q., Li, P., & Mao, J. (2023). Efficient inhibition of *Aspergillus flavus* to reduce aflatoxin contamination on peanuts over ag-loaded titanium dioxide. *Toxins (Basel)*, *15*(3), 216.
40. Sun, D., Mao, J., Wang, Z., Li, H., Zhang, L., Zhang, W., & Li, P. (2021). Inhibition of *Aspergillus flavus* growth and aflatoxins production on peanuts over α-Fe2O3 nanorods under sunlight irradiation. *Int J Food Microbiol*, *353*, 109296. doi:10.1016/j.ijfoodmicro.2021.109296
41. Yan, L., Song, W., Chen, Y., Kang, Y., Lei, Y., Huai, D., Wang, Z., Wang, X., & Liao, B. (2021). "Effect of non-aflatoxigenic strains of *Aspergillus flavus* on aflatoxin contamination of pre-harvest peanuts in fields in China. *Oil Crop Sci*, *6*(2), 81–86.
42. Norlia, M., Jinap, S., Nor-Khaizura, M. A. R., Radu, S., Samsudin, N. I. P., & Azri, F. A. (2019). Aspergillus section Flavi and Aflatoxins: Occurrence, Detection, and Identification in Raw Peanuts and Peanut-Based Products. *Front Microbol*, 22(10), 2602.

# 2.6 Association between Aflatoxigenic Fungi and Plant Nutrition

*I. Lavkor*

## INTRODUCTION

Mineral nutrients are essential for plants and microorganisms to grow and develop. Additionally, they significantly contribute to disease control by increasing resistance or tolerance to diseases [1, 2]. Adequate and balanced nutrient supply can notably decrease disease occurrence, thus reducing reliance on chemical pesticides. This is crucial for ensuring the safety of food, protecting human and environmental well-being, and thwarting pesticide resistance. The availability of nutrients is affected by several factors, including the use method and type of fertilizer, soil pH, tillage practices, moisture content, and hardness of the seedbed, crop rotations and the presence of mycorrhizal symbiosis [3]. Plant disease severity is closely related to the essential nutrients of plants [4]. Plants can become susceptible to diseases caused by pathogens due to excess or deficiency of nutrients [5]. However, the severity and incidence of other diseases may be increased by a particular nutrient that increases resistance to one disease [3]. The mode of nutrition of pathogens can determine the expression or suppression of diseases. For instance, in the case of obligate parasites, an increase in the nitrogen (N) level in plants leads to an increase in disease severity. On the other hand, high N reduces infection severity for facultative parasites. This prevents the efficient use of nutrients and can lead to nutrient deficiencies and over-accumulation or toxicity of nutrients [4]. Some organisms can consume a significant number of nutrients, which can lead to nutrient deficiencies in the plants and can increase the susceptibility of plants to nutrient deficiencies [6]. Nutritional adequacy leads to resistance in the host, mainly through the formation of mechanical obstacles like a cell wall thickness or the synthesis of natural defense compounds such as flavonoids, antioxidants, and phytoalexins, which provide resistance to pathogenic agents [7].

Nutrient levels can affect the development of disease in several ways. Nutrients have the potential to directly influence the physiology of both plants and pathogens. Nutrients play a pivotal role in influencing various aspects of pathogen behavior, such as sporulation rates and infection frequencies. They also influence the structural integrity of cell walls, membrane stability, and chemical makeup of host organisms. Furthermore, they can affect the growth rates of hosts, allowing seedlings to evade or withstand infection during their vulnerable developmental stages [2, 3].

The use of inorganic fertilizers has led to enhanced plant resistance, disease prevention, and altered pathogenicity or microbial interactions. Many pathogens can significantly alter the nutrition of plants, and differentiating between biotic and abiotic factors that cause nutrient deficiencies and excesses can be challenging. Hence, through changes in uptake, translocation, and distribution, all local and systemic symptoms of pathogenesis resemble abiotically induced nutrient deficiencies or excesses [8].

Diversity within pathogen populations is significant not only for plant protection but also for the potential industrial value of these fungi as producers of secondary metabolites [9, 10]. Mycotoxins are secondary metabolites produced by certain strains of fungi such as *Aspergillus, Fusarium*, and *Penicillium*. These fungi can infiltrate crops in the field and thrive on foods during storage when the temperature and humidity are conducive to their growth. Aflatoxins, fumonisins, ochratoxin


DOI: 10.1201/9781003487043-12

A, trichothecenes and zearalenone are the most common mycotoxins. Aflatoxins are particularly widespread and contaminate a wide range of tropical and subtropical foods/feeds [11]. Aflatoxigenic fungi are recognized for their production of secondary metabolites, including aflatoxin in *A. flavus* and *A. parasiticus* [12]. Four primary and naturally occurring aflatoxins ($B_1$, $B_2$, $G_1$ and $G_2$) are produced. *A. flavus* mainly produces aflatoxins $B_1$ and $B_2$, whereas nearly all isolates of *A. parasiticus* produce aflatoxins $B_1$, $B_2$, $G_1$, and $G_2$. The production of aflatoxins is influenced by several factors such as environment, genotype, insufficient pre-harvest and storage application, in addition to the presence of storage pests and the occurrence of other fungal diseases [12–17]. Peanuts, corn, wheat, soybeans, rice, pistachios, almonds, walnuts, tree nuts, and their products are widely used for human and animal consumption, and the combination of some or all of these factors ultimately increases the risk of aflatoxin production [13, 14, 16, 18–20].

Because abiotic factors affect the growth and production of secondary metabolites in fungi, mineral elements and different fertilization treatments can be used to prevent fungal contamination during plant cultivation. This chapter evaluates the relationship between plant nutrition, aflatoxigenic fungi, and aflatoxin contamination of crops and their products.

## RELATIONSHIP OF MACRONUTRIENTS TO AFLATOXIGENIC FUNGI

Researchers have studied the influence of several sources of macroelements on aflatoxin production in various species. The ability to produce high yields of aflatoxin using inorganic elements such as nitrogen (N) depends on the fungus strains used and the nutrient solution composition. Unless other environmental or biological stresses are extreme, nutrition, in combination with other production practices, has been shown to have a beneficial effect on suppressing *A. flavus* group infection and aflatoxin contamination in corn. However, research suggests that variations in aflatoxin levels at harvest may be influenced by N levels, particularly when treatment levels are low or high. N appears to play a significant role in determining aflatoxin concentration differences. Data indicate that inadequate irrigation, high population density or extreme N deficiency are associated with aflatoxin contamination. Importantly, these findings are objective, not factual, based on subjective assessments. Additionally, late or excessive application of N, phosphorus (P) and potassium (K) has led to increased concentrations [21].

Jones and Duncan [22] undertook a study to assess the impact of N fertilization, planting, and harvest date changes on aflatoxin contamination in corn fields. Regardless of the isolate used, treatments with a higher N rate (145.7 kg/ha) exhibited lower levels of aflatoxin B1 (AFB1) compared to treatments with a lower N rate (11.2 kg/ha), as indicated by the study. On average, across cultivar, planting date, and isolate, low-N plants had 2.4 times more AFB1 than high-N plants. Jones and Duncan [22] suggested that insufficient N fertilization changes the nutrition of corn before harvest, rendering it a more favorable medium for aflatoxin production.

Özer and Derici [23] identified a notable correlation between the N content in dried figs and the occurrence of AFB2 as well as total aflatoxin. It is important to note that these findings are objective and based on scientific evidence. Research indicates that this phenomenon occurs because an increase in N content leads to higher water content in plant cells. This heightened water content results in turgidity in the fruit, thereby promoting the growth rate of the pathogen. Additionally, N extends the time and the potential for soiling. According to Gül and Aksu [24], elevated levels of nitrogen contribute to the degradation of color in dried fig fruits. Additionally, a connection exists between the darker hue of fruit and the presence of aflatoxin in dried figs, reinforcing the association between N content and aflatoxin levels.

Sufficient provision of plant nutrients is among the potential beneficial agricultural practices for diminishing *A. flavus* presence. The risk of aflatoxigenic fungi colonization and aflatoxin production in corn grain is reduced using N and P application [25]. Studies have shown that insufficient application of N fertilization can result in higher levels of aflatoxin accumulation in grains [25, 26].

Applying appropriate nitrogen and phosphorus levels can accelerate corn seedling and root growth, resulting in a robust plant with greater resistance to pathogenic threats.

Corn plants that are deficient in nitrogen and under other stresses are more susceptible to infection by *Aspergillus* spp., which can lead to higher concentrations of aflatoxin in the grain. A study carried out at North Carolina State University determined that corn exposed to nitrogen deficiency experienced a 28% increase in aflatoxin contamination in ears receiving insufficient nitrogen compared to ears receiving sufficient nitrogen levels [26]. However, ideal N and P ratios in the corn plant can support the development of maize seedlings and roots, promote better crop viability and strengthen cell walls, potentially serving as a defense mechanism against *A. flavus* infiltration [27, 28].

Lavkor [29] studied the effects of nutrients on fungi and aflatoxin contamination in peanuts. A total of 40 peanut samples were collected from the Osmaniye province of Turkey during both the harvest and storage periods. The results showed that low N levels and other stresses could make peanuts susceptible to infection by *Aspergillus* species and increased aflatoxin concentration.

Applying calcium (Ca) to peanuts grown in Ca-deficient sandy soils can increase yield and quality while decreasing damage during harvest. In addition, the presence of Ca in peanuts may affect both the interactions of the *A. flavus* group and aflatoxin contamination in peanut kernels. Studies in Georgia have shown that applying calcium to certain soils has a beneficial effect on reducing aflatoxin contamination [30]. In a study by Wilson and Walker [21], the effect of gypsum, a form of Ca, on aflatoxin contamination in peanuts was investigated. The study found that applying gypsum at rates of 112, 224, and 332 kg/ha significantly reduced aflatoxin contamination compared to peanuts grown without applied calcium. Ca and irrigation practices influenced hull and kernel infection by the *A. flavus* group, which led to peanuts with high concentrations of aflatoxin. Therefore, mineral nutrition is just one of several production practices that may be used to help control *A. flavus* infection and aflatoxin contamination of crops. In a similar study, Mixon et al. [31] did not detect any aflatoxin in gypsum-treated plots. However, aflatoxin was found in peanuts without gypsum treatments, resulting in a highly significant difference between the treatments. For this result, gypsum application was found to reduce the incidence of seed colonization by A. *flavus* in peanuts.

A correlation has been found between certain nutrients and forms of aflatoxin. Generally, an increase in N levels leads to a significant increase in AFB2 and total aflatoxin. A positive correlation is also determined between Ca and AFB1. Furthermore, there appears to be a significant relationship between K and AFB1 as well as total aflatoxin and between Na and AFB1 [23].

## RELATIONSHIP OF MICRONUTRIENTS TO AFLATOXIGENIC FUNGI

Microelements, such as zinc (Zn), iron (Fe), and copper (Cu) may have an impact on aflatoxin. Most previous studies have focused solely on AFB1 production on crops by *A. flavus* and *A. parasiticus*. Nevertheless, the relationships between the different mineral nutrients of crops and the varying levels of AFB1 contamination have only been investigated in a few studies. Lillehoj et al. [32] indicated that the addition of 500 μg of Cu/g to corn germ resulted in a decrease in aflatoxin levels compared to non-contaminated corn germ. Zn has a strong impact on aflatoxin production, particularly by *Aspergillus* spp. [33]. The range of Zn concentrations that affect aflatoxin production is narrower than that affecting growth, similar to phytotoxins and other secondary metabolites. It is crucial to emphasize that this range falls within the essential, stimulatory, and inhibitory levels. At the same time, Tiwari et al. [34] detected that Tiwari et al. [35] detected that aflatoxin production was reduced in varying quantities by iron, copper, and manganese sulfate. Limiting Fe and other essential metals reduces aflatoxin production due to growth impediments. However, Zn deficiency limits the production of aflatoxin without affecting growth, demonstrating Zn's regulatory role [36]. Gupta and Venkitasubramanian [37] clarified that adding Zn to autoclaved and non-autoclaved soybeans increased the production of aflatoxins. According to Gupta et al. [38], Zn is necessary for the synthesis and integrity of *AflR*, the primary regulatory gene for the production of aflatoxins. Moreover, the production of AFB1 by *Aspergillus* is dependent on Zn [39]. They also declared that $Zn^{2+}$ is essential for the biosynthesis of aflatoxin by *A. flavus* [40, 41]. Moreover, a minimum of 20 enzymes rely on zinc [42], potentially contributing significantly to its pivotal function in AFB1 biosynthesis. In vitro studies have confirmed the role of Zn in aflatoxin biosynthesis in contaminated food and feed. Stössel [43] also concluded that Zn supplementation could promote AFB1 biosynthesis in soybeans. Analysis of microelements showed that defatted corn germ contained the highest amount of Zn. Following this, defatted peanut and soybean were introduced. The zinc content found in corn and corn endosperm was notably low, and these measurements showed a strong correlation with AFB1 contamination. Therefore, Zn satisfaction could potentially emerge as a primary factor influencing the diverse contamination levels of these grains by *A. flavus*. Zinc quantity in corn grain is associated with the growth and aflatoxin production of *A. flavus* and *A. parasiticus* [35, 44], soybeans [37], sunflower seed [45], and chicken feed [46]. There is a positive correlation between the Zn level in the soil, the saprophytic growth of *Aspergillus* in the soil, and the contamination of corn plants with aflatoxin when grown in that soil.

## CONCLUSION

The inhibitory effect of mineral nutrition on aflatoxin accumulation in *A. parasiticus* and *A. flavus* has been determined in various studies. Research into the suppressive impact of mineral nutrition on aflatoxin accumulation has revealed a correlation between different mineral nutrients present in crops and the levels of aflatoxin contamination across various crop types.

Mineral elements play a key role in aflatoxin biosynthesis and mycelial growth. The interactions between other mineral elements remain poorly understood in the aflatoxin production of aflatoxigenic *Aspergillus* spp. on various crops. The investigation of the effects of these interactions on aflatoxin production by aflatoxigenic *Aspergillus* spp. should be a priority for future research.

## REFERENCES


1. Agrios, G.N. 2005. *Plant Pathology* 5th ed. Elsevier Academic Press. Burlington, Mass, EU, 251–262.
2. Graham, R.D., and M.J. Webb. 1991. Micronutrients and disease resistance and tolerance in plants. *Micronutrients in Agriculture* 4: 329–370.
3. Dordas, C. 2008. Role of nutrients in controlling plant diseases in sustainable agriculture: A review. *Agronomy for Sustainable Development* 28: 33–46.
4. Huber, D.M., and R.D. Graham. 1999. The role of nutrition in crop resistance and tolerance to diseases. In *Mineral Nutrition of Crops: Fundamental Mechanisms and Implications*, ed. Z. Rengel. Boca Raton: CRC Press, 169–204.
5. Downer, A.J., D.R. Hodel, D.M. Mathews, and D.R. Pittenger. 2013. Effect of fertilizer nitrogen source on susceptibility of five species of field-grown palms to *Fusarium oxysporum* f. sp. canariensis. *Palms* 57: 89–92.
6. Timonin, M.I. 1965. Interaction of higher plants and soil microorganisms. In *Microbiology and Soil Fertility*. Oregon State University Press, Corvallis, OR, 135–138.
7. Gupta, N., S. Debnath, S. Sharma, P. Sharma, and J. Purohit. 2017. Role of nutrients in controlling the plant diseases in sustainable agriculture. In *Agriculturally İmportant Microbes for Sustainable Agriculture*. Springer, Singapore, 217–262.
8. Huber, D.M. 1993. Introduction. In *Soil borne Plant Pathogens: Management of Diseases With Macro and Microelements*, ed. A.W. Engelhard. APS Press, St. Paul, Minnesota, USA, 1–8.
9. Bills, G.F. 1995. Analyses of microfungal diversity from a user's perspective, *Canadian Journal of Botany* 73: 33–41.
10. Singh, S.B., R.G. Ball, G.F. Bills, C. Cascales, J.B. Gibbs, M.A. Goetz, K. Hoogsteen, R.G. Jenkins, et. al. 1996. Chemistry and biology of cylindrols: Novel inhibitors of Ras farnesyl-protein transferase from *Cylindrocarpon lucidum*. *The Journal of Organic Chemistry* 61: 7727–7737.
11. Katole, S.B., P. Kumar, and R.D. Patil. 2013. Environmental pollutants and livestock health: A review. *Veterinary Research International* 1: 1–13.

12. Wee, J., D.M. Day, and J.E. Linz. 2016. Effects of zinc chelators on aflatoxin production in *Aspergillus parasiticus*. *Toxins* 8: 1–14.
13. Cardwell, K.F., and S.H. Henry. 2004. Risk of exposure to and mitigation of effect of aflatoxin on human health: A West African example. *Journal of Toxicology: Toxin Reviews* 23: 217–247.
14. Hell, K.P., R. Fandohan, S. Bandyopadhyay, R. Kiewnick, J.P. Sikora, and P.J. Cotty. 2008. Pre- and postharvest management of aflatoxin in maize. An African perspective. In *Mycotoxins: Detection Methods, Management, Public Health, and Agricultural Trade*, eds. J.F. Leslie, R. Bandyopadhyay, and A. Visconti. Wallingford UK: CABI, 219–229.
15. Cotty, P.J., and R. Jaime-Garcia. 2007. Influences of climate on aflatoxin producing fungi and aflatoxin contamination. *International Journal of Food Microbiology* 119: 109–115.
16. Hell, K., C. Mutegi, and P. Fandohan. 2010. Aflatoxin control and prevention strategies in maize for sub-Saharan Africa. 10$^{th}$ international working conference on stored product protection *Julius-Kühn-Archiv* 425: 534–541.
17. Warburton, M.L., and W.P. Williams. 2014. Aflatoxin resistance in maize: What have we learned lately? *Advances in Botany* 2014: 1–10.
18. Ma, X., W. Wang, X. Chen, Y. Xia, W. Shijia, N. Duan, and Z. Wang. 2014. Selection, identification, and application of aflatoxin $B_1$ aptamer. *European Food Research and Technology* 238: 919–925.
19. Li, X.Y., L.H. Zhao, Y. Fan, Y. Jia, L. Sun, S. Ma, C. Ji, Q. Ma, and J. Zhang. 2014. Occurrence of mycotoxins in feed ingredients and complete feeds obtained from the Beijing region of China. *Journal of Animal Science and Biotechnology* 5: 37.
20. Bayman, P., J.L. Baker, and N.E. Mahoney. 2002. *Aspergillus* on tree nuts: Incidence and associations. *Mycopathologia* 155: 161–169.
21. Wilson, D.M., and M.E. Walker. 1981. Calcium potential aflatoxin foe. *Southeastern Peanut Farmer* 19: 6B.
22. Jones, R.K., and H.E. Duncan. 1981. Effect of nitrogen fertilizer, planting date and harvest date on aflatoxin production in corn inoculated with *Aspergillus flavus*. *Plant Disease* 65: 741–744.
23. Özer, B., and B. Derici. 1998. A research on the relationship between aflatoxin and ochratoxin a formation and plant nutrients. *ISHS Acta Horticulturae* 480: 199–207.
24. Gül, N., and U. Aksu. 2001. Sarılop incir çeşidinde bazı kuru meyve nitelikleri ile aflatoksin ilişkisi üzerinde araştırmalar, *Ege Üniversitesi Fen Bilimleri Enstitüsü Dergisi* 2001: 19–22.
25. Lillohoj, E.B., and M.S. Zuber. 1975. Aflatoxin problem in corn and possible solutions. In: Proceedings 30th Annual Corn and Sorghum Research Conference. pp. 230–250.
26. Payne, G.A., E.J. Kamprath, and C.R. Adkins. 1989. Increased aflatoxin contamination in nitrogen stressed corn. *Plant Disease* 73: 556–559.
27. Hubber, D.M., and S. Haneklaus. 2007. Managing nutrition to control plant disease. *Landbauforsch Volkenrode* 57: 313–322.
28. Zhao, Q., Z. De-Hui, F. Zhi-Ping, Y. Zhan-Yuan, H. Ya-Lin, and Z. Jianwei. 2009. Seasonal variations in phosphorus fractions in semi-arid sandy soils under different vegetation types. *Forest Ecology and Management* 258: 1376–1382.
29. Lavkor, I. 2017. The effects of nutrients on fungi and aflatoxin contamination in peanut. *International Journal of Scientific and Technological Research* 5: 165–175.
30. Wilson, D.M., M.E. Walker, and G.J. Gascho. 1989. Some effects of mineral nutrition on aflatoxin contamination of corn and peanuts. In *Soilborne Plant Pathogens: Management of Diseases With Macro and Microelements*, ed. A.W. Engelhard. APS Press, St. Paul, Minnesota, USA, 137–151.
31. Mixon, A.C., D.K. Bell, and D.M. Wilson. 1984. Effect of chemical and biological agents on the incidence of *Aspergillus flavus* and aflatoxin contamination of peanut seed, *Phytopatholgy* 74: 1440–1444.
32. Lillehoj, E.B., W.J. Garcia, and M. Lambrow. 1974. *Aspergillus flavus* infection and aflatoxin production in corn: Influence of trace elements. *Journal of Applied Microbiology* 28: 763–767.
33. Cuero, R., T. Ouellet, J. Yu, and N. Mogongwa. 2003. Metal ion enhancement of fungal growth, gene expression and aflatoxin synthesis in *Aspergillus flavus*: RT-PCR characterization. *Journal of Applied Microbiology* 94: 953–961.
34. Tiwari, R.P., V. Mittal, T.C. Bhalla, S.S. Saini, G. Singh, and D.V. Vadehra. 1986. Effect of metal ions on aflatoxin production by *Aspergillus parasiticus*. *Folia Microbiologica* 31: 124–128.
35. Failla, L., J. Lynn, and W.G. Niehaus Jr. 1986. Correlation of $Zn^{2+}$ content with aflatoxin content of corn. *Applied Environmental Microbiology* 52: 73–74.
36. Marsh, P.B., M.E. Simpson, and M.W. Trucksess. 1975. Effect of trace metals on the production of aflatoxin by *Aspergillus parasiticus*. *Applied Microbiology* 30: 52–57.
37. Gupta, S.K., and T.A. Venkitasubramanian. 1975. Production of aflatoxin on soybeans. *Journal of Applied Microbiology* 29: 834–836.

38. Gupta, S.K., K.K. Maggon, and T.A. Venkitasubramanian. 1977. Effect of zinc on tricarboxylic acid cycle intermediates and enzymes in relation to aflatoxin biosynthesis. *The Journal of General Microbiology* 99: 43–48.
39. Liu, B.H., and F.S. Chu. 1998. Regulation of *aflR* and its product, *AflR*, associated with aflatoxin biosynthesis. *Applied and Environmental Microbiology* 64: 3718–3723.
40. Mateles, R.I., and J.C. Adye. 1965. Production of aflatoxins in submerged culture. *Journal of Applied Microbiology* 13: 208–211.
41. Marsh, P.B., M.E. Simpson, and M.W. Trucksess. 1975. Effects of trace metals on the production of aflatoxins by *Aspergillus parasiticus. Journal of Applied Microbiology* 30: 52–57.
42. Parisi, A.F., and B.L. Vallee. 1969. Zinc metalloenzymes: Characteristics and significance in biology and medicine. *The American Journal of Clinical Nutrition* 22: 1222–1239.
43. Stössel, P. 1986. Aflatoxin contamination in soybeans: Role of proteinase inhibitors, zinc availability, and seed coat integrity, *Applied and Environmental Microbiology* 52: 68–72.
44. Aziz, N.H., A.A. Shahin, A.A. Abou-Zeid, and S.A. el-Zeany. 2000. Correlation of growth and aflatoxin production by *Aspergillus flavus* with some essential metals in gamma irradiated crushed corn. *Nahrung* 44: 354–359.
45. Chulze, S., S. Fusero, A. Dalcero, M. Etcheverry, and E. Varsavsky. 1987. Aflatoxin in sunflower seeds: Effect of zinc in aflatoxin production by two strains of *Aspergillus parasiticus. Mycopathology* 99: 91–94.
46. Jones, F.T., W.M. Hagler, and P.B. Hamilton. 1984. Correlation of aflatoxin contamination with zinc content of chicken feed. *Applied Environmental Microbiology* 47: 478–480.

# 2.7 Aflatoxins in Plant Pathogenesis

*Simran Rani, Anshu, Priyanka Dahiya, Chetna Rathi, A. Sankara Narayanan, Amita Suneja Dang, and Pooja Suneja*

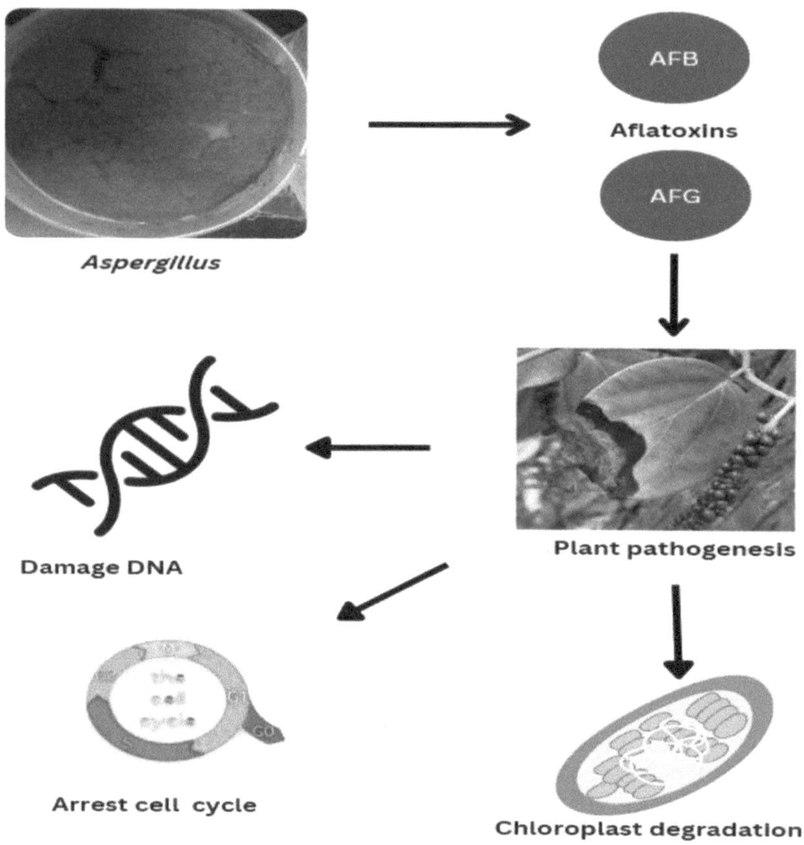


## INTRODUCTION

Aflatoxins are the group of mycotoxins produced by several fungal genera such as *Aspergillus, Alternaria*, *Fusarium,* and *Penicillium*, however, *Aspergillus flavus* and *Aspergillus parasiticus* have been reported to produce most toxic aflatoxins. These are the toxic secondary metabolites possessing mutagenic, teratogenic and immunosuppressive properties [1]. Aflatoxins were first detected in 1960 in association with Brazilian groundnut meal and found to be stable under extreme conditions, such as temperature and pH. Aflatoxins B1, B2, and G1, G2 are the most common aflatoxins produced by *A. flavus* and *A. parasiticus*, respectively. *A. flavus*, a yellow-green fungus, impairs plant growth and yield, particularly under abiotic stresses such as drought and heat [2]. Insects and


DOI: 10.1201/9781003487043-13

nematodes enter the plant host through surface lesions and compromise the defense mechanism, making it susceptible to fungal attack and enhances the production of aflatoxins [3]. *A. parasiticus* is a saprophyte found in soil and decayed plant litter. It can be distinguished from *A. flavus* by the presence of darker green conidial heads, rougher conidium surface ornamentation, and produces more aflatoxins than *A. flavus* [4, 5].

Around one-fourth of the world's crops remain affected by aflatoxins during harvesting, storage, and transport. Aflatoxins are associated with a variety of plants such as cereals (*Zea mays*, *Oryza sativa*, *Triticum aestivum*), oilseeds (*Arachis hypogea*, *Gossypium*), spices (*Capsicum annum*, *Piper nigrum*, *Curcuma longa*, *Coriandrum sativum*, *Zingiber officinale*), and nuts (*Prunus dulcis*, *Bertholletia excels*, *Juglans*, *Cocos nucifera*) growing in the humid tropics and subtropics [6]. Humans and animals are exposed to aflatoxins or the products of their metabolism by consuming the contaminated food [5]. Fungi, insects and weeds generally cause wounds on plants, making them susceptible to aflatoxigenic fungi [3].

Plants mitigate aflatoxin toxicity to an extent by the production of secondary metabolites, such as alkaloids, phenols, terpenoids and flavonoids, which have antifungal activity. However, several economically important plants succumb to this toxicity [7]. Therefore, there is a need to focus on inducing resistance in infected plants, improving detection techniques to identify the infection at the earliest and develop novel approaches to combat contamination. Understanding the interactions of aflatoxins in plant pathogenesis is essential for its prevention and control. This chapter is focused on the role of aflatoxins in plant pathogenesis, factors influencing their production, and mitigation strategies.

## TYPES AND CHEMISTRY OF AFLATOXINS

Mainly four types of aflatoxins—AFB1, AFB2, and AFG1, AFG2—are designated by strong bluish- and greenish-yellow fluorescence under UV light, respectively. Many studies have proved the ability of *A. flavus* to produce aflatoxins B1 and B2, whereas *A. parasiticus* can produce G1 and G2 as well [8]. Out of all, B1 is considered as the most toxic aflatoxin and identified as a potent hepatocarcinogen by the International Agency for Research on Cancer (IARC). The order of carcinogenicity amongst aflatoxins is as follows: B1>G1>B2>G2. Aflatoxins are further modified in the digestive tract of animals and mammals. M1 and M2, hydroxylated forms of B1 and B2, have been isolated from milk of cow fed with aflatoxin-contaminated feed. Aflatoxins are difuranocoumarin derivatives in which a bifuran group is attached to one side of the coumarin while the other side contains a lactone ring that is either 5-membered (B-group) or 6-membered (G-group) [9]. Synthesized from an elaborate pathway including approximately 27 enzymatic reactions, the expression of aflatoxins is regulated by aflR and aflS, two cluster-specific regulators. However, because they are secondary metabolites, their synthesis remains under the control of many other mechanisms, which are hitherto cryptic and expressed only under certain environmental stimuli such as temperature, light, pH, and nutrients [10].

The optimum range of pH favoring the production of aflatoxins is 3–7 [11, 12]. Lower pH reduces fungal growth, and a slight rise in pH induces the production of both fungus and aflatoxins [13]. In addition, lower (5) and higher pH (7) induce the production of aflatoxins B and G, respectively. Light inhibits the production of aflatoxin and fungal growth, whereas high moisture content favors aflatoxin production, with 85% humidity being the optimal value [14, 15]. *A. flavus* possess an astonishing ability to survive under temperatures ranging between 12 and 48°C, however, 28–37°C is an optimum range [16]. Temperature range has also been reported to influence the ratio of AFB and AFG production [17]. $CO_2$ and $O_2$ suppressed the aflatoxin production when present in their higher and lower concentrations, respectively [18]. Nutritional factor such as carbon source, nitrogen source, lipids, and some microelements influence the production of aflatoxins. Substrate having carbohydrates supports more production of aflatoxins as compared to oil [19]. Amino acids such as Gly (Glycine), Glu (Glutamate), and Ala (Alanine) with $Zn^{+2}$ and $Mg^{+2}$ endorse aflatoxin production while Trp (Tryptophan) inhibits the same [20, 21].

## AFLATOXINS IN PLANT PATHOGENESIS

Plants are infected with a variety of phytopathogenic fungi, primarily those belonging to Ascomycota (*Aspergillus* spp., *Cladosporium* spp., *Magnaporthe* spp., *Botrytis* spp.) and Basidiomycota (rusts and smuts) during their life cycles [22]. Attachment and penetration are the two key steps of fungal pathogenesis. Fungal spores come in contact with plants while dispersed by abiotic (wind, water) or biotic factors (insects). Usually, the infection cycle begins with the formation of a germ tube out of spore followed by growth of hyphae all along the plant surface. This step is mediated by several physical (hardness of surface, hydrophobicity) and chemical (cutin, waxes) stimuli. Many phytopathogenic fungi develop a dome-shaped appressorium that attaches firmly to the host surface and ruptures the cuticle to grow inside the epidermal cells of the host plant. A few of them also differentiate a hyphal tip swelling at the point of host penetration; nevertheless, this structure is not separated by a cell wall. Herein, penetration of fungi is facilitated by the activity of lytic enzymes rather than a mechanical force involved in case of appressorium. Pathogenic fungi also use stomata to penetrate the host and have evolved directional growth patterns on the host leaf surface to locate and recognize stomatal guard cells through thigmotropism [23–25].

Aflatoxins modulate the entire primary and secondary metabolism of the plant, hence affecting its growth and yield [26]. They have also been reported to alter the internal structures of the infected tissues, causing the size of various organelles to expand or reduce in an unusual manner. This leads to membrane disintegration (especially the vacuolar membrane [tonoplast]), which is essential for osmoregulation, rupture of organelles like the endoplasmic reticulum associated with protein and lipid synthesis, and loss of ribosomes, thus inhibiting the elongation of the root and shoot [27]. Aflatoxins also affect cell membrane permeability and inhibit the activity of many critical enzymes, such as those involved in biosynthesis of chlorophyll and photosynthesis. Several previous studies have demonstrated their phytotoxic effect. An inhibitory effect on the biosynthesis of photosynthetic pigments, reduction in germination percentage, and growth of lettuce seedling were observed in *Sesamum indicum* [28]. A drastic decrease in the physiological activity and chlorophyll content of rice following infection of *A. flavus* was observed [29]. Aflatoxins prevent the formation of grana in chloroplasts, inhibit chlorophyll, and carotenoid synthesis, thus making the plant susceptible to albinism. Reduction in these pigments reduce photosynthesis, ultimately affecting the mass and development of leaves. Aflatoxins bind to the DNA during its replication, causing various damages or inhibiting DNA-dependent polymerase activity, reducing the amount of DNA within a plant cell. Under such circumstances, the p53 gene plays a significant role in maintaining the cell in the G2/M phase to prevent DNA replication and allow it more time to undergo DNA repair. Nevertheless, too much damage to DNA pushes the cell to opt for apoptosis. Genotoxic effects and chromosomal aberrations during meiosis, including bridges, bivalents, and laggards, have also been noticed in various plants [30]. Table 2.7.1 cites various economically valuable plants infected with aflatoxin contamination.

## STRATEGIES TO MITIGATE AFLATOXIN CONTAMINATION IN PLANTS

Aflatoxin contamination poses a significant threat to the growth and development of economically important crops, especially in humid regions. Cleaning processes like dehulling, milling, and other physical agents such as microwave and ultraviolet radiation, are commonly used to eliminate contamination from the crop produce post-harvesting. Steam under pressure and roasting under dry conditions have also been reported effective in reducing aflatoxin [60]. *Arachis hypogea* seeds were found to undergo 40%–73% depletion in aflatoxin level when heated at 180°C [61]. Roasting the seeds of *A. hypogea* and *Zea mays* at 100°C with 30% moisture for 2 hours caused an 85% reduction in aflatoxins [62]. Sunlight exposure of 10–12 hours detoxified AFB1 in artificially infected *Z. mays* (80%) and *A. hypogea* (17%) [14]. Moisture leads to the production of aflatoxins, so the infection can be mitigated by drying for one to two days, as observed in *A. hypogea* (6.6%) and

**TABLE 2.7.1**
**Crops Infected with Aflatoxins**

| Fungus Species | Food Crop | Type of Mycotoxin | Reference |
|---|---|---|---|
| *A. flavus* | *Zea mays* | AFB1 | [31–41] |
| | *Arachis hypogaea* | | |
| | *Lens culinaris* | | |
| | *Helianthus* | | |
| | *Capsicum annum* L. | | |
| | *Prunus dulcis* | | |
| | *Lens culinaris* | | |
| | *Coriandrum sativum* | | |
| | *Cuminum cyminum* | | |
| | *Oryza sativa* | | |
| | *Digitaria exilis Stapf* | AFB1 | [42, 43] |
| | *Sorghum* | AFB2 | |
| | *Glycine max* | AFB1 | [44, 45] |
| | *Coffea arabica* | AFB2 | |
| | | AFG1 | |
| | | AFG2 | |
| *A. parasiticus* | *Psidium guajava* | AFB1 | [46] |
| | | AFG1 | |
| *A. flavus* | *Zea mays* | AFB1 | [47, 48] |
| *A. parasiticus* | *Oryza Sativa* | | |
| | *Cydonia oblonga* | AFB1 | [49] |
| | | AFB2 | |
| | *Capsicum* | AFB1 | [50] |
| | | AFB2 | |
| | | AFG1 | |
| | *Zea mays* | AFB1 | [51–56] |
| | *Arachis hypogaea* | AFB2 | |
| | *Triticum* | AFG1 | |
| | *Anacardium occidentale* | AFG2 | |
| | *Zingiber officinale* | | |
| | *Linum usitatissimum* | | |
| *A. flavus* | *Ficus carica* | AFB1 | [57–59] |
| *A. parasiticus* | *Fragaria* x *ananassa* | AFB2 | |
| *A. niger* | *Sesamum indicum* | AFG1 | |
| | | AFG2 | |

*Z. mays* (15%) [63–64]. Nevertheless, significant reduction in the nutritional value of the product and the highly expensive technology required for physical treatment have limited their widespread use to control aflatoxins [65].

Strong acids/bases, oxidizing agents, and aldehydes have the potential to mitigate aflatoxin contamination [66]. Though costly, ozone is a strong aflatoxin oxidizing agent that causes remarkable degradation of aflatoxin in cereals and legumes through electrophilic attack on the furan ring [67]. The ammoniation process has been found effective in reducing aflatoxin contamination in several crops like maize, wheat and barley because these are not stable under alkaline conditions [68]. $NaHSO_3$ (sodium bisulfite), $Ca(OH)_2$ (calcium hydroxide), NaClO (sodium hypochlorite), and sorbents are the potential reducing agents for aflatoxins. Activated charcoal mixed with aflatoxin-contaminated feed at different concentrations was used to prevent aflatoxicosis [69]. Chemicals used for the long-term storage of food, such as propionic acid ($C_3H_6O_2$), boric acid ($H_3BO_3$), and benzoic acid ($C_6H_5COOH$), inhibit the growth of aflatoxigenic fungi [70]. In fact, calcium propionates and sodium propionates decrease the production of aflatoxins in maize even under the conditions favorable for the fungus [71].

Amongst biological agents, bacteria, fungi and microscopic algae exhibit various degrees of aflatoxin degradation potential, mainly by absorption or enzymatic degradation [72]. Several microbes have been reported to directly absorb aflatoxins into their cell wall, mediated by internalization [73]. Even dead microflora can absorb aflatoxins, a property that is utilized to prepare aflatoxin-free fluids [74]. Both intracellular and extracellular enzymes can degrade aflatoxins [75]. Supernatant, derived from the microbial culture of *Bacillus velezensis*, exhibited 91.5% degradation activity against AFB1 [76]. Microbial consortium of *Geobacillus* and *Tepidimicrobium* resulted in more than 95% degradation of AFB1 within 72 hours at a temperature range of 55–60°C [77]. Many other bacterial and fungal strains, such as *Ralstonia, Pseudomonas, Lactobacilli, Bacillus*, and *Trichoderma*, have displayed significant antagonistic activity against *A. flavus* infecting crops prior to harvesting [78–79]. Other practices, such as effective weed and pest control, timely fertilization, irrigation, harvest, proper transportation and packaging, and the use of storage protectants, can also control the aflatoxin-led damage to a great extent. In addition, public awareness is required to sensitize the population regarding the spread and elimination of aflatoxigenic fungi.

## ACKNOWLEDGMENT

The authors acknowledge the infrastructural support from the Department of Microbiology and Centre for Medical Biotechnology (CMBT), Maharshi Dayanand University (MDU), Rohtak, India. SR and CR acknowledge the Council of Scientific & Industrial Research (CSIR), New Delhi, India, for the CSIR fellowship. PD is thankful to University Grants Commission (UGC), New Delhi, India, for fellowship support.

## REFERENCES

1. Bhatnagar-Mathur P., Sunkara S., Bhatnagar-Panwar M., Waliyar F., Sharma K. K. 2015. Biotechnological advances for combating *Aspergillus flavus* and aflatoxin contamination in crops. *Plant Science* 234, https://doi.org/10.1016/j.plantsci.2015.02.009
2. Kumar A., Pathak H., Bhadauria S., Sudan J. 2021. Aflatoxin contamination in food crops: Causes, detection, and management. *Food Production, Processing and Nutrition* 3: 1–9. https://doi.org/10.1186/s43014-021-00064-y
3. Kinyungu S. W. 2019. Efficacy of pre-harvest *Aspergillus flavus* biocontrol treatment on reducing aflatoxin accumulation during drying (Doctoral dissertation, Purdue University).
4. Chang P. K. 2021. Authentication of *Aspergillus parasiticus* strains in the genome database of the National Center for Biotechnology Information. *BMC Research Notes* 14: 111. https://doi.org/10.1186/s13104-021-05527-6
5. Kumar P., Mahato D. K., Kamle M., Mohanta T. K., Kang S. G. 2017. Aflatoxins: A global concern for food safety, human health, and their management. *Frontiers in Microbiology* 7: 2170. https://doi.org/10.3389/fmicb.2016.02170
6. Rajarajan P. N., Rajasekaran K. M., Devi N. K. 2013. Aflatoxin contamination in agricultural commodities. *Indian Journal of Pharmaceutical and Biological Research* 1(4): 148–151. https://doi.org/10.30750/ijpbr.1.4.25
7. Matrose N. A., Obikeze K., Belay Z. A., Caleb O. J. 2021. Plant extracts and other natural compounds as alternatives for post-harvest management of fruit fungal pathogens. *Food Bioscience* 41: 100840. https://doi.org/10.1016/j.fbio.2020.100840
8. Quadri S. H., Shantaram U., Hs E. 2012. An overview of chemistry, toxicity, analysis, and control of aflatoxins. *International Journal of Chemical and Life Sciences* 2: 1071–1078.
9. Nakai V. K., Rocha L. O., Goncalez E., Foneseca H., Ortega E. M., Correa B. 2008. Distribution of fungi and aflatoxins in stored peanut variety. *Food Chemistry* 106: 190–285. https://doi.org/10.1016/j.foodchem.2007.05.087

10. Caceres I., Al Khoury A., El Khoury R., et al. 2020. Aflatoxin biosynthesis and genetic regulation. *Toxins* 12(3): 150. https://doi.org/10.3390/toxins12030150
11. Mannaa M., Kim K. D. 2017. Influence of temperature and water activity on deleterious fungi and mycotoxin production during grain storage. *Mycobiology* 45(4): 240–254. https://doi.org/10.5941/MYCO.2017.45.4.240
12. Yoshinari T., Noda Y., Yoda K., Sezaki H., Nagasawa H., Sakuda S. 2010. Inhibitory activity of blasticidin A, a strong aflatoxin production inhibitor, on protein synthesis of yeast: Selective inhibition of aflatoxin production by protein synthesis inhibitors. *The Journal of Antibiotics* 63(6): 309–314. https://doi.org/10.1038/ja.2010.36
13. Eshelli M., Harvey L., Edrada-Ebe R., McNeil B. 2015. Metabolomics of the bio-degradation process of aflatoxin B1 by actinomycetes at an initial pH of 6.0. *Toxins* 7(2): 439–456. https://doi.org/10.3390/toxins7020439
14. Rushing B. R., Selim M. I. 2019. Aflatoxin B1: A review on metabolism, toxicity, occurrence in food, occupational exposure, and detoxification methods. *Food and Chemical Toxicology* 124: 81–100. https://doi.org/10.1016/j.fct.2018.11.047
15. Ding N., Xing F., Liu X., et al. 2015. Variation in the fungal microbiome (mycobiome) and aflatoxin in stored in-shell peanuts at four different areas of China. *Frontiers in Microbiology* 6: 1055. https://doi.org/10.3389/fmicb.2015.01055
16. Hawkins L. K., Windham G. L., Williams W. P. 2005. Effect of different postharvest drying temperatures on *Aspergillus flavus* survival and aflatoxin content in five maize hybrids. *Journal of Food Protection* 68(7): 1521–1524. https://doi.org/10.4315/0362-028X-68.7.1521
17. Matumba L., Sulyok M., Njoroge S. M., et al. 2015. Uncommon occurrence ratios of aflatoxin B 1, B 2, G 1, and G 2 in maize and groundnuts from Malawi. *Mycotoxin Research* 31(1): 57–62. https://doi.org/10.1007/s12550-014-0209-z
18. Mahbobinejhad Z., Aminian H., Ebrahimi L., Vahdati K. 2019. Reduction of aflatoxin production by exposing *Aspergillus flavus* to CO2. *Journal of Crop Protection* 8(4): 441–448.
19. Liu J., Sun L., Zhang N., et al. 2016. Effects of nutrients in substrates of different grains on aflatoxin B1 production by *Aspergillus flavus. BioMed Research International*. https://doi.org/10.1155/2016/7232858
20. Bolu S. A., Elelu N., Ahmed R. N., et al. 2014. Effect of vitamins, amino acids, and phyto-active biomolecules on *Aspergillus flavus* in poultry production. *Pharmacology and Therapeutics* 21(2): 23–44.
21. Chang P. K., Hua S. S. T., Sarreal S. B. L., Li R. W. 2015. Suppression of aflatoxin biosynthesis in *Aspergillus flavus* by 2-phenyl ethanol is associated with stimulated growth and decreased degradation of branched-chain amino acids. *Toxins* 7(10): 3887–3902. https://doi.org/10.3390/toxins7103887
22. Doehlemann G., Ökmen B., Zhu W., Sharon A. 2017. Plant pathogenic fungi. *Microbiology Spectrum* 5(10): 1128. https://doi.org/10.1128/microbiolspec.funk-0023-2016
23. Kleemann J., Rivera L. J., Takahara H., et al. 2012. Sequential delivery of host-induced virulence effectors by appressoria and intracellular hyphae of the phytopathogen *Colletotrichum higginsianum. PLoS Pathogens* 8: e1002643. https://doi.org/10.1371/journal.ppat.1002643
24. Mendoza A., Berndt P., Djamei A., Weise C., Linne U., Marahiel M., Vraneš M., Kämper J., Kahmann R. 2009. Physical-chemical plant-derived signals induce differentiation in *Ustilago maydis. Molecular Microbiology* 71: 895–911. https://doi.org/10.1111/j.1365-2958.2008.06567.x
25. Hoch H. C., Staples R. C., Whitehead B., Comeau J., Wolf E. D. 1987. Signaling for growth orientation and cell differentiation by surface topography in uromyces. *Science* 235:1659–1662. https://doi.org/10.1126/science.235.4796.1659
26. Samue A. T., Valentine I. T. 2014. Effect of total aflatoxin on the growth characteristics and chlorophyll level of sesame (*Sesamum indicum* L.). *New York Science Journal* 7: 8–13.
27. Sultana T., Javed B., Raja N. I., Mashwani Z.-U. 2021. Silver nanoparticles elicited physiological, biochemical, and antioxidant modifications in rice plants to control *Aspergillus flavus. Green Processing and Synthesis* 10: 314–324. https://doi.org/10.1515/gps-2021-0034
28. Loi M., Paciolla C., Logrieco A. F., Mulè G. 2020. Plant bioactive compounds in pre- and postharvest management for aflatoxins reduction. *Frontiers in Microbiology* 11: 243. https://doi.org/10.3389/fmicb.2020.00243
29. Fouad A. M., Ruan D., El-Senousey H. K., Chen W., Jiang S., Zheng C. 2019. Harmful effects and control strategies of aflatoxin B1 produced by *Aspergillus flavus* and *Aspergillus parasiticus* strains on poultry. *Toxins* 11: 176. https://doi.org/10.3390/toxins11030176
30. Ismaiel A. A., Papenbrock J. 2015. Mycotoxins: Producing fungi and mechanisms of phytotoxicity. *Agriculture* 5(3): 492–537. https://doi.org/10.3390/agriculture5030492

31. Mudili V., Siddaih C. N., Nagesh M. 2014. Mold incidence and mycotoxin contamination in freshly harvested maize kernels originated from India. *Journal of the Science of Food and Agriculture* 94(13): 2674–2683. https://doi.org/10.1002/jsfa.6608
32. Sirma A. J., Ouko E. O., Murithi G. 2015. Prevalence of aflatoxin contamination in cereals from Nandi County, Kenya. *International Journal of Agricultural Sciences and Veterinary* 3(3): 55–63.
33. Bumbangi N. F., Muma J. B., Choongo K., Mukanga M., Velu M. R., Veldman F., Mapatano M. A. 2016. Occurrence and factors associated with aflatoxin contamination of raw peanuts from Lusaka district's markets, Zambia. *Food Control* 68: 291–296. https://doi.org/10.1016/j.foodcont.2016.04.004
34. Magembe K. S., Mwatawala M. W., Mamiro D. P., Chingonikaya E. E. 2016. Assessment of awareness of mycotoxins infections in stored maize (*Zea mays* L.) and groundnut (*Arachis hypogea* L.) in Kilosa District, Tanzania. *International Journal of Food Contamination* 3(1): 12. https://doi.org/10.1186/s40550-016-0035-5
35. Baydan E., Küçükersan S., Yurdakök D. B. et al. 2016. Comparison of nutritional composition (moisture, ash, crude protein, nitrogen) and safety (aflatoxin, nitrate/nitrite) of organic and conventional rice and lentil samples consumed in Ankara. *Ankara Üniversitesi Veteriner Fakültesi Dergisi* 63(4): 365–370.
36. Mmongoyo J. A., Wu F., Linz J. E. et al. 2017. Aflatoxin levels in sunflower seeds and cakes collected from micro-and small-scale sunflower oil processors in Tanzania. *PLoS One* 12(4): e0175801. https://doi.org/10.1371/journal.pone.0175801
37. Singh P., Cotty P. J. 2017. Aflatoxin contamination of dried red chilies: Contrasts between the United States and Nigeria, two markets differing in regulation enforcement. *Food Control* 80: 374–379. https://doi.org/10.1016/j.foodcont.2017.05.014
38. Diella G., Caggiano G., Ferrieri F. et al. 2018. Aflatoxin contamination in nuts marketed in Italy: Preliminary results. *Ann Ig*, 30(5): 401–409.
39. Nazir A., Kalim I., Sajjad M., Usman M., Iqbal M. 2019. Prevalence of aflatoxin contamination in pulses and spices in different regions of Punjab. *Chemical International* 5(4): 274–280.
40. Anthony M. H., Ojochenemi A. D., Yemi A. H. R. 2014. Determination of aflatoxins in sesame, rice, millet, and acha from Nigeria using HPLC. *Chemical Science Transactions* 3(4): 1516–1524.
41. Iqbal S. Z., Asi M. R., Hanif U., Zuber M., Jinap S. 2016. The presence of aflatoxins and ochratoxin a in rice and rice products; and evaluation of dietary intake. *Food Chemistry* 210: 135–140. https://doi.org/10.1016/j.foodchem.2016.04.104
42. Ezekiel C. N., Udom I. E., Frisvad J. C. 2014. Assessment of aflatoxigenic *Aspergillus flavus* and other fungi in millet and sesame from plateau state, Nigeria. *Mycology* 5(1): 16–22. https://doi.org/10.1080/21501203.2014.889769
43. Chala A., Taye W., Ayalew A., et al. 2014. Multi mycotoxin analysis of sorghum (*Sorghum bicolor* L. Moench) and finger millet *(Eleusine coracana* L. Garten) from Ethiopia. *Food Control* 45: 29–35. https://doi.org/10.1016/j.foodcont.2014.04.018
44. Pratiwi C., Rahayu W. P., Lioe H. N., Herawati D., Broto W., Ambarwati S. 2015. The effect of temperature and relative humidity for *Aspergillus flavus* BIO 2237 growth and aflatoxin production on soybeans. *International Food Research Journal* 22(1): 82.
45. García-Moraleja A., Font G., Mañes J., Ferrer E. 2015. Analysis of mycotoxins in coffee and risk assessment in Spanish adolescents and adults. *Food and Chemical Toxicology* 86: 225–233. https://doi.org/10.1016/j.fct.2015.10.014
46. Embaby E., Hassan M. K. 2015. The decay of guava fruit (*Psidium guajava* Linn.) quality caused by some mold fungi. *International Journal of Agricultural Technology* 11(3): 713–730.
47. Dobolyi C. S., Sebők F., Varga J., Kocsubé S., Szigeti G., Baranyi N., Szoboszlay S. 2013. The occurrence of aflatoxin producing *Aspergillus flavus* isolates in maize kernel in Hungary. *ActaAlimentaria* 42(3): 451–459 https://doi.org/10.1556/aalim.42.2013.3.18
48. Lai X., Liu R., Ruan C., Zhang H., Liu C. 2015. The occurrence of aflatoxins and ochratoxin a in rice samples from six provinces in China. *Food Control* 50: 401–404. https://doi.org/10.1016/j.foodcont.2014.09.029
49. Bala P., Gupta D., Sharma Y. P. 2016. Mycoflora and natural aflatoxin contamination in dried quince seeds from Jammu, India. *Journal of Environmental Biology* 37(1): 101–106.
50. Golge O., Hepsag F., Kabak B. 2013. Incidence and level of aflatoxin contamination in chili commercialized in Turkey. *Food Control* 33(2):514–520. https://doi.org/10.1016/j.foodcont.2013.03.048
51. Kos J., Mastilović J., Hajnal E. J., Šarić B. 2013. Natural occurrence of aflatoxins in maize harvested in Serbia during 2009–2012. *Food Control* 34(1): 31–34. https://doi.org/10.1016/j.foodcont.2013.04.004

52. ChalaChala A., Mohammed A., Ayalew A., Skinnes H. 2013. Natural occurrence of aflatoxins in groundnut (*Arachis hypogaea* L.) from eastern Ethiopia. *Food Control* 30(2): 602–605. https://doi.org/10.1016/j.foodcont.2012.08.023
53. Al-Wadai A. S., Al-Othman M. R., Mahmoud M. A., et al. 2013. Molecular characterization of *Aspergillus flavus* and aflatoxin contamination of wheat grains from Saudi Arabia. *Genetics and Molecular Research* 12(3): 3335–3352.
54. Milhome M. A. L., Lima C. G., De Lima L. K. 2014. Occurrence of aflatoxins in cashew nuts produced in northeastern Brazil. *Food Control* 42: 34–37. https://doi.org/10.1016/j.foodcont.2014.01.033
55. LippolisLippolis V., Irurhe O., Porricelli A. C. R., et al. 2017. Natural co-occurrence of aflatoxins and ochratoxin a in ginger *(Zingiber officinale)* from Nigeria. *Food Control* 73: 1061–1067. https://doi.org/10.1016/j.foodcont.2016.10.026
56. Ting W. E., Chang C. H., Szonyi B., Gizachew D. 2020. Growth and aflatoxin B1, B2, G1, and G2 production by *Aspergillus flavus* and *Aspergillus parasiticus* on ground flax seeds (*Linum usitatissimum*). *Journal of Food Protection* 83(6):975–983. https://doi.org/10.4315/JFP-19-539
57. Kabak B. 2016. Aflatoxins in hazelnuts and dried figs: Occurrence and exposure assessment. *Food Chemistry* 211: 8–16. https://doi.org/10.1016/j.foodchem.2016.04.141
58. Saleem A. R. 2017. Mycobiota and molecular detection of *Aspergillus flavus* and *A. parasiticus* aflatoxin contamination of strawberry (*Fragaria ananassa* Duch.) fruits. *Archives of Phytopathology and Plant Protection* 50(19–20): 982–996. https://doi.org/10.1080/03235408.2017.1411155
59. Matthew E., Oluma H., Ochokwunu A., Eche D. I., Olasan C. O. 2021. Aflatoxin contamination levels in sesame seeds sold in Benue State, North Central Nigeria. *American Journal of Food Science and Health* 7(1): 14–23.
60. Peng W. X., Marchal J. L. M., Van der Poel A. F. B. 2018. Strategies to prevent and reduce mycotoxins for compound feed manufacturing. *Animal Feed Science and Technology* 237: 129–153. https://doi.org/10.1016/j.anifeedsci.2018.01.017
61. Opoku G. F. 2013. Process and product optimization and storage characteristics of canned Peanut Soup Base. (Doctoral dissertation,) University of Ghana.
62. Leong Y. H., Ismail N., Latif A. A., Ahmad R. 2010. Aflatoxin occurrence in nuts and commercial nutty products in Malaysia. *Food Control* 21(3): 334–338. https://doi.org/10.1016/j.foodcont.2009.06.002
63. Awuah R. T., Ellis W. O. 2002. Effects of some groundnut packaging methods and protection with Ocimum and syzygium powders on kernel infection by fungi. *Mycopathologia* 154(1): 29–36. https://doi.org/10.1023/A:1015259518448
64. Eziah V. Y., Afreh-Nuamah K. 2015. Traditional storage practices on the quality of maize a case study in the Shai Osudoku District in the Greater Accra region. (Doctoral dissertation), University of Ghana.
65. Sipos P., Peles F., Brassó D. L., Béri B., Pusztahelyi T., Pócsi I., Győri Z. 2021. Physical and chemical methods for reduction in aflatoxin content of feed and food. *Toxins* 13(3): 204. https://doi.org/10.3390/toxins13030204
66. Udomkun P., Wiredu A. N., NagleM., Müller J., Vanlauwe B., Bandyopadhyay R. 2017. Innovative technologies to manage aflatoxins in foods and feeds and the profitability of application–a review. *Food Control* 76: 127–138. https://doi.org/10.1016/j.foodcont.2017.01.008
67. Jalili M. 2016. A review of aflatoxins reduction in food. *Iranian Journal of Health, Safety, and Environment* 3(1): 445–459.
68. Karlovsky P., Suman M., Berthiller F., et al. 2016. Impact of food processing and detoxification treatments on mycotoxin contamination. *Mycotoxin Res* 32: 179–205. https://doi.org/10.1007/s12550-016-0257-7
69. Bedi P. S., Agarwal R. K. 2014. Detoxification of aflatoxin B1 by physical and chemical methods. *World Journal of Pharmacy and Pharmaceutical Sciences* 3(12): 995–1002.
70. Aiko V., Mehta A. 2015. Occurrence, detection, and detoxification of mycotoxins. *Journal of Biosciences* 40(5): 943–954. https://doi.org/10.1007/s12038-015-9569-6
71. Hassan R., El-Kadi S., Sand M. 2015. Effect of some organic acids on some fungal growth and the production of their toxins. *International Journal of Advances in Biology* 2(1): 1–11.
72. Jard G., Liboz T., Mathieu F., Guyonvarc'h A., Lebrihi A. 2011. Review of mycotoxin reduction in food and feed: From prevention in the field to detoxification by adsorption or transformation. *Food Additives & Contaminants* 28(11): 1590–1609. https://doi.org/10.1080/19440049.2011.595377
73. Motawe H. F. A., Salam A. A., El-Meleigy K. M. 2014. Reducing the toxicity of aflatoxin in broiler chickens' diet by using probiotics and yeast. *International Journal of Poultry Science* 13(7): 397–407. https://doi.org/10.3923/ijps.2014.397.407

74. Mwakinyali S. E., Ding X., Ming Z., Tong W., Zhang Q., Li P. 2019. The recent development of aflatoxin contamination biocontrol in agricultural products. *Biological Control* 128: 31–39. https://doi.org/10.1016/j.biocontrol.2018.09.012
75. Aliabadi M. A., Alikhani F. E., Mohammadi M., Darsanaki R. K. 2013. Biological control of aflatoxins. *European Journal of Experimental Biology* 3(2): 162–166.
76. Shu X., Wang Y., Zhou Q., Li M., Hu H., Ma Y., Wu L. 2018. Biological degradation of aflatoxin B1 by cell-free extracts of bacillus velezensis DY3108 with broad PH stability and excellent thermostability. *Toxins* 10(8): 330. https://doi.org/10.3390/toxins10080330
77. Wang Y., Zhao C., Zhang D., Zhao M., Zheng D., Lyu Y. 2017. Effective degradation of aflatoxin B1 using a novel thermophilic microbial consortium TADC7. *Bioresource Technology* 224: 166–173. https://doi.org/10.1016/j.biortech.2016.11.033
78. Akocak P. B., Churey J. J., Worobo R. W. 2015. Antagonistic effect of chitinolytic pseudomonas and bacillus on the growth of fungal hyphae and spores of aflatoxigenic *Aspergillus flavus. Food Bioscience* 10: 48–58. https://doi.org/10.1016/j.fbio.2015.01.005
79. Yang X., Zhang Q., Chen Z. Y., Liu H., Li P. 2017. Investigation of *Pseudomonas* fluorescens strain 3JW1 on preventing and reducing aflatoxin contaminations in peanuts. *PLoS One* 12(6): e0178810. https://doi.org/10.1371/journal.pone.0178810

# 2.8 Association between Aflatoxigenic Fungi and Plants

*D.M. Yashaswini, G. Sathvik, and Lokesh Ravi*

## INTRODUCTION

Secondary metabolites known as mycotoxins are produced by a variety of fungal species, primarily *Alternaria*, *Fusarium*, *Penicillium*, and *Aspergillus* [1]. Aflatoxin is the mycotoxin produced by *Aspergillus* species (aflatoxigenic fungi). *Aspergillus flavus* and *Aspergillus parasiticus* are the species that typically produce aflatoxins; however, certain other species have also been reported to produce them, They are *A. nomius*, *A. pseudotamarii*, *A. parvisclerotigenus* and *A. bombycis* of section *flavi*; *A. ochraceoroseus* and *A. rambellii* from section *Ochraceorosei*; and *Emericella astellata* and *E. venezuelensis* from *Nidulatans* [2]. *A. flavus*, a prevalent contaminant in agriculture, and other strains of *A. flavus* and *A. parasiticus* create aflatoxins, which are difuranocoumarin derivatives, via a polyketide route [3]. Contamination of economically important crops with these fungi is a major concern worldwide because it widely affects human and animal health [4]. They affect many plants and crops and more than 18 different types of aflatoxins have been reported. A summary of reported aflatoxin-producing fungi affecting various plants are summarized in Table 2.8.1.

Aflatoxin is a white crystalline solid, is optically active, and among the most toxic mycotoxins. They are known to be carcinogenic and genotoxic. AFG2 (aflatoxin G2) exhibits substantially less activity than AFG1 (aflatoxin G1) (Figures 2.8.1–2.8.4), which is linked to toxicity and hepatocarcinogenicity in both human and animal populations [5].

## PHYTOTOXICITY

Aflatoxins have an excessive inhibitory effect on chlorophyll and carotenoid synthesis as well as seedling germination and development in plants like lettuce, mung, gram, mustard, sesame, and cowpea. Electron microscopy investigations showed that aflatoxin-treated maize leaf chloroplasts inhibited the development of grana [3, 7]. AFB1's phytotoxic effects on in vitro cultures of *Nicotiana tabacum* regenerating plantlets revealed a dose-dependent, substantial inhibition of both root and leaf mass and development. Growth of roots was more severely impacted than that of leaves due to AFB1 [3, 8]. After germination, the growth rate of hypocotyls was maximally reduced by aflatoxin at concentrations more than 10 μg·mL–1. High aflatoxin accumulation in cells leads to DNA damage, which triggers apoptosis [3]. At increased doses of the toxin, the inhibition of elongation of the roots and shoots was accompanied ultrastructurally by the disappearance of the endoplasmic reticulum, loss of ribosomes, disruption of the organellar membrane, and derangement of cytoplasmic elements, especially the tonoplast. It has been proposed that there is a correlation between the increasing disruption of the organelles and the increased inhibition of root and shoot extension as AFB1 concentration increases [9].

DOI: 10.1201/9781003487043-14

**TABLE 2.8.1**
**Aflatoxin-Producing Fungal Species and the Affected Plant**

| Aflatoxin | Producer | Affected Host | Reference |
|---|---|---|---|
| B (B1, B2) | *A. arachidicola, A. bombycis, A. flavus, A. minisclerotigenes, A. nomius, A. ochraceoroseus, A. oryzae, A. parasiticus, A. parvisclerotigenus, A. pseudotamarii, A. rambellii, A. tamarii, A. toxicarius, A. versicolor, Emericella astellata, E. venezuelensis.* | Cottonseed, dairy products, figs, fruit juices (apple, guava), maize, maize flour, meat, oilseed rape, peanuts, peanut butter, pea, pistachio, rice, sorghum, sunflower seed, spices | [4, 6] |
| G (G1, G2) | *A. minisclerotigenes, A. bombycis, A. arachidicola, A. pseudotamarii, A. nomius, A. parasiticus, A. terreus, A. toxicarius, A. versicolor* | Figs, cottonseed, dairy goods, and fruit juices (guava and apple)<br>Meat, rice, sorghum, sunflower seed, spices, peas, peanut butter, peanut butter, oilseed rape, and maize | [4, 6] |
| B3 - Parasiticol *(Aflatoxin G1 Metabolite)* | *A. flavus, A. mottae, A. nomius, A. novoparasiticus, A. parasiticus* | The same hosts as B1 and G1 | [4] |

## MODE OF ACTION

Aflatoxin suppresses the production of protochlorophyllide, chlorophyll a, and chlorophyll b in mung bean seeds (*Vigna radiata*). It also suppresses the amounts of protein and nucleic acid in germinating maize and mung seeds. Aflatoxin prevents DNA-dependent polymerase activity linked to chromatin. Although the lack of m-RNA is thought to be the cause of the reduction of protein synthesis, aflatoxin binding to DNA during replication or DNA polymerase inhibition is thought to be the cause of the inhibition of DNA synthesis. AFB1 has been immunolocalised in the stem cell nucleus of tobacco plantlets. Apoptosis caused by DNA damage buildup in cells may cause a decrease in tillers in plants treated with aflatoxin. In cultured cells, aflatoxin has been shown to halt the cell cycle and trigger apoptosis. When it comes to DNA repair, a cell's only choice is the apoptotic pathway. Systems have too much damage, which overloads them. According to research by Vogelstein and Kinzler, the p53 gene typically stops DNA replication in cells that have suffered DNA damage by keeping the cell in the G2/M phase, which increases the chance of DNA repair. Therefore, cells with inactivated p53 may live abnormally and permit more DNA damage to accumulate, a condition that promotes the development of cancer. In wheat plants, AFB1 showed a genotoxic effect. Following application, various chromosomal abnormalities were found during

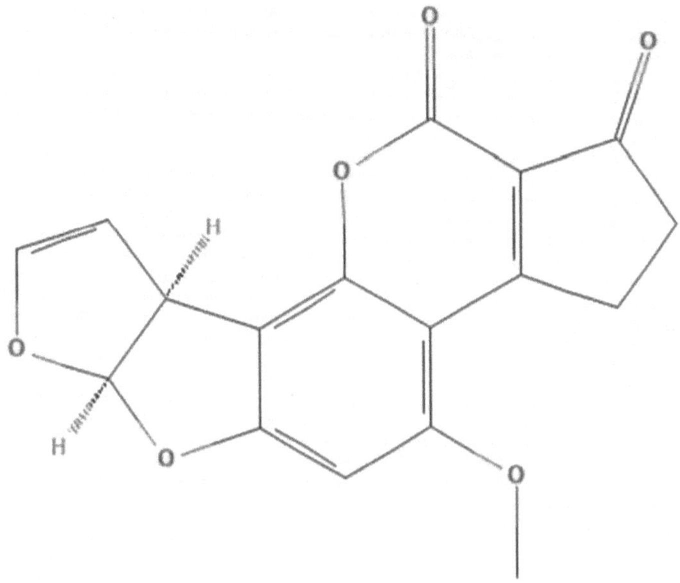


**FIGURE 2.8.1** Aflatoxin B1.

meiosis; these aberrations included chromosome stickiness, outside bivalents, bridges, laggards, unequal division and micronuclei [3].

## ENVIRONMENTAL CONDITIONS

Aflatoxigenic molds can infect a variety of agricultural commodities, contaminating them with aflatoxins. Important crops, including rice, cottonseed and maize, have also been discovered to be contaminated in many nations by aflatoxigenic molds. In tropical and subtropical areas, where

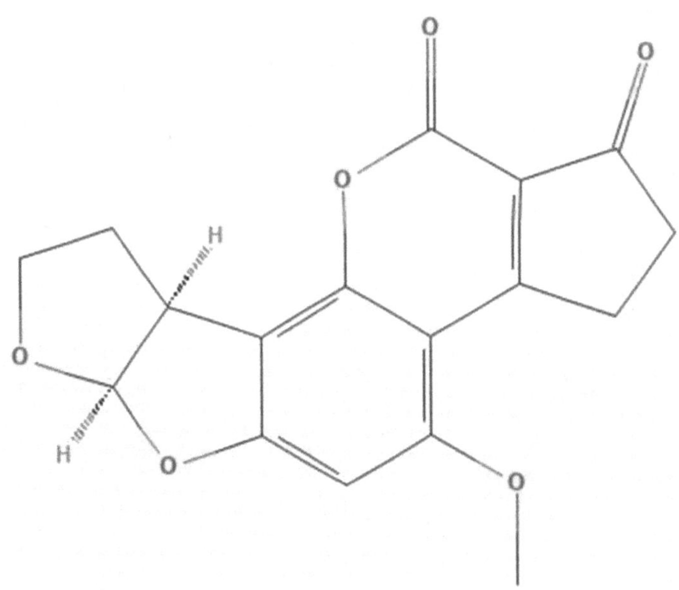


**FIGURE 2.8.2** Aflatoxin B2.

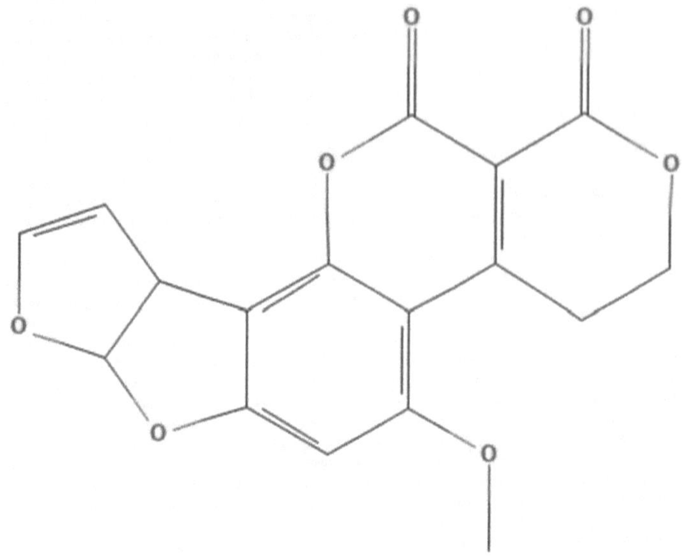


**FIGURE 2.8.3** Aflatoxin G1.

the warm, humid weather creates ideal circumstances for the growth of molds, aflatoxins are more common in foods and feeds. The type of substrate, fungus species, moisture content, minerals, temperature, humidity and physical damage to the kernels all affect *A. flavus* development and aflatoxin production in the natural substrate [10, 11]. For *A. parasiticus* to grow, the ideal temperature ranges are 25–35°C, the maximum is 44–46°C, and the minimum is 6–8°C. *A. flavus* can manufacture aflatoxin at temperatures as low as 12°C and 34°C and stop the formation of aflatoxin at 36°C, with the highest production occurring between 28 and 30°C. Aflatoxin contamination in food and the growth of *Aspergillus* fungi are influenced by various circumstances. Contamination from preharvest to storage can occur at any point in food production. The climate of the area, the genotype

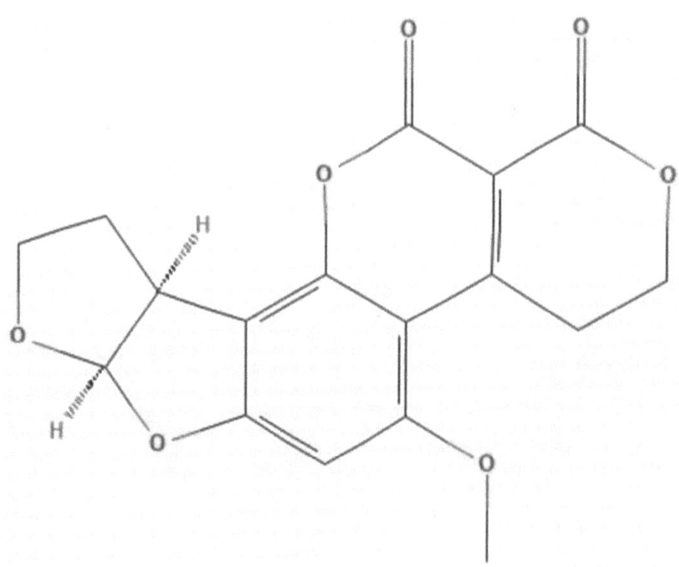


**FIGURE 2.8.4** Aflatoxin G2.

of the crop planted, the type of soil, the minimum and maximum daily temperatures, and the daily net evaporation are some of the factors that affect aflatoxin contamination [10].

## ETIOLOGY

Aflatoxin, the most common type of mycotoxin, affects about 25% of the world's crops. Inadequate storage of cassava, cottonseed, chili peppers, maize, wheat, millet, peanuts, rice, sesame, sunflower seeds and several spices are common places to find them. Two stages of crop contamination are possible: Crops become infected with *Aspergillus* species during growth and development. When products are stored or transported in warm, humid environments or during a prolonged dry spell, contamination may develop. Humans may be exposed when animals given tainted feed produce products of aflatoxin metabolism into eggs, milk products, and meat [12, 13].

## CONCLUSION

Curbing the effects of aflatoxins is a serious effort all around the globe due to economic losses and due to their adverse effects on plant and animal species. These molds grow easily because the conditions are favorable, so it is critical to adopt efficient and systematic pre- and post-harvest processing methods, develop and use resistant plant varieties, and amend the safety regulations to minimize exposure to these toxins.

## REFERENCES


1. Nazhand A, Durazzo A, Lucarini M, Souto EB, Santini A. Characteristics, occurrence, detection and detoxification of aflatoxins in foods and feeds. Foods 2020;9. https://doi.org/10.3390/foods9050644
2. Ahmad MM, Ahmad M, Ali A, Hamid R, Javed S, Abdin MZ. Detection of *Aspergillus flavus* and *Aspergillus parasiticus* from aflatoxin-contaminated peanuts and their differentiation using PCR-RFLP. Ann Microbiol 2014;64:1597–605. https://doi.org/10.1007/s13213-014-0803-5
3. Ismaiel AA, Papenbrock J. Mycotoxins: Producing fungi and mechanisms of phytotoxicity. Agriculture (Pol'nohospodarstvo) 2015;5:493–537. https://doi.org/10.3390/agriculture5030492
4. Shabeer S, Asad S, Jamal A, Ali A. Aflatoxin contamination, its impact and management strategies: An updated review. Toxins (Basel) 2022;14. https://doi.org/10.3390/toxins14050307
5. Popescu RG, Rădulescu AL, Georgescu SE, Dinischiotu A. Aflatoxins in feed: Types, metabolism, health consequences in swine and mitigation strategies. Toxins (Basel) 2022;14. https://doi.org/10.3390/toxins14120853
6. Abdin MZ, Ahmad MM, Javed S. Advances in molecular detection of aspergillus: An update. Arch Microbiol 2010;192:409–25. https://doi.org/10.1007/s00203-010-0563-y
7. Crisan, E. V. Effects of aflatoxin on germination and growth of lettuce. Appl. Microbiol 1973;25(3):342–45.
8. McLean M, Dutton MF, Watt MP, Berjak P. Aflatoxin b1—its effects on an in vitro plant system. Food Addit Contam 1995;12:435–43. https://doi.org/10.1080/02652039509374327
9. Mclean M, Beajak P, Watt MP, Dutton MF. The Effects of Aflatoxin B1 on Immature Germinating Maize (*Zea Mays*) Embryos. vol. 119. Kluwer Academic Publishers; 1992.
10. Abrar M, Anjum FM, Butt MS, Pasha I, Randhawa MA, Saeed F, et al. Aflatoxins: Biosynthesis, occurrence, toxicity, and remedies. Crit Rev Food Sci Nutr 2013;53:862–874. https://doi.org/10.1080/10408398.2011.563154
11. Molina MM, Giannuzzi L. Modelling of aflatoxin production by *Aspergillus parasiticus* in a solid medium at different temperatures, pH and propionic acid concentrations. 2001.
12. Kumar P, Mahato DK, Kamle M, Mohanta TK, Kang SG. Aflatoxins: A global concern for food safety, human health and their management. Front Microbiol 2017;7. https://doi.org/10.3389/fmicb.2016.02170
13. Thakur S, Singh RK, De PS, Dey A. Aflatoxins in feeds: Issues and concerns with safe food production. Indian J Anim Health 2022. https://doi.org/10.36062/ijah.2022.14421

# 2.9 Aflatoxigenic Fungi and Groundnut

*Hamadou Mamoudou, Difo Voukang Harouna, Jairus Olumasai Nandwa, Vandi Yonas, and Martin Alain Mune Mune*

## INTRODUCTION

Groundnuts (*Arachis hypogaea*) are an important crop in many parts of the world, providing essential nutrients and income. They also play a crucial role in crop rotation and soil fertility management due to their ability to fix nitrogen [1–4]. However, the presence of aflatoxigenic fungi poses a threat to the benefits of this grain legume. It is important to note that the use of objective language is necessary in discussing the potential risks associated with aflatoxigenic fungi. These fungi, specifically *Aspergillus flavus* and *A. parasiticus*, are responsible for contaminating peanuts with aflatoxins [5, 6].

Aflatoxins, a group of toxic secondary metabolites produced primarily by the fungal species *A. flavus* and *A. parasiticus*, pose a significant challenge to the food industry, particularly to groundnut crops. These mycotoxins are recognised for their teratogenic, carcinogenic, hepatotoxic and immunosuppressive properties. Aflatoxin contamination is mainly associated with *A. flavus* and *A. parasiticus*, which are prolific producers of various aflatoxin subtypes, including B1, B2, G1, and G2 [5, 7, 8]. These aflatoxins are highly resistant to degradation processes, allowing them to persist throughout the food chain. Consequently, aflatoxin contamination presents a serious public health risk to both humans and animals, with additional economic ramifications at local and international scales [5, 7, 8].

The contamination of groundnut cultivation by aflatoxins due to aflatoxigenic fungi is a complex phenomenon influenced by multiple interacting factors. Environmental conditions play a critical role, with aflatoxigenic fungi thriving under specific parameters such as elevated temperatures, high relative humidity and substrates rich in organic matter [4, 9, 10]. Groundnuts are highly susceptible to aflatoxin contamination due to their nutritional composition, which provides a favourable growth medium for these fungi. Aflatoxin contamination can occur as early as harvest and may be worsened by inadequate storage practices and unfavourable environmental conditions during storage [5, 6, 11]. The prevalence of aflatoxigenic fungi in groundnut cultivation poses a multifaceted challenge. Climatic conditions, agricultural practices and post-harvest handling procedures all contribute to the risk of aflatoxin contamination [5, 6, 11]. Therefore, it is crucial to have a thorough understanding of aflatoxigenic fungi and their interactions within the groundnut agroecosystem to develop integrated management strategies that effectively minimise contamination throughout the groundnut food chain.

Furthermore, aflatoxin contamination is a global issue, primarily affecting tropical and subtropical regions where climatic conditions favour the growth of aflatoxigenic fungi. Developing countries are particularly vulnerable because they often rely on groundnuts as a source of income and staple food. The presence of aflatoxins in peanut-based products can result in significant economic losses, food safety issues and adverse health effects for exposed populations [5, 12, 13].

Ensuring food safety by detecting and preventing aflatoxins in peanut products is a major challenge. To quantify aflatoxins in food samples, sophisticated analytical methods are used for effective


DOI: 10.1201/9781003487043-15

monitoring of contamination and corrective action where necessary. Additionally, various prevention strategies have been implemented to reduce contamination, including preventive farming practices, appropriate storage methods and post-harvest treatments.

This work aims to present a thorough overview of the current knowledge regarding aflatoxigenic fungi and their interactions with groundnuts. The investigation includes a detailed examination of the fungal species responsible for aflatoxin production, the environmental conditions and agricultural practices that promote contamination, and the harmful effects of aflatoxins on human and animal health. Additionally, this work explores the methods used to detect aflatoxin in peanut-based products, as well as strategies for preventing and reducing contamination. The goal of this chapter is to provide a solid understanding of the mechanisms behind aflatoxin contamination in groundnut crops and suggest solutions to minimise it. The significance of this lies in its potential to enhance food safety and protect public health. It also informs future research and regulatory policies across relevant fields.

## RISK FACTORS FOR CONTAMINATION OF PEANUTS BY AFLATOXIGENIC FUNGI

### Peanuts: An Overview

Groundnuts, also known by the scientific name *Arachis hypogaea*, are plants of the *Fabaceae* family. The peanut, also known as the groundnut, is a leguminous plant belonging to the *Fabaceae* family (bean family) and the Papilionaceae subfamily [4, 10, 12]. Despite its name and appearance, the peanut is not a true nut but a legume, similar to peas, beans and lentils. Native to South America, peanuts are grown in many parts of the world for their edible seeds, which are rich in protein, fat and essential nutrients [4, 10, 12].

#### Botany and Cultivation

The peanut is an annual herbaceous plant that typically grows to a height of 30–60 cm (12–24 inches). It has two main growth forms, erect and spreading. Erect peanuts have short branches, whereas spreading peanuts have long branches that trail along the ground. The peanut plant produces yellow butterfly-shaped flowers that self-pollinate. Following fertilization, a unique characteristic of the peanut is that a peg-like structure develops from the base of the flower, pushing the developing a peanut pod (technically a fruit) underground where it matures [4, 9, 10, 12].

Groundnuts are annual herbaceous plants that grow best in warm, sunny climates with well-drained soil. They produce yellow flowers that grow at the base of the plant and, once fertilised, produce pods containing peanut seeds. Peanuts have a unique flowering habit. After pollination, the flower grows towards the ground on an elongated peduncle, where the fruit ripens as a hypocotyl, commonly known as a peanut pod, which develops underground. This botanical peculiarity classifies peanuts as a legume, but with a unique method of fruiting known as geocarpy [4, 9, 10, 12].

Optimal conditions for growing groundnuts include a warm climate between 20 and 30°C and a dry season for harvesting. They require light, well-drained soil that is rich in calcium. Groundnuts play an important role in phytosanitation by fixing atmospheric nitrogen in the soil, thus benefiting successive crops [4, 9, 12].

#### Composition and Nutritional Value

Groundnuts are a valuable legume crop that serves as a primary source of protein, particularly in developing nations where meat may be scarce or expensive. Peanuts are recognised for their abundance of essential nutrients, including unsaturated fatty acids, dietary fibre, B vitamins and minerals such as magnesium, phosphorus and zinc [4, 9, 12]. Peanuts can be consumed in various forms, including raw, roasted and processed peanut butter. In addition, peanuts are used as ingredients in various culinary dishes, sauces, confections and snack products. They also have industrial applications, such as the production of edible oils, cosmetics and biofuels [4, 9, 12].

Groundnut seeds are known for their exceptional nutritional profile. They have a protein content of approximately 25%–30%, making them a valuable source of dietary protein. Additionally, peanut seeds are rich in lipids (45%–55%), with the majority of these lipids being unsaturated fatty acids [1, 4, 9, 12]. This unique fatty acid composition contributes to the seeds' role as a source of energy and a beneficial component of a well-balanced diet.

### Uses and Applications

Peanuts are an important crop for global agriculture and food security due to their high-protein and nutrient content, especially in developing countries. Additionally, peanuts contribute to soil health by fixing nitrogen through a symbiotic relationship with nitrogen-fixing bacteria in their root nodules [9], which improves soil fertility and reduces the need for synthetic fertilizers.

Peanuts have a wide range of uses, including direct consumption, peanut oil, peanut butter and various snacks. Peanut oil, extracted from the seeds, is valued for its stability at high temperatures and neutral taste in cooking. Furthermore, the peanut market extends beyond food sectors, with the oil being used in cosmetics and the residues from oil production serving as animal feed [9].

### Agricultural and Health Challenges

Peanut crops are vulnerable to diseases and pests, including root rot and mould caused by fungi, as well as contamination by aflatoxins, which are carcinogenic to humans. Additionally, peanut allergies are a significant and widespread food allergy, posing a major public health concern [9, 14].

## Biological and Ecological Characteristics of the Primary Aflatoxigenic Fungi

### Aspergillus Flavus

#### *Taxonomy and Morphology*

*A. flavus* is a filamentous fungus belonging to the *Aspergillus* genus, which is a diverse group of fungi commonly found in soil, decaying organic matter and various agricultural products. It is an important species within this genus due to its ability to produce a group of toxic compounds known as aflatoxins [11].

Classified within the kingdom Fungi, phylum Ascomycota, *A. flavus* is a member of the order Eurotiales and family Aspergillaceae. This species exhibits a characteristic filamentous growth pattern, forming predominantly septate mycelia. The conidiophores of *A. flavus* are specialised aerial hyphae that give rise to conidia, the asexual spores responsible for its widespread distribution. These conidia are borne on phialides and encapsulated in structures known as conidial heads, which range in colour from green to yellow-green, facilitating identification [13, 15].

#### *Mycotoxin Production*

*A. flavus* is distinguished by its capacity to produce aflatoxins, secondary metabolites of high toxigenic and carcinogenic potency. Aflatoxins are toxic metabolites that contaminate crops such as groundnuts, maize and tree nuts, posing a serious threat to food safety and human health. These mycotoxins primarily include aflatoxins B1, B2, along with lesser quantities of G1 and G2 [11, 13, 15]. The biosynthesis of aflatoxins involves a complex enzymatic pathway, encoded by a gene cluster within the fungal genome. The expression of these genes and resultant aflatoxin production are influenced by environmental factors such as substrate composition, temperature and humidity, with optimal conditions centring around warm and moist environments [11, 15].

#### *Ecology and Pathogenicity*

*A. flavus* is a fungus that is commonly found in soil, decaying vegetation various agricultural commodities. It thrives in warm and humid climates, with optimal growth temperatures ranging from

25 to 35°C. This species has a versatile metabolic capability, enabling it to degrade complex organic compounds and produce a wide range of metabolites, including aflatoxins [6, 11, 16].

*A. flavus* is a saprophytic fungus that contributes to the decomposition of organic matter, playing a critical role in nutrient cycling within ecosystems. However, it can also colonise and contaminate agricultural commodities, such as nuts, grains and legumes, posing significant challenges for food safety and public health [11, 13]. *A. flavus* is capable of infecting crops both before and after harvesting, by penetrating plant tissues through wounds or natural openings. It can also act as a plant pathogen, causing diseases such as Aspergillus ear rot in maize and aflatoxin contamination in crops. The fungus produces spores, called conidia, which are dispersed through the air and can easily contaminate food and feed sources [11, 16].

*A. flavus* has a high genetic variability, with different strains possessing varying levels of aflatoxin-producing capability.

#### *Mitigation Strategies*

Management of *A. flavus* and reduction of aflatoxin contamination involve an integrated approach encompassing agricultural practices, such as crop rotation and timely harvesting, alongside biotechnological interventions, such as the development of biological control agents and genetically engineered crops with enhanced resistance to fungal colonization [11, 17].

### Aspergillus Parasiticus

*A. parasiticus* is a species of fungus that is critically important due to its role in the production of aflatoxins. This fungus, along with *A. flavus*, is the primary contributor to the contamination of a broad spectrum of agricultural commodities with aflatoxins, affecting crops such as peanuts, maize, cottonseed and various nuts [15, 18, 19].

#### *Biological Characteristics*

*A. parasiticus* exhibits a filamentous growth form, characterised by dense mycelia. The conidia (spores) are borne on conidiophores and exhibit a green to yellow-green coloration. Its colonies can be woolly to velvety in texture, displaying a greenish hue on the surface with a gold to yellow reverse, depending on the growing conditions [15, 18, 19].

#### *Aflatoxin Production*

This species is a prolific producer of aflatoxins, specifically aflatoxins B1, B2, G1 and G2, among others. The ability to produce these toxins is genetically determined, with environmental factors influencing the extent of production [15].

#### *Genetics*

The genome of *A. parasiticus* includes specific genes and gene clusters responsible for aflatoxin biosynthesis. This complex pathway involves more than 25 genes, which encode enzymes that catalyse the various steps leading to aflatoxin formation [16].

#### *Ecological Characteristics*

*A. parasiticus* is predominantly found in soil but can colonise a wide range of organic substrates. Its distribution is global, with a predilection for warm and humid environments that are conducive to aflatoxin production. *A. parasiticus* can infect and proliferate on a variety of crop plants, especially during pre-harvest and post-harvest phases. The fungus enters crops through wounds and continues to grow and produce aflatoxins in storage under favourable conditions of temperature and humidity [15, 18, 19].

In addition to being a plant pathogen, *A. parasiticus* acts as a saprophyte, decomposing dead organic material. This ecological role contributes to nutrient cycling in ecosystems but also presents challenges for managing aflatoxin contamination in agricultural settings [15].

### Aspergillus Nomius

#### *Taxonomy*

*Aspergillus nomius* is located within the kingdom Fungi, in the phylum Ascomycota, characterised by the production of ascospores within a special sac-like structure known as an ascus [13, 20]. This species is enshrined within the genus *Aspergillus*, a diverse grouping that has garnered attention for its industrial, environmental and medical significance. As part of the section Flavi, *A. nomius* shares close phylogenetic ties with several aflatoxin-producing species [13, 18, 20].

#### *Morphological Insights*

The morphology of *A. nomius* exhibits the quintessential features of *Aspergillus* species, including filamentous, branched hyphae that form a dense network known as the mycelium. This species produces conidiophores and conidia (asexual spores), mechanisms through which it achieves widespread dispersion. Moreover, *A. nomius* can produce sclerotia under certain conditions, which are structures that facilitate survival in adverse environments [20].

#### *Pathogenicity and Toxigenesis*

The propensity of *A. nomius* to produce aflatoxins, potent mycotoxins known for their carcinogenicity, marks it as a significant concern in agriculture and food safety. Aflatoxins B1, B2, G1 and G2 have been identified among the metabolites produced by this species, emphasizing its similarity to *A. flavus* and *A. parasiticus*. The presence of *A. nomius* in crops such as nuts, pulses and legumes underscore the necessity for vigilant monitoring and control measures to safeguard food supplies from contamination [8, 15, 20].

## Environmental Factors Favouring Aflatoxin Production

*Aspergillus* species thrive and produce aflatoxins in large part due to environmental variables. Under some circumstances, *A. flavus, A. parasiticus* and *A. nomius* may all produce aflatoxins. Temperature, moisture content, pH, oxygen availability, carbon and nitrogen sources, light exposure and the presence of mycotoxin precursors all affect aflatoxins' capacity to proliferate [5, 11, 13]. Environmental variables must be closely monitored and controlled to reduce the danger of aflatoxin contamination in food and feed products. Maintaining the right temperature and moisture content during handling and storage can assist stop the growth of *Aspergillus* species and reduce the volume of aflatoxins produced [5, 11, 13].

The generation of mycotoxins by *A. flavus, A. parasiticus* and *A. nomius* is significantly influenced by several environmental parameters, including temperature, the groundnut's moisture content, its chemical makeup, and several other factors that are described in Table 2.9.1 [5, 8, 11, 13, 15, 17].

## Agronomic Factors Influencing Contamination of Groundnuts

The presence of aflatoxins produced by *A. flavus, A. parasiticus* and *A. nomius* in peanuts presents a serious risk to consumer safety and the stability of the economy. Aflatoxin contamination in peanuts can vary greatly in frequency and severity depending on the agronomic measures used (Table 2.9.2) [11, 13, 17].

Peanut producers can reduce the danger of aflatoxin by using a holistic strategy by knowing how agronomic variables impact *Aspergillus* contamination. This involves choosing resistant cultivars, putting pre-harvest stress-reduction measures into place, and embracing sensible cultural customs [13, 18]. Agronomic management optimisation is essential to guaranteeing the production of safe, premium peanuts.

**TABLE 2.9.1**
**Environmental Factors Favouring Aflatoxin Production [5, 8, 11, 13, 15, 17]**

| Environmental Factor | Effect on Aflatoxin Production | Optimal Range for Aflatoxin Production | Notes |
|---|---|---|---|
| Water Activity ($a_w$) | Directly affects fungal growth and toxin production. | 0.95–0.99 (High) | Lower aw reduces aflatoxin production but may not completely inhibit fungal growth. |
| Temperature | Aflatoxin production peaks at temperatures slightly lower than those optimal for growth, typically in the range of 28 to 32°C, depending on the species and strain. | 25–37°C | Optimal fungal growth for *A. flavus* and *A. parasiticus* generally occurs between 25 and 35°C. Lower temperatures slow growth and toxin production. *A. nomius* may have a slightly higher optimal temperature range. |
| Storage Time | Longer storage under favourable conditions increases risk. | Extended Periods | Aflatoxin production increases as *A. flavus* colonises the substrate over time. |
| Substrate Composition | Nutrients influence growth and toxin production. | High Carbohydrates, Low Protein/Fat | *Aspergillus* spp. prefers readily available carbohydrates for energy. |
| Carbon & Nitrogen Source | Specific sources can influence toxin production. | Varies | Availability of specific carbon (carbohydrates) and nitrogen (protein) sources can affect the growth rate and biomass production. |
| pH | Acidic conditions, particularly pH values between 4 and 6, are favourable for aflatoxin production, reflecting the enzymatic activity required for toxin biosynthesis. | pH 4–7.5 | Extremely acidic or alkaline environments can inhibit growth and toxin production. *Aspergillus* spp. can grow in a broad pH range, with optimal growth observed in slightly acidic to neutral conditions (pH 4–8). |
| Light | Limited impact, but some studies suggest light inhibition. | Darkness | Light exposure might influence aflatoxin production in certain situations, but darkness generally promotes toxin formation. Studies indicating that darkness or ultraviolet (UV) light may enhance toxin synthesis in some strains. |
| Oxygen ($O_2$) & Carbon Dioxide ($CO_2$) | Balance is crucial. Aflatoxin production is also an aerobic process, necessitating ample oxygen supply for the expression of genes involved in aflatoxin biosynthesis. | Aerobic ($O_2$ Required), Moderate $CO_2$ | *A. flavus* is aerobic, but elevated $CO_2$ can sometimes stimulate aflatoxin production. |
| Physical Damage | Creates entry points for fungal growth. | — | Insect damage or mechanical injury to crops increases aflatoxin contamination risk. |
| Microbial Interactions | Competition or antagonism can influence growth and toxin production. | Varies | Presence of competing or beneficial microorganisms can affect *A. flavus* growth and aflatoxin production. Some microbes may suppress fungal growth. |

**TABLE 2.9.2**
**Agronomic Factors Influencing Contamination of Groundnuts [11, 13, 17]**

| Agronomic Factor | Impact on Aflatoxin Contamination/Management Strategies |
|---|---|
| Climatic Conditions | Temperature: Warm and humid conditions favour fungal growth and aflatoxin production.<br>Rainfall: Excessive moisture during pod development and maturation increases susceptibility to fungal infection. |
| Soil Type and Quality | Well-Drained Soil: Proper drainage reduces waterlogged conditions that promote fungal growth.<br>Nutrient Balance: Soil fertility affects peanut health and resistance to fungal infection. |
| Crop Rotation | Diverse Rotation: Regularly rotating crops reduces the buildup of fungal spores in the soil.<br>Avoidance of Continuous Peanut Cultivation: Monoculture increases the risk of contamination. |
| Irrigation Practices | Implement irrigation practices to maintain adequate soil moisture during critical peanut development stages, particularly during pod filling. Avoid overwatering because this can also create favourable conditions for fungal growth. |
| Planting Density | Optimal Spacing: Proper spacing allows for better air circulation and reduces humidity around plants. |
| Nutrient Management | Ensure balanced fertilization practices that meet the specific needs of peanuts at different growth stages. Avoid excessive nitrogen application, which can promote fungal growth. |
| Harvest Timing | Early Harvest: Timely harvesting minimises exposure to wet conditions and fungal growth.<br>Avoid Delayed Harvest: Delayed harvest increases the risk of fungal contamination. |
| Storage Practices | Prompt Drying: Rapid drying after harvest prevents fungal growth.<br>Proper Storage Conditions: Low temperature and humidity inhibit fungal activity during storage.<br>Insect Control: Manage insect pests effectively to minimise damage to developing peanuts. Prevent insect infestations during storage. |
| Pest and Disease Management | Integrated Pest Management: Effective pest control reduces stress on peanut plants and lowers susceptibility to fungal infection. |
| Weed Management | Weed Competition: Weeds can harbour fungal spores and compete for nutrients. Effective weed control is essential. |
| Fertilization | Balanced Nutrient Application: Proper fertilization supports healthy plant growth and resistance to pathogens. |
| Irrigation Practices | Avoid Over-Irrigation: Excess moisture promotes fungal growth.<br>Drip Irrigation: Targeted irrigation reduces leaf wetness. |
| Harvest and Post-Harvest Handling | Gentle Harvesting: Minimise mechanical damage to pods.<br>Prompt Drying: Rapid drying after harvest prevents fungal growth.<br>Storage Hygiene: Clean storage facilities to prevent contamination. |

## BIOCHEMICAL ASPECTS OF AFLATOXINS: TYPES AND MECHANISM OF ALTERATION OF GROUNDNUT QUALITY

### Types of Aflatoxins

Peanuts are one of the crops most vulnerable to contamination by aflatoxins, a class of mycotoxins produced by fungus of the *Aspergillus* genus. Aflatoxins represent a significant concern for food safety and public health globally. Among the most studied aflatoxins are B1, B2, G1 and G2, each exhibiting unique chemical structures and biological effects (Table 2.9.3). Among these aflatoxins, AFB1 is the most powerful immune system suppressor, mutagenic, hepatocarcinogenic and teratogenic agent. It is AFB1 > AFG1 > AFG2 > AFB2 in order of toxicity [6, 8, 15, 18].

#### Aflatoxin B1 (AFB1)

Aflatoxin B1 (AFB1) is considered the most toxic and prevalent member of the aflatoxin family, posing a significant threat to human and animal health [15, 18, 21]. AFB1 is a polyketide heterocyclic molecule with a complex chemical structure. It possesses a difuran ring system and a lactone group,

**TABLE 2.9.3**
**Biochemical Characteristics of Aflatoxins Found in Groundnuts**

| Feature | Aflatoxin B1 | Aflatoxin B2 | Aflatoxin G1 | Aflatoxin G2 |
|---|---|---|---|---|
| 2D structure | 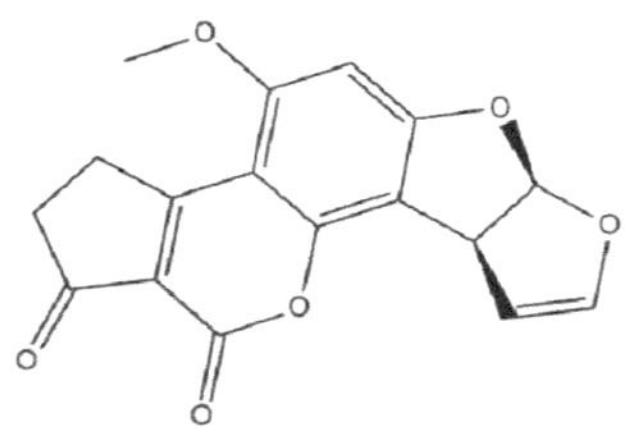 | 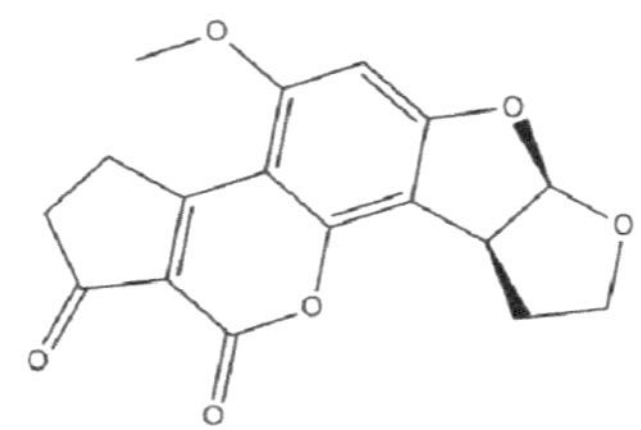 | 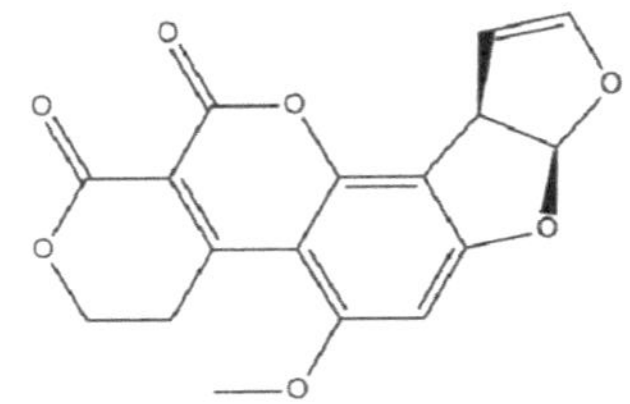 | 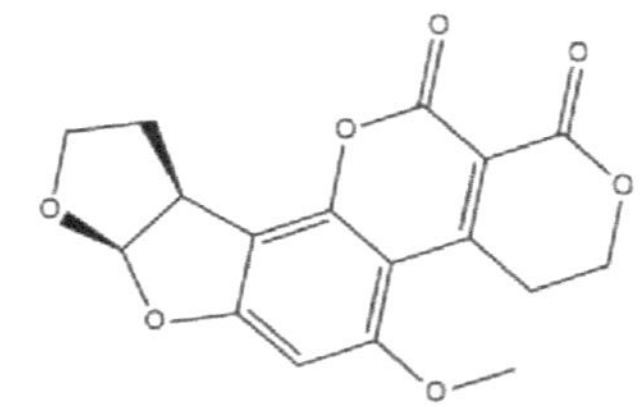 |
| Chemical Formula | $C_{17}H_{12}O_6$ | $C_{17}H_{14}O_6$ | $C_{17}H_{12}O_7$ | $C_{17}H_{14}O_7$ |
| Molecular Weight | 312.27 g/mol | 314.29 g/mol | 328.27 g/mol | 330.29 g/mol |
| CAS Number | 1162-65-8 | 7220-81-7 | 1165-39-5 | 7241-98-7 |
| Structure | Difuranocoumarin with a double bond at C8=C9 | Difuranocoumarin with a saturated C8–C9 bond | Difuranocoumarin with a hydroxyl group at C8 | Difuranocoumarin with a hydroxyl group at C8 and a saturated C8–C9 bond |
| Toxicity | Most potent carcinogen among the four | Less toxic than B1 | Less potent carcinogen than B1 | Least toxic among the four |
| Occurrence | Most abundant aflatoxin | Common, but less abundant than B1 | Less common than B1 and B2 | Least common |
| Regulatory Limits | Strict regulatory limits worldwide due to its toxicity; limits vary by country but are generally low (parts per billion levels). | Subject to regulatory scrutiny; often, limits for B1 and B2 are combined due to their co-occurrence. | Although specific limits may vary, regulatory bodies often impose stringent controls similar to those for B1 and B2. | Regulated alongside G1, with oversight by food safety authorities to minimise exposure in the human diet |

which contribute to its chemical stability and biological activity [15, 18]. AFB1 exerts its detrimental effects through various mechanisms, primarily targeting the liver:

- Hepatotoxicity: AFB1 binds to DNA, inducing mutations and genotoxicity. This can lead to hepatocellular carcinoma (HCC), the most well-documented consequence of chronic AFB1 exposure [15, 18, 21]. Additionally, AFB1 disrupts liver function by causing cell death (necrosis) and inflammation [15, 21, 22].
- Carcinogenesis: The aforementioned DNA adduct formation and genotoxicity are the primary drivers of AFB1's carcinogenicity. Chronic exposure is strongly linked to HCC development, particularly in individuals co-infected with hepatitis B virus [18, 23, 24].
- Immunotoxicity: AFB1 can suppress the immune system, making individuals more susceptible to infections and hindering the body's ability to fight cancer [18, 23, 24].
- Other effects: Beyond the liver, AFB1 exposure has been linked to adverse effects in various organs, including the kidneys, spleen and reproductive system [18, 23, 24].

### Aflatoxin B2 (AFB2)

AFB2 has a chemical structure comparable to AFB1, however the double bond inside the difuran ring is configured somewhat differently [8, 16, 20]. This small distinction results in variations in their chemical characteristics and biological consequences.

AFB2 is generally considered less toxic than AFB1. However, it exhibits similar detrimental effects, albeit at a lower potency [8, 16, 20]:

- Hepatotoxicity: AFB2 can also bind to DNA and induce liver damage, although to a lesser extent compared to AFB1.
- Carcinogenesis: AFB2 demonstrates carcinogenic potential, but its contribution to HCC development is believed to be lower than AFB1.
- Genotoxicity: Similar to AFB1, AFB2 can induce DNA damage, raising concerns about its long-term effects.
- Metabolism: A critical aspect of AFB2 is its in vivo conversion to AFB1 by the body's enzymes. This metabolic transformation can contribute to the overall toxicity associated with AFB2 exposure.

### Aflatoxin G1 (AFG1)

AFG1 shares a similar structure to AFB1 but lacks an epoxide group on the cyclohexene ring [8, 15, 25]. Because of this structural difference, AFG1 is intrinsically less hazardous and less reactive than AFB1. AFG1, despite its lower intrinsic toxicity, is a major contributor to the total health risk posed by aflatoxin exposure [20, 26].

### Aflatoxin G2 (AFG2)

Aflatoxin G2 (AFG2) is another member of the aflatoxin family produced by *Aspergillus* fungi, alongside its frequently discussed counterparts AFB1 and AFG1. Similar to AFG1, AFG2 is categorised as a precursor toxin. It requires metabolic activation within the body to exert its toxic effects [5, 20, 25]. AFG2 belongs to the aflatoxin B class, characterised by a coumarin lactone structure with a difuran ring.

## Biochemical Mechanisms of Aflatoxin-Induced Changes in Peanut Quality

Aflatoxins, primarily produced by *A. flavus* and *A. parasiticus*, exert profound biochemical impacts on groundnuts (peanuts), significantly impairing their quality. These secondary fungal metabolites, among which aflatoxin B1 is the most potent, interfere with essential biochemical pathways in groundnuts, leading to diminished nutritional value, safety and marketability [6, 13, 14, 17, 18].

The mechanisms through which aflatoxins affect the quality of groundnuts encompass a range of biochemical and molecular interactions.

### Interference with Nucleic Acid Synthesis

Aflatoxin B1 is metabolised in plant tissues (and animal liver) to aflatoxin B1-8,9-epoxide, which can bind to DNA and RNA, forming adducts. This binding interrupts the synthesis of nucleic acids, essential for cell replication and function [27, 28]. In groundnuts, this can translate to stunted growth or compromised defence mechanisms against further fungal invasion or environmental stressors. The primary mechanism by which aflatoxins influence nucleic acid synthesis is through the formation of aflatoxin-DNA adducts, which can interfere with DNA and RNA synthesis and function in several fundamental ways [13, 18, 29]:

- Formation of aflatoxin-DNA adducts
  After groundnuts are infected and aflatoxins are produced, these compounds can be bioactivated by plant and perhaps microbial enzymes in the groundnut matrix, converting aflatoxin B1 into its reactive form, aflatoxin B1-8,9-epoxide. This reactive form has a high affinity for nucleophilic sites in DNA, leading to the formation of covalent bonds with guanine bases, creating aflatoxin-DNA adducts [13, 18, 29].
- Interference with DNA and RNA synthesis
  The presence of aflatoxin-DNA adducts disrupts the normal processes of replication and transcription. Adduct formation can lead to miscoding during DNA replication, potentially resulting in mutations. These mutations can accumulate and interfere with the genetic integrity of the groundnut cells, affecting their proper growth and development. Additionally, by altering the transcription process, these adducts can hinder RNA synthesis, compromising the synthesis of essential proteins required for various physiological processes in the groundnuts [13, 18, 29].
- Disruption of gene expression
  The modification of DNA by aflatoxin-DNA adducts can also affect gene expression. Critical genes that regulate cellular growth, apoptosis (programmed cell death), and repair mechanisms can be dysregulated, leading to abnormal cellular development or an increased rate of cell death. This dysregulation can severely impact the development of groundnuts, leading to reduced yield and quality [7, 13, 18, 29].
- Implications for groundnut health and quality

The interference of aflatoxins with nucleic acid synthesis in groundnuts has direct implications for their health and quality. Impairment in DNA and RNA integrity can lead to reduced seed germination and plant vigour, compromised immune response against various stressors, and disruptions in metabolic pathways essential for nutrient accumulation in the nuts. Consequently, the nutritional and market value of groundnuts can be significantly diminished [13, 18, 29, 30].

### Protein Synthesis Disruption

Aflatoxins have a high affinity for binding to proteins. They can form covalent bonds with amino acids, particularly cysteine and lysine residues, in proteins. This binding can lead to the formation of aflatoxin-protein adducts, which can alter the structure and functionality of proteins in groundnuts [7, 13, 18, 29, 31, 32]. This can result in reduced nutritional value and impaired protein functionality, affecting the overall quality of groundnuts. The aflatoxin epoxide can also form complexes with cellular proteins, including enzymes critical for metabolic pathways [7, 13, 18, 29, 31]. This disruption hampers protein synthesis, affecting the nutritional quality of groundnuts by decreasing their protein content or altering the amino acid composition. Additionally, essential enzymatic activities in groundnuts necessary for healthy plant development and immune response can be compromised.

The implications of protein synthesis disruption in groundnuts are multifaceted. Protein content and quality are paramount to the nutritional value of groundnuts, influencing their marketability and suitability for consumption [3, 18, 26]. Furthermore, proteins are essential for the development, stress response and defence mechanisms of groundnut plants. Disruption in protein synthesis can lead to stunted growth, increased susceptibility to diseases and environmental stress, and ultimately, reduced yield and quality of groundnut crops. Aflatoxins, particularly aflatoxin B1, affect the cellular functions of the groundnuts themselves, including the vital process of protein synthesis [7, 13, 18, 29–31]:

- Aflatoxin B1 metabolism and reactive epoxide formation
  Upon contamination of groundnuts by aflatoxigenic fungi, aflatoxin B1 can undergo metabolic transformation within the plant tissues to generate a highly reactive intermediate, aflatoxin B1-8,9-epoxide. This transformation is crucial as the epoxide form of aflatoxin B1 possesses high reactivity towards nucleophilic groups in biomolecules, including those in proteins [16, 25, 31].
- Interaction with ribosomal RNA and proteins
  The disruption of protein synthesis primarily stems from the interaction of aflatoxin B1-8,9-epoxide with components of the ribosome, the cellular machinery responsible for protein synthesis. This interaction can lead to the modification of ribosomal RNA (rRNA) and ribosomal proteins, impairing their function. Given that the ribosome plays a central role in decoding mRNA and forming peptide bonds between amino acids, any interference with its components can significantly hinder the assembly of proteins [7, 23, 24, 32].
- Modification of tRNA and aminoacyl-tRNA synthetase
  Transfer RNA (tRNA) molecules and their associated enzymes, aminoacyl-tRNA synthetases, are also potential targets of aflatoxin-induced damage. By forming adducts with tRNA or altering the enzymatic activity of aminoacyl-tRNA synthetases, aflatoxins can disrupt the accurate translation of mRNA into the corresponding protein sequence [7, 23, 24, 32]. This disruption can lead to the misincorporation of amino acids or premature termination of protein synthesis, resulting in faulty proteins that may fail to function properly or degrade rapidly [7, 23, 24, 32].
- Alteration of signal transduction pathways

Aflatoxins can also interfere with protein synthesis by modulating signal transduction pathways that regulate this process. For instance, aflatoxins have been shown to influence pathways involving mitogen-activated protein kinases (MAPKs), which play significant roles in cell growth, division and response to stress. Altered signalling due to aflatoxin exposure can dysregulate the initiation and elongation phases of protein synthesis, affecting overall protein production [7, 23, 24, 32].

### Oxidative Stress Induction

Aflatoxins generate reactive oxygen species (ROS), leading to oxidative stress within groundnut tissues. This stress can damage cellular components, including lipids, proteins and nucleic acids, through peroxidation and oxidation reactions. The lipid content in groundnuts, vital for their energy value and flavour, can be particularly affected, leading to rancidity and decreased shelf life [33–35].

Aflatoxins produce ROS and free radicals, which can initiate lipid peroxidation reactions in groundnut kernels. This process can lead to the breakdown of lipids, resulting in the production of volatile compounds and off-flavours. Lipid peroxidation can also cause changes in the texture, colour and aroma of groundnuts, negatively impacting their quality [33–36].

### Impairment of Nutrient Absorption and Transport

The impairment of nutrient absorption and transport in groundnuts due to aflatoxin contamination can result in nutrient deficiencies, reduced growth and compromised yield. It is crucial to implement measures to prevent aflatoxin contamination, such as proper storage, drying and post-harvest handling practices, to ensure the nutritional quality and safety of groundnuts [15, 35, 37, 38].

The interaction of aflatoxins with cellular components may hinder nutrient absorption and transport within the groundnut plant. By disrupting cellular membranes and metabolic pathways, aflatoxins can limit the efficient uptake and distribution of nutrients essential for the development of quality groundnut kernels. The presence of aflatoxins in groundnuts can lead to the impairment of nutrient absorption and transport through various mechanisms [15, 35, 37, 38].

- Disruption of cell membranes
  Aflatoxins can disrupt the integrity and functionality of cell membranes in groundnut tissues. This disruption can affect the selective permeability of membranes, including those of root cells involved in nutrient absorption. As a result, the uptake of essential nutrients, such as minerals and water, may be compromised, leading to impaired nutrient absorption [15, 35, 37, 38].
- Inhibition of enzyme activity
  Aflatoxins can inhibit the activity of enzymes involved in nutrient metabolism and transport. For example, aflatoxins can inhibit the activity of ATPases, which are essential for energy-dependent nutrient transport across cell membranes. This inhibition can disrupt the normal functioning of nutrient transporters, impairing the uptake and transport of nutrients within groundnut plants [15, 35, 37, 38].
- Alteration of root architecture
  Aflatoxins can affect root development and architecture in groundnuts. They can inhibit root elongation and branching, reducing the surface area available for nutrient absorption. This reduction in root surface area can limit the contact between roots and the soil, leading to decreased nutrient uptake [15, 35, 37, 38].
- Impaired nutrient transport in xylem and phloem
  Aflatoxins can interfere with nutrient transport in the xylem and phloem, the vascular tissues responsible for the movement of water, minerals and organic compounds within plants. Aflatoxins can disrupt the integrity of vascular tissues, impairing the flow of nutrients from roots to shoots and vice versa. This disruption can result in nutrient imbalances and deficiencies in different parts of the groundnut plant [15, 35, 37, 38].
- Induction of oxidative stress

Aflatoxins can induce oxidative stress in groundnut tissues. Oxidative stress occurs when there is an imbalance between the production of ROS and the antioxidant defence system. Aflatoxins generate free radicals and ROS, which can damage cellular components, including proteins and membranes involved in nutrient absorption and transport [22, 33–35]. This oxidative damage can further impair the efficiency of nutrient uptake and transport in groundnuts.

### Inhibition of Fatty Acid Metabolism

Groundnuts are valued for their lipid content, predominantly unsaturated fatty acids, which contribute significantly to their nutritional profile. Aflatoxins can inhibit enzymes involved in lipid metabolism, affecting the synthesis and accumulation of fatty acids in groundnuts. This inhibition can alter the fatty acid composition, potentially reducing the content of beneficial unsaturated fatty acids [39–41].

### Reduction in Germination Rate

The accumulation of aflatoxins in groundnut seeds can reduce their germination rate. Aflatoxins compromise the integrity and vigour of the seed, which is essential for germination and the establishment of seedlings [15]. This reduction in germination rate not only affects the quality of groundnuts for planting purposes but also their overall market value.

## DETECTION METHODS FOR AFLATOXINS IN GROUNDNUTS

Aflatoxins are a group of toxic secondary metabolites produced by certain *Aspergillus* fungi. Contamination of peanuts with aflatoxins is a significant concern due to their carcinogenic properties. Several methods are available for detecting aflatoxins in peanuts as presented in Table 2.9.4.

**TABLE 2.9.4**
**Detection Methods for Aflatoxins in Peanuts [20, 45–51]**

| Method | Principles | Characteristics |
|---|---|---|
| High-performance liquid chromatography (HPLC) | Utilises liquid chromatography to separate and quantify aflatoxins. | Shows good recoveries for the levels of interest. Used in combination with mass spectrometry or with post-column derivatization and ultraviolet detection for enhanced sensitivity. |
| Enzyme-linked immunosorbent assay (ELISA) | Utilises antibodies to specifically bind to aflatoxins. | Utilised for both qualitative and quantitative analysis, it demonstrates good sensitivity and specificity for aflatoxin detection. |
| Immunochromatographic rapid tests | Uses antibodies specific to aflatoxins to capture them on a membrane strip. A coloured signal appears if aflatoxin is present, indicating a positive test. | Offer rapid and efficient detection suitable for on-site analysis. |
| Solid-phase extraction (SPE) | Isolates aflatoxins from groundnut samples using a specialised sorbent material. | Found to have better recovery of aflatoxins in peanuts compared to immuno-affinity columns (IACs). |
| Methanol/water extraction | A simple method that uses a methanol/water solution to extract aflatoxins from groundnuts. | Used for extracting aflatoxins from peanut samples followed by statistical analysis for quantification. |
| Thin-layer chromatography (TLC) | Separates aflatoxins based on their chemical properties using a thin layer of adsorbent material. Aflatoxins are visualised under UV light or by spraying with specific reagents. | Initially used for qualitative assessment, it provides a basis for further quantitative analysis. |
| Fluorometric spectroscopy | Measures the fluorescence emitted by aflatoxins under UV light. | Developed for quantitative analysis, yielding precise results with a lower limit of quantification. |
| Small particle column-high performance liquid chromatography (SPLC) | Similar to HPLC, but uses smaller particles in the column for higher resolution and faster separation. Aflatoxins are detected based on their retention time and can be quantified using a calibration curve. | Shows improved accuracy and efficiency compared to the national standard for aflatoxin detection in peanuts. |
| Reversed phase high-performance liquid chromatography fluorescence detection (RP-HPLC-F) | A specific type of HPLC where aflatoxins are separated using a reversed-phase column. Aflatoxin detection relies on their inherent fluorescence properties when exposed to ultraviolet light. | Optimised for detecting aflatoxins and ochratoxin A in peanuts, demonstrating high sensitivity and specificity. |
| Holaday–Velasco method | Not a commonly used method today. It involved chemical reactions to derivatise aflatoxins, enhancing their fluorescence for detection. | Showed better performance in detecting aflatoxins in raw peanuts with no missed detections in a certain aflatoxin level range and no false positive reports from uncontaminated peanuts. |

*(Continued)*

**TABLE 2.9.4** ***(Continued)***
**Detection Methods for Aflatoxins in Peanuts [20, 45–51]**

| Method | Principles | Characteristics |
|---|---|---|
| Real-time polymerase chain reaction (PCR) | Detects the DNA of aflatoxin-producing fungi present in groundnuts. This method does not directly detect aflatoxins but indicates the potential for contamination. | Standardised for detecting *A. flavus* in stored peanuts, showing a significant agreement with traditional methods and useful for distinguishing toxigenic from atoxigenic isolates. |
| Mid-infrared spectral characterization | Analyses the infrared light absorption pattern of the sample. Aflatoxins have a unique spectral signature that can be identified for detection. | Focused on distinguishing aflatoxin contamination levels in peanuts through spectral analysis. |
| Polymerase chain reaction (PCR) | Detects the DNA of aflatoxin-producing fungi present in groundnuts. | Optimised alongside RP-HPLC-F for aflatoxin detection, showing potential for quick identification of toxigenicity. |
| Rapid methods using acoustic techniques and fermentation | An emerging method that explores using sound waves to detect aflatoxins based on their interaction with the sample. Still under development. | Proposed for rapid detection and elimination of aflatoxins in contaminated corn and nuts. |

## AFLATOXIN MANAGEMENT STRATEGIES IN GROUNDNUTS

Aflatoxins pose a significant threat to groundnut safety and economic viability. A multi-pronged approach is necessary to effectively manage aflatoxin contamination throughout the peanut production chain.

Strategies for managing aflatoxins in peanuts require a multifaceted approach that includes measures from pre-harvest to post-harvest. This approach involves both biological and technological interventions to protect consumer health and reduce economic losses. Recent scholarly discussions have identified effective strategies for managing aflatoxin [5, 20, 25, 42–44]. These include the use of biocontrol agents, near-infrared spectrometric sorting, biofertilizers, improved tillage, predictive forecasting, post-harvest rapid testing, Decision Support Systems, and meticulous risk monitoring during drying, storage and transportation processes [5, 20, 25, 42–44]. It is important to note that these strategies should be implemented objectively and with precision to ensure successful management of aflatoxin.

### Pre-harvest Practices

The pre-harvest management tactics are practices used by producers to improve crop health, production and quality before harvest. These techniques address numerous areas of agricultural production that affect plant growth, development and defence mechanisms. Effective pre-harvest management strategies can help to reduce crop losses, increase output and guarantee that fruits, vegetables and other commodities meet quality requirements [5, 20, 25, 42–44].

The pre-harvest stage emphasises the importance of biological control as an eco-friendly substitute for chemical pesticides. This involves using native biocontrol agents to combat toxigenic fungi. Additionally, integrating biofertilizers and implementing improved agricultural practices can further help to reduce aflatoxin contamination [5, 20, 25, 42–44]. Furthermore, the risk of aflatoxin contamination can be significantly reduced by using resistant cultivars and implementing specific crop production enhancements.

#### Crop Rotation and Intercropping

Crop rotation involves planting different crops in sequence on the same land over time. Rotating peanuts with non-susceptible crops disrupts the life cycle of *Aspergillus* fungi in the soil, minimizing infection risk [5, 25, 52, 53]. Intercropping, on the other hand, refers to the practice of cultivating two or more crops simultaneously within the same field. Both techniques offer a range of pre-harvest benefits [5, 25, 52, 53]:

- Disease and pest control: Continuous monoculture (planting the same crop repeatedly) can lead to a buildup of soilborne pathogens and pests adapted to that specific crop. Rotation disrupts pest and disease life cycles, reducing their populations and minimizing reliance on chemical controls [5, 25, 52, 53].
- Nutrient management: Different crops have varying nutrient requirements. Rotation allows the soil to replenish nutrients depleted by the previous crop, promoting healthy growth of the subsequent crop. Stress from drought or nutrient deficiency can increase aflatoxin risk. Maintaining optimal soil moisture and providing adequate nutrients through irrigation and fertilization promotes healthy plant growth and reduces susceptibility to fungal infection [5, 25, 52, 53].
- Weed suppression: Certain crops are allelopathic, meaning they release chemicals that inhibit the growth of weeds. Strategic rotation can help suppress weed populations and reduce competition for resources. Certain weeds act as alternate hosts for *Aspergillus* species. Proper weed management reduces the overall fungal population in the field.
- Improved soil structure: Root systems of different crops vary in depth and structure. Rotation with deep-rooted crops can improve soil aeration and drainage, benefiting subsequent shallow-rooted crops. Applying farmyard manure and lime can improve soil quality, potentially reducing pathogen infection and aflatoxin production.

### Soil Management Practices

Soil health is paramount for successful crop production. Pre-harvest management strategies often involve practices that improve soil quality [5, 25]:

- Tillage practices: Tillage methods influence soil structure, aeration and water infiltration. Conservation tillage practices minimise soil disturbance, promoting organic matter retention and beneficial soil biota [5, 25].
- Nutrient management: Soil testing helps identify nutrient deficiencies and excesses. Applying amendments like compost, manure, or fertilizers based on test results ensures optimal nutrient availability for crops [5, 25].
- Organic matter management: Organic matter addition improves soil fertility, structure and water holding capacity. Practices like cover cropping and compost application can increase soil organic matter content [5, 25].
- Irrigation management: Proper irrigation practices ensure crops receive adequate moisture for growth without waterlogging, which can lead to root rot and nutrient leaching [5, 25].

### Pest and Disease Control Measures

Pre-harvest pest and disease control strategies aim to minimise crop damage and ensure food safety. These strategies can be categorised into preventive and reactive measures.

- Preventive measures [5, 25, 44]:
  - Selection of resistant varieties: Planting cultivars that are genetically resistant to common diseases and pests lessens the need for chemical treatments.
  - Sanitation and hygiene: Reducing pest and disease harbourage sites can be achieved by clearing away agricultural trash, controlling weeds and practicing excellent field hygiene.
  - Scouting and monitoring: Consistently keeping an eye out for indications of pest or disease infestation in crops enables prompt identification and remediation.

- Reactive measures [5, 25, 44]:
  - Biological control, which involves using beneficial organisms such as insects or microbes to manage pest populations, can be a sustainable alternative to chemical pesticides.
  - Chemical control, on the other hand, may be necessary to manage pest outbreaks or severe disease infestations, but responsible application practices are crucial to minimise environmental impact and ensure food safety.

## Harvest and Post-Harvest Practices

Groundnuts are susceptible to deterioration after harvest if not handled appropriately. Effective post-harvest management strategies are crucial to minimise losses, preserve quality and ensure groundnuts reach consumers in a safe and desirable state. Timely harvesting at peanut maturity avoids prolonged exposure to favourable conditions for fungal growth and aflatoxin production.

### Proper Drying Techniques

Groundnuts are usually sun-dried after harvest. However, it is essential to use proper drying techniques to prevent spoilage and maintain nut quality. Rapid drying is crucial to prevent fungal growth and aflatoxin contamination, which can pose health risks. Good air circulation and avoiding overcrowding during drying accelerates the process. In addition, to ensure uniform drying, groundnuts should be spread in a single layer and turned regularly to promote even moisture loss throughout the pod [42].

Furthermore, to determine when drying is complete, use a moisture meter to monitor the moisture content. Additionally, it is important to protect the groundnuts from direct sunlight as it can have a negative impact on their quality [42, 44]. It is recommended to maintain a final moisture content of around 6–8% for storage purposes. Covering drying pods with a shade cloth can prevent excessive heat and light exposure [42, 44].

### Storage Conditions and Facilities

Preserving groundnut quality and minimising post-harvest losses need ideal storage conditions. To avoid aflatoxin development and insect pest activity, groundnuts should be stored in a cold, dry environment. The recommended range for storage temperature is 10–15°C, with a relative humidity of no more than 65% [42, 44]. Sufficient ventilation is also required. It's critical to keep the storage facility's air circulation system functioning properly to avoid spoiling. Hermetic bags or sealed metal silos are examples of airtight, moisture-proof containers that may dramatically lower storage losses. Jute bags and other conventional storage techniques could be less successful in keeping moisture and pest infestation under control [42, 44].

### Sorting and Grading Practices

Grading and sorting are necessary to guarantee that only premium groundnuts are sold and kept in storage [43]. To avoid contamination and spoiling, foreign elements including as leaves, trash and immature or broken pods should be removed. In addition, size, homogeneity of the kernel, and lack of flaws may all be used to grade groundnuts, allowing for the most effective marketing and pricing strategies depending on quality variances.

## Aflatoxin Mitigation Approaches

Biological control is a highly effective pre-harvest method for reducing aflatoxin levels. It involves managing pest populations using other living organisms. Competing fungi can reduce seed infection, thereby limiting the growth of *A. flavus*. *A. parasiticus*, another fungus that produces aflatoxins, can be out-competed, resulting in a reduction of aflatoxin production. Aflatoxin B1, which has negative health implications for livestock, is also affected [10, 43, 54, 55]. Aflatoxin can cause a reduction in protein content in contaminated seeds and an increase in poor quality by-products. This is especially evident in groundnuts, where groundnut meal is a significant livestock food product. A biocontrol approach has been used to reduce aflatoxin levels in both pre- and post-harvest situations by competitively displacing toxigenic strains of *A. flavus* with atoxigenic strains, which are less virulent towards the host [10, 43, 54–56].

The primary approach to biologically controlling aflatoxins at the post-harvest stage is to decrease the level of Aspergillus infection in commodities intended for human or animal consumption. This is typically accomplished by targeting specific points where infection by *A. flavus* or *A. parasiticus* can occur during storage. Periods of moisture stress during drying, inadequate drying due to unevenness or prolonged drying, and the high-risk period for infection, which occurs when the moisture content and temperature of the commodity are both high, are the main points to consider. To reduce infection by *A. flavus*, it is important to maintain a competitive growth of non-toxigenic strains of *A. flavus* [10, 43, 54–56].

## Biological Control Methods

Various saprophytic and parasitic moulds and bacteria are employed in soil ecology approaches to eliminate aflatoxin toxicity caused by competing microorganisms. The most common saprophyte used is Trichoderma species, which employs mycoparasitism to combat aflatoxin-producing strains of *Aspergillus* [10, 43, 54–56]. This group of fungi is capable of degrading the germination and mycelial growth of A. flavus by competing for nutrients and space. Another effective method of mycobiocontrol involves introducing parasitic microorganisms that can kill or inactivate the host organism. Biocontrol agents can be effective in the microbial detoxification of aflatoxins. Certain bacteria and fungi can use aflatoxins as a nutrient source and break them down into less toxic metabolites [10, 43, 54–56]. This method of microbial detoxification can be used in grains and feeds. However, there is currently no authorization for the use of such detoxified products in edible commodities.

Recently, biological control has emerged as a promising method for controlling pre-harvest and post-harvest aflatoxin contamination [13, 44]. This method is eco-friendly, effective and sustainable compared to other chemical methods. Biological control involves the use of atoxigenic strains of *A. flavus* and *A. niger*, competing microorganisms and microbial degrading enzymes for aflatoxin detoxification. Atoxigenic strains of *A. flavus* and *A. niger* can be used as biological control agents to competitively exclude indigenous toxigenic aflatoxin-producing strains [13, 44, 55]. These atoxigenic strains multiply in the soil and extensively compete with native aflatoxin strains at various ecological niches, eventually reducing the overall level of aflatoxins in groundnuts. The effectiveness of this atoxigenic bio-control depends on the compatibility of bio-control strains with indigenous strains of groundnut. Additionally, it relies on the persistence and spore dispersal of bio-control strains in the soil, and it must provide equal or greater adaptability than native aflatoxin strains [13, 44, 55].

## Physical and Chemical Treatments

A double-extraction procedure with a food-grade solvent can be used as an alternative approach. The peanuts are soaked in the solvent to dissolve the aflatoxins. However, it is important to dissolve as little of the oil or protein as possible. Since aflatoxins have polar groups, a polar solvent is required. However, peanut oil and proteins also have polar groups, making it difficult to separate the two. Another chemical method that was tested is the use of sodium hypochlorite. A complete detoxification of an artificially contaminated peanut sample was achieved. However, this method significantly affects the flavour and nutritive value of the peanuts and is therefore not a suitable option [44, 57, 58].

Chemical detoxification methods have been explored for treating peanuts contaminated with aflatoxins. Ammonia reacts with aflatoxins to produce various products, including non-toxic ammoniation products. Treating peanuts with a 1.5% ammonia solution for up to 2 hours at 45°C has been found to reduce the amount of B1 aflatoxin by 70% and G1 aflatoxin by 25% [44, 57, 58]. However, controlling the reaction of ammonia with peanuts to produce non-toxic products without damaging the peanuts is difficult. There are concerns about the impact of ammoniation on the peanuts' nutritional value.

### Genetic Modification Techniques

Genetic modification techniques are being explored to develop groundnut varieties with inherent resistance to aflatoxin contamination [54, 55]. One approach being investigated is the introduction of genes from other organisms that encode enzymes capable of detoxifying aflatoxin. Another approach is the silencing of aflatoxin production pathways. Genetic modification can be employed to silence genes within the aflatoxin biosynthesis pathway in *A. fungi*, thus preventing toxin production.

## CONCLUSION

Aflatoxigenic fungi can pose a significant threat to food safety, especially in crops like groundnuts, where conditions are favourable for fungal growth and toxin production. Contamination of groundnuts with aflatoxins can negatively affect human and animal health, leading to problems such as liver damage and an increased risk of cancer. To reduce the risks associated with aflatoxigenic fungi, it is crucial for farmers, processors and regulators to implement strict control measures throughout the production and supply chain. This involves monitoring and managing environmental conditions, implementing good agricultural practices and using effective storage and processing techniques to prevent fungal growth and toxin production. The relationship between aflatoxigenic fungi and groundnuts is a complex and challenging issue that requires a coordinated effort from all stakeholders involved in the food production chain. By working together to address the risks posed by these moulds, we can ensure the safety and quality of our food supply for generations to come.

## REFERENCES


1. De Ron AM (2015) Grain Legumes. Springer New York.
2. FAO (2016) Légumineuses : des graines nutritives pour un avenir durable
3. Hamadou M, Martin Alain MM, Obadias FV, et al (2022) Consumption of underutilised grain legumes and the prevention of type II diabetes and cardiometabolic diseases: Evidence from field investigation and physicochemical analyses. Environ Challenges 9:100621. https://doi.org/10.1016/j.envc.2022.100621
4. Toomer OT (2018) Nutritional chemistry of the peanut (*Arachis hypogaea*). Crit Rev Food Sci Nutr 58:3042–3053. https://doi.org/10.1080/10408398.2017.1339015
5. Kumar D, Desai LJ, Bhanu C, et al (2023) Pre-Harvest Management of Aflatoxin Contamination in Groundnut Through Biocontrol Products. In: Applied Mycology for Agriculture and Foods. Apple Academic Press, New York, pp 217–246
6. Nsabiyumva G, Mutegi CK, Wagacha JM, et al (2023) Aflatoxin contamination of maize and groundnut in Burundi: Distribution of contamination, identification of causal agents and potential biocontrol genotypes of *Aspergillus flavus*. Front Microbiol 14. https://doi.org/10.3389/fmicb.2023.1106543
7. Jose AM, Tejaswi A, Kokiladevi E, et al (2024) Genome-wide identification of the Glycine-rich RNA-binding protein genes and their expression analysis upon *Aspergillus flavus* infection in groundnut (*Arachis hypogaea*). Agronomy 14:165. https://doi.org/10.3390/agronomy14010165
8. Ráduly Z, Szabó L, Madar A, et al (2019) Toxicological and medical aspects of aspergillus-derived mycotoxins entering the feed and food chain. Front Microbiol 10:2908. https://doi.org/10.3389/fmicb.2019.02908
9. Chang AS, Sreedharan A, Schneider KR (2013) Peanut and peanut products: A food safety perspective. Food Control 32:296–303. https://doi.org/10.1016/j.foodcont.2012.12.007
10. Lu Q, Huang L, Liu H, et al (2024) A genomic variation map provides insights into peanut diversity in China and associations with 28 agronomic traits. Nat Genet 56:530–540. https://doi.org/10.1038/s41588-024-01660-7
11. Tai B, Chang J, Liu Y, Xing F (2020) Recent progress of the effect of environmental factors on *Aspergillus flavus* growth and aflatoxins production on foods. Food Qual Saf 4:21–28. https://doi.org/10.1093/fqsafe/fyz040

12. Singh B, Singh U (1991) Peanut as a source of protein for human foods. Plant Foods Hum Nutr 41: 165–177. https://doi.org/10.1007/BF02194085
13. Lare N, Bodjona BPT, Gambogou B, Karou DS (2023) Molecular characterization of aflatoxigenic and non-aflatoxigenic *Aspergillus flavus* strains isolated from peanut sold in Northern Togo. Food Sci Appl Biotechnol 6:77. https://doi.org/10.30721/fsab2023.v6.i1.227
14. Geng Q, Zhang Y, Song M, et al (2023) Allergenicity of peanut allergens and its dependence on the structure. Compr Rev Food Sci Food Saf 22:1058–1081. https://doi.org/10.1111/1541-4337.13101
15. Zahija I, Jeršek B, Demšar L, et al (2022) Production of aflatoxin B1 by *Aspergillus parasiticus* grown on a novel meat-based media. Toxins (Basel) 15:25. https://doi.org/10.3390/toxins15010025
16. Bharose AA, Hajare ST, Narayanrao DR, et al (2024) Whole genome sequencing and annotation of *Aspergillus flavus* JAM-JKB-B HA-GG20. Sci Rep 14:18. https://doi.org/10.1038/s41598-023-50986-5
17. Madilo FK, Glover RLK, Islam MN, et al (2023) Microbiological assessment of groundnut (*Arachis hypogaea* L.) Sold for consumption in Ghana. J Food Qual 2023:1–10. https://doi.org/10.1155/2023/7836774
18. Bukhari SH, Asghar MA, Ahmed F, Jabeen S (2023) The production of aflatoxin B 1 by Aspergillus parasiticus in peanuts and walnuts under the influence of controlled temperature and water activity. Int J Food Eng 19:551–560. https://doi.org/10.1515/ijfe-2023-0116
19. Boukaew S, Petlamul W, Yossan S, et al (2024) Biofumigant potential and inhibition mechanism of *Trichoderma asperelloides* SKRU-01 volatile organic compounds for controlling aflatoxigenic *Aspergillus parasiticus* and *Aspergillus flavus* in stored peanuts. Food Control 157:110194. https://doi.org/10.1016/j.foodcont.2023.110194
20. Musangi CR, Juma BS, Mukhebi DW, et al (2024) Aspergillus population diversity and its role in aflatoxin contamination of cashew nuts from coastal Kenya. PLoS One 19:e0292519. https://doi.org/10.1371/journal.pone.0292519
21. Zaineldin AI, Elsebaey E, Habotta OA, et al (2024) Mitigating aflatoxin B1-induced growth impairment and hepatic stress in Nile Tilapia (*Oreochromis niloticus*): Comparative efficacy of saccharomyces cerevisiae and silicate-based detoxifiers. Probiotics Antimicrob Proteins. https://doi.org/10.1007/s12602-023-10210-2
22. Gao Y, Bao X, Meng L, et al (2021) Aflatoxin B1 and aflatoxin M1 induce compromised intestinal integrity through Clathrin-Mediated endocytosis. Toxins (Basel) 13:184. https://doi.org/10.3390/toxins13030184
23. Colin Garner R (1973) Microsome-dependent binding of aflatoxin B1 to DNA, RNA, polyribonucleotides and protein in vitro. Chem Biol Interact 6:125–129. https://doi.org/10.1016/0009-2797(73)90079-3
24. Yu F-L, Bender W, Geronimo IH (1988) The binding of aflatoxin B 1 to rat liver nuclear proteins and its effect on DNA-dependent RNA synthesis. Carcinogenesis 9:533–540. https://doi.org/10.1093/carcin/9.4.533
25. Akale Z, Mohammed A, Kebede A, Abady S (2023) Pre-harvest host-resistance to *Aspergillus* infection and aflatoxin B1 contaminations in groundnut (*Arachis hypogaea* L.) genotypes. Heliyon 9:e23034. https://doi.org/10.1016/j.heliyon.2023.e23034
26. Kortei NK, Osei S, Addae AA, et al (2024) Impact of some Ghanaian local roasting methods on aflatoxin reduction in groundnuts (*Arachis hypogaea* L). Sci African 23:e02022. https://doi.org/10.1016/j.sciaf.2023.e02022
27. Posey S, Magnan N, McCullough EB, et al (2024) Challenges to groundnut value chain development: Lessons from an (attempted) experiment in Ghana. J Dev Eff 16: 468–484. https://doi.org/10.1080/19439342.2024.2319657
28. Saha Turna N, Comstock SS, Gangur V, Wu F (2023) Effects of aflatoxin on the immune system: Evidence from human and mammalian animal research. Crit Rev Food Sci Nutr 64:9955–9973. https://doi.org/10.1080/10408398.2023.2219336
29. Zhu FH, Chen XY, Hou LL, et al (2024) *Limosilactobacillus reuteri* peptidoglycan alleviates aflatoxin B 1 -induced toxicity through adsorbing toxins and improving growth, antioxidant status, immunity and liver pathological changes in chicks. Br Poult Sci 65:352–360. https://doi.org/10.1080/00071668.2024.2316228
30. Salano EN, Mulwa RM, Obonyo MA (2024) Peanut (Arachis hypogea) accessions differentially accumulate aflatoxins upon challenge by *Aspergillus flavus*: Implications for aflatoxin mitigation. J Agric Food Res 15:100923. https://doi.org/10.1016/j.jafr.2023.100923
31. Wanniarachchi PC, Uthpala TGG, Ranaweera KKDS (2023) Aflatoxin occurrence, contamination, detection, and decontamination with special emphasis on coconut oil: A review. J Agric Sci – Sri Lanka 18:101–128. https://doi.org/10.4038/jas.v18i1.10101

32. Yang C, Wu D, Lin H, et al (2024) Role of RNA modifications, especially m6A, in aflatoxin biosynthesis of aspergillus flavus. J Agric Food Chem 72:726–741. https://doi.org/10.1021/acs.jafc.3c05926
33. Marin DE, Taranu I (2012) Overview on aflatoxins and oxidative stress. Toxin Rev 31:32–43. https://doi.org/10.3109/15569543.2012.730092
34. Alpsoy L, Yalvac ME (2011) Key roles of Vitamins A, C, and E in Aflatoxin B1-induced oxidative stress. Vitam Horm 86:287–305. https://doi.org/10.1016/B978-0-12-386960-9.00012-5
35. Cao W, Yu P, Yang K, Cao D (2022) Aflatoxin B1: Metabolism, toxicology, and its involvement in oxidative stress and cancer development. Toxicol Mech Methods 32:395–419. https://doi.org/10.1080/15376516.2021.2021339
36. Nganso Y, Sidjui Sidjui L, Ngnoung Amang AGA, et al (2020) Identification of peptides in the leaves of *Bauhinia rufescens Lam* (Fabaceae) and evaluation of their antimicrobial activities against pathogens for aquaculture. Sci J Chem 8:81. https://doi.org/10.11648/j.sjc.20200804.12
37. Chen X, Naehrer K, Applegate TJ (2016) Interactive effects of dietary protein concentration and aflatoxin B1 on performance, nutrient digestibility, and gut health in broiler chicks. Poult Sci 95:1312–1325. https://doi.org/10.3382/ps/pew022
38. Urugo MM, Teka TA, Lema TB, et al (2024) Dietary aflatoxins exposure, environmental enteropathy, and their relation with childhood stunting. Int J Food Sci Nutr 75: 241–254. https://doi.org/10.1080/09637486.2024.2314676
39. Zhou J, Tang L, Wang J, Wang J-S (2018) Aflatoxin B1 disrupts gut-microbial metabolisms of short-chain fatty acids, long-chain fatty acids, and bile acids in male F344 rats. Toxicol Sci 164:453–464. https://doi.org/10.1093/toxsci/kfy102
40. Song C, Shi W, Yang J, et al (2024) Mechanism study on Monarda didyma essential oil inhibiting *Aspergillus flavus* infection and aflatoxins accumulation in peanuts. Ind Crops Prod 212:118311. https://doi.org/10.1016/j.indcrop.2024.118311
41. Gong A, Song M, Liu J, et al (2024) Acetyl-CoA carboxylase as potential molecular target of dimethyl trisulfide on inhibiting the *Aspergillus flavus* growth and aflatoxins production. LWT 198:115924. https://doi.org/10.1016/j.lwt.2024.115924
42. Chulze SN, Alaniz Zanon MS, Taniwaki MH, et al (2024) Aflatoxins in the nut chains: Strategies to reduce their impact on consumer's health and economic losses. World Mycotoxin J 17:33–56. https://doi.org/10.1163/18750796-bja10003
43. Danso JK, Mbata GN, Holton RL (2024) Preharvest insect pests of peanuts and associated aflatoxin contaminants in Georgia, USA. J Econ Entomol. https://doi.org/10.1093/jee/toae074
44. Gong A, Song M, Zhang J (2024) Current strategies in controlling *Aspergillus flavus* and aflatoxins in grains during storage: A review. Sustainability 16:3171. https://doi.org/10.3390/su16083171
45. Norlia M, Jinap S, Nor-Khaizura MAR, et al (2019) *Aspergillus* section Flavi and aflatoxins: Occurrence, detection, and identification in raw peanuts and peanut-based products along the supply chain. Front Microbiol 10. https://doi.org/10.3389/fmicb.2019.02602
46. Holaday CE (1968) Rapid method for detecting aflatoxins in peanuts. J Am Oil Chem Soc 45:680–682. https://doi.org/10.1007/BF02541256
47. Fu Z, Huang X, Min S (2008) Rapid determination of aflatoxins in corn and peanuts. J Chromatogr A 1209:271–274. https://doi.org/10.1016/j.chroma.2008.09.054
48. Waliyar F, Reddy SV, Lava-Kumar P (2009) Review of immunological methods for the quantification of aflatoxins in peanut and other foods. Peanut Sci 36:54–59. https://doi.org/10.3146/AT07-007.1
49. Gilbert J, Shepherd MJ (1985) A survey of aflatoxins in peanut butters, nuts and nut confectionery products by HPLC with fluorescence detection. Food Addit Contam 2:171–183. https://doi.org/10.1080/02652038509373541
50. Chen P, Xu Y, Cao J, et al (2024) A smartphone-enabled visual platform for detecting aflatoxins in peanut oil using colorimetric analysis coupled with magnetic imprinted solid-phase extraction. Food Chem 456:139294. https://doi.org/10.1016/j.foodchem.2024.139294
51. Almusa MA, Al-Otibi FO, Alreshoodi FM, et al (2024) Detection and molecular characterization of aflatoxin and ochratoxin produce aspergillus species in capsicum spices in Saudi Arabia. Food Control 162:110377. https://doi.org/10.1016/j.foodcont.2024.110377
52. Batty M, Bennett MR, Yu E (2022) The role of oxidative stress in atherosclerosis. Cells 11. https://doi.org/10.3390/cells11233843
53. Harb HE, El-Tabakh MAM, Khattab AM, et al (2024) Recent Advances of Using Innovative Strategies in Management of Millet Plant Pathogens. In: Genetic Improvement of Small Millets. Springer Nature Singapore, Singapore, pp 297–328.

54. Krska R, Leslie JF, Haughey S, et al (2022) Effective approaches for early identification and proactive mitigation of aflatoxins in peanuts: An EU–China perspective. Compr Rev Food Sci Food Saf 21:3227–3243. https://doi.org/10.1111/1541-4337.12973
55. Pandey MK, Kumar R, Pandey AK, et al (2019) Mitigating aflatoxin contamination in groundnut through a combination of genetic resistance and post-harvest management practices. Toxins (Basel) 11:315. https://doi.org/10.3390/toxins11060315
56. Guo B, Yu J, Holbrook CC, et al (2009) Strategies in prevention of preharvest aflatoxin contamination in peanuts: Aflatoxin biosynthesis, genetics and genomics. Peanut Sci 36:11–20. https://doi.org/10.3146/AT07-001.1
57. Castro-Ríos K, Montoya-Estrada CN, Martínez-Miranda MM, et al (2021) Physicochemical treatments for the reduction of aflatoxins and *Aspergillus niger* in corn grains (*Zea mays*). J Sci Food Agric 101:3707–3713. https://doi.org/10.1002/jsfa.11001
58. Miller JD (2002) Relevance of mycotoxins in the food supply and in the built environment: Introduction. In: DeVries JW, Trucksess MW, Jackson LS. (eds) Mycotoxins and Food Safety. Advances in Experimental Medicine and Biology, vol 504. Springer, Boston, MA. https://doi.org/10.1007/978-1-4615-0629-4_1.

# 2.10 Aflatoxigenic Fungi and Tuber Crops

*Harita Parmar and Roshni Patel*

## INTRODUCTION

Root and tuber crops have held a significant role in human nutrition since the earliest days of civilization, offering indispensable carbohydrates, proteins, and vitamins essential for our well-being[1]. Tuber crops are plants that yield starchy roots, tubers, rhizomes, corms, and stems. Their significance stems from a myriad of agronomic advantages, allowing them to thrive in a wide range of soil and environmental conditions while demonstrating remarkable adaptability to various farming systems[2]. Beyond their resilience, these crops stand out as highly efficient sources of edible carbohydrates, surpassing many other food crops in nutritional value[3]. In this section, we further explore the pivotal role played by tuber crops in particular, as they emerge as the second most crucial group of crop plants in providing food energy[3]. As agricultural landscapes evolve and agro-ecologies diversify, their importance continues to grow. This surge is propelled by several factors, including the expanding global population, rapid urbanization trends, increasing consumer awareness of dietary choices, and the escalating prices of animal products. This continuity of importance is evidenced by their consistent inclusion in dietary patterns worldwide. According to Abe's (2019) report, the global production figures for potato (*Solanum tuberosum*), sweet potato (*Ipomoea batatas*), and cassava (*Manihot esculenta*) were approximately 368, 91, and 277 megatons, respectively, highlighting their immense importance in sustaining populations worldwide[4]. The potato is cultivated as an annual crop. With proper irrigation and grown in a temperate climate, potato plants exhibit high productivity, often yielding over 40 tons per hectare within 4 months after planting[1]. It boasts a rich nutritional profile, abundant in carbohydrates, proteins, minerals, and vitamins such as B6 and C. It has risen to become the world's third most important food crop for human consumption, trailing only behind wheat and rice. It holds a central position in the diets of developed countries, contrasting with its consumption in developing nations. The daily energy intake from potatoes per individual in developed countries amounts to 130 kcal/day, whereas in developing countries, it stands at 41 kcal/day[5]. Additionally, in certain European countries like the Netherlands, Norway, and Finland, potatoes contribute to approximately 10% of the total folate intake[6]. Approximately two-thirds of the potato harvest is consumed fresh, with over 1.3 billion people worldwide relying on potatoes as a staple food, with consumption exceeding 50 kg per person per year in regions such as India and China[7]. Cassava is Sub-Saharan Africa's second most crucial staple food crop. It has the potential for remarkable yields, ranging from 50 to 82 metric tons per hectare. Notably, it thrives in marginal soil conditions where other crops struggle to yield economically viable results. Its highest per capita consumption, reaching approximately 800 grams per person per day, is observed in sub-Saharan Africa, where it serves as the primary energy source for nearly 40% of the population[8]. Sweet potatoes, globally, thrive in tropical, subtropical, and warm temperate regions across the world. They boast a unique resilience, capable of year-round cultivation under suitable climatic conditions with minimal risk of complete crop loss even in adverse weather. This resilience has earned them the title of an "insurance crop." Notably, their exceptional features and nutritional value have led the National Aeronautics and Space Administration (NASA) to select them as a candidate crop for cultivation and incorporation into the diets of astronauts during space missions[9]. Yams are not only a staple food, but also have important cultural and social functions. Taro (*Colocasia esculenta*) has

DOI: 10.1201/9781003487043-16

also been recognized as a promising smart food in the category of underutilized roots and tuber crops, with the potential to address chronic malnutrition and hidden hunger in Asia[10]. The native inhabitants of Nigeria's middle belt perceive Polynesian arrowroot tubers (*Tacca leontopetaloides*) as a divine blessing because they naturally grow abundantly in the wild[11]. These crops also serve as important income sources in rural and marginalized areas, prized for their ability to yield stable harvests even under adverse conditions such as severe drought and salinity, circumstances where other crops may falter.

They serve as staple foods, cash crops, and increasingly, fodder for livestock, raw materials for industrial processes, and ingredients for human consumption[12]. Moreover, these crops offer a virtually unlimited source of macronutrients, micronutrients, and phytonutrients, all of which exert positive effects on human health. Macronutrients comprise carbohydrates, lipids, and proteins. Omega-3 fatty acids have positive effects on cardiovascular health, brain function, and anti-inflammatory properties. Bioactive peptides are effective for promoting human health, affecting systems such as digestion, hormones, heart health, immunity, and the nervous system. They hold promise for enhancing the treatment of numerous diseases and disorders, ultimately leading to better health and overall well-being. Micronutrients include vitamins and minerals, essential to sustain healthy body function. Phytochemicals are a heterogeneous group of minor components, including terpenes, alkaloids, phenolics, and organosulfur compounds[13]. Flavonoids, lignans, stilbenes, and phenolic acids classified as polyphenols act as defenders of the body against ultraviolet radiation, oxidative stress, and pathogens. Incorporating dietary polyphenols into one's diet can lead to improvements in lipid profiles, blood pressure, insulin resistance, and systemic inflammation. Particularly, flavonoids have the potential to alter risk factors associated with chronic human conditions, including unresolved inflammation, high blood pressure, dyslipidemia, insulin resistance, and other oxidative stress-related ailments(Table 2.10.1).

## MYCOTOXINS

Mycotoxins are fungal metabolites produced on various organic materials, which can cause severe or adverse effects when consumed by humans or animals. These effects can include vomiting, weight loss, tumor growth, and even death. Many of the toxins produced by fungi are particularly prevalent in grain crops. Most mycotoxins target specific organs in the body, posing risks to the blood, kidneys, skin, or central nervous system, with some identified as carcinogens[29]. These toxins are mainly hazardous secondary metabolites synthesized by fungi in agricultural commodities prone to mold contamination and can be categorized based on their fungal origin, chemical composition, and biological impact. Mycotoxicosis represents a multidisciplinary concern, requiring the involvement of professionals from various fields such as environmental science, medicine, veterinary medicine, agricultural engineering, plant pathology, analytical chemistry, and agronomy[29]. Consuming foods tainted with mycotoxins can result in a condition known as mycotoxicosis, characterized by various diseases or disorders.

Mycotoxigenic fungi are commonly categorized into two groups: Field fungi and Storage fungi. Field fungi, including species from *Fusarium, Alternaria, Cladosporium, and Helminthosporium* genera, typically infect crops before harvest, thriving in high moisture conditions ranging from 18% to 30%. Conversely, storage fungi, comprising genera such as *Aspergillus, Eurotium*, *Penicillium*, and *Mucor*, are adept at thriving in dry environments with moisture content between 14% and 16%, thus preferring post-harvest infection of food crops[30]. Aflatoxins, trichothecenes, fumonisins, ochratoxins, and zearalenone are prevalent in food grains and various crucial agricultural commodities. Among these toxins, aflatoxins are particularly concerning due to their exceptionally high toxicity, carcinogenic properties, and frequent occurrence across various food and feed products[30].

Mycotoxins, such as those produced by *Claviceps purpurea* causing ergot of rye, have long been recognized for their severe impact. Throughout history, outbreaks of diseases associated with these mycotoxins, known as 'St. Anthony's Fire' or 'Holy Fire' in medieval Europe, resulted in numerous

**TABLE 2.10.1**
**Presents a Comprehensive Overview of the Nutraceutical Values, and Essential Medicinal Properties**

| Sr. No | Tuber Crop | Nutritional Qualities | Medicinal Properties | References |
|---|---|---|---|---|
| 1 | *Amorphophallus paeoniifolius* Common name: Elephant foot yam | Tubers contain 18.0% starch, 1%–5% protein, and up to 2% fat. Leaves contain 2%–3% protein, 3% carbohydrates, and 4%–7% crude fiber, Omega-3 fatty acids, and vitamin B6 | Antioxidant, wound healer, immune booster, cardio-protective effects | 14,15 |
| 2 | *Colocasia esculenta L.* Common name: Taro | Starch (70–80 g/100 g dry taro), fiber (0.8%), ash (1.2%), thiamine, riboflavin, iron, phosphorus, and zinc, vitamin B6, vitamin C, niacin, potassium, copper, manganese, and β-carotene | Antioxidant, anti-carcinogenic, anti-compulsive, anti-hyperglycemic, anti-hypertensive, anti-inflammatory, hepatoprotective, immunoprotective, and neuroprotective properties | 14,10 |
| 3 | *Pachyrhizus erosus* (L) Urban Common name: Yam bean | The root contains 32% soluble sugars and 15% starch. Seeds contain high levels of oil (20%–28%), protein (23%–34%), and phytochemicals such as flavonoids, retinoids, and phenyl furanocoumarin. | Antitumor activity and antifungal activity, Anti-anxiety, and Anti-aggressive activity. | 14 |
| 4 | *Plectranthus rotundifolius* Common name: Chinese potato | Rich in minerals like calcium and iron and certain vitamins, including thiamine, riboflavin, niacin, and ascorbic acid | Antifungal, Anti-inflammatory | 16 |
| 5 | *Alocasia macrorrhiza* Common name: Giant Taro | Protein, 0.1 grams of lipid fat, 23 grams of carbohydrates, and 1.9 grams of dietary fiber. Calcium, 0.8 mg of Iron, 52 mg of magnesium, 267 mg of potassium, 30 mg of sodium, and 1.6 mg of zinc, along with trace amounts of vitamins | Antioxidant, anti-inflammatory, antimicrobial, antidiarrheal, free radical scavenging, and antiprotozoal properties | 17 |
| 6 | *Tacca leontopetaloides* Common name: Tacca | Flavonoids, phenols, alkaloids, tannins, coumarins, polysaccharides, glycosides, gums, terpenes, and terpenoids | Analgesic, antipyretic, and anti-inflammatory activities; leprosy, hemorrhagic diathesis, and skin diseases | 11 |
| 7 | *Dioscorea sp.* Common name: Yam species | Phenolic compounds like caffeic acid, chlorogenic acid, p-hydroxybenzaldehyde, and methyl ester of protocatechuic acid (methyl protocatechuate), etc. | Rheumatoid arthritis, stomach cramps, and pain from gallstones, menstrual cramps, and discomfort, relieves labor pains; dysmenorrhea, and intestinal colic, treats scabies and boils | 14,18 |
| 8 | *Psophocarpus tetragonolobus* Common name: Winged beans | Proteins, B-complex vitamins contain thiamin, pyridoxine (vitamin B6), niacin, and riboflavin). | Antimicrobial activities, cures skin sores such as boils and ulcers | 14,19 |
| 9 | *Smallanthus sonchifolius* Common name: Yacón | Minerals are calcium and potassium, 46% inulin, 11%–17% protein | Cytoprotective and anti-hyperglycemic active | 20 |

*(Continued)*

**TABLE 2.10.1** *(Continued)*
**Presents a Comprehensive Overview of the Nutraceutical Values, and Essential Medicinal Properties**

| Sr. No | Tuber Crop | Nutritional Qualities | Medicinal Properties | References |
|---|---|---|---|---|
| 10 | *Solanum tuberosum* Common name: Potatoes | Carbohydrates, potassium, and ascorbic acid | Folate protects against oxidative breakdown | 6 |
| 11 | *Ipomoea batatas* Common name: Sweet Potato | Rich in dietary fiber, minerals, carotenoids, vitamins, and bioactive compounds such as phenolic acids and anthocyanins | Improves vision, wound healer, and has anti-ulcer activity | 21,22,23 |
| 12 | *Manihot esculenta* Common name: Cassava | Carbohydrate content, vitamins and minerals, proteins, fiber, and essential amino acids, | Diabetes, celiac diseases, bone and neurological health, cardiovascular diseases, prostate problems, allergies, preventing anemia, etc. | 24,25 |
| 13 | *Zingiber officinale* Common name: Ginger | Phenolic compounds, terpenes, vitamins, and minerals | Antioxidant, anti-inflammatory, antidiabetic, and antimicrobial effects against a few gram-positive and gram-negative bacteria | 26,27 |
| 14 | *Dolomiaea costus* Common name: Costus | Bioactive components, essential oils | Treatment of diabetes, abdominal pains, gallbladder pain, chest pains, liver problems and jaundice, fever, skin diseases, worm infection, leprosy, burning sensation, bronchitis, anthelmintic, astringent, expectorant properties, antibacterial and antifungal activities | 14 |
| 15 | *Canna edulis* Common name: Achira | Rhizomes contain 70% carbohydrate, 12% protein, and 22% fiber on a dry-weight basis | Antimicrobial, anti-oxidant, gastro protectant | 28 |
| 16 | *Oxalis tuberosa* Common name: Oca | Carbohydrates | Anti-oxidant, anti-hyperglycemic active | 20 |

deaths, underscoring the gravity of the issue. In India, an outbreak of enteric ergotism has been linked to ergot consumption[31]. These outbreaks typically arise from the consumption of contaminated cereals grown under drought conditions and stored in humid environments conducive to mold growth and proliferation. The toxicity of mycotoxins can range from acute to chronic, influenced by factors such as the specific type of toxin and dosage, as well as the age and susceptibility of individuals based on their immune system[31]. Nevertheless, even in industrialized nations, approximately 4.5 billion individuals are subjected to unregulated and unmonitored levels of mycotoxins, largely attributable to the importation of contaminated food[32]. As per the World Health Organization, more than 500 million individuals, primarily in sub-Saharan Africa, are subjected to hazardous levels of mycotoxin contamination. Childhood exposure to such unsafe levels leads to elevated mortality and morbidity rates. The prevalence of mycotoxin contamination, particularly in Africa, poses a significant barrier to achieving 15 out of the 17 Sustainable Development Goals outlined by the United Nations[33]. This revised estimate challenges the previously widely cited figure of 25%, as attributed to the Food and Agricultural Organization (FAO) of the United Nations.

## AFLATOXIGENIC FUNGUS

*Aspergillus* species are groups of molds widely distributed all over the world. They can easily contaminate foodstuffs and animal feeds when conditions are favorable, especially in tropical and subtropical countries where relative humidity and temperature are very high. Several *Aspergillus* species grew in foods with low water activity and produced mycotoxins. *Aspergillus* is an important mold in food processing and industrial food[34].

*A. flavus* is common in soil, air, and decaying plant matter. Aflatoxin production and contamination can occur in the field, during transportation, or in storage. *A. flavus*, a significant producer of aflatoxin, demonstrates a wide host range as both an opportunistic pathogen and a saprobe and is also rarely produced by *A. nominus* and *A. pseudotamari*. It exists in soil in the form of conidia or sclerotia and as mycelia within plant tissues. Although it is isolated across various climate zones, it is most commonly found in warm climate zones between latitudes of 16° and 35° and is less prevalent above latitudes of 45°[35]. *A. flavus* exhibits a complex morphology and is categorized into two groups based on the size of its sclerotia: L strains (Group I), characterized by sclerotia larger than 400 mm in diameter, and S strains (Group II), with sclerotia smaller than 400 mm in diameter[36]. Both *A. flavus* S and L strains produce aflatoxins B1 and B2, while S strains can additionally produce aflatoxins G1 and G2. S strains of *A. flavus* are found distributed across the globe. It has been reported in India across a range of consumables, including cereals, oilseeds, spices, vegetables, dried fruits, pulses, areca nuts, coconuts, and fast foods[37].

Under harsh environmental conditions, sclerotia endure within the soil and generate conidia, potentially giving rise to ascospores[36]. The dissemination of conidia through the air is linked to the infection of crops above the ground, such as maize and tree nuts, whereas the spread through soil movement and rain splashes may play a more significant role in the infection of peanut and cotton seeds[38]. The retrieval of viable conidia from buried sclerotia can persist for up to three years. This process contributes to an escalation in population during hot and dry weather. Colonization is facilitated not only by favorable conditions like warmth and dryness, but also by damage from insects and birds, which create entry points for the fungus[39]. Aflatoxin contamination primarily impacts agricultural produce in African and Southeast Asian nations due to their specific climatic environments[40]. These regions, characterized by hot and humid tropical or subtropical climates with annual rainfall exceeding 700 mm, offer ideal conditions for mold growth. Post-harvest products stored in environments with high relative humidity and limited aeration further promote fungal proliferation[41]. Nevertheless, the escalating effects of global warming are extending the threat of aflatoxin contamination to regions previously unaffected, including Europe[42]. According to research conducted by the FAO, around a quarter of crops are impacted in Europe due to climate change[43]. Droughts also served as a significant catalyst for aflatoxins biosynthesis, particularly noticeable in tropical regions where acute aflatoxicosis from consuming compromised crops is frequent. Central Europe faced severe droughts in both 2018 and 2019, coupled with unprecedented summer heat across the continent[44]. The study found that *A. flavus*, the most abundant fungal species identified in cassava (*Manihot esculenta*) and yams (*Dioscorea* spp.), had the highest colony-forming units per gram (cfu/g) recorded at 8950 and 6030, respectively, among others[45]. In Nigeria, aflatoxin was found in detectable amounts in 80.9% of *Zingiber officinale* (Ginger) samples, ranging from 1.92 to 10.72 μg/kg[27]. Another study also reported that 80% of whole, unpeeled ginger samples exceeded Tanzania's legal limit of 10 μg/kg aflatoxin[26]. Dried yam chips (*Dioscorea rotundata)* that were procured from the Nigerian market revealed that 51% of them surpassed the internationally accepted threshold for aflatoxin levels (20 ppb)[46]. The Food and Drug Administration (FDA) in the United States sets regulations on permissible aflatoxin contamination, allowing up to 20 parts per billion (ppb) in crops or 0.5 ppb in milk for human consumption[35].

Similar aflatoxin regulatory laws exist in over 100 countries, but acceptable threshold levels vary. For instance, in Europe, peanuts with more than 2 ppb of aflatoxin B1 or 4 ppb of total aflatoxins are rejected, whereas any food exceeding 30 ppb is prohibited in India[47]. The connection between mycotoxins and the duration of storage was also examined by Manjula (2009), who revealed aflatoxin B1 ranged from 0.3 to 4.4 ppb in cassava chip flour with elevated contamination levels after 4 months of storage. Testing found Aflatoxin B1 levels at various intervals in Gbodo (derived from fermented yam), and the concentration 32.33 μg/kg compared to 25.17 μg/kg after 6 months and one month of storage, respectively[17]. However, under unfavorable storage conditions, it can proliferate and generate aflatoxin in almost any crop seed. Additionally, it can exhibit pathogenicity towards animals and insects.

## AFLATOXIN HEALTH THREATS

Aflatoxins are carcinogenic in various animal species, including mice, fish, rats, marmosets, ducks, tree shrews, and monkeys. In humans, *A. flavus* is an opportunistic pathogen, particularly affecting immunocompromised individuals. Its economic impact is extensive because it affects plant health and contributes to storage rot in various crops[34]. At the same time; its production of the regulated mycotoxin aflatoxin B1 further complicates matters. As human pathogens, *Aspergillus* species, including *A. flavus*, have gained significance, especially considering the heightened susceptibility of immunocompromised individuals to infection. Aflatoxin B1 stands out as one of the most potent naturally occurring carcinogens, with notable concerns regarding its potential use as a biological weapon. Ingestion of aflatoxin has been linked to hepatocellular carcinoma, a form of liver cancer[34]. It is classified as a Group 1 carcinogen in the International Agency for Research on Cancer (IARC) classification of carcinogenic substances[48].

In the early stages of aflatoxicosis, initial signs and symptoms include fever, loss of appetite, and general discomfort. As the disease advances, affected animals may experience abdominal pain, vomiting, and enlargement of the liver (hepatomegaly). It also poses significant health risks, including aflatoxicosis, cancer, infertility, hepatocellular carcinoma, liver cirrhosis, nephropathy, immunodeficiency, anemia, and stunting, underweight, and nutritional interference. The immunosuppressive effect of aflatoxin B1 (AFB1) stems from its ability to reduce the functionality of antigen-presenting cells (APCs), specifically dendritic cells[49].

Its presence in the diet poses a significant risk factor for hepatocellular carcinoma. Among them, aflatoxin B1 stands out as a genotoxic hepatocarcinogen, presumably initiating cancer by inducing DNA adducts, which lead to genetic alterations in liver cells[50]. The underlying molecular mechanism of AFB1-induced mutagenesis in hepatocellular carcinoma reveals that the lesion is remarkably mutagenic, resulting in a high replication error rate of 97%. The primary base substitution observed is a G-to-T transversion[51]. This exposure during pregnancy can directly or indirectly contribute to maternal anemia and adverse birth outcomes such as low birth weight and impaired fetal growth[52].

In parts of the world where aflatoxins are not regulated, outbreaks of aflatoxicosis and subsequent fatalities among the human population occur periodically. As reported by the African Union Commission (AUC) through the Partnership for Aflatoxin Control in Africa (PACA), Africa suffers a loss exceeding US$100 million annually due to cases of cancer linked to aflatoxin exposure[31]. Experimental studies with animals have documented prenatal effects, indicating that aflatoxin B1 exhibits teratogenic properties. It emerges as a potent teratogen due to its capacity to bind DNA, consequently impeding protein synthesis. Specifically, skeletal anomalies have been observed in the offspring of animals exposed to aflatoxin during pregnancy[53]. It is linked to roughly 25% of liver cancer cases worldwide, equating to approximately 172,000 new liver cancer diagnoses annually[54].

## POSSIBLE MITIGATION STRATEGIES

Various methods, both physical and chemical, such as field management, physical separations, moisture control, and the use of fungicides and chemical absorbents, have been employed to control aflatoxin and the molds that produce it. However, these approaches often prove inefficient due to factors such as nutritional loss in processed foods, difficulty in eliminating toxic compound residues, or the emergence of resistant pathogen strains. Nevertheless, once agro-foods and feeds are contaminated, complete removal of contaminants like aflatoxins becomes unfeasible. The lack of resistant genotypes to mycotoxin contamination underscores the need for strategies to minimize aflatoxin presence, including adopting less susceptible crop varieties, employing biological control methods, integrating risk prediction systems, and improving pre- and post-harvest management practices to mitigate losses of essential dietary staples. Biological control methods have been recognized as environmentally friendly and safer alternatives[55], typically implemented during pre-and/or post-harvest stages. Post-harvest strategies primarily target the elimination of aflatoxin. Utilizing essential oils, such as peppermint oil, as safe alternatives to synthetic fungicides offers a cost-effective approach to reducing aflatoxins in peanut seeds[56,57]. Additionally, secondary metabolites from various microorganisms, including nontoxigenic *Aspergillus*, *Trichoderma*, *Penicillium* spp., and *yeast strains*, have shown promise as biocontrol agents against aflatoxin-producing fungi, highlighting their potential in biological control[58]. Financial and technical assistance from economically advanced nations and international organizations is vital, particularly to support education, research, and initiatives aimed at reducing contamination in products[59].

## REFERENCES

1. Zierer W, Rüscher D, Sonnewald U, Sonnewald S. Tuber and tuberous root development. *Annu Rev Plant Biol.* 2021;72:551–580. doi:10.1146/annurev-arplant-080720-084456
2. Chandrasekara A, Josheph Kumar T. Roots and tuber crops as functional foods: A review on phytochemical constituents and their potential health benefits. *Int J Food Sci.* 2016;2016(1):3631647. doi:10.1155/2016/3631647
3. Nanbol KK, Namo O. The contribution of root and tuber crops to food security: A review. *J Agric Sci Technol B.* 2019;9(10.17265):2161–6264.
4. Abe M, Kosaka S, Shibuta M, et al. Transient activity of the florigen complex during the floral transition in *Arabidopsis thaliana. Development.* 2019;146(7):dev171504. doi:10.1242/dev.171504
5. Burlingame B, Mouillé B, Charrondière R. Nutrients, bioactive non-nutrients and anti-nutrients in potatoes. *J Food Compos Anal.* 2009;22(6):494–502. doi:10.1016/j.jfca.2009.09.001
6. Hale AL, Reddivari L, Nzaramba MN, Bamberg JB, Miller JC. Interspecific variability for antioxidant activity and phenolic content among *Solanum* species. *Am J Potato Res.* 2008;85(5):332–341. doi:10.1007/s12230-008-9035-1
7. Devaux A, Goffart JP, Kromann P, Andrade-Piedra J, Polar V, Hareau G. The potato of the future: Opportunities and challenges in sustainable Agri-food systems. *Potato Res.* 2021;64(4):681–720. doi:10.1007/s11540-021-09501-4

8. Adebayo WG. Cassava production in Africa: A panel analysis of the drivers and trends. *Heliyon.* 2023;9(9). doi:10.1016/j.heliyon.2023.e19939
9. Bovell-Benjamin AC. Sweet Potato: A Review of Its Past, Present, and Future Role in Human Nutrition. In: *Advances in Food and Nutrition Research.* Vol 52. Academic Press; 2007:1–59. doi:10.1016/S1043-4526(06)52001-7
10. Aditika, Kapoor B, Singh S, Kumar P. Taro (*Colocasia esculenta*): Zero wastage orphan food crop for food and nutritional security. *South Afr J Bot.* 2022;145:157–169. doi:10.1016/j.sajb.2021.08.014
11. Amadi CO, Olojede OA, Ironkwe A, Ebeniro CN, Nwokocha CC, Amadi G. Prospects of domestication of Polynesian arrowroot (*Tacca leontopetaloides* L.) in Nigeria. *Niger Agric J.* 2021;52(3):184–196.
12. Ferraro V, Piccirillo C, Tomlins K, Pintado ME. Cassava (*Manihot esculenta* Crantz) and Yam (*Dioscorea spp.*) crops and their derived foodstuffs: Safety, security and nutritional value. *Crit Rev Food Sci Nutr.* Published online December 9, 2016;56:2714–2727. Accessed November 8, 2024. https://www.tandfonline.com/doi/abs/10.1080/10408398.2014.922045
13. Kussmann M, Abe Cunha DH, Berciano S. Bioactive compounds for human and planetary health. *Front Nutr.* 2023;10:1193848. doi:10.3389/fnut.2023.1193848
14. Chauhan V, Mallick S, Pati K, Arutselvan R, Nedunchezhiyan, M. Status and importance of underexploited tuber crops in relation to nutritional security and economic prosperity. *Compendium for Winter School on "Unexpected Vegetables: Unexplored Treasure Trove for Food, Nutritional and Economic Security*; 2022:246–264.
15. Singh B, Kaur S, Kaur A. Bioactive Chemicals and Biological Activities of Elephant Foot Yam (*Amorphophallus Paeoniifolius* (Dennst.) Nicolson). In: Murthy HN, Paek KY, Park SY, eds. *Bioactive Compounds in the Storage Organs of Plants.* Springer Nature Switzerland; 2023:1–23. doi:10.1007/978-3-031-29006-0_8-1
16. Madhavan M, Vishnu VR, Shanavas S, Jyothi DAN. Phenolic content and in vitro bioactivities of Chinese potato (*Plectranthus rotundifolius*) tuber extracts. *J Root Crops.* 2021;47(1 & 2):16–23.
17. Müller JV, Guzzon F. The forgotten giant of the Pacific: A review on giant taro (*Alocasia macrorrhizos* (L.) G.Don). *Genet Resour Crop Evol.* 2024;71(1):519–527. doi:10.1007/s10722-023-01664-y
18. Epping J, Laibach N. An underutilized orphan tuber crop—Chinese Yam: A review. *Planta.* 2020;252(4):58. doi:10.1007/s00425-020-03458-3
19. Bassal H, Merah O, Ali AM, Hijazi A, El Omar F. *Psophocarpus tetragonolobus*: An underused species with multiple potential uses. *Plants.* 2020;9(12):1730. doi:10.3390/plants9121730
20. Eugenia Jimenez M, Rossi A, Sammán N. Health properties of oca (*Oxalis tuberosa*) and yacon (*Smallanthus sonchifolius*). *Food Funct.* 2015;6(10):3266–3274. doi:10.1039/C5FO00174A
21. van Jaarsveld PJ, Marais DW, Harmse E, Nestel P, Rodriguez-Amaya DB. Retention of $\beta$-carotene in boiled, mashed orange-fleshed sweet potato. *J Food Compos Anal.* 2006;19(4):321–329. doi:10.1016/j.jfca.2004.10.007
22. Chimkode R, Patil MB, Jalalpure SS. Wound healing activity of tuberous root extracts of Ipomoea batatas. *Adv Pharmacol Toxicol.* 2009;10(3):69–72.
23. Panda V, Sonkamble M. Anti-ulcer activity of Ipomoea batatas tubers (sweet potato). *Funct Foods Health Dis.* 2012;2:48–61. doi:10.31989/ffhd.v2i3.99
24. Blagbrough IS, Bayoumi SAL, Rowan MG, Beeching JR. Cassava: An appraisal of its phytochemistry and its biotechnological prospects. *Phytochemistry.* 2010;71(17):1940–1951. doi:10.1016/j.phytochem.2010.09.001
25. Tsige TZ, Basa B, Herago T. Medicinal, nutritional and anti-nutritional properties of cassava (*Manihot esculenta*): A review. *Acad J Nut.* Published online January 1, 2019;8:34–46.
26. Makundi I, Ngoyinde N. Assessment of aflatoxin levels in whole and peeled ginger marketed in Dar es Salaam, Tanzania. *J Food Saf Hyg.* 2023;9(4):299–305. doi:10.18502/jfsh.v9i4.15005
27. Negedu A, Apeh DO, Ogara IM, et al. Natural occurrence of major mycotoxins across the Ginger Value Chain in Nigeria. *J Food Qual Hazards Control.* Published online September 13, 2022;9:68–63. doi:10.18502/jfqhc.9.2.10642
28. Najini R, Bachri MS, Yuliani S. The gastroprotective effects of canna edulis ker. tuber starch on peptic ulcer and the histopathological profile of rat stomach. *Pharmaciana.* 2018;8(1):43.
29. Imran M, Cao, S., Wan, S.F., Chen, Z., Saleemi, M.K., Wang, N., Naseem, M.N., Munawar. Mycotoxins - A global one health concern: A review. *Agrobiol Rec.* 2020;2:1–16. https://doi.org/10.47278/journal.abr/2020.008
30. Gurikar C, Shivaprasad DP, Sabillón L, Nanje Gowda NA, Siliveru K. Impact of mycotoxins and their metabolites associated with food grains. *Grain Oil Sci Technol.* 2023;6(1):1–9. doi:10.1016/j.gaost.2022.10.001

31. Adeyeye SAO. Fungal mycotoxins in foods: A review. Yildiz F, ed. *Cogent FoodAgric.* 2016;2(1):1213127. doi:10.1080/23311932.2016.1213127
32. Marroquín-Cardona AG, Johnson NM, Phillips TD, Hayes AW. Mycotoxins in a changing global environment – A review. *Food Chem Toxicol.* 2014;69:220–230. doi:10.1016/j.fct.2014.04.025
33. Eskola M, Kos G, Elliott CT, Hajšlová J, Mayar S, Krska R. Worldwide contamination of food-crops with mycotoxins: Validity of the widely cited 'FAO estimate' of 25%. *Crit Rev Food Sci Nutr.* 2020;60(16):2773–2789. doi:10.1080/10408398.2019.1658570
34. Adejumo TO, Adejoro DO. Incidence of aflatoxins, fumonisins, trichothecenes and ochratoxins in Nigerian foods and possible intervention strategies. *Food Sci Qual Manag.* 2014;31:127–147.
35. Klich MA. Aspergillus flavus: The major producer of aflatoxin. *Mol Plant Pathol.* 2007;8(6):713–722. doi:10.1111/j.1364-3703.2007.00436.x
36. Horn BW. Ecology and population biology of aflatoxigenic fungi in soil. *J Toxicol Toxin Rev.* 2003;22 (2–3):351–379. doi:10.1081/TXR-120024098
37. Singh U, Gupta S, Gupta M. A review study on biological ill effects and health hazards of aflatoxins. *Asian J Adv Med Sci.* 2021;3(1)1–8..
38. Cotty PJ. Cotton Seed Losses and Mycotoxins. In: Kirkpatrick TL, Rothrock CS, eds. *Compendium of Cotton Diseases.* Infectious Diseases; Published online 2001.
39. Amaike S, Keller NP. Aspergillus flavus. *Annu Rev Phytopathol.* 2011;49:107–133. doi:10.1146/annurev-phyto-072910-095221
40. Jallow A, Xie H, Tang X, Qi Z, Li P. Worldwide aflatoxin contamination of agricultural products and foods: From occurrence to control. *Compr Rev Food Sci Food Saf.* 2021;20(3):2332–2381. doi:10.1111/1541-4337.12734
41. Benkerroum N. Aflatoxins: Producing-molds, structure, health issues and incidence in Southeast Asian and Sub-Saharan African Countries. *Int J Environ Res Public Health.* 2020;17(4):1215. doi:10.3390/ijerph17041215
42. Leggieri MC, Toscano P, Battilani P. Predicted aflatoxin B1 increase in Europe due to climate change: Actions and reactions at global level. *Toxins.* 2021;13(4):292. doi:10.3390/toxins13040292
43. Moretti A, Pascale M, Logrieco AF. Mycotoxin risks under a climate change scenario in Europe. *Trends Food Sci Technol.* 2019;84:38–40. doi:10.1016/j.tifs.2018.03.008
44. Hari V, Rakovec O, Markonis Y, Hanel M, Kumar R. Increased future occurrences of the exceptional 2018–2019 Central European drought under global warming. *Sci Rep.* 2020;10(1):12207. doi:10.1038/s41598-020-68872-9
45. Gnonlonfin GJB, Hell K, Fandohan P, Siame AB. Mycoflora and natural occurrence of aflatoxins and fumonisin B1 in cassava and yam chips from Benin, West Africa. *Int J Food Microbiol.* 2008;122(1): 140–147. doi:10.1016/j.ijfoodmicro.2007.11.047
46. Bankole S, Adebanjo A. Aflatoxin contamination of dried yam chips marketed in Nigeria. *Trop Sci.* 2003;43(4):201–203. doi:10.1002/ts.119
47. Manjula K, Hell K, Fandohan P, Abass A, Bandyopadhyay R. Aflatoxin and fumonisin contamination of cassava products and maize grain from markets in Tanzania and republic of the Congo. *Toxin Rev.* 2009;28(2–3):63–69. doi:10.1080/15569540802462214
48. Ostry V, Malir F, Toman J, Grosse Y. Mycotoxins as human carcinogens—the IARC monographs classification. *Mycotoxin Res.* 2017;33(1):65–73. doi:10.1007/s12550-016-0265-7
49. Mehrzad J, Devriendt B, Baert K, Cox E. Aflatoxin B1 interferes with the antigen-presenting capacity of porcine dendritic cells. *Toxicol In Vitro.* 2014;28(4):531–537. doi:10.1016/j.tiv.2013.11.015
50. Hamid AS, Tesfamariam IG, Zhang Y, Zhang ZG. Aflatoxin B1-induced hepatocellular carcinoma in developing countries: Geographical distribution, mechanism of action and prevention. *Oncol Lett.* 2013;5(4):1087–1092. doi:10.3892/ol.2013.1169
51. Lin YC, Li L, Makarova AV, Burgers PM, Stone MP, Lloyd RS. Molecular basis of aflatoxin-induced mutagenesis—Role of the aflatoxin B1-formamidopyrimidine adduct. *Carcinogenesis.* 2014;35(7):1461–1468. doi:10.1093/carcin/bgu003
52. Smith LE, Prendergast AJ, Turner PC, Humphrey JH, Stoltzfus RJ. Aflatoxin exposure during pregnancy, maternal anemia, and adverse birth outcomes. *Am J Trop Med Hyg.* 2017;96(4):770. doi:10.4269/ajtmh.16-0730
53. El-Nahla SM, Imam HM, Moussa EA, Ibrahim AM, Ghanam AR. Teratogenic effects of aflatoxin in rabbits (*Oryctolagus cuniculus*). *J Vet Anat.* Published online October 1, 2013;6:67–85. doi:10.21608/jva.2013.45024
54. Liu Y, Wu F. Global burden of aflatoxin-induced hepatocellular carcinoma: A risk assessment. *Environ Health Perspect.* 2010;118(6):818–824. doi:10.1289/ehp.0901388

55. Zhu Y, Hassan YI, Watts C, Zhou T. Innovative technologies for the mitigation of mycotoxins in animal feed and ingredients—A review of recent patents. *Anim Feed Sci Technol.* 2016;216:19–29. doi:10.1016/j.anifeedsci.2016.03.030
56. Kodape A, Lama A, Vivek Babu CS. Metagenomic insights of fungal diversity of peanuts under storage conditions and mitigation of aflatoxigenic fungi through competitive exclusion and phytochemicals. *Food Biosci.* 2024;58:103711. doi:10.1016/j.fbio.2024.103711
57. Abd El-Hack ME, Kamal M, Altaie HAA, et al. Peppermint essential oil and its nano-emulsion: Potential against aflatoxigenic fungus *aspergillus flavus* in food and feed. *Toxicon.* 2023;234:107309. doi:10.1016/j.toxicon.2023.107309
58. Ren X, Zhang Q, Zhang W, Mao J, Li P. Control of aflatoxigenic molds by antagonistic microorganisms: Inhibitory behaviors, bioactive compounds, related mechanisms, and influencing factors. *Toxins.* 2020;12(1):24. doi:10.3390/toxins12010024
59. Sowley ENK. Aflatoxins: A silent threat in developing countries. *Afr J Biotechnol.* 2016;15(35):1864–1870. doi:10.5897/AJB2016.15305

# 2.11 Aflatoxigenic Fungi and Beverage Crops

*Priyanka Dahiya, Simran Rani, Chetna Rathi, Vajiha Bhanu Habeeb Mohammed, Amita Suneja Dang, and Pooja Suneja*

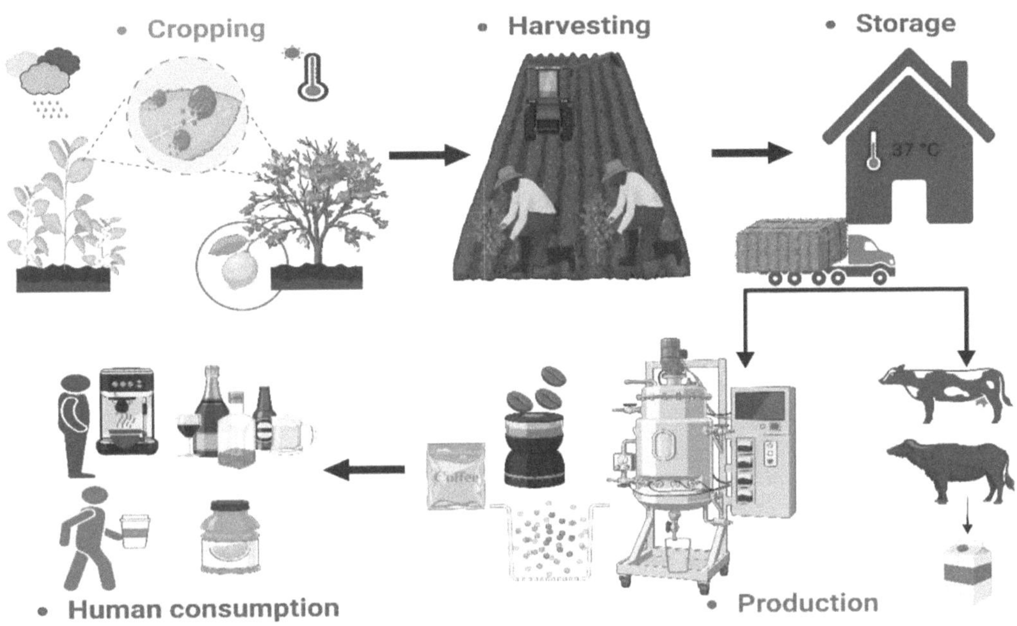


## MYCOTOXINS: THE LETHAL SIDE OF FUNGAL METABOLISM

Secondary metabolites synthesized by fungi are utilized for different purposes, such as antibiotics, anticancer agents, and combatting infectious diseases (1). However, secondary metabolites, such as mycotoxins, are low molecular weight harmful substances that negatively impact animal and human health. Hepatotoxicity (liver damage), nephrotoxicity (kidney damage), neurotoxicity (nervous system damage), and immunotoxicity (immune system suppression) are among the reported adverse effects of aflatoxins (2). Mycotoxins are synthesized by aerobic, mycelial fungi mainly belonging to the genera *Aspergillus, Fusarium,* and *Penicillium* (3).

## AFLATOXINS AND THEIR HARMFUL EFFECTS

Among mycotoxins, aflatoxins are primarily synthesized by the genera *Aspergillus*, specifically via species *A. parasiticus* and *A. flavus.* The fungal infection of crops occurs in the field and synthesizes aflatoxins at pre- or post-harvest phases. Aflatoxin production by fungi is closely linked to their development. The fungi's optimal growth occurs around 35–37°C and for toxin production, the

DOI: 10.1201/9781003487043-17

ideal temperature is between 28 and 33°C. *Aspergillus* species grow well in temperate, and moist environments. Aflatoxin B1 (AFB1), B2 (AFB2), G1 (AFG1), and G2 (AFG2) are among the major types of aflatoxins. These are categorized into two groups based on the color of fluorescence they emit under ultraviolet light. AFB1 and AFB2 appear blue, while AFG1 and AFG2 appear green (2). Other aflatoxins include aflatoxin M1 (AFM1) and M2 (AFM2) which are hydroxy derivatives of AFB1 and AFB2 respectively. Aflatoxins B1, B2, G1, and G2 contamination occurs in food crops or their products, while dairy products like milk contain M1 and M2 (4).

Aflatoxins pose significant hazards to humans and livestock such as carcinogenic, teratogenic (causing birth defects), and mutagenic effects. They can lead to fetal maldevelopment, and miscarriages, directly affect the structure of DNA and suppress the immune system (5). The International Agency for Research on Cancer (IARC) has categorized AFB1 as a class I carcinogen because it is 10 times more toxic than potassium cyanide and a highly polluting toxin (6). It can damage DNA through lipid peroxidation or oxidation. It is closely linked to liver cancer in humans. In the liver, AFB1 is converted to AFB1 8,9-epoxide via cytochrome p- 450 activation, responsible for various cancer-causing effects (4).

## THE ECONOMIC IMPACT OF AFLATOXIN CONTAMINATION

Aflatoxin contamination, apart from its impact on public health, poses a significant economic and trade challenge for developed and developing countries. Several countries have implemented aflatoxin levels limits in food and feed due to its high toxicity. Strict regulations, such as those established by the United States Food and Drug Administration (USFDA) (20 parts per billion [ppb] and 0.5 ppb in food and milk, respectively) and the European Union (EU) (2–4 ppb), restrict the sale of contaminated crops in premium markets (7). The Food Safety and Standards Regulations, 2011, has imposed a limit of 30 μg/kg for all food commodities in India. Several lakh tons of harvested products fail to meet export standards yearly due to aflatoxin contamination and go to waste, leading to serious negative economic effects (4). In the United States, maize farmers suffer an estimated loss of $160 million annually, whereas developing countries, like Saharan Africa, face even greater losses (8). It is necessary to implement appropriate pre- and post-harvest control practices, such as proper drying and storage techniques, to reduce aflatoxin contamination and protect farmers' livelihoods and national economies.

## GLOBAL TRENDS IN TROPICAL BEVERAGE CONSUMPTION

Tea is a more popular drink than coffee globally, especially outside of Europe and America. For example, black tea is highly preferred in countries like Iran (9). Regional variations in the type of tea consumed can be noted. Iced tea is a popular choice in the United States, fruit-infused in Europe, foamy in Taiwan, and herbal in China and Japan, whereas ready-to-drink tea is preferred in Japan, the United States, and Taiwan. Coffee is consumed more in Nordic countries like Finland, where consumption can exceed 12 kg per person annually. Chocolate consumption varies significantly across different regions. It is relatively low in China and exceptionally high in Switzerland (3). The trend in beverage consumption has shifted from nutrient-deficient, high-sugar drinks to natural fruit and herbal juices because of obesity issues (10). Pineapple juice is one of the major players in the tropical beverage market, driven by high demand in the United States and Europe (11).

## BEVERAGE CROPS AT RISK OF AFLATOXIN CONTAMINATION

Aflatoxins contaminate an array of food commodities such as cereals, seeds, oilseeds, spices, nuts, and beverages, as well as milk and milk products (8). Aflatoxin contamination is especially problematic for beverages like coffee (*Coffea* spp.), tea (*Camelia sinensis*), cocoa (*Theobroma cacao*), and fruits. Toxin accumulation is significant in these commodities because/as they are usually grown in warm and humid conditions ideal for fungal growth. Even though detoxification techniques such as chemical, biological, and physical methods (e.g., roasting and shelling) are used. Toxins can still reach consumers because they resist harsh processing conditions and may concentrate on specific parts of the crop used in beverages. As a result, contamination of beverages like tea, coffee, and juices is common (3).

## Tea

Tea is a popular beverage and an essential product globally. Within the middle class and the urbanized population, the trend for tea consumption is rising (3). It is preferred due to its beneficial properties such as improved antioxidant status and contains essential natural bioactive compounds, such as carotenoids and phenols (12). Previous research on tea primarily focused on its health benefits, but overlooked safety risks like aflatoxin contamination (13). Aflatoxins are among the top mycotoxins detected in raw herbs worldwide and various studies have documented their contamination (Table 2.11.1) (3). Recent research has shown that even contaminated leaves can transfer aflatoxins to the final tea beverage. A research study revealed that during the processing of contaminated raw leaves, approximately 50% of the aflatoxin can be transferred to the beverage (14).

## Coffee

Coffee ranks second among the world's most popular beverages and is prepared by brewing roasted coffee beans. The beans and berries are susceptible to fungal contamination during cultivation and processing, despite being made from roasted seeds (32). Although earlier research suggested caffeine might prevent aflatoxin growth, studies have found aflatoxin in the ground and whole coffee beans (19). The level of aflatoxin contamination in coffee is relatively low and sporadic compared to other mycotoxins. Two different studies conducted in Spain and Pakistan found that 53% and 50%

**TABLE 2.11.1**
**Aflatoxins Contamination in Different Beverages**

| Beverage Type | Country | Contamination Rate | Maximum Level (µg/kg) | References |
|---|---|---|---|---|
| Tea | Pakistan | 5–50 µg\kg | 0.11–16.17 | 14 |
| Black tea | Portugal | 65% | 28.15 | 15 |
| Tea beverages | Different countries | 65% | 14.4–332.2 | 16 |
| Coffee beans | Yemen | 14–27 ppb | 14.225–23.231 | 17 |
| Cocoa beans | Brazil | 4.9% | 0.35–30 | 18 |
| Green coffee | Not specified | Low | 1.2–19.8 | 19 |
| Milk and cheese | Costa Rica | Not specified | 0.019–0.629 | 20 |
| Milk and milk products | Serbia | 56.3%–76% | <0.025–1 | 21 |
| Breast milk | Mexico | 89% | 0.0030–0.03424 | 22 |
| Raw milk | Kenya | 30%–60% | 0.0154–4.563 | 23 |
| Fruit juice (mixed) | Pakistan | Not specified | Not specified | 24, 25 |
| Mango juice | | | | |
| Berry juice | Not specified | 57% | Not specified | 26 |
| Red wine | Italy | 57.9% | 0.02–0.05 | 27 |
| Beer | China | 97.8% | 0.1–0.03 | 28 |
| | Ethiopia | Unspecified | 1.23–12.47 | 29 |
| | Nigeria | 16.7%–17.9% | 3.43–38 | 30 |
| | South Africa | Low | 5.8–7 | 31 |

of coffee samples, respectively, were contaminated with aflatoxin. However, the highest level of contamination measured was low, that is, 1.2 μg/kg (33, 34). Researchers have developed a method for detecting multiple aflatoxins in coffee simultaneously, providing a more comprehensive picture of potential aflatoxin contamination (35). Roasting can be a preventive measure because it can reduce aflatoxin levels in coffee beans by up to 20% (17).

## Cocoa-based Beverages

Cocoa beans (*T. cacao*) are the source of chocolate. They are primarily cultivated in warm, humid regions of West Africa, Asia, and Central and South America. Improper handling and drying practices expose cocoa beans to contamination by filamentous fungi. The fungal genera, mainly *Aspergillus*, have been identified as associated with cocoa bean contamination. The fungi not only degrade the quality of cocoa beans, but pose a significant health risk due to aflatoxins production ability (3). Different *Aspergillus* species, that is, *A. flavus, A. parasiticus,* and *A. nomius*, all well-known for aflatoxin production, were identified in cocoa beans ( 36). Aflatoxins exhibit thermal stability, and therefore can persist through various stages of chocolate manufacturing (3).

## Fruits and Fruit Drinks

Fruit juices are a popular beverage choice due to their high antioxidant, vitamin, and mineral content that provides health benefits, including the prevention of cardiovascular disease, diabetes, and cancer (37). However, fruits used to make juice are prone to fungal decay due to their high sugar content and low pH levels, offering an ideal environment for fungal growth. Fruits with fungal growth, such as citrus fruits, can also contain mycotoxins that persist throughout processing. Citrus fruits, like oranges, lemons, and tangerines cultivated in tropical and subtropical climates with high temperatures and humidity, are at major risk. In an early study, Alderman and Marth noted that *A. parasiticus* grows and produces significant levels of aflatoxin (up to 220 μg/kg) even under low humidity conditions (13%–20%) in grapefruit, oranges, and limes (3). Similarly, other fruits like guava are susceptible to aflatoxins, with reported concentrations of up to 120 μg/kg (38). Sugarcane is also at risk. Aflatoxin B1 and G1 have been detected in sugarcane, with concentrations exceeding 300 μg/kg (39). Juan and García reported in 2017 that the most significant risk in different berry juices is posed by aflatoxins (57% contaminated samples). Moreover, 26% of the samples surpassed the EU aflatoxins legal limit for dried fruits (4–10 μg/kg) (26).

## Alcoholic Beverages

Aflatoxin contamination in wines is rare, but it has been detected occasionally in both traditional and industrially produced beer (8). A survey in Ethiopia revealed that out of the 12 sampled domestic beer brands, 11 tested positive for total aflatoxin ranging between 1.23 and 12.47 μg/L (29). None of the imported brands had detectable levels of aflatoxin, suggesting local manufacturing practices are a matter of concern. Similarly, in a Nigerian study, it was found that nearly 18% of locally brewed beers were contaminated with aflatoxin B1. Interestingly, a small percentage of imported beers also contained detectable aflatoxin, highlighting a need for further investigation (30). Although contaminated raw materials are a source of aflatoxin in beer, research suggests the brewing process can be a mitigation alternative. The brewing process reduced aflatoxin contamination by up to 80% (40). It was also observed that wine fermentation offers around a 30% reduction in aflatoxin levels compared to beer fermentation (41). Despite these findings, the use of contaminated raw materials remains a significant risk factor. Therefore, to ensure consumer safety and produce aflatoxin-free beer, it is important to implement stricter regulations, improve quality control, and provide educational programs for manufacturers.

## AFLATOXIN-CONTAMINATION PREVENTION METHODS

Good Agricultural Practices (GAPs) are essential in preventing aflatoxin contamination in crops. These practices include planting and harvesting at appropriate times, using resistant crop varieties, crop rotation, insect control, and irrigation. In addition, Good Manufacturing Practices (GMPs) such as sorting raw materials, washing, and dehulling. The preventive measures for aflatoxin contamination are biological and chemical methods, along with plant breeding (42).

### Biological Methods for Aflatoxin Prevention

Biological control offers promising avenues for preventing aflatoxin contamination by utilizing various microorganisms to combat aflatoxin-producing fungi. Currently, the most successful method involves introducing atoxigenic (non-toxin-producing) strains of *A. flavus* (Table 2.11.2). These strains outcompete toxigenic strains in the field, altering the fungal community and reducing aflatoxin production. However, concerns exist regarding potential ecological disruptions and mutations enabling aflatoxin production in atoxigenic strains. Researchers are actively exploring other microorganisms like *Trichoderma* (fungus) and *Bacillus* (bacteria) for their potential in aflatoxin biocontrol. These methods show promise, but further research is needed to optimize their effectiveness and ensure long-term safety. Also, other biological control agents and plant extracts present exciting possibilities for preventing aflatoxin. However, further research is needed to bridge the gap between promising laboratory results and real-world applications, including field efficacy, cost considerations, and long-term safety.

### Crop Breeding

Researchers are actively investigating the development of aflatoxin-resistant crops. Although there are no commercially available resistant crops yet, significant progress has been made in understanding the genes that are responsible for aflatoxin resistance in crops such as peanuts and maize. Studies conducted on peanuts (49) and (50, 51) maize, using biotechnological tools that include genomics and breeding methods, offer promising prospects for developing resistant varieties. This will help to identify genes and integrate resistance traits to achieve inheritable resistance in these crops, which will offer a long-term solution for controlling aflatoxin.

Aflatoxin contamination in beverage crops is a growing concern due to its potentially harmful impact on public health. Applying stricter regulations, improving quality control measures, setting aflatoxin contamination levels, and adopting preventive strategies like GAP and MAP are necessary

**TABLE 2.11.2**
**Different Biological Approaches to Mitigate Aflatoxin Contamination**

| Microorganism / Bioactive Compounds | Mechanism | Pros | References |
|---|---|---|---|
| Atoxigenic *A. flavus* strains | Outcompete toxigenic strains | Commercially available, reduces contamination | 43, 44, 45 |
| *Trichoderma* species (*T. harzianum*, *T. viride*) | Reduces aflatoxin production via the production of extracellular enzymes | Suppresses mold growth and aflatoxin production | 46 |
| *Bacillus* species (*B. subtilis, B. amyloliquefaciens*) | Competition, antibiotic production | Strong competitive ability | 47 |
| Complex essential oils like carvacrol, cinnamaldehyde, eugenol | Disrupt fungal cell membranes, suppress enzyme secretion, weaken metabolism | Reduce aflatoxin production | 48 |

to minimize aflatoxin contamination. Exploring the potential of biological control methods using atoxigenic strains and developing aflatoxin-resistant crop varieties offer promising long-term solutions for mitigating this threat.

## ACKNOWLEDGMENT

The authors acknowledge the infrastructural support provided by the Department of Microbiology and the Centre for Medical Biotechnology (CMBT) at Maharshi Dayanand University (MDU), Rohtak, India. PD is grateful to the University Grants Commission (UGC), New Delhi, India, for the UGC fellowship. SR and CR acknowledge the Council of Scientific & Industrial Research (CSIR), New Delhi, India, for their CSIR fellowships. Biorender was used to prepare graphical abstract.

## REFERENCES

1. Adekoya, I., Njobeh, P., Obadina, A., et al. (2018) Investigation of the metabolic profile and toxigenic variability of fungal species occurring in fermented foods and beverages from Nigeria and South Africa using UPLC-MS/MS. *Toxins*, *11*(2), 85.
2. Gholami-Shabani, M., Shams-Ghahfarokhi, M., & Razzaghi-Abyaneh, M. (2017) Aflatoxins and aflatoxigenic fungi in Iran: A systematic review of the past, present, and future. *Mycologia Iranica*, *4*(2), 65–84.
3. Granados-Chinchilla, F., Redondo-Solano, M., & Jaikel-Víquez, D. (2018) Mycotoxin contamination of beverages obtained from tropical crops. *Beverages*, *4*(4), 83.
4. Kumar, A., Pathak, H., Bhadauria, S., & Sudan, J. (2021) Aflatoxin contamination in food crops: Causes, detection, and management: A review. *Food Production, Processing, and Nutrition*, *3*, 1–9.
5. Razzaghi-Abyaneh, M., Chen, Z. Y., Shams-Ghahfarokhi, M., & Rai, M. (2022) Research efforts, challenges, and opportunities in mitigating aflatoxins in food and agricultural crops and its global health impacts. *Frontiers in Microbiology*, *13*, 881858.
6. Peng, Z., Zhang, Y., Ai, Z., et al. (2023) Current physical techniques for the degradation of aflatoxins in food and feed: Safety evaluation methods, degradation mechanisms and products. *Comprehensive Reviews in Food Science and Food Safety*, *22*(5), 4030–4052.
7. Gurtler, J. B., & Keller, S. E. (2019) Microbiological safety of dried spices. *Annual Review of Food Science and Technology*, *10*, 409–427.
8. Jallow, A., Xie, H., Tang, X., Qi, Z., & Li, P. (2021) Worldwide aflatoxin contamination of agricultural products and foods: From occurrence to control. *Comprehensive Reviews in Food Science and Food Safety*, *20*(3), 2332–2381.
9. Rezaee, E., Mirlohi, M., Hassanzadeh, A., & Fallah, A. (2016) Factors affecting tea consumption pattern in an urban society in Isfahan, Iran. *Journal of Education and Health Promotion*, *5*(1), 13.
10. Ng, S. W., Ostrowski, J. D., & Li, K. P. (2017) Trends in added sugars from packaged beverages available and purchased by US households, 2007–2012. *The American Journal of Clinical Nutrition*, *106*(1), 179–188.
11. Reinhardt, A., & Rodriguez, L. V. (2007) Industrial processing of pineapple–trends and perspectives. *VI International Pineapple Symposium*, *822*, 323–328.
12. Chandrasekara, A., & Shahidi, F. (2018) Herbal beverages: Bioactive compounds and their role in disease risk reduction - A review. *Journal of Traditional and Complementary Medicine*, *8*(4), 451–458.
13. Sedova, I., Kiseleva, M., & Tutelyan, V. (2018) Mycotoxins in tea: Occurrence, methods of determination and risk evaluation. *Toxins*, *10*(11), 444. https://doi.org/10.3390/toxins10110444
14. Ismail, A., Akhtar, S., Riaz, M., et al. (2020) Prevalence and exposure assessment of aflatoxins through black tea consumption in the Multan City of Pakistan and the impact of tea making process on aflatoxins. *Frontiers in Microbiology*, *11*, 446. https://doi.org/10.3389/fmicb.2020.00446
15. Duarte, S. C., Salvador, N., Machado, F., et al. (2020) Mycotoxins in teas and medicinal plants destined to prepare infusions in Portugal. *Food Control*, *115*, 107290.
16. Pallares, N., Font, G., MañEs, J., & Ferrer, E. (2017) Multimycotoxin LC–MS/MS analysis in tea beverages after dispersive liquid– liquid microextraction (DLLME). *Journal of Agricultural and Food Chemistry*, *65*(47), 10282–10289. https://doi.org/10.1021/acs.jafc.7b03507
17. Humaid, A. A., Alghalibi, S. M., & Al-Khalqi, E. A. A. (2019) Aflatoxins and ochratoxin a content of stored Yemeni coffee beans and effect of roasting on mycotoxin contamination. *International Journal of Molecular Microbiology*, *2*(1), 11–21.

18. Pires, P. N., Vargas, E. A., Gomes, M. B., et al. (2019) Aflatoxins and ochratoxin A: Occurrence and contamination levels in cocoa beans from Brazil. *Food Additives & Contaminants. Part A, Chemistry, Analysis, Control, Exposure & Risk Assessment, 36*(5), 815–824. https://doi.org/10.1080/19440049.2019.1600749
19. Bessaire, T., Perrin, I., Tarres, A., et al. (2019) Mycotoxins in green coffee: Occurrence and risk assessment. *Food Control, 96*, 59–67. https://doi.org/10.1016/j.foodcont.2018.08.033
20. Chavarría, G., Granados-Chinchilla, F., Alfaro-Cascante, M., & Molina, A. (2015) Detection of aflatoxin M1 in milk, cheese and sour cream samples from Costa Rica using enzyme-assisted extraction and HPLC. *Food Additives & Contaminants, 8*(2), 128–135. https://doi.org/10.1080/19393210.2015.1015176
21. Tomašević, I., Petrović, J., Jovetić, M., et al. (2015) Two year survey on the occurrence and seasonal variation of aflatoxin M1 in milk and milk products in Serbia. *Food Control, 56*, 64–70. https://doi.org/10.1016/j.foodcont.2015.03.017
22. Cantú-Cornelio, F., Aguilar-Toalá, J., de León-Rodríguez, C., et al. (2016) Occurrence and factors associated with the presence of aflatoxin M1 in breast milk samples of nursing mothers in central Mexico. *Food Control, 62*, 16–22. https://doi.org/10.1016/j.foodcont.2015.10.004
23. Kuboka, M. M., Imungi, J. K., Njue, L., et al. (2019) Occurrence of aflatoxin M1 in raw milk traded in peri-urban Nairobi, and the effect of boiling and fermentation. *Infection Ecology & Epidemiology, 9*(1), 1625703. https://doi.org/10.1080/20008686.2019.1625703
24. Iqbal, S. Z., Malik, S., Asi, M. R., Selamat, J., & Malik, N. (2018) Natural occurrence of patulin in different fruits, juices and smoothies and evaluation of dietary intake in Punjab, Pakistan. *Food Control, 84*, 370–374.
25. Hussain, S., Asi, M. R., Iqbal, M., et al. (2020) Mycotoxin in mango and orange fruits, juices, pulps, and jams marketed in Pakistan. *Toxins, 12*, 52.
26. Juan, C., Mañes, J., Font, G., & Juan-García, A. (2017) Determination of mycotoxins in fruit berry by-products using QuEChERS extraction method. *LWT, 86*, 344–351.
27. Di Stefano, V., Avellone, G., Pitonzo, R., et al. (2015) Natural co-occurrence of ochratoxin A, ochratoxin B and aflatoxins in Sicilian red wines. *Food Additives & Contaminants: Part A: Chemistry, Analysis, Control, Exposure & Risk Assessment, 32*(8), 1343–1351. https://doi.org/10.1080/19440049.2015.1055521
28. Zhao, Y., Huang, J., Ma, L., Liu, S., & Wang, F. (2017) Aflatoxin B1 and sterigmatocystin survey in beer sold in China. *Food Additives & Contaminants, 10*(1), 64–68. https://doi.org/10.1080/19393210.2016.1247919
29. Nigussie, A., Bekele, T., Gemede, H. F., & Woldegiorgis, A. Z. (2018) Level of aflatoxins in industrially brewed local and imported beers collected from Ethiopia market. *Cogent Food & Agriculture, 4*(1), 1453317. https://doi.org/10.1080/23311932.2018.1453317
30. Salami Oluwafemi, M., Onyemelukwe Ngozi, F., Olowu Frederick, A., Ibrahim Hadizat, H., & Tok Pam, D. (2019) Detection of aflatoxins from foreign and locally made beer. *IOSR Journal of Environmental Science, Toxicology and Food Technology, 13*(2), 09–14.
31. Adekoya, I., Obadina, A., Adaku, C. C., et al. (2018) Mycobiota and co-occurrence of mycotoxins in South African maize-based opaque beer. *International Journal of Food Microbiology, 270*, 22–30. https://doi.org/10.1016/j.ijfoodmicro.2018.02.001
32 Galarce-Bustos, O., Alvarado, M., Vega, M., & Aranda, M. (2014) Occurrence of ochratoxin A in roasted and instant coffees in Chilean market. *Food Control, 46*, 102–107.
33. García-Moraleja, A., Font, G., Mañes, J., & Ferrer, E. (2015) Analysis of mycotoxins in coffee and risk assessment in Spanish adolescents and adults. *Food and Chemical Toxicology, 86*, 225–233.
34. Azam, K., Akhtar, S., Gong, Y. Y., et al. (2020) Evaluation of the impact of activated carbon-based filtration system on the concentration of aflatoxins and selected heavy metals in roasted coffee. *Food Control, 121*, 107583. https://doi.org/10.1016/j.foodcont.2020.107583
35. García-Moraleja, A., Font, G., Mañes, J., & Ferrer, E. (2015) Simultaneous determination of mycotoxin in commercial coffee. *Food Control, 57*, 282–292.
36. Copetti, M. V., Iamanaka, B. T., Pereira, J. L., Fungaro, M. H., & Taniwaki, M. H. (2011) Aflatoxigenic fungi and aflatoxin in cocoa. *International Journal of Food Microbiology, 148*(2), 141–144.
37. Mandappa, I. M., Basavaraj, K., & Manonmani, H. K. (2018) Analysis of mycotoxins in fruit juices. In *Fruit Juices* (pp. 763–777). Academic Press.
38. Abdel-Sater, M. A., Zohri, A. A., & Ismail, M. A. (2001) Natural contamination of some Egyptian fruit juices and beverages by mycoflora and mycotoxins. *Journal of Food Science and Technology-Mysore, 38*(4), 407–411.
39. Abdallah, M. F., Krska, R., & Sulyok, M. (2016) Mycotoxin contamination in sugarcane grass and juice: First report on detection of multiple mycotoxins and exposure assessment for aflatoxins B1 and G1 in humans. *Toxins, 8*(11), 343.

40. Inoue, T., Nagatomi, Y., Uyama, A., & Mochizuki, N. (2013) Fate of mycotoxins during beer brewing and fermentation. *Bioscience, Biotechnology, and Biochemistry*, *77*, 130027. https://doi.org/10.1271/bbb.130027
41. Inoue, T., Nagatomi, Y., Uyama, A., & Mochizuki, N. (2013) Degradation of aflatoxin B1 during the fermentation of alcoholic beverages. *Toxins*, *5*(7), 1219–1229. https://doi.org/10.3390/toxins5071219
42. Mahato, D. K., Lee, K. E., Kamle, M., et al. (2019) Aflatoxins in food and feed: An overview on prevalence, detection and control strategies. *Frontiers in Microbiology*, *10*, 2266. https://doi.org/10.3389/fmicb.2019.02266
43. Gasperini, A. M., Rodriguez-Sixtos, A., Verheecke-Vaessen, C., et al. (2019) Resilience of biocontrol for aflatoxin minimization strategies: Climate change abiotic factors may affect control in non-GM and GM-maize cultivars. *Frontiers in Microbiology*, *10*, 2525. https://doi.org/10.3389/fmicb.2019.02525
44. Weaver, M., & Abbas, H. (2019) Field displacement of aflatoxigenic *Aspergillus flavus* strains through repeated biological control applications. *Frontiers in Microbiology*, *10*, 1788. https://doi.org/10.1080/02652039409374266
45. Mauro, A., Garcia-Cela, E., Pietri, A., Cotty, P. J., & Battilani, P. (2018) Biological control products for aflatoxin prevention in Italy: Commercial field evaluation of atoxigenic *Aspergillus flavus* active ingredients. *Toxins*, *10*(1), 30. https://doi.org/10.3390/toxins10010030
46. Gachomo, E. W., & Kotchoni, S. O. (2008) The use of *Trichoderma harzianum* and *T. viride* as potential biocontrol agents against peanut microflora and their effectiveness in reducing aflatoxin contamination of infected kernels. *Biotechnology*, *7*(3), 439–447. https://doi.org/10.3923/biotech.2008.439.447
47. Siahmoshteh, F., Siciliano, I., Banani, H., et al. (2017) Efficacy of *Bacillus subtilis* and *Bacillus amyloliquefaciens* in the control of *Aspergillus parasiticus* growth and aflatoxins production on pistachio. *International Journal of Food Microbiology*, *254*, 47–53. https://doi.org/10.1016/j.ijfoodmicro.2017.05.011
48. Wang, L., Liu, B., Jin, J., et al. (2019) The complex essential oils highly control the toxigenic fungal microbiome and major mycotoxins during storage of maize. *Frontiers in Microbiology*, *10*, 1643. https://doi.org/10.3389/fmicb.2019.01643
49. Sharma, K. K., Pothana, A., Prasad, K., et al. (2018) Peanuts that keep aflatoxin at bay: A threshold that matters. *Plant Biotechnology Journal*, *16*(5), 1024–1033. https://doi.org/10.1111/pbi.12846
50. Pandey, M. K., Kumar, R., Pandey, A. K., et al. (2019) Mitigating aflatoxin contamination in groundnut through a combination of genetic resistance and post-harvest management practices. *Toxins*, *11*(6), 315. https://doi.org/10.3390/toxins11060315
51. Warburton, M. L., & Williams, W. P. (2014) Aflatoxin resistance in maize: What have we learned lately? *Advances in Botany*, *2014*, 10. https://doi.org/10.1155/2014/352831

# 2.12 Aflatoxigenic Fungi and Corn

*I. Lavkor*

## INTRODUCTION

Corn (*Zea mays* L.) is the first of the most widely grown cereals in the world. It surpassed rice, and wheat in terms of the quantity of grain. Notably, corn has demonstrated remarkable production levels, reaching 1.2 billion tons in 2021, and has experienced the most rapid growth over the past two decades, soaring by 104% since 2000 [1]. This surge can be attributed to corn's extensive applications in biofuel, animal feed industries, snack foods and oil production [2]. Like any other crop, corn is susceptible to contamination by toxigenic fungi, which can produce mycotoxins harmful to animal and human health and cause financial loss to the world's food production [3].

Mycotoxins, produced by a diverse group of toxigenic fungi, are estimated to contaminate approximately 25% of the world's grains [4, 5]. Aflatoxins, mainly produced by *Aspergillus* section *Flavi*, have received highly significant notice among naturally occurring mycotoxins due to their high toxicity for humans and domestic animals, and their ability to contaminate a vast array of agricultural commodities in optimum control [6].

Aflatoxins are potent secondary metabolic products produced by certain fungi, which grow in various substrates such as soil, cereals, nuts, fruit, rotting plants, straw, and grain [7]. They are highly toxic for humans and domestic animals, and can contaminate a wide range of agricultural commodities under suitable conditions [6]. Many nations have implemented or suggested regulations to manage aflatoxins in food and feed [8].

The US Food and Drug Administration (FDA) has established regulations stipulating that total aflatoxin levels must not exceed 20 parts per billion (ppb) in interstate trade of food and feed. Additionally, for milk sales, the allowable limit for aflatoxin M1 is set at 0.5 ppb. In numerous developing countries, contamination of locally produced goods occurs significantly. A study demonstrated a clear link between exposure to aflatoxin and the prevalence of stunting (a sign of chronic malnutrition) and underweight conditions (indicative of acute malnutrition) among children in West Africa, highlighting the severity of the issue [9]. Moreover, an epidemic of acute aflatoxicosis occurred in Kenya in 2004, leading to 125 fatalities as a result of consuming contaminated corn [10].

Aflatoxins can emerge in corn while it is growing in the field and during storage, rendering the grains unsafe for consumption. Insufficient drying, high relative humidity and heat, agricultural practices like intercropping with aflatoxin-contaminated crops, early or late harvest, and poor storage facilities increase the risk of infection. Corn is susceptible to certain factors and is at increased risk of fungal infection, particularly by *Aspergillus* spp., potentially leading to heightened formation of aflatoxin [11].

The latest report indicates that aflatoxins are produced by thirteen species classified into three sections within the genus *Aspergillus*. These sections include: *Aspergillus* section *Flavi*, which encompasses species such as A. *flavus, A. parasiticus, A. nomius, A. pseudotamarii, A. arachidicola, A. bombycis, A. parvisclerotigenus*, and *A. minisclerotigenes; Aspergillus* section *Nidulantes*, which includes species like *Emericella olivicola, E. astellata, and E. venezuelensis;* and *Aspergillus* section *Ochraceorosei,* which contains *A. rambellii* and *A. ochraceoroseus* [12]. The predominant and widely recognized producers of aflatoxin in foodstuffs within *Aspergillus* section *Flavi* are *A.*

DOI: 10.1201/9781003487043-18

*flavus* [13, 14], *A. parasiticus* [15–17], and *A. nomius* [18, 19]. $AFB_1$, $AFB_2$, $AFG_1$, and $AFG_2$ can be produced by *A. parasiticus* and *A. nomius*, while $AFB_1$ and $AFB_2$ can be produced by *A. flavus* [20].

## AFLATOXIGENIC FUNGI IN FIELD

In the field, corn can be contaminated by aflatoxigenic *Aspergillus* spp. in three stages. At the outset, conidia transmitted through the air or insects contaminate the silk and subsequently proliferate within the developing ear. Afterward, the corn kernels within the ear, likely harmed by insects or birds, become invaded by aflatoxigenic *Aspergillus* spp. and contaminated with aflatoxins. The plant's susceptibility to fungal contamination increases due to stress caused by climatic and cultural factors, such as drought and nutrient deficiency. Multiple field studies concur on the primary influence of drought and high temperatures in the increased production of aflatoxin in corn [21–23]. An examination of 3000 corn samples collected in Serbia between 2012 and 2016 [24] indicated that increased levels of aflatoxin were primarily attributed to extreme climatic conditions, notably severe drought and high temperatures during the summer. The hindered growth of corn plants resulting from heat and drought stress creates favorable conditions for pathogenic fungi infection. Additionally, the hot and dry atmospheric conditions are ideal for the proliferation of *A. flavus* [25–37]. Figure 2.12.1 indicates that aflatoxigenic *A. parasiticus* isolated from corn plant tissues in the field.

The spread of *A. flavus* and its ability to infect corn ears rely on the presence of airborne spores in the surrounding environment [38]. Moisture on leaves and soil has a detrimental effect on the extent of infection, as does precipitation, which washes away conidia and spores from the air [38, 39]. Elevated temperatures, leading to dry conditions on plant and soil surfaces, are often linked to severe outbreaks of infection [28, 29, 40].

Kernel infection could also occur through the scar left by silk shedding, as observed by [41]. Fennel et al. [42] indicated that 68% of *A. flavus* sporulation occurred in the kernel, 48% in the tip, and 12% in the entire endosperm in corn. According to Mycock et al. [43], *A. flavus* internally infected in corn kernels, when allowed to germinate, could give rise to plants that were internally infected with the same fungus. It has been determined that fungi might be transmitted through seeds, while it has been proposed that *A. flavus* primarily resides in soil and can persist for extended durations in the form of sclerotia [44, 45]. Figure 2.12.2 shows aflatoxigenic *A. flavus* infected corn cob in the field.

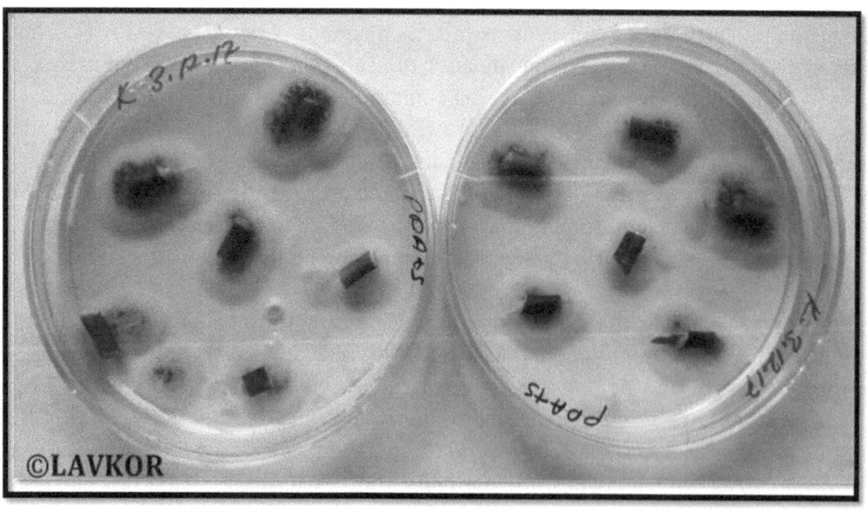


**FIGURE 2.12.1** *A. parasiticus* isolated from corn plant tissues.

**FIGURE 2.12.2** Aflatoxigenic *A. flavus* infected corn cob.

## FACTORS OF AFLATOXIGENIC FUNGI INFECTION AND AFLATOXIN PRODUCTION IN CORN

Aflatoxigenic fungi grow in corn and environmental conditions, with optimal growth typically occurring in hot and humid conditions. However, they can increase and produce toxins across humidity levels, temperatures, and pH. Mold growth on organic materials occurs under optimal conditions of warm temperatures and high moisture levels, typically 7% or more. The ideal growth temperatures for optimal development range between 35 and 37°C. The preferred temperature is within the range of 28–33°C for the production of fungal toxins. *Aspergillus* species exhibit growth capability within a water activity range of 0.78 to 1, with the most favorable condition being around 0.95. Aflatoxigenic strains producing aflatoxins can generate these toxins within a water activity range of 0.83 to 0.97. The optimal conditions for aflatoxin production occur within a narrower water activity range of 0.90 to 0.95 [3].

As relative humidity surpasses 85%, the proliferation of *A. flavus* experiences a significant surge. Within this threshold, even a slight uptick in moisture content can substantially elevate the likelihood of aflatoxin contamination [46]. Aflatoxins may produce at relative humidity levels of 88%, 90%, and 99% [47]. Moreover, Hell [48] stated that goods kept in environments with humidity ranging from 75% to 85% are prone to fungal infestation within the usual storage duration. The growth of storage fungi in a post-harvest commodity is impacted by the duration it spends in storage [49]. Bhatti et al. [50] indicated that in Pakistan, seed samples stored for 8 to 12 months at elevated moisture levels (17%) and those obtained from mud-plastered storage facilities showed a greater ratio of aflatoxin production strains of *A. flavus*.

On the other hand, Griffin et al. [51] found an increase in *A. flavus* population in the soil over successive years when corn was cultivated in rotation with peanuts. Barry et al. [52] also demonstrated that corn varieties resistant to ear-infesting insects also exhibited reduced aflatoxin levels in pre-harvest kernels. Zuber et al. [53] detected that insects consuming corn ears in the field can increase the susceptibility of kernels to *A. flavus* infection due to the physical damage inflicted by their feeding activity.

## AFLATOXIN CONTAMINATION

Corn is a perfect substrate for mold growth and aflatoxin contamination, which has been reported by several authors. According to Hell et al. [54], 9.9% to 32.2% of corn samples prepared for storage from different ecological zones of Benin contained more than 5 ppb of aflatoxins, which increased to 15% and 32.2% after six months of storage. Lavkor [55] found the fungal agents responsible for ear rot and assessed the levels of aflatoxin contamination in primary and secondary corn crops during harvest seasons in Adana, from 2015 to 2016. During the survey studies, a total of 134 cornfields were sampled. Fungal isolates including *A. flavus, A. terreus*, and *A. niger* were identified from infected corn cobs. The 48 corn samples revealed aflatoxin concentrations ranging from 0.3 to 72.6 μg/kg. Chehri et al. [56] investigated studied the prevalence of *Aspergillus* species in corn kernels in Kermanshah province from 2006 to 2013. They collected 100 samples, primarily from diseased corn kernels intended for human and animal consumption, from various geographic locations within the province. They isolated and identified 100 *Aspergillus* strains, categorizing them into six species. Among these species, *A. flavus* comprised 27% of the isolates. The researchers also investigated $AFB_1$ in 66 samples. The levels of $AFB_1$ ranged from 0.046 to 10.776 $\mu g.kg^{-1}$.

Setamou et al. [57] observed a direct correlation between the extent of insect damage in pre-harvest corn in Benin and the percentage of kernels infected with *A. flavus*, the prevalence of aflatoxin contamination in samples, and the average aflatoxin content. Hell et al. [58] discovered that corn without insect damage showed no aflatoxin contamination, whereas corn with 70% of its cobs damaged by insects exhibited aflatoxin contamination in 30.3% of cobs. According to Payne [59], the production of aflatoxin in corn before harvest was influenced by the weather conditions prevailing during the maturation of the crop. Moreover, aflatoxin contamination before harvest increases significantly in conditions of moisture deficit combined with elevated temperatures.

## HARVESTING AND DRYING

Following the harvest, it is essential to adequately dry corn kernels to a safe moisture level to prevent the growth of fungi. Improper drying of corn kernels can lead to a tenfold increase in aflatoxin contamination within three days [60]. Rapidly drying harvested field crops to attain moisture levels between 10% and 13% is strongly advocated to ensure their safety. To keep fungal contamination to a minimum during the drying process, farmers need to dry their kernels in the field and on the ground. Dried kernels have a lasting shelf life, are seldom susceptible to insect infestation, and typically do not foster mold growth due to the unavailable moisture necessary for such development. In developing countries, drying typically relies on solar power, often resulting in a longer time to achieve an acceptable moisture level deemed safe. In the event of heavy rainfall during harvest, farmers often resort to various strategies to protect their produce. These include stacking cobs next to stalks to protect against rain, storing kernels in covered areas in their home garden, or threshing wet and dried kernels to reduce moisture levels. It should be ensured that the product must dry before storing [61]. Figure 2.12.3 shows aflatoxigenic *A. parasiticus* infected corn cobs at harvest.

## STORAGE CONDITIONS AND AFLATOXIN CONTAMINATION OF CORN

The most effective approach is to control aflatoxin in the standing crop because the level of contamination in the harvested cobs tends to increase over time. Several factors exacerbate aflatoxin contamination, such as high temperatures, humidity, inadequate ventilation in storage areas, and

**FIGURE 2.12.3** Aflatoxigenic *A. parasiticus* infected corn cobs.

damage caused by insects and rodents. Reducing the number of aflatoxins by separating damaged cobs, cobs that have been infested by insects, cobs that have not been fully hulled, or cobs that have gone moldy from the rest of the kernels is an essential step for ensuring the safety and quality of the final product. Sorting stored corn is an efficient method to reduce aflatoxin levels.

Corn cobs should be stored in a well-ventilated drying bin after sorting to reduce aflatoxin contamination. The quality of the corn kernel should be checked, and any insect infestation must be controlled. If significant insect infestation is detected, it is advisable to husk the corn cobs, discard any damaged kernels, and store the intact kernels in jute bags. Although polypropylene bags are a popular choice among African farmers, it is important to note that they may not provide adequate aeration. Therefore, it is recommended that alternative storage methods be used that allow for better air circulation, especially for kernels that have not been dried to a safe level [58, 62].

The presence of aflatoxigenic fungi in corn can be influenced by whether the corn is stored as cobs or shelled kernel. Mora and Lacey [63] determined that corn kernels left on the cob through drying had lower levels of aflatoxigenic fungi compared to those that were shelled immediately after harvest. To prevent damage to the kernels and reduce the risk of fungal infection, it is recommended to avoid shelling corn by beating the cobs in a bag with a stick.

A more effective method is recommended. Husking corn-breaking the cob with a stick in a bag can damage the kernels and increase the kernel's susceptibility to fungal attack. Damaged corn kernels and corn cobs that have been threshed with mechanical shellers are highly susceptible to contamination with aflatoxins [64].

The prevalence of *A. flavus* was particularly high during corn kernel storage, reaching a peak of 32% infected bagged kernels and 30% infected kernels in clay stores after four months of storage period. Conversely, corn kernels stored on the cob with husks showed a low incidence of *A. flavus*, consistently measuring below 1.3% across different storage structures [48]. Fungus is one of the leading causes of corn harm and crop yield loss during storage. The study by Lavkor and Ay [65] was conducted in Adana as a survey of 56 corn warehouses in 2015–2016. In this study, fungi in stored corn kernels were detected and identified using blotter and agar techniques. In the agar and blotter methods, *A. flavus, A. parasiticus*, and *A. niger* were determined as the most common species. Therefore, a need to choose the levels of aflatoxin-producing fungi in stored corn kernels for food and feed safety and quality control was concluded. Lavkor and Ay's [66] other study investigated fungal diseases in 32 corn seed samples collected from various storage locations in Adana. In this research, *A. flavus, A. terreus*, and *A. niger* were identified as the predominant species.

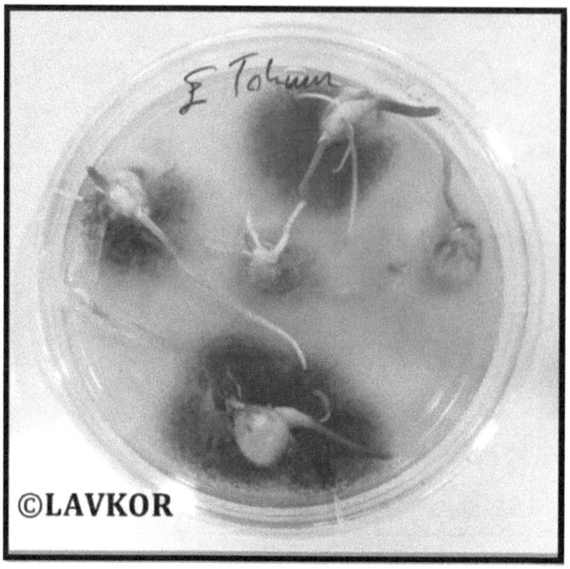


**FIGURE 2.12.4** Aflatoxigenic *A. parasiticus* isolated from corn kernels.

The study was conducted in the Adana and Osmaniye provinces of Turkey in 2018; the aim was to analyze aflatoxin contamination in 32 corn samples collected from different warehouses. High-performance liquid chromatography (HPLC) was used to analyze the total aflatoxin in the corn samples. It was found that the total aflatoxin concentration in the contaminated 32 samples ranged from 0.50 to 596.33 μg/kg. All corn samples indicated 100% aflatoxin contamination. Additionally, it was determined that 11 out of 32 samples exceeded the legal limit set by the Turkish government, which is 10 μg/kg [67]. Figure 2.12.4 shows aflatoxigenic *A. parasiticus* isolated from corn kernels in storage.

## EFFECTIVE CONTROL METHODS

The presence of insects is correlated with aflatoxin contamination in both pre-harvest [32] and post-harvest [1] stages. Certain insect species, notably *Coleopteran* and *Lepidopteran* species, have been found to be associated with high levels of aflatoxin contamination. Among these, specific species such as *Mussidia nigrivinella* (Lepidoptera: Pyralidae) have been identified as playing a role in the transmission of fungal spores [68]. Efforts to mitigate insect infestation post-harvest involve various measures, including the use of commercial insect control products in warehouse facilities and implementing barriers to shield cobs from infestation, both in the field and during storage. Utilizing prophylactic pesticides, particularly at the onset of storage, has proven to be effective in combating infestation. These proactive steps are crucial for safeguarding harvested crops and preserving their quality during storage.

Numerous farmers utilize indigenous plant extracts, either in their natural state or in the form of an oil or water extract, for insect protection purposes. *Monodera myristica, Zingiber officinalis, Ocimum basilicum, Tetrapleura tetraptera, Ocimum gratissimum, Xylopia aethiopica, Aframonium* spp., and *Piper guineense* have all undergone examination regarding their potential to impede the mycelial growth of *A. flavus* [69].

The aqueous extracts derived from a combination of dried fruits from *Xylopia aethiopica* and *Piper guineense* have demonstrated inhibitory effects on the growth of all evaluated corn pathogens. Etheric oils extracted from *Morinda lucida* and *Azadirachta indica* have inhibited the growth of aflatoxigenic *A. flavus* and notably decreased the production of aflatoxins in inoculated corn kernels [70]. Conidia germination and growth of *A. flavus, A. fumigatus* and *F. verticillioides* have been inhibited by the essential oils of *Ocimum gratissimum, Thymus vulgaris,* and *Cymbopogon citratus* [71]. Further investigations are necessary to elucidate the inhibition mechanism(s) and identify the active ingredients present in natural products that hinder fungal growth. Only through such thorough examinations can definitive conclusions be drawn regarding the effectiveness of natural botanical products in mitigating post-harvest aflatoxin contamination.

## ELIMINATING AFLATOXIN THROUGH PHYSICAL SEGREGATION AND SANITATION PRACTICES

The distribution of aflatoxin within a lot of corn cobs or kernels exhibits a high degree of heterogeneity, wherein substantial quantities of the toxin are concentrated in only a select few kernels or a minor percentage of the overall batch. Severely moldy or damaged kernels typically have the highest concentrations of aflatoxin. Physically damaged and infected kernels should be sorted based on their characteristics such as discoloration, irregular shapes, shriveled appearance, and reduced size, aflatoxin levels can be reduced by 40%–80% [72]. This process effectively removes contaminated kernels from the intact commodity, thereby mitigating aflatoxin contamination to a significant extent [73].

Basic sanitation practices, including removal of the previous crop's remnants and destruction of infective plant debris, can go a long way toward reducing grain spoilage both in the field and during storage. Studies have shown that cleaning storage areas before refilling them with the new harvest can lead to reduced aflatoxin levels [54]. Additionally, maintaining cleanliness near storage facilities helps minimize infestations by insects that often seek refuge in host plants nearby. Another effective measure is the storage of healthy maize cobs after segregating heavily damaged ones, particularly those with over 10% ear damage due to insects because this practice has been demonstrated to lower aflatoxin levels [68]. Furthermore, the levels of mycotoxins in contaminated commodities can be reduced before consumption through various food processing methods. These methods include wet and dry milling, grain purification, autoclave, roast, bake, fry, nixtamalize, and extruder cooking. Traditional food processing techniques have been identified in different parts of Africa, which significantly decrease aflatoxin levels in corn and peanuts-based food products. However, to identify the methods that expose consumers to the lowest levels of aflatoxin, additional evaluations of these processing techniques are not required [64].

## CONCLUSION

The natural occurrence of aflatoxigenic fungi and contamination of corn risk for public health all over the world. Aflatoxin contamination of corn is a problem that mostly occurs from field to storage stages. Aflatoxin, a secondary metabolite, in corn and corn products is a major public health concern. This toxin is highly resistant to most physical and chemical inactivation methods, making it a formidable challenge. Prioritize accurate, sensitive, specific, and rapid procedures for detecting and determining aflatoxin contamination in corn. Develop effective strategies for preventing contamination and safe decontamination procedures. Networking at both national and global levels is crucial for gathering information and developing effective strategies to prevent future aflatoxin exposure issues.

## REFERENCES

1. FAO. 2022. Agricultural production statistics. 2000–2021. FAOSTAT Analytical Brief Series No 60. Rome:1–17. https://www.fao.org/3/cc3751en/cc3751en.pdf (accessed March 25, 2024).
2. Mayfield, K.L. 2006. Preharvest aflatoxin in maize genotypes under inoculation with *Aspergillus flavus*. A Thesis master of science, 1–148. Texas A&M University.

3. Razzaghi-Abyaneh, M., P.K. Chang, M. Shams-Ghahfarokhi, and M. Rai. 2014. Global health issues of aflatoxins in food and agriculture: Challenges and opportunities. *Frontiers in Microbiology* 12: 420.
4. Bhat, R., R.V. Rai, and A.A. Karim. 2010. Mycotoxins in food and feed: Present status and future concerns. *Comprehensive Reviews in Food Science and Food Safety* 9: 57–81.
5. Park, D.L., H. Njapau, and E. Boutrif. 1999. Minimizing risks posed by mycotoxins utilizing the HACCP concept. *Food, Nutrition and Agriculture* 3: 49–54.
6. Gholami-Shabani, M., M. Shams-Ghahfarokhi, and M. Razzaghi-Abyaneh. 2017. Aflatoxins and aflatoxigenic fungi in Iran: A systematic review of the past, present, and future. *Mycologia Iranica* 4: 65–84.
7. Iram, W., T. Anjum, M. Iqbal, A. Ghaffar, and M. Abbas. 2016. Structural elucidation and toxicity assessment of degraded products of aflatoxin $B_1$ and $B_2$ by aqueous extracts of *Trachyspermum ammi*. *Frontiers in Microbiology* 7: 346.
8. Haumann, F. 1995. Eradicating mycotoxins in food and feeds. *Inform* 6: 248–256.
9. Gong, Y.Y., K. Cardwell, A. Hounsa, S. Egal, P.C. Turner, A.J. Hall, et. al. 2002. Dietary aflatoxin exposure and impaired growth in young children from Benin and Togo: Cross sectional study. *British Medical Journal* 325: 20–21.
10. Probst, C., H. Njapau, and P.J. Cotty. 2007. Outbreak of an acute aflatoxicosis in Kenya in 2004: Identification of the causal agent. *Applied and Environmental Microbiology* 73: 2762–2764.
11. Akrobortu, E.D. 2008: Aflatoxin contamination of maize of maize from different locations in Ghana. MPhil Thesis. Kwame Nkrumah University of Science and Technology, Kumasi. 45 pp.
12. Varga, J., J.C. Frisvad, and R.A. Samson. 2015. Two new aflatoxin producing species, and an overview of *Aspergillus* section *Flavi*. *Studies in Mycology* 69: 57–80.
13. Nesbitt, B.F., J. O'Kelly, K. Sargeant, and A.N.N. Sheridan. 1962. *Aspergillus flavus* and Turkey X disease. Toxic metabolites of *Aspergillus flavus*. *Nature* 195: 1062–1063.
14. Varga, J., J. Frisvad, and R.A. Samson. 2009. A reappraisal of fungi producing aflatoxins. *World Mycotoxin Journal* 2: 263–277.
15. Codner, R.C., K. Sargeant, and R. Yeo. 1963. Production of aflatoxin by the culture of strains of *Aspergillus flavus-oryzae* on sterilized peanuts. *Biotechnology and Bioengineering* 5: 185–192.
16. Buchanan, R.L., and J.C. Ayres. 1976. Effect of sodium acetate on growth and aflatoxin production by *Aspergillus parasiticus* NRRL 2999. *Journal of Food Science* 41: 128–132.
17. Schmidt-Heydt, M., C.E. Rüfer, A. Abdel-Hadi, N. Magan, and R. Geisen. 2010. The production of aflatoxin $B_1$ or $G_1$ by *Aspergillus parasiticus* at various combinations of temperature and water activity is related to the ratio of *AflS* to *AflR* expression. *Mycotoxin Research* 26: 241–246.
18. Kurtzman, C.P., B.W. Horn, and C.W. Hesseltine. 1987. *Aspergillus nomius*, a new aflatoxin-producing species related to *Aspergillus flavus* and *Aspergillus tamarii*. *Antonie Leeuwenhoek* 53: 147–158.
19. Yunes, N.B.S., R.C. Oliveira, T.A. Reis, A.C. Baquião, L.O. Rocha, and B. Correa. 2020. Effect of temperature on growth, gene expression, and aflatoxin production by *Aspergillus nomius* isolated from Brazil Nuts. *Mycotoxin Research* 36: 173–180.
20. Pickova, D., V. Ostry, and F. Malir. 2021. A recent overview of producers and important dietary sources of aflatoxins. *Toxins* 13: 136.
21. Leggieri, C., T. Bertuzzi, A. Pietri, and P. Battilani. 2015. Mycotoxin occurrence in maize produced in Northern Italy over the years 2009–2011: Focus on the role of crop related factors. *Phytopathologia Mediterranea* 54: 212–221.
22. Pleadin, J., A. Vuli´c, N. Perši, M. Škrivanko, B. Capek, and Ž. Cvetni´c. 2013. Aflatoxin $B_1$ occurrence in maize sampled from Croatian farms and feed factories during. *Food Control* 40: 286–291.
23. Hajnal, E.J., J. Kos, J. Krulj, S. Krstovic, I. Jajic, L. Pezo, et al. 2017. Aflatoxins contamination of maize in Serbia: The impact of weather conditions in 2015. *Food Additives and Contaminants: Part A* 34: 1999–2010.
24. Kos, J., E. Jani´c Hajnal, B. Šari´c, P. Jovanov, A. Mandi´c, O. Đuragi´c, et al. 2018. Aflatoxins in maize harvested in the Republic of Serbia over the period 2012–2016. *Food Additives and Contaminants: Part B* 11: 246–255.
25. Cotty, P.J., and R. Jaime-Garcia. 2007. Influences of climate on aflatoxin producing fungi and aflatoxin contamination. *International Journal of Food Microbiology* 119: 109–115.
26. Diener, U., R. Cole, T. Sanders, G. Payne, L. Lee, and M. Klich. 1987. Epidemiology of aflatoxin formation by *Aspergillus flavus*. *Annual Review of Phytopathology* 25: 249–270.
27. McMillian, W., D. Wilson, and N. Widstrom. 1985. Aflatoxin contamination of pre-harvest corn in Georgia: A six-year study of insect damage and visible *Aspergillus flavus*. *Journal of Environmental Quality* 14: 200–202.

28. Payne, G.A., D.K. Cassel, and C.R. Adkins. 1986. Reduction of aflatoxin contamination in maize by irrigation and tillage. *Phytopathology* 76: 679–684.
29. Payne, G.A., W.M. Hagler, and C.R. Adkins. 1988. Aflatoxin accumulation in inoculated ears of field grown maize. *Plant Disease* 72: 422–424.
30. Payne, G.A., D.L. Thompson, E.B. Lillehoj, M.S. Zuber, and C.R. Adkins. 1988. Effect of temperature on the pre-harvest infection of maize kernels by *Aspergillus flavus*. *Phytopathology* 78: 1376–1380.
31. Herrera, M., R. Anadón, S.Z. Iqbal, J.D. Bailly, and A. Ariño. 2016. Climate change and food safety. In *Food Safety: Basic Concepts, Recent Issues, and Future Challenges*, 149–160. Springer, Berlin/Heidelberg.
32. Sétamou, M., K. Cardwell, F. Schulthess, and K. Hell. 1997. *Aspergillus flavus* infection and aflatoxin contamination of pre-harvest maize in Benin. *Plant Disease* 81: 1323–1327.
33. Thompson, D.L., G.A. Payne, M.S. Zuber, and E.B. Lillehoj. 1983. Early appearance of aflatoxin in developing corn kernels after inoculation with *Aspergillus flavus*. *Plant Disease* 67: 1321–1322.
34. Widstrom, N.W. 1992. Aflatoxin in developing maize: Interactions among involved biota and pertinent econiche factors. In *Handbook of Applied Mycology*, eds. D. Bhatnagar, E.B. Lillehoj, and D.K. Arora, 23–57. Marcel Dekker, New York.
35. Widstrom, N.W. 1996. The aflatoxin problem with corn grain. In *Advances in Agronomy*, ed. D.L. Sparks, 219–280. Academic Press, New York.
36. Widstrom, N.W., B.Z. Guo, and D.M. Wilson. 2003. Integration of crop management and genetics for control of pre-harvest aflatoxin contamination of corn. *Toxin Reviews* 22: 195–223.
37. Widstrom, N.W., W. McMillian, R. Beaver, and D.M. Wilson. 1990. Weather associated changes in aflatoxin contamination of pre-harvest maize. *Journal of Agricultural Production* 3: 196–199.
38. Jones, R., H. Duncan, G. Payne, and K. Leonard. 1980. Factors influencing infection by *Aspergillus flavus* in silk-inoculated corn. *Plant Disease* 64: 859–863.
39. Jones, R., H. Duncan, and P. Hamilton. 1981. Planting date, harvest date, and irrigation effects on infection and aflatoxin production by *Aspergillus flavus* in field corn. *Phytopathology* 71: 810–816.
40. Shearer, J., L. Sweets, N. Baker, and L. Tiffany. 1992. A study of *Aspergillus flavus/parasiticus* in Iowa crop fields: 1988–1990. *Plant Disease* 76: 19–22.
41. Marsh, S.F., and G.A. Payne. 1984. Scanning EM studies on the colonization of dent corn by *Aspergillus flavus*. *Phytopathology* 74: 1284–1289.
42. Fennell, D.I., R.J. Bothast, E.B. Lillehoj, and R.E. Peterson.1993. Bright greenish-yellow fluorescence and associated fungi in white corn naturally contaminated with aflatoxin. *Cereal Chemistry* 50: 504–414.
43. Mycock, D.J., F.H.J. Rijkenberg, and P. Berjak. 1992. Systemic transmission of *A. flavus* var. columnaris from one maize seed generation to the next. *Seed Science and Technology* 20: 1–13.
44. Cotty, P.J. 1994. Influence of field application of an atoxigenic strain of *Aspergillus flavus* on the population of *A. flavus* infecting cotton bolls and on the aflatoxin content of cottonseed. *Phytopathology* 84: 270–1277.
45. Saito, M., O. Tsurata, P. Siriacha, S. Kawasugi, and M. Manabe. 1989. A typical strains if *Aspergillus flavus* isolated in maize fields - Aflatoxin producing ability and distribution in Thailand. *Japan Agricultural Research Quarterly* 23: 151–154.
46. Christensen, C.M., and C.J. Mirocha. 1976. Relationship of relative humidity to the invasion of rough rice by *Aspergillus parasiticus*. *Pythopathology* 66: 204–205.
47. Lillehoj, E.B. 1983. Effect of environmental and cultural factors on aflatoxin contamination of developing maize kernels. In *Aflatoxin and Aspergillus flavus in Corn*, eds. U.L. Diener, R.L. Asquith, and J.W. Dickens, 27–34. Alabama Agricultural Experiment Station, Auburn University, Auburn, Ala.
48. Hell, K. 1997. Factors contributing to the distribution and incidence of aflatoxin producing fungi in stored maize in Benin. An unpublished PhD diss., 1–182, Hannover University.
49. Lillehoij, E.B., and M.S. Zuber. 1988. Distribution of toxin-producing fungi in mature maize kernels from diverse environments. *Tropical Science* 28: 19–24.
50. Bhatti, N.N., M.A. Nasir, and M.A. Randhawa.1990. Association of aflatoxin producing strains of *Aspergillus* with some stored cereals. *Pakistan Journal of Phytopathology* 2: 74–82.
51. Griffin, G.J., K.H. Garren, and J.D. Taylor.1981. Influence of crop rotation and minimum tillage on the population of *Aspergillus flavus* group in peanut field soil. *Plant Disease* 65: 898–900.
52. Barry, D., N.W. Widstrom, L.L. Darrah, W.W. McMillian, T.J. Riley, G.E. Scott et. al. 1992. Maize ear damage by insects in relation to genotype and aflatoxin contamination in preharvest maize grain. *Journal of Economic Entomology* 85: 2492–2495.
53. McMillian, W.W. 1986. Relation of insects to aflatoxin contamination in maize grown in the southeastern USA. In *Aflatoxin in Maize*, eds. M.S. Zuber, E.B. Lillehoj, and B.L. Renfro, Proceedings of the Workshop 194–200., CIMMYT, Mexico.

54. Hell, K., K.F. Cardwell, M. Setamou, and H. Poehling. 2000. The influence of storage practices on aflatoxin contamination in maize in four agroecological zones of Benin, West Africa. *Journal of Stored Products Research* 34: 1–2.
55. Lavkor, I. 2019. Fungal species causing ear rot in corn and occurred mycotoxins in corn. *The Journal of Food* 44: 1–16.
56. Chehri, K., E. Azami, and A. Mosaber.2015. *Aspergillus* and aflatoxin B contamination of stored corn grains in Western Iran. *Global Veterinaria* 14: 39–42.
57. Setamou, M., K.F. Cardwell, F. Schulthess, and K. Hell. 1997. *Aspergillus flavus* infection and aflatoxin contamination of preharvest maize in Benin. *Plant Disease* 81: 1323–1327.
58. Hell, K., K. Cardwell, M. Setamou, and F. Schulthess.2000. Influence of insect infestation on aflatoxin contamination of stored maize in four agroecological regions in Benin. *African Entomology* 6: 169–177.
59. Payne, G.A. 1992. Aflatoxin in maize. *Critical Reviews in Plant Science* 10: 423–440.
60. Tanboon-ek, P. 1989. Control of aflatoxin in maize. In *Mycotoxin Prevention and Control in Food Grains*, eds. R.L. Semple, A.S. Frio, P.A. Hicks, and J.V. Lozare, 166–172. FAO, Rome, Italy.
61. Hell, K., P. Fandohan, R. Bandyopadhyay, S. Kiewnick, R. Sikora, and P.J. Cotty. 2008. Pre- and post-harvest management of aflatoxin in maize: An African perspective. In *Mycotoxins: Detection Methods, Management, Public Health and Agricultural Trade*, eds. J.F. Leslie, R. Bandyopadhyay, and A. Visconti, 219–229. CABI, Wallingford UK.
62. Udoh, J.M., K.F. Cardwell, and T. Ikotun. 2000. Storage structures and aflatoxin content of maize in five agro-ecological zones of Nigeria. *Journal of Stored Products Research* 36: 187–201.
63. Mora, M., and J. Lacey. 1997. Handling and aflatoxin contamination of white maize in Costa Rica. *Mycopathologia* 138: 77–89.
64. Fandohan, P., R. Ahouansou, P. Houssou, K. Hell, W.F.O. Marasas, and M.J. Wingfield. 2006. Impact of mechanical shelling and dehulling on *Fusarium* infection and fumonisin contamination in maize. *Food Additives and Contaminants* 23: 451–421.
65. Lavkor, I., and T. Ay. 2019. *Detection of Major Storage Fungal Pathogens of Maize*, Proceedings Book 66–72. 1st International Gobeklitepe Agricultural Congress November 25–27 Sanlıurfa/Turkey.
66. Lavkor, I., and T. Ay. 2019. *Fungal Diseases of Corn Grains During Storage*, Proceedings Book 666. 4th International Mediterranean Science and Engineering Congress, Alanya/Turkey.
67. Lavkor, I. 2020. *Aflatoxin Contamination on Stored Corn Kernels in Cukurova Province*, 74. 5th Internaional Scientific Research Congress 1-2 September, Istanbul/Turkey.
68. Sétamou, M., K.F. Cardwell, F. Schulthess, and K. Hell. 1998. Effect of insect damage to maize ears, with special reference to *Mussidia nigrivenella* (*Lepidoptera: Pyralidae*), on *Aspergillus flavus* (*Deuteromycetes: Monoliales*) infection and aflatoxin production in maize before harvest in the Republic of Benin. *Journal of Economic Entomology* 91: 433–438.
69. Cardwell, K.F., and L. Dongo. 1994. Effects of extracts from nine plant species found in Africa on the mycelia growth of *A. flavus* Link ex Fries. In *Stored Product Protection, Proceedings of the 6th International Working Conference on Stored-Product Protection*, 17–23 April, Canberra, Australia..
70. Bankole, S.A. 1997. Effect of essential oil from two Nigerian medicinal plants (*Azadirachta indica* and *Morinda lucida*) on growth and aflatoxin $B_1$ production in maize grain by a toxigenic *Aspergillus flavus*. *Letters in Applied Microbiology* 24: 190–192.
71. Nguefack, J., V. Leth, P.H. Amvam Zollo, and S.B. Mathur. 2004. Evaluation of five essential oils from aromatic plants of Cameroon for controlling food spoilage and mycotoxin producing fungi. *International Journal of Food Microbiology* 94: 329–334.
72. Park, D.L. 2002. Effect of processing on aflatoxin. *Advances in Experimental Medicine and Biology* 504: 173–179.
73. Whitaker, T.B. 2003. Standardization of mycotoxin sampling procedures: An urgent necessity. *Food Control* 14: 233–237.

# 2.13 Aflatoxigenic Fungi and the Agricultural Economy

*Yutika Narzary and S. Uma Devi*

## INTRODUCTION

Mycotoxins are toxic secondary metabolites produced by fungus, posing a substantial risk to human and animal health. Mycotoxins can induce both acute poisoning and long-term health issues like cancer and immune system deficits [1]. There are more than 400 mycotoxins, but the major ones in terms of health and the economy are aflatoxins, fumonisins, ochratoxins, trichothecenes, zearalenones, and deoxynivalenol. Aflatoxins are the mycotoxins produced by several strains of *Aspergillus* section *Flavi*, section *Ochraceorosei*, and section *Nidulantes* [2]. Scientists have identified more than 18 different forms and metabolites of aflatoxins, but among all aflatoxins, the four main types of aflatoxin B1 (AFB1), aflatoxin B2 (AFB2), aflatoxin G1 (AFG1), and aflatoxin G2 (AFG2) are the most well-studied, prevalent, and pose serious risks to human as well as animal's health [3]. Aflatoxin B1 is the most potent and toxic of the four. Aflatoxin, the most common type of mycotoxin, affects about 25% of the world's crop. Inadequate storage of cassava, cottonseed, chili peppers, maize, wheat, millet, peanuts, rice, sesame, sunflower seeds, and various spices are the common places to find them [4]. As per a survey conducted by the Food and Agriculture Organization (FAO), mycotoxins and aflatoxins during storage cause contamination of 1000 million metric tons of plant-based products every year [5]. The market value of contaminated crops, due to their concerning health risk to humans and animals, is reduced or they are rendered completely unsellable. This affects not only farmers' incomes, but also the entire supply chain, which includes distributors, merchants, and consumers [6].

## TYPES OF AFLATOXIN

The aflatoxins—aflatoxin B1 (AFB1), aflatoxin B2 (AFB2), aflatoxin G1 (AFG1), and aflatoxin G2 (AFG2)—can be differentiated according to their fluorescence under UV light (green or blue) and comparative chromatographic movement during thin-layer chromatography [7]. Aflatoxin B1 ($C_{17}H_{12}O_6$) is a potent hepatotoxic, mutagenic, teratogenic, and hepatocarcinogenic mycotoxin and causes immunosuppression in animals. It is found as a contaminant in peanuts, cottonseed meal, corn, and other grains. Aflatoxin B1 has a tetrahydrocyclopenta[c]furo[3′,2′:4,5]furo[2,3-h]chromene skeleton with oxygen functionality at positions 1, 4 and 11. It is aflatoxin, aromatic ether, and an aromatic ketone [8]. Aflatoxin B2 ($C_{17}H_{14}O_6$) is aflatoxin having a hexahydrocyclopenta[c]furo[3′,2′:4,5]furo[2,3-h]chromene skeleton with oxygen functionality at positions 1, 4 and 11 [9]. Aflatoxin B2 shows weak mutagenicity, hepatoxicity, and carcinogenicity as compared to AFB1 [10]. Aflatoxin G (G1 and G2) belongs to the family of Difurocoumarolactone Series. These are polycyclic aromatic compounds containing a delta-valerolactone ring fused to the coumarin moiety of the difurocoumarin skeleton. Difurocoumarolactones are a subgroup of the aflatoxins and related compounds [11]. Aflatoxin B2a, aflatoxin M1, and aflatoxin G2a are some of the derivatives of the above four aflatoxins that are equally significant in terms of contaminants and toxicity.

DOI: 10.1201/9781003487043-19

## AFLATOXIGENIC FUNGI

Aflatoxins are secondary metabolites produced by *Aspergillus* species distributed on three main sections of the genus (section *Flavi*, section *Ochraceorosei*, and section *Nidulantes*). *A. ochraceoroseus* and *A. rambellii* are two species from section *Ochraceorosei*, and *A. stellatus*, *A. miraensis*, *A. olivicola*, and *A. venezuelensis* are from section *Nidulantes*, which produce aflatoxin B1. However, the maximum species producing aflatoxin belongs to section *Flavi* [10]. *Aspergillus* section *Flavi* contains both hazardous and beneficial *Aspergillus* sp., such as *Aspergillus oryzae*, which is employed in food fermentation and enzyme manufacturing, as well as the notorious *A. flavus* and *A. parasiticus*, known for production of mycotoxins [12]. *Aspergillus* section *Flavi* produces both aflatoxins and ochratoxins, which are among the most important mycotoxins. However, they are not produced by the same species. There are 33 species included in *Aspergillus* section *Flavi*, from which, *A. pseudotamarii* and *A. togoensis* are the two species that produce aflatoxins B1 and B2. Fourteen species of *Aspergillus* section *Flavi* (*A. arachidicola*, *A. aflatoxiformans*, *A. austwickii* and *A. cerealis*, *A. minisclerotigenes*, *A. mottae*, *A. luteovirescens* (formerly *A. bombycis*), *A. nomius*, *A. novoparasiticus*, *A. parasiticus*, *A. pseudocaelatus*, *A. pseudonomius*, *A. sergii*, *A. transmontanensis*) are able to produce aflatoxins B1, B2, G1, and G2 [13]. Although it is widely believed that *A. flavus* cannot make aflatoxins of type G, Frisvad et al. reported two strains from Korea that can produce aflatoxins G1 and G2. The mycotoxin, ochratoxin A is produced by three of the four species in the *A. alliaceus* clade: *A. alliaceus* s. str. and two newly described species, *A. neoalliaceus* and *A. vandermerwei* [13].

## AGRICULTURE LOSS DUE TO AFLATOXIN

Agricultural losses occur when aflatoxins contaminate crops, resulting in reduced production and quality. The following is a list of factors linked to economic loss caused by aflatoxin: (1) decreased product value; (2) yield loss; (3) decreased animal productivity as a result of health problems associated with aflatoxin; (4) detrimental effects on human health; (5) enormous costs connected with associated disorders; and (6) methods related to aflatoxin management, mitigation, and prevention. When crops become contaminated, it can lead to decreased yield and quality, causing significant economic losses for farmers. Aflatoxins also affect the yield of the crop due to inhibition of plant growth and development and leading to lower yields. Contamination might also occur after harvest during storage if suitable mold prevention procedures are not followed and leads to extra crop losses [14]. Additionally, the animals fed with mycotoxin contaminated feed can experience reduced feed intake, reduced growth rates, digestive disorders, decreased milk production in dairy animals, immune suppression, increased susceptibility to diseases and low fertility which causes economic loss to farmers engaged in livestock rearing business. The impact of aflatoxins extends beyond crop and livestock losses because it can also result in increased expenditures for farmers and agricultural businesses as they implement measures to prevent fungal contamination and mitigate the presence of aflatoxins [15]. Aflatoxin causes liver cancer, child malnutrition, food poisoning, increased mortality and morbidity due to their effect on nutritional absorption, and cause of aggravated health conditions of major diseases such as malaria and kwashiorkor that are associated with mycotoxins and aflatoxins [16].

## ENVIRONMENT REQUIRED FOR AFLATOXIN CONTAMINATION IN FOOD AND FEED COMMODITIES

Mycotoxigenic molds are found all over the world; therefore, mycotoxins such as aflatoxin infect a wide range of crops and foods, particularly those grown in tropical locations. It has been discovered that 30% to 50% of food commodities are wasted due to inadequate storage facilities, particularly in developing nations with hot and humid climates [17]. Aflatoxin generation in pre-harvest and post-harvest crops is influenced by various factors, including geographical location, favorable climate conditions such as temperature, relative humidity, aeration, moisture content, and storage conditions. The interaction of colonizing fungal species, their biotic potential, substrate composition, and duration of fungal growth, stress factors such as water scarcity, insect infestation, and nutritional aspects of the substrate can further promote toxin production [18, 19]. According to an FAO survey, aflatoxins and other mycotoxins during storage contaminate 1000 million metric tons of plant-based products per year [5]. Aflatoxigenic fungi and other fungal contamination destroy both vital calorie crops (rice, wheat, maize, and soybeans) and commodity crops (bananas, coffee, and barley) [20]. Since the middle of the 20th century, crop-damaging fungi have become more severe and widespread, posing a significant threat to global food security [21]. Despite disease interventions, still 10%–23% of our crop is lost before harvest and another 10%–20% after harvest [21].

It has also been seen that the nutritional composition of crops affects their susceptibility to aflatoxin infection. Higher concentration of fats, oils and soluble sugars, pH, temperature, and water activity significantly had a substantial influence on *Aspergillus flavus* growth and aflatoxin generation [22–24]. In their study, Majeed et al. [25] revealed that the content of tocopherol negatively correlates with aflatoxin formation in maize. Generally, low water activity is required to minimize the post-harvest accumulation of aflatoxin in cereals [26, 27]. In one study, three parameters-water activity, temperature, and incubation period- were used to study the mycotoxin formation in barley rootlets [28]. The experiment found that the maximum Aflatoxin B1 accumulation was recorded at 0.80 aw (water activity), 25°C, and on the $14^{th}$ day of culture. Furthermore, it was noted that an increase in water activity levels and temperature to 30°C reduced the Aflatoxin B1 production [28]. Such experiments and related literature attest that aflatoxin incidence occurs because of a combination of environmental and crop-intrinsic variables.

## CONTAMINATION OF FOOD COMMODITIES BY AFLATOXINS

According to the World Health Organization (WHO), cereals such as corn, wheat, sorghum and rice, oil-seeds, peanuts, sunflower seeds, cotton seeds, black peppers, turmeric, ginger, chili pepper, pistachio, almond, coconut, Brazil nut, walnuts, and even milk fed with contaminated feed have been found to contain aflatoxin [3].

Among the cereals, maize-induced human and animal aflatoxin exposure is the most significant food safety concern. Maize is staple of the diet in many countries, and the environmental conditions of those countries are favorable for fungal growth and aflatoxin production. In a study of aflatoxin contamination in Congo, 32% of maize was found to be contaminated before harvest, and nearly 100% infection was recorded during post harvest stages [29]. In a survey conducted in Nigeria, it was found that 52% of maize harvests were contaminated with aflatoxin levels exceeding the Nigerian regulatory limit [30]. Surveys in Kenya, Togo, and Burundi also reported similar contamination [31, 32, and 33]. Several studies in Latin America have also reported maize contamination with aflatoxin. In Costa Rica, 38.6% of maize samples and 27.8% of maize products were contaminated with aflatoxin [34]. In Mexico, aflatoxin B1 contamination exceeded Mexican and, European Union regulatory limits in 26% of the samples [35]. Similarly, in a survey in Brazil, 25.6% of maize was contaminated with aflatoxin B1, and in 7.4%, aflatoxin G1 was detected [36].

Rice is another significant cereal and staple diet of half of the population of the world [37]. Rice is a food staple in most Asian countries, particularly in South, South East, and East Asia. Asia is also a leading continent in production and consumption of rice. Rice is cultivated in subtropical environments with hot and humid climates, which stimulate the fungal growth and production of secondary metabolites. Rice may be contaminated in the field and in the pre-harvest, post-harvest, and storage stages, especially when the climatic conditions become favorable for their growth. Aflatoxin B1 was found in 63.5% of the sample studied from six different regions of China [38]. In another survey, aflatoxin was detected in rice samples from India, Pakistan, the United States, Egypt, and Australia [39]. The rice samples contaminated with aflatoxin B1 above the permissible amount were reported from other Asian countries including Vietnam, Sri Lanka, Bangladesh, Malaysia, Pakistan, Thailand, Indonesia, Iran, and the United Arab Emirates [40]. Many of these countries have tropical and subtropical environments, which are conducive to the growth of aflatoxin-producing fungi. Similar cases of contamination of other cereals like sorghum, wheat, and barley by aflatoxin are reported from several countries of the world [41].

Among oil crops, peanuts are the most susceptible to aflatoxin contamination. The invasion of peanut plants by hazardous fungus, followed by aflatoxin contamination of the nuts, is a grave risk to food security in peanut-producing countries around the world [42]. Peanuts, a commercial and staple crop in West Africa, contribute significantly to aflatoxin dietary exposure [43].

Other oil-seed crops, such as sunflowers, flax-seed, mustard seed, canola, rape-seed, sesame, cotton seed, soy, and their products, are susceptible to aflatoxin contamination [44]. Other nuts, such as almonds, pistachios, walnuts, chestnuts, apricots, and Brazil nuts, are equally susceptible to aflatoxin poisoning [45].

Spices are widely used in the food and culinary industries around the world as organoleptic enhancers, preservatives, and medicines in some cultures, but unfortunately, spices and herbs have been associated as a source of human aflatoxin exposure. Aflatoxin contamination was reported in spices such as cumin, turmeric, black pepper, cloves, saffron, coriander, cinnamon, nutmeg, capsicum, mint, and thyme [46]. Among alcoholic beverages, the occurrence of aflatoxin in wine was not frequent, but traces were found in traditionally and industrially prepared beer [47]. Sporadic and lower (maximum level of 1.2 μg/kg) levels of aflatoxin exposure through coffee were detected in samples from nine countries, despite the preceding supposition that caffeine can inhibit fungal toxin production [48].

Many reports have suggested the existence of aflatoxins in foods derived from animals. This has resulted in the development of regulations to restrict the control of aflatoxin in animal feeds [49]. Aflatoxin has also been found in human milk from nursing women in aflatoxin hotspots around the world [50]. The study concluded that about 14% of 121 samples of human breast milk were contaminated with aflatoxin M1 at levels between 2 and 187 ng/L. European and North American countries were previously unaffected by aflatoxins, but climate change and rising temperatures have made them vulnerable.

According to the DSM Global Mycotoxin Survey of 2023 [51], the highest aflatoxin contamination in several food and feed items was found in South Asia and South East Asia, zearalenones in Southern Europe, deoxynivalenol in Central Asia, MYCOTOXIN(T2) in South Asia, fumonisins in South East Asia, and ochratoxins in South Asia. According to this survey, the risk associated with aflatoxin contamination in South Asia is 77%, followed by South East Asia at 60%, South Europe at 49% and Sub-Saharan Africa at 40% [51]. Malaysia, Nepal, and Bangladesh are among the Asian countries with the highest prevalence of aflatoxin exposure [52]. East and West African countries are also favorable for the growth of *Aspergillus*, and intake of susceptible crops such as maize or groundnuts as staple foods is high. The countries of East and African countries like Uganda, Tanzania, and Kenya had higher percentages of biomarker AF-alb in collected samples [52].

## HEALTH IMPLICATIONS ASSOCIATED WITH AFLATOXIN

Aflatoxin has both mutagenic and carcinogenic characteristics. Hepatocellular carcinoma (HCC) was identified as the most frequent cancer worldwide, with 83% of cases occurring in developing nations. Africa and South Asia have a significant number of cases of hepatitis B and C, in addition to HCC. In fact, studies have indicated that aflatoxin and hepatitis B might interact synergistically, raising the risk of HCC [53]. The P53 gene hotspot mutation at codon 249 was linked to aflatoxin exposure [54]. Aflatoxin has also been linked to the development of various liver conditions, including cirrhosis and hepatomegaly [55]. The in utero exposure of aflatoxin causes lower birth weights and stunted child growth [52]. The co-relations between the detection of AFM1 was in breast milk samples and growth impairment in infants under 6 months old in Northern Tanzania was established in a study [56]. It has been proposed that aflatoxin exposure could impact the insulin-like growth factor (IGF) pathway via toxicity to the liver. Other reasons for aflatoxin-induced impairment in children include the immunosuppressive effect of aflatoxin exposure, which may increase infection susceptibility and thus affect nutritional status through appetite suppression and lower nutrient absorption [57]. Aflatoxin exposure may cause intestinal damage by inhibiting protein synthesis, resulting in reduced nutrition absorption and stunted development [58]. Aflatoxin's immunosuppressive effects, such as decreased antibody production, increased susceptibility to infectious illnesses, and diminished cell-mediated immunity, have been intensively examined in numerous animal species [59].

The degree of individual exposure to aflatoxin can be estimated by analyzing the contamination level of food samples and with food intake surveys, such as 24-hour recalls and food frequency questionnaires [52]. However, biomarkers are considered more accurate indicators of exposure to aflatoxins. Aflatoxin-$N^7$-guanine adducts excreted in urine (sign of previous day's exposure); AFM1, the hydroxylated metabolite of AFB1 (found in breast milk); and the aflatoxin-albumin adduct (AF-alb) in plasma or serum (half-life period of approximately 2 months) are some of the biomarkers used to detect aflatoxin exposure [60]. The enzyme-linked immunosorbent assay (ELISA) method is used to detect the AF-alb biomarker and is expressed as pg AF-alb/mg albumin or pg AF-lysine equivalent/mg albumin [61]. The aflatoxin-albumin adduct (AF-alb) can also be measured using high-performance liquid chromatography (HPLC)-fluorescence or liquid chromatography-tandem mass spectrometry (LC-MS/MS) chromatographic methods, which involve quantifying the aflatoxin-lysine adduct (AFB1-lysine) after proteolytic digests of the serum [62, 63].

## PREVENTION AND MITIGATION OF AFLATOXIN CONTAMINATION

The WHO has emphasized that an integrated approach is needed to control aflatoxins and associated problems [41]. Applicable strategies include pre-harvest and post-harvest blockage of the toxin. There are various ways for a competent authority to contribute to the control of aflatoxin contamination, including removing sources of contamination, promoting better agricultural and storage techniques, and enforcing strict food safety standards [64]. At least 199 countries have instituted regulations and legislation to monitor the trade of food and feed and transfer/migration of aflatoxins and the fungus from place to place [65]. These regulations specify maximum acceptable aflatoxin levels in feeds and foods, which vary for different commodities in various countries [66, 67]. Aflatoxins are regulated by maximum permissible levels in food. Several national and international agencies, such as the FAO, WHO, European Union Commission, and Codex Alimentarius Commission have set regulations regarding different mycotoxins and aflatoxins [68]. The absence of such constraints or even associated guidelines and legislation has had serious effects in some countries.

Industrious and low-income countries encounter different economic losses. The price of crops is reduced in industrialized nations, resulting in the disposal of a significant number of crops; nevertheless, low-income countries face an additional burden due to the health hazards linked with aflatoxin poisoning [69]. Low-income countries continue to face major health consequences of excessive aflatoxin levels in staple foods such as maize and peanuts [70]. These nations lack the resources necessary for regular food monitoring and management. The problem is particularly detrimental in Sub-Saharan African and Asian nations, where hepatitis B virus (HBV) incidence is high, leading to greater cancer rates [71]. More than 100 countries have established regulatory limits on aflatoxin levels in human and animal feeds. However, appropriate aflatoxin standards and regulations vary over the world. Because of the disparities in rigorous standards, the best quality food is exported to rich nations, while lower-grade contaminated food is retained locally, increasing the health risk in poor African, South Asian, and South East Asian countries. Sometimes the best-quality goods produced in a few nations are refused for export because they exceed the permissible limits, resulting in a monetary loss of millions of dollars [72, 73].

There are several methods or approaches to prevent and decontaminate commodities that are contaminated by aflatoxin. (1) Physical approaches to decontamination of mycotoxins include use of gamma irradiation; pulsed light treatment; cold plasma technology, in which reactive species (e.g., O, $O_3$, OH, NO, $NO_2$) are used to decontaminate agricultural products. (2) Biological methods of mycotoxin detoxification including use of beneficial microbes like lactic acid bacteria, *Bacillus licheniforms*, *B. subtilis*, and *Saccharomyces cerevisiae*, and botanicals like *Cassia fistula* and *Ocimum basilicum*. (3) Use of essential oils of botanicals such as *Azadirachta indica*, *Morinda lucida*, and *Satureja hortensis*. (4) Use of permissible chemical decontaminants [74].

## CONCLUSION

Aflatoxins are a major global concern because they contaminate crops and lead to decreased yield and quality, resulting in significant economic losses. It has been shown that one-third to one-half of food commodities are wasted due to poor storage facilities and fungal contamination, especially in underdeveloped countries with hot and humid climates. The contamination can also affect animal productivity and human health, causing increased expenditures for farmers and agricultural businesses. Several biomarker-based studies report that human exposure to aflatoxins causes liver cancer, child malnutrition, suppressed immunity, food poisoning, and increased mortality and morbidity. To minimize aflatoxin contamination, applying effective strategies can be the precautionary measures against fungal growth and aflatoxigenic fungi in agricultural products. Implementation of good agricultural and manufacturing practices during harvesting, storage, and distribution of agricultural products may ensure a lower level of aflatoxin in the final product. Physical, chemical, and microbiological approaches can be used to reduce the aflatoxin levels in the agricultural commodities. Preventive measures are preferred over curative measures, and a set of laws play a vital role in ensuring food safety. Most countries have enacted aflatoxin-related regulations and regulatory limits on permissible aflatoxin levels in human food or animal feed. However, many nations lack the resources necessary for regular food monitoring and management. Accurate, affordable, and simple aflatoxin detection methods are required to determine the effectiveness of control measures. Current methods are effective, but improving the speed, detection, and accuracy of aflatoxin analysis at a low cost would lead to safer food and agricultural products. Management, mitigation, and prevention methods are necessary to combat the aflatoxin problem.

## REFERENCES


1. World Health Organization. 2023. Mycotoxins. https://www.who.int/news-room/fact-sheets/detail/mycotoxins. Accessed on March 2024.
2. Payne, G. A., G. J. Nystrom, D. Bhatnagar, T. E. Cleveland, and C. P. Woloshuk. 1993. Cloning of the afl-2 gene involved in aflatoxin biosynthesis from *Aspergillus flavus*. *Applied and Environmental Microbiology* 59, no. 1: 156–162.
3. World Health Organization, Food. 2007. *Safety Evaluation of Certain Food Additives and Contaminants*. WHO.
4. Dhakal, A., Hashmi, M. F., & Sbar, E. (2023). *Aflatoxin toxicity*. In StatPearls. StatPearls Publishing. (PMID: 32491713)
5. Moosavi, M. H., A. M. Khaneghah, F. Javanmardi, et al. 2021. A review of recent advances in the decontamination of mycotoxin and inactivation of fungi by ultrasound. *Ultrasonics Sonochemistry* 79: 105755.
6. Ashiq, S. 2015. Natural occurrence of mycotoxins in food and feed: Pakistan Perspective. *Comprehensive Reviews in Food Science and Food Safety* 14, no. 2: 159–175.
7. Singh, J., and A. Mehta. 2020. Rapid and sensitive detection of mycotoxins by advanced and emerging analytical methods: A review. *Food Science & Nutrition* 8, no. 5: 2183–2204.
8. National Center for Biotechnology Information. 2024. PubChem Compound Summary for CID 186907, Aflatoxin B1. https://pubchem.ncbi.nlm.nih.gov/compound/Aflatoxin-B1. Accessed March 20, 2024.
9. National Center for Biotechnology Information. 2024. PubChem Compound Summary for CID 2724360, Aflatoxin B2. https://pubchem.ncbi.nlm.nih.gov/compound/Aflatoxin-B2. Accessed March 20, 2024.

10. Benkerroum, N. 2020. Aflatoxins: Producing-molds, structure, health issues and incidence in Southeast Asian and Sub-Saharan African Countries. *International Journal of Environmental Research and Public Health* 17, no. 4: 1215.
11. National Center for Biotechnology Information. 2024. PubChem Compound Summary for CID 14421, Aflatoxin G1. https://pubchem.ncbi.nlm.nih.gov/compound/Aflatoxin-G1. Accessed March 20, 2024.
12. Kjærbølling, I., T. Vesth, J. C. Frisvad, et al. 2020. A comparative genomics study of 23 *Aspergillus* species from section *Flavi*. *Nature Communications* 11, no. 1: 1106.
13. Frisvad, J. C., V. Hubka, C. N. Ezekiel, et al. 2019. Taxonomy of *Aspergillus* section *Flavi* and their production of aflatoxins, ochratoxins and other mycotoxins. *Studies in Mycology* 93, no. 1: 1–63.
14. Gruber-Dorninger, C., T. Jenkins, and G. Schatzmayr. 2019. Global mycotoxin occurrence in feed: A ten-year survey. *Toxins* 11, no. 7: 375.
15. Leite, M., A. Freitas, A. S. Silva, J. Barbosa, and F. Ramos. 2020. Maize (*Zea mays* L.) and mycotoxins: A review on optimization and validation of analytical methods by liquid chromatography coupled to mass spectrometry. *Trends in Food Science & Technology* 99: 542–565.
16. Gbashi, S., N. E. Madala, S. De Saeger, M. De Boevre, I. Adekoya, O. A. Adebo, and P. B. Njobeh. 2018. The Socio-Economic Impact of Mycotoxin Contamination in Africa. In *Mycotoxins-Impact and Management Strategies*. ed. P. B. Njobeh and F. Patman. IntechOpen.
17. Fimeche. 2013. https://catalogue.unccd.int/1482_IME-Global-Food-Report.pdf. Accessed on March 20, 2024.
18. Tola, M., and B. Kebede. 2016. Occurrence, importance and control of mycotoxins: A review. *Food Science and Technology* 2: 1191103. https://doi.org/10.1080/23311932.2016.1191103
19. Deligöz, E., and N. Bilge. 2017. Sütle Gelen Tehdit: Aflatoksin M1. Türk Tarım. *Gıda Bilim Ve Teknoloji Dergisi* 5, no. 8: 846–857.
20. Fones, H. N., D. P. Bebber, T. M. Chaloner, W. T. Kay, G. Steinberg, and S. J. Gurr. 2020. Threats to global food security from emerging fungal and oomycete crop pathogens. *Nature Food* 1, no. 6: 332–342.
21. Fisher, M. C., D. A. Henk, C. J. Briggs, J. S. Brownstein, L. C. Madoff, S. L. McCraw, and S. J. Gurr. 2012. Emerging fungal threats to animal, plant and ecosystem health. *Nature* 484, no. 7393: 186–194.
22. Liu, J., L. Sun, N. Zhang, J. Zhang, J. Guo, C. Li, et al. 2016. Effects of nutrients in substrates of different grains on aflatoxin B 1 production by *Aspergillus flavus*. *BioMed Research International* 10: 7232858. https://doi.org/10.1155/2016/7232858
23. Rajasekaran, K., G. Ford, K. Sethumadhavan, C. Carter-Wientjes, J. Bland, H. Cao, and D. Bhatnagar. 2017. *Aspergillus flavus* growth and aflatoxin production as influenced by total lipid content during growth and development of cottonseed. *Journal of Crop Improvement* 31, no. 1: 91–99.
24. Casquete, R., M. J. Benito, M. de Guía Córdoba, S. Ruiz-Moyano, and A. Martín. 2017. The growth and aflatoxin production of *Aspergillus flavus* strains on a cheese model system are influenced by physicochemical factors. *Journal of Dairy Science* 100, no. 9: 6987–6996. https://doi.org/10.3168/jds.2017-12865
25. Majeed, S., M. Iqbal, M. R. Asi, S. Z. Iqbal, and J. Selamat. 2017. Analysis of nutritional traits and aflatoxin contamination in selected maize varieties from Pakistan. *Journal of Food Protection* 80, no. 12: 1993–1998. https://doi.org/10.4315/0362-028X.JFP-17-117
26. Lahouar, A., I. Jedidi, S. Saïd, and A. V. Sanchís. 2018. Incidence, legislations and strategies of control of mycotoxins in North African countries. *International Food Research Journal* 25, no. 6: 2229–2247.
27. Dorner, J. 2008. Management and prevention of mycotoxins in peanuts. *Food Additives and Contaminants* 25, no. 2: 203–208. https://doi.org/10.1080/02652030701658357
28. Ribeiro, J., L. Cavaglieri, M. Fraga, G. Direito, A. Dalcero, and C. Rosa. 2006. Influence of water activity, temperature and time on mycotoxins production on barley rootlets. *Letters in Applied Microbiology* 42, no. 2: 179–184. https://doi.org/10.1111/j.1472-765X.2005.01830.x
29. Kamika, I., and M. Tekere. 2016. Occurrence of aflatoxin contamination in maize throughout the supply chain in the democratic Republic of Congo. *Food Control* 69: 292–296.
30. Liverpool-Tasie, L. S. O., N. S. Turna, O. Ademola, A. Obadina, and F Wu. 2019. The occurrence and co-occurrence of aflatoxin and fumonisin along the maize value chain in southwest Nigeria. *Food and Chemical Toxicology* 129: 458–465.
31. Udomkun, P., C. Wossen, T. Atehnkeng, J. Nabahungu, N. L. Nabahungu, E. Njukwe, B. Vanlauwe, and R. Bandyopadhyay. 2018. Incidence and farmers' knowledge of aflatoxin contamination and control in Eastern Democratic Republic of Congo. *Food Science & Nutrition* 6, no. 6: 1607–1620. https://doi.org/10.1002/fsn3.735

32. Nabwire, W. R., J. Ombaka, C. P. Dick, C. Strickland, L. Tang, K. S. Xue, and J.-S. Wang. 2020. Aflatoxin in household maize for human consumption in Kenya, East Africa. *Food Additives & Contaminants: Part B* 13, no. 1: 45–51. https://doi.org/10.1080/19393210.2019.1690053
33. Hanvi, D. M., P. Lawson-Evi, M. De Boevre, C. Goto, S. De Saeger, and K. Eklu-Gadegbeku. 2019. Natural occurrence of mycotoxins in maize and sorghum in Togo. *Mycotoxin Research* 35: 1–7. https://doi.org/10.1007/s12550-019-00351-1
34. Granados-Chinchilla, F., A. Molina, G. Chavarría, M. Alfaro-Cascante, D. Bogantes-Ledezma, and A. Murillo-Williams. 2017. Aflatoxins occurrence through the food chain in Costa Rica: Applying the one health approach to mycotoxin surveillance. *Food Control* 82: 217–226. https://doi.org/10.1016/j.foodcont.2017.06.023
35. Zuki-Orozco, B. A., L. E. Batres-Esquivel, M. D. Ortiz-Pérez, B. I. Juárez-Flores, and F. Díaz-Barriga. 2018. Aflatoxins contamination in maize products from rural communities in San Luis Potosi, Mexico. *Annals of Global Health* 84, no. 2: 300. https://doi.org/10.29024/aogh.918
36. Oliveira, M. S., A. Rocha, M. Sulyok, R. Krska, and C. A. Mallmann. 2017. Natural mycotoxin contamination of maize (*Zea mays* L.) in the South Region of Brazil. *Food Control* 73: 127–132. https://doi.org/10.1016/j.foodcont.2016.07.033
37. Food and Agriculture Organization. 2004. Worldwide Regulations for Mycotoxins in Food and Feed in 2003. Rome. FAO Food and Nutrition Paper no. 81, pp. 1–180.
38. Lai, X., R. Liu, C. Ruan, H. Zhang, and C. Liu. 2015. Occurrence of aflatoxins and ochratoxin A in rice samples from six provinces in China. *Food Control* 50: 401–404. https://doi.org/10.1016/j.foodcont.2014.09.029
39. Al-Zoreky, N. S., and F. A. Saleh. 2019. Limited survey on aflatoxin contamination in rice. *Saudi Journal of Biological Sciences* 26, no. 2: 225–231.
40. Ali, N. 2019. Aflatoxins in rice: Worldwide occurrence and public health perspectives. *Toxicology Reports* 6: 1188–1197.
41. Jallow, A., H. Xie, X. Tang, Z. Qi, and P. Li. 2021. Worldwide aflatoxin contamination of agricultural products and foods: From occurrence to control. *Comprehensive Reviews in Food Science and Food Safety* 20, no. 3: 2332–2381.
42. Waliyar, F., M. Osiru, B. Ntare, K. V. K. Kumar, H. Sudini, A. Traore, and B. Diarra. 2015. Post-harvest management of aflatoxin contamination in groundnut. *World Mycotoxin Journal* 8, no. 2: 245–252. https://doi.org/10.3920/WMJ2014.1766
43. N'dede, C. B., C. Jolly, S. D. Vodouhe, and P. Jolly. 2012. Economic risks of aflatoxin contamination in marketing of peanut in Benin. *Economics Research International* 2012: 1–12. https://doi.org/10.1155/2012/230638
44. Filazi, A., and U. T. Sireli. 2023. Occurrence of Aflatoxins in Food. In *Recent Advances and Future Prospects*, ed. A. Filazi and U. T. Sireli, 143–170. Rijeka: IntechOpen. https://doi.org/10.5772/51031
45. Diella, G., G. Caggiano, F. Ferrieri, et al. 2018. Aflatoxin contamination in nuts marketed in Italy: Preliminary results. *Annali Di Igiene* 30: 401–409. https://doi.org/10.7416/ai.2018.2240
46. Kabak, B., and A. D. Dobson. 2017. Mycotoxins in spices and herbs: An update. *Critical Reviews in Food Science and Nutrition* 57, no. 1: 18–34. https://doi.org/10.1080/10408398.2013.772891
47. Scott, P. M. 2008. Mycotoxins in Alcoholic Beverages and Fruit Juices: Occurrence and Analysis. In *Food Contaminants, ACS Symposium Series*, eds. D. P. Siantar, M. W. Trucksess, P. M. Scott and E. M. Herman, 1001, 170–191. ACS Publishers. https://doi.org 10.1021/bk-2008-1001.ch009
48. Bessaire, T., I. Perrin, A. Tarres, A. Bebius, F. Reding, and V. Theurillat. 2019. Mycotoxins in green coffee: Occurrence and risk assessment. *Food Control* 96: 59–67. https://doi.org/10.1016/j.foodcont.2018.08.033
49. Fink-Gremmels, J., and D. van der Merwe. 2019. Mycotoxins in the Food Chain: Contamination of Foods of Animal Origin. In *Chemical Hazards in Foods of Animal Origin*, eds. J. M. Frans, R. M. C. M. Smulders and R. Martin, 241–261. Wageningen Academic. https://doi.org/10.3920/978-90-8686-877-3_10
50. Oluwafemi, F. T. 2012. Aflatoxin M1 levels in lactating mothers in two Nigerian cities. *Archives of Clinical Microbiology* 3, no. 4: 10–3823.
51. DSM-Firmenich. 2023. World Mycotoxin Survey. https://www.dsm.com/anh/products-and-services/tools/mycotoxin-contamination/mycotoxin-survey.html. Accessed on March 03, 2024
52. Gong, Y. Y., S. Watson, and M. N. Routledge. 2016. Aflatoxin exposure and associated human health effects, a review of epidemiological studies. *Food Safety* 4, no. 1: 14–27.
53. Qian, G. S., R. K. Ross, and M. C. Yu, et al. 1994. A follow-up study of urinary markers of aflatoxin exposure and liver cancer risk in Shanghai, People's Republic of China. *Cancer Epidemiology, Biomarkers & Prevention* 3: 3–10.

54. Gouas, D., H. Shi, and P. Hainaut. 2009. The aflatoxin-induced TP53 mutation at Codon 249 (R249S): Biomarker of exposure, early detection and target for therapy. *Cancer Letters* 286: 29–37. https://doi.org/10.1016/j.canlet.2009.02.057
55. International Agency for Research on Cancer. 2012. IARC monographs on the evaluation of carcinogenic risks to humans; Vol 56, Chemical Agents and Related Occupations. IARC, Lyon, 100F: 1–599.
56. Magoha, H., M. Kimanya, B. De Meulenaer, D. Roberfroid, et al. 2014. Association between aflatoxin M 1 exposure through breast milk and growth impairment in infants from Northern Tanzania. *World Mycotoxin Journal* 7: 277–284. https://doi.org/10.3920/WMJ2014.1705
57. Gong, Y. Y., P. C Turner, A. J. Hall, and C. P Wild. 2008. Aflatoxin exposure and impaired child growth in West Africa: An unexplored international public health burden? In, *Mycotoxins: Detection methods, management, public health and agricultural trade*, eds. J. F. Leslie, R. Bandyopadhyay, and A. Visconti, 53–66. CABI North American Office.
58. Smith, L. E., R. J. Stoltzfus, and A. Prendergast. 2012. Food chain mycotoxin exposure, gut health, and impaired growth: A conceptual framework. *Advances in Nutrition* 3: 526–531. https://doi.org/10.3945/an.112.002188
59. Bondy, G. S., J. J. Pestka, G. S. Bondy, and J. J. Pestka. 2000. Immunomodulation by fungal toxins. *Journal of Toxicology and Environmental Health, Part B* 3: 109–143. https://doi.org/10.1080/109374000281113
60. Routledge, M. N., and Y. Y. Gong. 2011. Developing Biomarkers of Human Exposure to Mycotoxins. In *Determining Mycotoxins and Mycotoxigenic Fungi in Food and Feed*, eds. M. Magan and M. Olsen, 225–244. Woodhead Publishing.
61. Chapot, B., and C. P. Wild. 1991. ELISA for quantification of aflatoxin-albumin adducts and their application to human exposure assessment. *Techniques in Diagnostic Pathology* 2: 135–155.
62. Sabbioni, G., S. Ambs, G. N. Wogan, and J. D. Groopman. 1990. The aflatoxin—lysine adduct quantified by high-performance liquid chromatography from human serum albumin samples. *Carcinogenesis* 11, no. 11: 2063–2066.
63. Scholl, P. F., L. McCoy, T. W. Kensler, and J. D. Groopman. 2006. Quantitative analysis and chronic dosimetry of the aflatoxin B1 plasma albumin adduct lys-AFB1 in rats by isotope dilution mass spectrometry. *Chemical Research in Toxicology* 19, no. 1: 44–49.
64. Gonçalves, A., J. L. A. Gkrillas, C. J. L. Dorne, R. C. Dall'Asta, R. Palumbo, N. R. Lima, P. Battilani, A. Venâncio, and V. P. Giorni. 2019. Pre- and postharvest strategies to minimize mycotoxin contamination in the rice food chain. *Comprehensive Reviews in Food Science and Food Safety* 18: 441–454.
65. de Freitas, C. H., C. L. Gonçalves, and P. d. S. Nascente. 2018. Aflatoxins B 1 and M 1: risks related to milk produced in Brazil. *Annals of Microbiology* 68: 793–802.
66. Ajani J, Chakravarthy DVS, Tanuja P, Pasha KV (2014). Aflatoxins. *Indian J. Adv. Chem. Sci.* 3:49–60.
67. FDA (US Food and Drug Administration). 2019. Compliance Policy Guides–CPG Sec. 683.100. Action Levels for Aflatoxins in Animal Food: Guidance for FDA Staff. https://www.fda.gov/ICECI/ComplianceManuals/CompliancePolicyGuidanceManual/default.htm. Accessed on April 3, 2024.
68. Adeyeye, S. A. 2016. Fungal mycotoxins in foods: A review. *Cogent Food & Agriculture* 2, no. 1: 1213127.
69. Wu, F. 2004. Mycotoxin risk assessment for the purpose of setting international regulatory standards. *Environmental Science & Technology* 38, no. 15: 279A–284A. https://doi.org/10.1021/es035353n.
70. Groopman, J. D., T. W. Kensler, and C. P. Wild. 2008. Protective interventions to prevent aflatoxin-induced carcinogenesis in developing countries. *Annual Review of Public Health* 29: 187–203.
71. Liu, Y., and F. Wu. 2010. Global burden of aflatoxin-induced hepatocellular carcinoma: A risk assessment. *Environmental Health Perspectives* 118, no. 6: 818–824.
72. Wu, F., and H. Guclu. 2012. Aflatoxin regulations in a network of global maize trade. *PLoS ONE* 7, no 9: e45151.
73. Meneely, J. P., O. Kolawole, S. A. Haughey, S. J. Miller, R. Krska, and C. T. Elliott. 2023. The challenge of global aflatoxins legislation with a focus on peanuts and peanut products: A systematic review. *Exposure and Health* 15, no. 2: 467–487.
74. Pandey, A. K., M. K. Samota, A. Kumar, A. S. Silva, and N. K. Dubey. 2023. Fungal mycotoxins in food commodities: Present status and future concerns. *Frontiers in Sustainable Food Systems* 7: 1162595.

# 2.14 Aflatoxigenic Fungi and Animal Feed

*S. Meenakshi, R. Saravanan, S. Umayaparvathi, and S. Kumaran*

## INTRODUCTION

Of all the mycotoxins, aflatoxins are the most common and well-researched type. They are found in warm, humid climates, such as those found in India and many other Asian nations. Dairy feeds and human food products include aflatoxins, which are mostly generated by *Aspergillus* fungus (*A. flavus, A. parasiticus,* and *A. nomius*). As with other fungi, the taxonomy of Aspergillus is complex and constantly changing. Probably the most prevalent and well-known mycotoxin contamination is aflatoxin. The molds *A. flavus* and *A. parasiticus* are the ones that create it. Actually, "*A. flavus* toxin" is the composite term from which the name is formed. Commonly impacted foods include rice, maize, wheat, sorghum, beans, figs, oilseed cakes, cottonseed, copra, tree nuts like pistachio and Brazil nuts, and all nuts, particularly groundnuts. Aflatoxin is typically present in unrefined vegetable oils derived from infected seeds or nuts. However, the refining process eliminates aflatoxin, making refined oils safe.

Toxins that glow blue (B) or green (G) when exposed to UV light are known as aflatoxin B1, B2, G1, and G2. These toxins may be separated using thin-layer chromatography (TLC). The presence of an oxygen atom in the cyclopentanone ring is the only structural difference between toxins B and G. The toxins B1 and B2, which have been metabolized inside the body of a nursing animal, are represented by aflatoxin M1 and M2. They were given the moniker "M" after discovering milk. The hydroxyl group's inclusion distinguishes B and M's structures clearly. TLC, high-performance liquid chromatography (HPLC), and enzyme-linked immunosorbent assay (ELISA) are the analytical techniques used to detect aflatoxin. The most often employed technique has been TLC. In many species, aflatoxins are incredibly powerful carcinogens. Aflatoxins degrade in the presence of light. Thus, it is imperative that all analytical materials be sufficiently shielded from light. Standard aflatoxin solutions should be maintained in amber-colored vials or aluminum foil [1].

Aflatoxin B1 is the most prevalent and harmful type of aflatoxin, with other major forms detected in feeds including aflatoxin B2, B2, G1, and G2. Aflatoxin M1 can be discovered in dairy products, including milk. To reduce the amount of aflatoxin M1 in milk and milk products, there must be strict control over the amount of aflatoxin B1 in the feeds fed to nursing dairy cattle and buffaloes. The Food Safety and Standards Authority of India [2] and the United States Food and Drug Administration [3].

The contamination of animal feed with aflatoxigenic fungi and their associated toxins poses a significant threat to animal health, food safety, and economic stability within the livestock industry. Acute intake of aflatoxins in high amounts may cause serious poisoning (aflatoxicosis), which can be life-threatening, usually through damage to the liver. Several outbreaks of aflatoxicosis have been observed in animals and in human populations since the 1960s. Aflatoxins may also lead to health problems in animals; furthermore, the destruction of contaminated crops poses a severe burden on the economy. Several factors have been investigated in connection with aflatoxin contamination of the milk value chain. A promising direction to address the challenge may be the breeding of crops that are resistant to mycotoxins. Another direction is the utilization of nontoxigenic *Aspergillus* strains that outcompete the toxigenic strains, thereby limiting toxin production. There are also different


DOI: 10.1201/9781003487043-20

postharvest practices (cleaning, sorting, chemical treatments, etc.) to eliminate contamination or mitigate the adverse effects on animals [4].

Aflatoxigenic fungi in the feed supply chain occur at multiple points, from pre-harvest contamination of crops such as maize, peanuts, and tree nuts to post-harvest storage and processing. Factors influencing fungal growth and aflatoxin production include environmental conditions (temperature, humidity), agronomic practices, and interactions between fungal and plant physiology. Understanding these factors is crucial for implementing effective preventive strategies and mitigating the risk of aflatoxin contamination in animal feed.

## AFLATOXINS: NATURE AND CHARACTERISTICS

All animals are affected by aflatoxins, but some species are more so than others. Rabbits, ducks, and pigs are highly susceptible; dogs, calves, turkeys, and sheep are moderately susceptible; and chickens and cattle are relatively resistant. Fish vary from highly susceptible to resistant, and honey bees are relatively resistant. The aflatoxin problem has been reported to be more serious in tropical and subtropical regions of the world, where climatic conditions of temperature and relative humidity favor the growth of *A. flavus and A. parasiticus* Table 2.14.1 shows types of aflatoxin, producing fungal species, and commodities affected by aflatoxin.

## AFLATOXIGENIC FUNGI: IDENTIFICATION AND BEHAVIOR

Although immunochemical techniques that employ antibodies are increasingly popular as aflatoxins' analytical instruments, chemical approaches are still the primary means of analyzing aflatoxins. Because flavonoids are formed by certain molds and have a complicated chemical structure that requires accurate quantification, chemical techniques are the primary means of analysis for this category of very hazardous chemicals. These procedures frequently combine mass spectrometry and chromatography, which can precisely separate and detect aflatoxins in a variety of food and feed samples.

### METHODOLOGY

To generate tenfold serial dilutions, 5 g of each feed sample was first individually placed aseptically into sterile tubes and mixed with 45 ml of sterile distilled water. Subsequently, 1 milliliter of the previously made serial dilutions was inoculated individually into sterile Petri dish plates and combined with either Sabouraud's dextrose agar (SDA) medium containing 0.05 milligrams of chloramphenicol per milliliter or Dichloran Rose-Bengal chloramphenicol agar. The plates were allowed to dry and harden.

Subsequently, it was incubated for five days at 25°C ± 1°C in an aerobic environment. After two to five days of incubation, the plates were read. The morphological traits of the fungal cultures, such as surface, texture, and colony size (millimeter, diameter), were used to differentiate them. Throughout the incubation phase, the fungal cultures were routinely inspected. After seven days of incubation at 28°C, the culture properties and sporulation on various culture mediums were noted. Each fungal isolate's morphological features were identified using a light microscope (OPTO-EDU, China). Following the sporulation of fungal colonies on various culture mediums, a description of the microscopic analysis of fungal isolates was provided. For this aim, distilled water was used to mount a tiny portion of mycelia from the center and periphery of the developing colony onto a microscope slide, which was then covered with a cover slip. It was established what the features of the vegetative and reproductive structures were, including the size, shape, and color of the hyphae, as well as their structure [6, 7]. The virulent genes of the tested *Aspergillus* sp. that were extracted from the animal feeding samples were determined by polymerase chain reaction (PCR) [8, 9].

**TABLE 2.14.1**
**Types of Aflatoxin, Producing Fungal Species, and Commodities Affected by Aflatoxin (Data from Benkerroum) [5]**

| Aflatoxin | Aflatoxin Producing *Aspergillus* spp. | Host/Affected Entity |
|---|---|---|
| B (B1, B2) | *A. arachidicola, A. bombycis, A. flavus, A. minisclerotigenes, A. nomius, A. ochraceoroseus, A. oryzae, A. parasiticus, A. parvisclerotigenus, A. pseudotamarii, A. rambellii, A. tamarii, A. toxicarius, A. versicolor, Emericella astellata, E. venezuelensis.* | Cottonseed, dairy products, figs, fruit juices (apple, guava), maize, maize flour, meat, oilseed rape, peanuts, peanut butter, pea, pistachio, rice, sorghum, sunflower seed, spices |
| B2a | Hydroxylated metabolite of aflatoxin B1 | — |
| B3 (Parasiticol) | Aflatoxin G1 metabolite, naturally produced by: *A. flavus, A. mottae, A. nomius, A. novoparasiticus, A. parasiticus* | Same as aflatoxin B1 and G1 |
| G (G1, G2) | *A. arachidicola, A. bombycis, A. minisclerotigenes. A. nomius, A. parasiticus, A. pseudotamarii, A. terreus, A. toxicarius, A. versicolor* | Cottonseed, dairy products, figs, fruit juices (apple, guava), maize, maize flour, meat, oilseed rape, peanuts, Peanut butter, pea, pistachio, Rice, sorghum, sunflower seed, spices |
| G2a | Hydroxylated metabolite of aflatoxin G1, also naturally produced by *A. flavus* | — |
| M (M1, M2) | Hydroxylated metabolite of aflatoxin B1 and B2, respectively | Dairy products, milk, meat |
| M2a | Aflatoxin M1 derivative | Dairy products, milk |
| GM1 | Hydroxylated metabolite of aflatoxin G1, naturally produced by *A. flavus*, also produced by *A. parasiticus* in vitro | Dairy products, milk |
| GM2 | Naturally produced by *A. flavus* and *A. parasiticus* and yeast, derived from aflatoxin G2 | Dairy products, milk |
| GM2a | Aflatoxin GM1 metabolite | Dairy products, milk |
| P1 | Metabolite of aflatoxin B1 (demethylated) | Dairy products, excreted in animals and human urine |
| Q1 | Metabolite of aflatoxin B1 (hydroxylated) | Present in the meat of cattle that feeds on aflatoxin-contaminated food |
| Q2a | Acid hydration of aflatoxin Q1 | — |
| Aflatoxicol R0 | Metabolite of aflatoxin B1, also naturally produced by *A. flavus* and *A. parasiticus* | Present in bird feed, also exists in birds that feed on aflatoxin-contaminated food |
| Aflatoxicol M1 | Aflatoxin B1, aflatoxin R0, or aflatoxin M1 metabolite | Dairy products, milk |
| Aflatoxicol H1 | Aflatoxin B1 and aflatoxin Q1 metabolite | Dairy products, milk |
| Aspertoxin | *A. flavus* and *A. parasiticus* | Crops and plants |

Aflatoxin levels in animal feed need to be investigated to ensure feed safety and, therefore, animal health. Many methods may be used to analyze aflatoxin in animal feed. Here are a few of the most often used methods: The most often used quantitative techniques for AFB1, AFB2, AFG1, and AFG2 analysis are near-infrared spectroscopy (NIRS), ELISA, and liquid chromatography (LC) with mass spectrometry (MS) or fluorescence detection (FD).

## High-Performance Liquid Chromatography (HPLC)

Aflatoxin in animal feed may be analyzed using HPLC, a method that is often employed. Because of their great sensitivity and selectivity, aflatoxins can be accurately quantified. HPLC analysis of AFB1. Samples are usually extracted with methanol or combinations of methanol and water or hexane for LC-FD analysis. When it comes to oil samples, the latter is employed. Aflatoxin-specific immunoassay columns can be used to clean samples before ending on the aflatoxin and matrix, limits of quantification (LOQs) and limits of detection (LODs) are usually reported to be in the range of 0.001–0.20 μg/kg.LC separation, post-column derivatization, and FD quantification. Depending on the aflatoxin and matrix, LOQs and LODs are typically reported to be in the range of 0.001–0.20 μg/kg [4].

## Liquid Chromatography-Mass Spectrometry (LC-MS)

LC-MS has the capacity to simultaneously detect and quantify many aflatoxin species with great sensitivity and selectivity. When handling complicated sample matrices or confirmatory analysis, LC-MS is especially helpful. The benefit of using mass spectrometry to determine aflatoxins is that post-column derivatization is not required. Before being analyzed by (LC-MS) or LC coupled with tandem mass spectrometry (LC-MS/MS), aflatoxins are usually extracted using acetonitrile, and occasionally mixed with water, formic acid, or hexane. The development of multi-mycotoxin techniques was significantly influenced by the use of LC-MS or LC-MS/MS. In a single investigation, these approaches have been developed to determine over 100 mycotoxins. Nevertheless, a heightened quantity of chemicals added to an analytical procedure might result in diminished sensitivity.

## Enzyme-Linked Immunosorbent Assay (ELISA)

ELISA is an immunoassay-based technique that looks for aflatoxins in animal feed by using certain antibodies. It can be completed quickly and easily, which makes it appropriate for high-throughput analysis. ELISA is the most widely used analytical technique for AFM1 measurement that has been documented in the literature. The LOD of commercial kits designed specifically to measure AFM1 in milk is usually around 0.005 μg/L. Additionally, LC-MS and LC-FD techniques are employed to determine AFM1. Commercial ELISA kits are also available for measuring aflatoxin total (AFT) and AFB1. Additional immunochemical-based techniques have been developed to analyze aflatoxins, as have the benefits and drawbacks of the various approaches [10].

## Near-Infrared Spectroscopy (NIRS)

An effective and non-destructive method for detecting aflatoxin contamination in animal feed is NIRS. Every AFB1-free rice sample was split into two subsamples, each weighing 115 g. To achieve the appropriate final concentration, the AFB1 standard solution was then added to each 115-g AFB1-free rice subsample. A diode array NIR spectrometer (DA 7200, Perten Instruments, Stockholm, Sweden) with a wavelength range of 950–1,650 nm in diffuse reflection mode and 5 nm of resolution was used to scan each spiking subsample. A 115-g sample of rice was put in an open cup that was 75 mm wide and 25 mm deep for each scan. The samples that were naturally infected underwent the same procedure. Every sample was split into two subsamples weighing 115 g apiece, and the NIR spectrometer was used to scan each subsample to collect spectral data. Three scans were performed on each subsample. The NIR scanning findings have a repeatability of 0.003. Using the Unscrambler 9.8 program (Camo, Norway), all spectral data from 210 subsamples were examined. The average reflection spectra were employed in the analysis process. The

AFB1 concentration in the scanned samples was then measured using HPLC and used as reference data [11]. There are options for verified reference materials and proficiency testing in several matrices. There are additional established techniques (EN-methods) for determining aflatoxins. The techniques employed for official control are subject to performance and quality assurance criteria, as outlined in the legislation.

## AFLATOXIN CONTAMINATION IN ANIMAL FEED

Because aflatoxins are not uniformly dispersed throughout meals and feeds, they are challenging to identify. As a result, it is challenging to get representative and trustworthy samples and identify low contamination levels. There are several sample techniques that have to be adhered to whenever feasible. Aflatoxin accumulation in animal tissues and products (including milk and eggs) can occur when animals eat contaminated feed, endangering the health of the animals as well as the safety of food items made from them. Animals exposed to aflatoxin may have a range of health problems, such as immunosuppression, liver damage, slowed growth rates, and, in extreme situations, even death.

### ALTERNATIVE USES OF AFLATOXIN-CONTAMINATED FEED

Because aflatoxins are not uniformly dispersed throughout meals and feeds, they are challenging to identify. As a result, it is challenging to get representative and trustworthy samples and identify low contamination levels. There are several sample techniques that have to be adhered to whenever feasible.

Aflatoxin accumulation in animal tissues and products (including milk and eggs) can occur when animals eat contaminated feed, endangering the health of the animals as well as the safety of food items made from them. Animals exposed to aflatoxin may have a range of health problems, such as immunosuppression, liver damage, slowed growth rates, and, in extreme situations, even death. Most aflatoxin-contaminated grains are probably best used when fed to suitable livestock, provided that there is no excessive danger to the animals' health. Numerous animals, especially older animals, can withstand aflatoxins effectively, although there are now no set quantities at which they can be assured to be safe for cattle. In fact, low aflatoxin levels do not appear to have statistically significant effects in several experimental trials. At lower aflatoxic dosages, there is a continuous pattern of fewer, less severe, or no indications, and at greater doses, there is an increase in effects.

### REGULATIONS FOR THE EXPORT OF CONTAMINATED FEEDS

Regulations in several nations allow feed and feed additives that don't fit national limitations to be exported. When grains are sent to nations where their final application cannot be strictly regulated, this becomes problematic [12]. Different people have defined quality in different ways, such as "fitness for use," "meeting an expectation," "degree of excellence," or "conforming to a standard." Many makers of feed do not examine components before usage, although many employ NIRS to quickly ascertain the moisture, fat, protein, and fiber content of an ingredient sample. Consequently, most firms rely on "supplier histories." Predictability is crucial when it comes to feed components, and it follows that high-quality products have consistent and predictable quality. Put another way, premium ingredients fulfill expectations most of the time, not just occasionally.

Three elements of the feed mill are involved in the process of turning premium materials into premium feeds: staff, equipment, and protocols. It is doubtful that high-quality feeds will be consistently produced if any one of these three quality components is deficient. But it's also critical to make sure that staff, equipment, and processes are "blended" to provide high-quality meals as efficiently as possible. Feeds are frequently utilized quickly after production, and animals eat the feeds before any tests can be completed. Nevertheless, tests for finished feed are essential and crucial as they give the mill a "final report card" on how effectively quality was managed. As a general guideline, one sample of each formula should be collected every week or one sample per 100 tons of output, whichever is higher. As soon as a problem is identified, it has to be addressed and fixed [13].

## REGULATORY GUIDELINES AND CONTROL MEASURES

Aflatoxin management is deemed necessary due to the harmful effects of aflatoxins on animals across the world. Several control methods, including chemical, biological, and physical management techniques, are employed globally for this reason and are covered in the sections that follow.

By using naturally existing atoxigenic strains through competitive exclusion, aflatoxin may be effectively controlled in the field. Strains of *A. flavus* are separated into S and L strains according to their morphology. Aflatoxigenic strains that are partially unable to manufacture aflatoxins due to genetic changes are included in L strains, which produce big sclerotia but a lower amount of aflatoxin. S strains produce small sclerotia but a high level of aflatoxin. [14]. Aflatoxin synthesis in the field has been effectively reduced by using atoxigenic strains of *Aspergillus* against toxigenic strains, which in turn has controlled aflatoxin contamination in storage settings. Aflatoxin has been controlled globally by utilizing the method of utilizing atoxigenic *Aspergillus* strains against toxigenic strains, which was initially used by Cotty and Bayman [15]. Table 2.14.2 shows list of nations where aflatoxin biopesticides are being used. For more than 20 years, toxic strains have been employed as biocontrol agents against toxic strains.

**TABLE 2.14.2**
**List of the Nations where Aflatoxin Biopesticides Are Being Investigated or Used (Cotty) [15]**

| S. No | Product/Strain Name | Country | References |
|---|---|---|---|
| 1. | AF36 | United States | [16] |
| 2. | Afla-Guard (strain RRL21882) | United States | [15] |
| 3. | CT3 (unregistered) | Southern United States | [17] |
| 4. | K49 (unregistered) | Southern United States | [18] |
| 5. | AF-X1 | Italy | [19] |
| 6. | Aflasafe SN01 | Senegal and The Gambia | [20] |
| 7. | Aflasafe GH01 | Ghana | [21] |
| 8. | Aflasafe GH02 | Ghana | [21] |
| 9. | Aflasafe | Nigeria | [21] |
| 10. | Aflasafe KE01 | Kenya | [21] |
| 11. | AR27 (unregistered) | Northern Argentina | [22] |
| 12. | AR100G (unregistered) | Northern Argentina | [22] |
| 13. | AFCHG2 (unregistered) | Northern Argentina | [22] |
| 14. | FS10 (unregistered) | China | [23] |
| 15. | AF051 (unregistered) | China | [24] |
| 16. | BN30 (unregistered) | Africa | [24] |
| 18. | Aflasafe TZ01 | Tanzania | [25] |
| 19. | Aflasafe TZ02 | Tanzania | [25] |
| 20. | Aflasafe ZM01 and ZM02 | Zambia | [25] |
| 21. | Aflasafe MW01 and MWMZ01 | Malawi | [25] |
| 22. | Aflasafe MZ01 and MWMZ01 | Mozambique | [25] |

## FUTURE DIRECTIONS AND RESEARCH OPPORTUNITIES

The first studies on aflatoxins date back to the late 1950s and early 1960s. From that time until now, various studies and discoveries have been made regarding aflatoxins. There are 18 different types of aflatoxins that have long been discovered. New and advanced technology has also enabled mankind to study the structure of aflatoxins and their biosynthesis pathways utilizing different methods, which can be used to detect them at early stages.

## CONCLUSION

This chapter provided insights of aflatoxins and their biosynthesis pathways, shedding light on their mechanisms of action and potential health effects. Advanced analytical techniques, such as MS and NIRS, have enabled researchers to study aflatoxins at a molecular level, enhancing our ability to detect and quantify them accurately. These methods have also contributed to the development of rapid and sensitive detection assays, allowing for the early identification of aflatoxin contamination in food and feed products. As a result, ongoing research continues to refine our understanding of aflatoxins and improve strategies for their detection, prevention, and control, ultimately safeguarding public health and food safety.

## REFERENCES


1. Krska R., Schubert P., Molinelli A., Sulyok M., MacDonald S., & Crews C. 2008. Mycotoxin analysis: An update. *Food Additives and Contaminants*, 25(2):152–163.
2. FSS. 2011. Food Safety and Standards (Contaminants, Toxins and Residues) Regulation.
3. USFDA. 2018. United States Food and Drug Administration.
4. World Health Organization. 2018. Food Safety Digest–Aflatoxins.
5. Benkerroum N. 2020. Aflatoxins: Producing-molds, structure, health issues and incidence in Southeast Asian and Sub-Saharan African Countries. *International Journal of Environmental Research and Public Health*, 17:1215.
6. ISO. 2008. Microbiology of food and animal feeding stuffs: Horizontal method for the enumeration of yeasts and molds.
7. Pitt J. I., & Hocking A. D. 2009. Fungi and food spoilage, 3rd Edition. Springer, New York, pp. 519–388.
8. Hassan A. A., Oraby N. H., El-Mesalamy M. M., & Sayed-ElAhl R. M. H. 2022. Effect of hybrid nanomaterial of copper-chitosan against aflatoxigenic fungi in poultry feed. *The Journal of World's Poultry Research*, 12(3):157–164.
9. Isalar O. F., Ogbuji N. G., Okungbowa F. I., & Ataga A. E. 2021. Fungal contaminants associated with groundnut (arachis hypogaea) seeds. *Journal of Bioinformatics and Systems Biology*, 4:182–193.
10. Matabaro E., Ishimwe N., Uwimbabazi E., & Lee B. H. 2017. Current immunoassay methods for the rapid detection of aflatoxin in milk and dairy products. *Comprehensive Reviews in Food Science and Food Safety*, 16(5):808–820.
11. Putthang R., Sirisomboon P., & Sirisomboon C. D. 2019. Shortwave near-infrared spectroscopy for rapid detection of aflatoxin B1 contamination in polished rice. *Journal of Food Protection*, 82(5):796–803.
12. Grace D., Lindahl J. F., Atherstone C., Kang'ethe E. K., Nelson F., Wesonga T., & Manyong V. M. 2015. Aflatoxin standards for feed in East Africa. International Livestock Research Institute (ILRI), Nairobi, Kenya.
13. Sohrabi N., & Taghizadeh M. 2018. Molecular identification of aflatoxigenic aspergillus species in feedstuff samples. *Current Medical Mycology*, 4(2):1.
14. Ren X., Zhang Q., Zhang W., Mao J., & Li P. 2020. Control of aflatoxigenic molds by antagonistic microorganisms: Inhibitory behaviors, bioactive compounds, related mechanisms, and influencing factors. *Toxins*, 12:24.
15. Cotty P. J. 2005. *Development of Aspergillus flavus AF36*. Plant Disease.
16. Okun D. O., Khamis F. M., Muluvi G. M., Ngeranwa J. J., Ombura F. O., Yongo M. O., & Kenya E. U. 2015. Distribution of indigenous strains of atoxigenic and toxigenic *Aspergillus flavus* and *Aspergillus parasiticus* in maize and peanuts agro-ecological zones of Kenya. *Agriculture & Food Security*, 4:1–10.
17. Dorner J., & Lamb M. C. 2006. Development and commercial use of Afla-Guard®, an aflatoxin biocontrol agent. *Mycotoxin Research*, 22:33–38.

18. Abbas H. K., Zablotowicz R. M., Bruns H. A., & Abel C. A. 2006. Biocontrol of aflatoxin in corn by inoculation with non-aflatoxigenic *Aspergillus flavus* isolates. *Biocontrol Science and Technology*, 16:437–449.
19. Mauro A., Garcia-Cela E., Pietri A., Cotty P. J., & Battilani P. 2018. Biological control products for aflatoxin prevention in Italy: Commercial field evaluation of atoxigenic *Aspergillus flavus* active ingredients. *Toxins*, 10:30.
20. Senghor A., Ortega-Beltran A., Atehnkeng J., Jarju P., Cotty P., & Bandyopadhyay R. 2021. Aflasafe SN01 is the first biocontrol product approved for aflatoxin mitigation in two nations, Senegal and the Gambia. *Plant Disease*, 105:1461–1473.
21. Agbetiameh D., Ortega-Beltran A., Awuah R. T., Atehnkeng J., Elzein A., Cotty P. J., & Bandyopadhyay R. 2020. Field efficacy of two atoxigenic biocontrol products for mitigation of aflatoxin contamination in maize and groundnut in Ghana. *Biological Control*, 150:104351.
22. Zanon M. S. A., Barros G. G., & Chulze S. N. 2016. Non-aflatoxigenic *Aspergillus flavus* as potential biocontrol agents to reduce aflatoxin contamination in peanuts harvested in Northern Argentina. *International Journal of Food Microbiology*, 231:63–68.
23. Xu D., Wang H., Zhang Y., Yang Z., & Sun X. 2013. Inhibition of non-toxigenic *Aspergillus niger* FS10 isolated from Chinese fermented soybean on growth and aflatoxin B1 production by *Aspergillus flavus*. *Food Control*, 32:359–365.
24. Yin Y.-N., Yan L.-Y., Jiang J.-H., & Ma Z. H. 2008. Biological control of aflatoxin contamination of crops. *Journal of Zhejiang University Science B*, 9:787–792.
25. Mahuku G. S. 2017. AR-NAFAKA project aflatoxin management: 2016–2017 progress. In Proceedings of the Africa RISING—NAFAKA scaling project end-of-project phase review meeting; Dar es Salaam, Tanzania, 3–4 July 2017, International Institute of Tropical Agriculture: Ibadan, Nigeria.

# 2.15 Effect of Aflatoxins on Animals of Veterinary Interest

*Mario E. Cancino-Díaz, Miguel A. De la Rosa-Ramos, Gabriel Betanzos-Cabrera, and Juan Carlos Cancino-Díaz*

## INTRODUCTION

Recently, contamination by metabolites produced by fungi in feedstuffs of production animals such as chickens, pigs, and cattle has been studied. Feed contaminated with these metabolites is introduced into the animals through the feed and, within the organism, can cause physiological damage to the animal and, in some cases, can lead to death. Therefore, these metabolites are considered toxic and, produced by fungi, are classified as mycotoxins. There are several types of mycotoxins, one of which is aflatoxin. These are classified as aflatoxin B1 (AFB1), AFB2, AFG1, AFG2, AFM1, and AFM2.

The physiological damage caused by AFB1 in animals is caused by the continuous ingestion of feed contaminated with mycotoxin, resulting in problems that impair proper growth and development (Sarker et al., 2021). Areas with hot, humid, or tropical climates provide suitable conditions for fungal growth in food, and stored food further facilitates fungal growth, allowing the mycotoxin to accumulate and reach high concentrations. For example, AFB1 has been reported in broiler feed at 0.3–0.7 μg/kg feed. This concentration range is constant worldwide, possibly due to variations or changes in environmental and climatic conditions affecting fungal growth.

Toxic effects caused by AFB1 in animals are generally hepatotoxic, teratogenic, mutagenic, and gastrointestinal (Monson et al., 2014). Regarding the digestive system, AFB1 is a compound that is readily absorbed (>80%) in the small intestine of poultry (Grenier & Applegate, 2013) and metabolized in the liver. Poultry diets are not free of AFB1, with approximately 20% of diets contaminated with AFB1 and, in some cases, approximately 5% showing toxic levels three times the estimated risk. This highlights the importance of implementing a feed control program to address this contamination problem.

Broiler chickens are high consumers of cereal grains, which makes them highly susceptible to mycotoxins, especially AFB1. They are unable to degrade AFB1 before it reaches their intestinal tract. Consumption of food contaminated with AFB1 poses a high risk of damage to intestinal tissue and other organs. Thus, addressing the problem of AFB1 contamination in food is crucial for intestinal protection and overall good health. Furthermore, AFB1 is commonly found in feed. It significantly affects broiler farming, causing disease and decreased broiler meat quality (Hussain et al., 2010).

This chapter addresses the detrimental effects of AFB1 on poultry and other animals of veterinary interest.

## AFLATOXIN B1 (AFB1)

Aflatoxins are a group of mycotoxins characterized by a difuran ring in their chemical structure and are considered highly toxic to animals. They are diverse, with an estimated number of 20 derivatives, classified into six groups according to their fluorescent properties under UV light and


DOI: 10.1201/9781003487043-21

chromatographic mobility: AFB1, AFB2, AFG1, AFG2, AFM1, and AFM2 (Filazi & Sireli, 2013). Among them, AFB1 is considered the most toxic, with strong pathogenic effects on animals. Its toxic properties were discovered in 1960 in London, England, when "turkey X disease" was identified. Since then, the World Health Organization (WHO) has classified AFB1 as a Class I natural carcinogen.

AFB1 is produced by *Aspergillus flavus* and *Aspergillus parasiticus* and is highly prevalent (Gündüz & Oznurlu, 2014). Its chemical formula is $C_{17}H_{12}O_6$, with a molecular weight of 312.27 g/mol. AFB1 is considered mutagenic, teratogenic, carcinogenic, and capable of producing DNA adducts (Woo et al., 2011). It is found in corn feed, the primary source of energy for poultry broilers, and other feedstuffs (Abbasi et al., 2018). In addition, AFB1 may be present in meat from edible animals fed contaminated feed (Yilmaz et al., 2017). The US Food and Drug Administration (FDA) considers AFB1 an important food contaminant (Williams et al., 2004).

## METABOLISM OF AFB1 IN THE LIVER

After ingesting AFB1-contaminated feed, the mycotoxin is absorbed in the duodenum of the small intestine and reaches the liver, where it accumulates and is metabolized. In the bovine intestine, detoxification of AFB1 can occur, particularly in the rumen, where microorganisms detoxify it. However, continued ingestion of AFB1 can damage the intestine in animals without a rumen.

The liver metabolizes foreign compounds, including AFB1 (Li et al., 2022). AFB1 is metabolized into less toxic and soluble compounds for excretion in urine or feces. The detoxification process occurs mainly in phases (Phase I and II), present in most animals, including birds, pigs, cattle, and humans. Phase I reactions involve oxidative, reductive, and hydrolytic reactions, producing intermediate metabolites conjugated with liver compounds to become soluble and excretable in Phase II.

### PHASE I: ENZYMATIC METABOLISM OF AFB1

AFB1 itself is not toxic, but it forms intermediate metabolites with increased toxicity during metabolism. In birds, AFB1 is mainly metabolized in the liver by enzymes catalyzing oxidation, hydroxylation, and demethylation reactions, which result in different intermediate metabolites. In the oxidation reaction, AFB1 is transformed into epoxy compounds such as AFB1-8, 9-epoxide (AFBO), subsequently converted into aflatoxin alcohol (AFL). In hydroxylation and demethylation reactions, AFB1 is transformed into intermediate metabolites such as aflatoxin Q1 (AFQ1), aflatoxin M1 (AFM1), and aflatoxin P1 (AFP1). All these intermediate metabolites exhibit high reactivity. Cytochrome P-450 (CYP-450) enzymes are involved in the biotransformation of AFB1, predominantly producing highly reactive intermediate metabolites (Akcam et al., 2013). Generally, CYP-450 enzymes are found in the endoplasmic reticulum and mitochondria, metabolizing AFB1 to reactive oxygen species.

In AFB1-susceptible avian species, such as poultry, AFB1 is metabolized by enzymes including CYP-1A1, 1A2, 1A4, 1A5, CYP2A4, 2A6, CYP3A4, and CYP3A37, forming AFBO, a metabolite highly sensitive to DNA and capable of generating DNA-damaging DNA adducts (Wu & Khlangwiset, 2010). Isoforms such as CYP1A2 and CYP3A4 participate in the transformation of AFB1 to AFBO and are present in liver microsomes (Kamdem et al., 2006). In addition, CYP1A2 converts AFB1 to AFM1, while CYP1A2, 1A5, CYP3A4, and CYP3A37 participate in the biotransformation of AFQ1, and CYP1A and CYP3A metabolize the intermediate metabolite AFP1.

Among the CYPs, CYP3A4 shows the highest activity in forming AFBO from AFB1 to produce a less toxic metabolite. CYP1A2 also shows significant activity in the bioactivation of AFB1, even at low concentrations. However, the biotransformation of AFB1 by CYPs aims to produce intermediate metabolites with low toxicity, but these metabolites can become more reactive and react with essential biomolecules in the cell. AFBO formation is highly reactive and capable of binding to proteins, causing acute toxicity, damaging DNA, inducing mutations, and causing cancer (Wu & Khlangwiset, 2010). AFBO can generate the guanine epoxide AF-N7, which interacts with DNA guanines, forming DNA adducts and mutations, and can bind to proteins, altering their function (Guengerich, 2001; Wild & Turner, 2002).

The intermediate metabolite AFL is produced in poultry, with reports indicating that AFB1-sensitive birds, such as ducks and turkeys, produce higher AFL levels than chickens and quail (Gregorio et al., 2015). In contrast, the non-toxic intermediate metabolites AFQ1, AFM1, and AFP1 are excreted directly via bile and urine.

In the human liver, CYP3A4 is the primary enzyme responsible for converting AFB1 to AFBO, whereas in birds such as turkeys, ducks, quail, and chickens, CYP2A6 is the predominant enzyme (Wang et al., 2019).

### Phase II: Detoxification of AFB1

In Phase II, some intermediate metabolites of AFB1 are mainly conjugated with glutathione to generate less toxic and soluble compounds that can be eliminated in feces and bile. The enzyme involved in Phase II metabolism of AFB1 is glutathione-S-transferase (GST), which has three superfamily branches: cytosolic GST, mitochondrial GST, and microsomal membrane-bound GST. Glutathione-S-transferase catalyzes the conjugation of glutathione with AFBO (Sherratt & Hayes, 2002). The glutathione (GSH) pathway is well known for its vital role in AFB1 detoxification because it facilitates the conjugation of hydrophobic substances (AFB1 metabolites) with GSH to produce more soluble metabolites (Oakley, 2011). In contrast, the intermediate metabolite AFL, derived from AFBO, is eliminated directly in the urine or as bile in the feces.

## AFB1 DAMAGE IN BIRDS

Although AFB1 can be metabolized in the liver for elimination, continuous consumption of AFB1-contaminated feed causes AFB1 levels to increase, and reactive intermediate metabolites such as AFBO to accumulate. This continuous consumption of contaminated feed can cause damage to poultry; however, it has been observed that the extent of damage caused by AFB1 is not uniform and depends on several factors. For example, poultry is susceptible to AFB1, and the impact of AFB1 on poultry varies depending on factors such as age, sex, type, and other characteristics of the birds.

Among poultry species, turkeys are the most susceptible to AFB1, followed by quail. Chickens are relatively tolerant of AFB1, whereas ducklings are more sensitive to AFB1 than turkeys and chicks. The lethal dose differs between bird species, with the median lethal dose (LD50) of AFB1 being 0.24–0.36 mg/kg for ducklings and 0.5–1.0 mg/kg for turkeys at ages over 9, 41, and 65 months. Age significantly affects AFB1 sensitivity; young poultry are more sensitive to AFB1 than older poultry, which is attributed to the more efficient microsomal activation of AFB1 and lower enzymatic activity for detoxification in younger birds.

Regarding the bird's sex, females have been reported to be more tolerant to AFB1 than males (Lozano & Diaz, 2006) (Figure 2.15.1).

## DIGESTION AND ABSORPTION OF NUTRIENTS

The function of the avian intestine is to digest food to form molecules or nutrients that the intestinal epithelium can readily absorb, while unwanted compounds are eliminated through feces. Digested nutrients are transported into the bloodstream by diffusion within the intestinal epithelium. Considering the length of the intestine, various criteria or indicators are used to assess its function and structural integrity. These indicators include villus height, crypt depth, and villus-to-crypt ratio, which serve as markers of intestinal health. Alterations in villus height and crypt depth suggest intestinal damage caused by specific compounds or toxins.

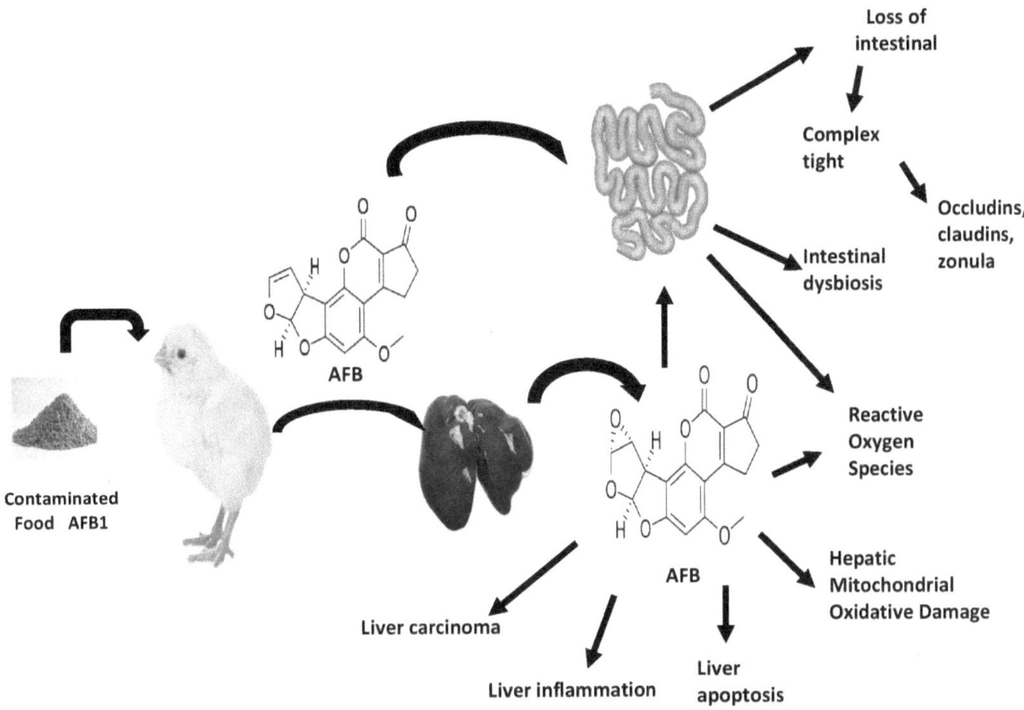


**FIGURE 2.15.1** Effects of AFB1 on birds. Feed contaminated with AFB1 ingested by chickens has an effect on the small intestine, altering its functions. Additionally, AFB1 is metabolized in the liver to form AFBO (AFB1-8, 9-epoxide), which exerts various effects on this organ.

Bird growth and development are critically dependent on the absorption of nutrients through the crypts and villi (Hernández et al., 2006). Changes in the development of enterocytes and the villus network determine the degree of digestion and the absorptive capacity of the intestine. As for aflatoxins, there has been a consensus in recent decades on their negative effects on the vulnerability of intestinal villi and the impairment of intestinal integrity (Bouhet & Oswald, 2005).

## INTESTINAL IMMUNE SYSTEM

The intestinal epithelial barrier serves as a vital defense mechanism for the overall health and stability of birds, functioning at mechanical (intestinal movement), chemical (pH, enzymes), biological (microbiota), and immune levels. The immune system is closely associated with intestinal protection because nearly 70% of the cells comprising the gut are immune cell types. The gastrointestinal tract and the immune system are intricately linked, with the intestinal immune system being essential for intestinal function, health, well-being, and disease prevention (Celi et al., 2017). Enteroendocrine enterocytes and goblet cells reside in the villi. In contrast, Paneth cells reside beneath the crypts, forming a protective barrier against harmful substances such as infections, toxins, and foreign antigens.

When hazardous substances compromise the intestinal epithelium's integrity, the gut may experience disturbances such as increased permeability, which can facilitate the diffusion of toxins (Grenier & Applegate, 2013), leading to intestinal inflammation.

Evidence indicates that AFB1 adversely affects broiler production performance by suppressing the immune system. Broilers fed with 40 μg/kg of AFB1 exhibited significant reductions in the production of antibodies, including IgM, IgG, and IgG, and decreased numbers of T and B cells (Liu et al., 2018). T-cell subsets (CD3, CD4, and CD8), antibodies (immunoglobulin [Ig]A, IgM, and IgG), and cytokines (interleukin [IL]-2, IL-10, and tumor necrosis factor [TNF]-alpha) were all found to decrease in the cecal tonsils of broilers following consumption of feed contaminated with AFB1 (Wang et al., 2019). Additionally, after 42 days of consuming feed contaminated with AFB1, Grozeva et al. (2020) observed mononuclear cell infiltrates in the small intestine of broilers fed with 0.05 mg/kg AFB1 in their feed, whereas broilers fed AFB1-free food did not exhibit cellular infiltrates, indicating an inflammatory process in the intestine due to the presence of AFB1.

## EFFECT OF AFB1 ON THE BROILER INTESTINE

Studies indicate that AFB1 adversely affects the viability intestinal (Zhang et al., 2022). AFB1 can diminish the surface area of the small intestine in broilers, leading to impaired nutrient assimilation. Broilers exposed to a diet contaminated with 2.0 g/kg of AFB1 exhibited a significant reduction in villus height, an increase in crypt depth, and a decline in jejunal villus/crypt ratio (Jahanian et al., 2017). Moreover, AFB1 can directly damage bird organs and disrupt normal physiological functions, including alterations in intestinal bacterial abundance (Wang et al., 2016). Broilers fed a diet contaminated with 0.02 to 0.7 mg/kg of AFB1 for 3 weeks exhibited lower total intestinal density (weight/length) (Yunus et al., 2011). AFB1 was also found to impact the gastrointestinal barrier function in broilers (Chen et al., 2016). Numerous epidemiological studies have confirmed the detrimental effects of consuming AFB1-contaminated feed on bird growth and health, leading to various metabolic disorders (Lewis et al., 2005). Specifically, regarding the intestine, epidemiological evidence indicates that AFB1 negatively affects the viability of cells, impeding cell proliferation and rendering cells unable to repair and recover from these effects. Consequently, the adverse effects of AFB1 on gut structure susceptibility and intestinal integrity in birds have gained increased recognition.

At the molecular level, the toxic effect of AFB1 primarily involves the loss of intestinal epithelial permeability by tight junction complexes (TJs) (Ghareeb et al., 2015). TJs consist of occludins, claudins, and the zonula occludens (ZO) proteins ZO-(1, -2, and -3) (Li et al., 2015). TJs play a crucial role in epithelial cell attachment, maintaining epithelial cell structure and preserving biological activity. The gene expression in TJ regulation in chickens is a biomarker of intestinal permeability. AFB1 decreases the expression of the ZO-1 protein, a component of TJs, compromising the integrity of the intestinal mechanical barrier.

Additionally, AFB1 may disrupt the mechanical barrier's activity by impeding gut mucus movement. AFB1 has been observed to disrupt the gastrointestinal epithelial structure and tissue integrity (Ren et al., 2019). Moreover, AFB1 affects cellular integrity by disrupting connections between actin and the cytoskeleton, resulting in erroneous intestinal permeability, which allows for the translocation of bacteria, toxins, and antigens, ultimately leading to intestinal inflammation.

Another mechanism of AFB1-induced damage involves the generation of excessive reactive oxygen species (ROS) in cells and tissues, causing significant harm (Souto et al., 2018). However, intestinal cells and tissues possess control mechanisms against oxidative damage, primarily mediated by antioxidant enzymes such as glutathione peroxidase (GPX), glutathione reductase (GR), glutathione-S-transferase (GST), catalase, and superoxide dismutase (SOD) (Duan et al., 2015).

## AFB1 DAMAGE IN THE LIVER OF BIRDS

AFB1 causes physiological changes in intoxicated birds. Among the various organs that AFB1 can affect in birds, the liver is one of the organs in which pathological changes are observed, such as blunt and rounded edges, brittle and fragile, earthy yellow color, and sometimes an increased liver weight/body weight ratio, and gray-white foci on the liver surface. In chickens affected by AFB1 over a long period of the disease, histopathological examination of the liver shows that cells are denatured, and necrotic, hepatic cords are disordered, hyperplasia in the bile duct, moderate hyperplasia of liver tissue and fibrosis of the liver (Liu et al., 2020). The following are the different biological processes by which AFB1 affects the liver.

## AFB1 CAUSES OXIDATIVE DAMAGE IN POULTRY

During the bioactivation process of AFB1 by CYP450 enzymes, ROS groups are produced in the liver, which trigger oxidative stress responses. Studies indicate that oxidative stress has a role in the AFB1 response, inducing DNA damage in cells through the different pathways outlined below.

AFB1 induces mitochondrial dysfunction in hepatocytes and is mainly manifested in the following aspects:

First, AFB1 affects the mitochondria by uncoupling the respiratory chain. Oxidative phosphorylation of the electron transport chain in hepatocyte mitochondria is a critical energy transfer process. AFB1 disrupts oxidative phosphorylation in the liver (Sekine & Ichijo, 2015). The AFB1 exposure to primary liver cells of broilers enhances respiration in mitochondrial complexes III and IV, while significantly reducing the ratio of complex III to complex IV components. This phenomenon indicates that AFB1 strongly uncouples the mitochondrial respiratory chain (Sekine & Ichijo, 2015), resulting in an imbalance in the oxidation/antioxidation balance in the hepatocyte, leading to oxidative damage.

Furthermore, AFB1 can decrease mitochondrial membrane potential and cause mitochondrial swelling, resulting in increased mitochondrial permeability transition pore (Liu & Wang, 2016), which subsequently induces mitochondrial dysfunction with an imbalance of oxidation and antioxidation.

## AFB1 INDUCES ROS PRODUCTION

The ingestion of AFB1 by poultry generates a significant volume of ROS through the metabolic action of the liver. The electron transport chain in mitochondrial oxidative phosphorylation is an essential energy source for liver cells; however, mitochondria are the primary site of ROS production. A significant volume of ROS is generated when AFB1 induces failure in oxidative phosphorylation by uncoupling the mitochondrial respiratory chain. Additionally, during Phase II of the AFB1 detoxification and metabolism process, high amounts of GSH are consumed for conjugation with AFBO. GSH is an important antioxidant. Excessive consumption of GSH (in conjugation reaction) leads to increased levels of ROS in the body and, subsequently, oxidative damage (Liu & Wang, 2016). The uptake of AFB1 by poultry can increase ROS levels in the liver, leading to oxidative damage.

Various antioxidant enzymes, such as catalase and SOD, can prevent damage caused by excessive ROS (Yang et al., 2012). However, it is reported that SOD viability is reduced by exposure of primary broiler hepatocytes to AFB1, indicating that AFB1 causes oxidative damage and reduces the antioxidant capacity of liver cells by inhibiting SOD activity and inducing excessive accumulation of ROS.

A molecular mechanism controlling the expression of antioxidant genes is based on the signaling pathway of the antioxidant nuclear transcription factor erythrocyte line-2 (Nrf2). Nrf2 is a transcriptional activating antioxidant enzyme that regulates the expression of antioxidant genes by binding to antioxidant response elements (AREs) in the promoter and altering the gene expression profile. The Nrf2/ARE pathway is one of the most important defense systems against oxidative damage (Liu & Wang, 2016). Nrf2 is controlled by Kelch-like ECH-associated protein 1 (Keap-1), which binds to Nrf2 to inactivate it. When ROS levels increase, Keap-1 dissociates from Nrf2 to exert antioxidant activity. AFB1 can alter the Nrf2 signaling pathway. When the body is exposed to AFB1, the expression of Keap-1 protein in liver cells increases, inhibiting Nrf2 activity and resulting in an accumulation of ROS (Chatterjee & Chatterji, 2017).

## AFB1 INDUCES APOPTOSIS IN POULTRY HEPATOCYTES

AFB1 primarily causes hepatocellular apoptosis through the mitochondrial pathway by producing high levels of ROS. The B-cell lymphoma-2 (Bcl-2) protein family regulates this mitochondrial pathway of apoptosis. Studies have demonstrated that exposure of primary chicken liver cells to AFB1 at a concentration of 2.5 μmol/L leads to a significant generation of ROS in the cell. The pro-apoptotic protein Bax, a member of the Bcl-2 family, translocates to the mitochondria, where it interacts with Bcl-2, forming Bax-Bax dimer pores in the mitochondrial outer membrane. This pore formation releases pro-apoptotic molecules such as cytochrome C, cysteinyl aspartate specific proteinase (Caspase)-3, and Caspase-9. Immunofluorescence analysis reveals the translocation of Bax, caspase-3, and caspase-9 to the cell nucleus, causing characteristic changes in apoptotic cells, including reduced size, rounded morphology, crescent-shaped nuclear membrane, and condensed chromatin (Liu & Wang, 2016).

Moreover, AFB1 upregulates the expression of death receptors (extrinsic apoptosis) such as Fas, TNFR1, and FADD in hepatocytes. Broiler livers exposed to AFB1 for 7, 14, and 21 days show significant upregulation of Fas, TNFR1, and FADD, whereas the anti-apoptotic protein Bcl-2 exhibits significant downregulation on days 14 and 21 (Mughal et al., 2017). Thus, AFB1 can induce apoptosis in chicken liver cells through the death receptor pathway.

Another cellular process affected by AFB1 is autophagy, an intracellular adaptive protective mechanism involving the degradation and recycling of cellular components under stress conditions. A significant decrease in the expression of LC3a (a marker of autophagy induction) and the ratio of LC3b-II/LC3b-I (proteins involved in autophagic vesicles) is observed in the livers of broiler chickens exposed to AFB1 (Wang et al., 2019). This indicates that AFB1 may inhibit autophagic activity and promote apoptosis, hindering recovery from hepatocyte damage.

## AFB1 IN HEPATOCELLULAR CARCINOMA

AFB1 is a potent liver carcinogen. Contamination of rapeseed oil with AFB1 dramatically increases the risk of carcinogenesis because AFB1 is involved in the formation of DNA adducts and mutations in the tumor suppressor protein P53. Ducks are the only poultry that is reported to develop liver tumors when chronically and sub-chronically exposed to AFB1 (Gregorio et al., 2015). Increased expression of P53 in the nucleus and cytoplasm of hepatocytes is observed in broilers exposed to 5 mg/kg of AFB1 (Wang et al., 2019). AFB1 and its metabolites can cause methylation in the p53 gene, increasing sensitivity to mutations in the P53 protein, and thereby promoting the carcinogenic process (Narkwa et al., 2017). This suggests that AFB1 induces mutations in P53, leading to the activation of carcinogenesis.

## AFB1 CAUSES INFLAMMATORY DAMAGE IN HEPATOCYTES

The inflammatory response in the liver due to AFB1 is another observed element of the toxicity of this mycotoxin. AFB1 can increase the expression of COX-2 in the liver, which in turn promotes the activation of the NLRP3 inflammasome complex (Zhang et al., 2019). Once activated, NLRP3 forms a complex with ASC and pro-Caspase-1 proteins, activating Caspase-1. Active Caspase-1 cleaves and activates IL-18 and IL-1beta and binds to the cell membrane to produce and enlarge membrane pores, resulting in cellular pyroptosis. Additionally, AFB1 can promote the activation of nuclear factor kappa-B (NF-kB) in chicken liver cells (Ma et al., 2015). Activated NF-kB regulates many interleukins and TNF genes, contributing to inflammation in the broiler liver. Furthermore, AFB1 upregulates the expression of RIPK1, RIPK3, and MLKL levels in the liver of broilers, along with reduced ATPase activity. These molecules are involved in the necroptosis process, as evidenced by the rupture of the cell membrane, cytoplasmic content flow, mitochondrial swelling, and vacuolization, all characteristic of necroptosis (Li et al., 2022).

## EFFECT OF AFB1 ON OTHER ANIMALS OF VETERINARY INTEREST

Pigs and cattle are interested in the meat production industry because they provide protein to the population. These animals are often fed processed feeds that may be contaminated with the fungi *A. flavus* and *A. parasiticus*, which can produce AFB1. There is evidence of the detrimental effects of AFB1 in these animals, which is described below (Figure 2.15.2).

## SWINE

The damage caused by AFB1 in pigs mainly affects the liver. The major intermediate metabolite of AFB1 produced in pig liver microsomes is AFBO, and there is a correlation between the level of AFBO and the protein levels and activities of cytochromes CYP3A, CYP2A, and CYP2A19, with higher metabolic activity for CYP2A19. Furthermore, CYP2A19 is more involved in DNA adduct formation and DNA damage through AFB1 metabolization, indicating that CYP2A19 is the primary mediator of AFB1-induced cytotoxicity through the formation of active AFBO (Wang et al., 2024).

To determine whether AFB1 affects the porcine intestine, as in poultry, porcine jejunal epithelial cells (IPEC-J2) were treated with different concentrations of AFB1 (10–60 mg/L). It was observed that cell viability decreased and that this effect was most pronounced at an AFB1 concentration of 30 mg/L. Another observation was that AFB1 can repress the expression of proteins involved in tight junctions and, on the other hand, AFB1 increases the expression level of Caspase-3 and the Bax/Bcl-2 ratio in these cells, suggesting that AFB1 is cytotoxic to IPEC-J2 cells by disrupting tight junctions and inducing apoptosis. In an in vivo model, mice fed AFB1 at a concentration of 0. 3 mg/kg for 28 days showed significant effects on villus height/crypt depth ratio, intestinal wall thickness, number of intestinal villi per 1000 μm in jejunum, expression levels of ZO-1, Claudin-3, Occludin, MUC2, and Caspase-3, and the Bax/Bcl-2 ratio, indicating that the in vitro results with IPEC-J2 cells are very similar to those obtained in live animals, demonstrating that AFB1 affects the intestine (jejunum) of pigs (Zhang et al., 2022).

In another study in pigs, it was shown that a diet contaminated with AFB1 at a concentration of 500 ppb caused pigs to have lower body weight and weight gain compared to pigs receiving a regular diet without AFB1. Pigs fed an AFB1-contaminated diet had lower feed intake between days 1 and

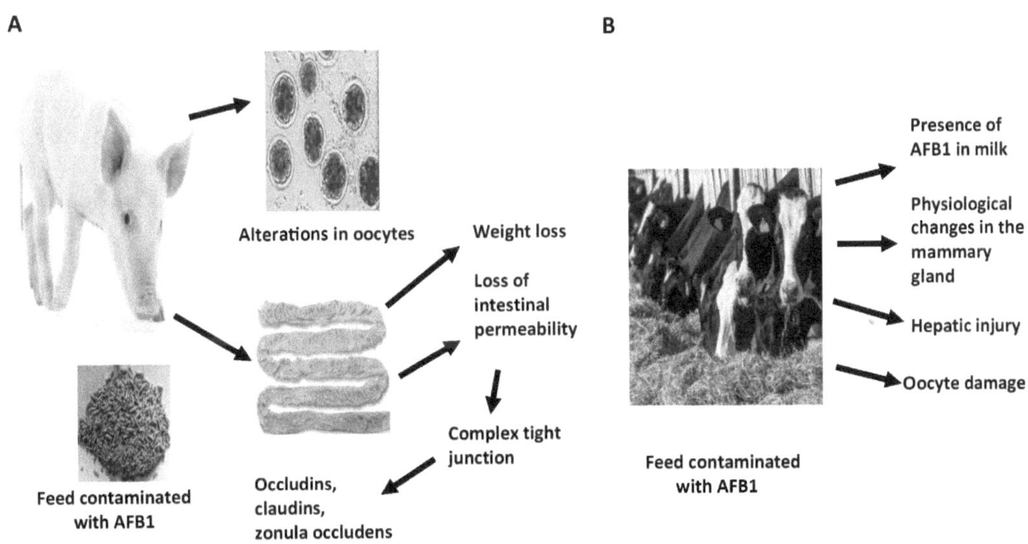


**FIGURE 2.15.2** Effects of AFB1 in pigs and cattle. Feed contaminated with AFB1 ingested by pigs (a) has effects on the small intestine and alters the reproductive system. In cattle, (b) AFB1 is presumed to be able to damage the mammary, hepatic, and reproductive systems.

20 and higher levels of the liver enzymes aspartate aminotransferase and alanine aminotransferase. Histological analysis of the intestine showed that pigs fed the AFB1-contaminated diet had higher villus height, and crypts were more profound in the intestine. These results indicate that feeding AFB1-contaminated diets to pigs negatively impacts animal health and growth (Tarasconi et al., 2024).

Another potential damage that AFB1 causes in pigs is to the lungs. One study showed that exposure of AFB1 to porcine alveolar macrophages (3D4/2 cell) at a concentration of 40 μg/mL for 24 hours in vitro resulted in inhibition of cell proliferation and induced apoptosis, and the damage caused by AFB1 occurred through the mechanism of non-coding RNA-mediated pathway (Chao et al., 2022).

The reproductive system of pigs may also be affected by AFB1. In vitro, exposure of AFB1 to porcine oocytes showed altered cumulus cell expansion and oocyte polar body extrusion. There was an alteration in the endoplasmic reticulum (ER) distribution and elevated expression of GRP78 protein, in addition to calcium storage in the ER, indicating an ER stress effect by AFB1. In addition, the structure of the cis-Golgi apparatus was affected because there was decreased expression of GM130. Oocytes showed aberrant accumulation of lysosomes and high expression of LAMP2, possibly due to aberrant mitochondrial function with low adenosine triphosphate (ATP) production and increased apoptosis due to increased expression of Bax. This indicates that AFB1 alters the intracellular membrane system, such as ER, Golgi, lysosome, and mitochondrial function, to affect the quality of porcine oocyte maturation (Hu et al., 2023).

Another study showed that AFB1 can induce testicular damage in pigs, manifested by decreased expression of GJA1, ZO1, and Occludin in mice and porcine Sertoli cell proliferation inhibition. The mechanism of action is suggested to be affecting the Ras/PI3K/Akt signaling pathway related to the cell cycle (Zhang et al., 2023). This demonstrates that AFB1 may have detrimental effects on the reproductive system of pigs.

## CATTLE

In the case of cattle, they seem to be resistant to the effects of AFB1, mainly because AFB1 can be detoxified in the rumen due to the participation of microorganisms in this site. However, studies on the effect of AFB1 have shown cellular damage, as observed in other animals. Studies on the toxicity of AFB1 have been conducted mainly using cell lines; however, an indication of the effect of AFB1 in cattle is the presence of this mycotoxin in cow's milk.

Apart from AFBO, the most toxic and carcinogenic aflatoxin derivative is aflatoxin M1. This hydroxylated derivative, AFB1, and varying amounts of aflatoxin can be found in milk and cheese from cows exposed to the mycotoxin daily. Of all agricultural species, cattle are relatively more resistant to aflatoxins due to the ability of the rumen microbiota to convert AFB1 into non-toxic or less toxic derivatives. However, contamination of daily milk with aflatoxins poses a serious food safety concern because it could pose health risks to humans who consume these products daily. In milk samples, AFG2, AFB1, and AFB2 have been detected as contaminants with incidences of 8.7%, 2.0%, and 0.67%, respectively (Ben Hassouna et al., 2023), indicating that consumption of mycotoxin-contaminated feed by cattle may contaminate milk.

To determine whether AFB1 affects the mammary gland of cows, AFB1 and AFM1 were tested for toxicity in bovine mammary epithelial (BME) cells at concentrations of 0–10 mg/L for 24 or 48 hours of exposure. AFB1 and AFM1 inhibited cell proliferation, apoptosis, and cell cycle arrest. There was a change in the expression of genes related to apoptosis, oxidative stress, cell cycle, binding, and signaling pathways, implying that these metabolites are toxic to BME cells (Wu et al., 2021). These results suggest that AFB1 may affect the bovine mammary gland and secrete AFB1 into milk, which may be a source of mycotoxins for humans.

Like other animals, cattle are also affected by AFB1 in the liver. A cell line of fetal bovine hepatocytes treated with AFB1 showed alterations in the expression of several genes involved in different pathways or processes such as cancer, cell damage, apoptosis, inflammation, and drug metabolism and transport. Moreover, these expression changes are very similar to those occurring in human hepatocyte cells, suggesting that the effect of AFB1 is similar in animals and humans (Pauletto et al., 2020). This indicates that AFB1 behaves similarly, affecting the liver in animals and humans.

Finally, damage to the bovine reproductive system due to AFB1 has also been observed. Cumulus oocyte complexes (COCs) from bovine embryos exposed to 32 μM AFB1 or 60 nM AFM1 showed a reduced rupture rate, while blastocyst formation was reduced to 3.2 or 32 μM AFB. In zygotes were observed erroneous nuclear and cytoplasmic maturation in COCs and increased mitochondrial membrane potential to AFB1 exposition. At the blastocyst stage, the in mt-ND2 and STAT3 gene expression was altered, suggesting an erroneous effect of the oocyte on embryonic development (Yaacobi-Artzi et al., 2023). This result suggests that AFB1 may have a significant detrimental effect on bovine reproduction.

## CONCLUSIONS

AFB1 has effects on broilers that include reduced growth, production, feed efficiency, bird weight loss, weakened immune function, increased susceptibility to infectious diseases, intestinal and hepatic injury. It is estimated that each part per million of AFB1 introduced into the bird reduces the growth rate by 5%; 300 ppm of AFB1 reduces the broiler's body weight by 21%. Feedstuffs containing AFB1 are frequently found in broiler farming, which is the primary reason for the reduction in chicken meat quality and profitability, along with other effects such as incubation period, chicken growth rate, vaccination efficiency, feed intake, and disease susceptibility in broilers. These results lead to significant losses in the broiler farming industry every year.

AFB1 poses a severe problem for the poultry breeding industry. Currently, there are some strategies for detoxifying AFB1 in foods. However, the results could be more encouraging and have defects such as destruction of nutritional components, poor effectiveness in eliminating AFB1, toxicity, and high cost.

Over time, food safety has become a significant concern for several industries. Thus, AFB1 is the primary contaminant, and the incidence of AFB1-contaminated feed in poultry has been occurring for a long time, often resulting in substantial losses to poultry production and economic benefits.

## REFERENCES


Abbasi F, Liu J, Zhang H, Shen X, Luo X. Effects of feeding corn naturally contaminated with aflatoxin on growth performance, apparent ileal digestibility, serum hormones levels and gene expression of Na+, K+-ATPase in ducklings. Asian-Australasian J Animal Sci 2018;31:91.

Akçam M, Artan R, Yilmaz A, Ozdem S, Gelen T, Nazıroğlu M. Caffeic acid phenethyl ester modulates aflatoxin B1-induced hepatotoxicity in rats. Cell Biochem Funct 2013;31(8):692–697.

Ben Hassouna K, Hamed AM, Salah-Abbès JB, Chaieb K, Abbès S, García-Campaña AM, Gámiz-Gracia L. Mycotoxin occurrence in milk and durum wheat samples from Tunisia using dispersive liquid-liquid microextraction and liquid chromatography with fluorescence detection. Toxins (Basel) 2023;15(11):633.

Bouhet S, Oswald IP. The effects of mycotoxins, fungal food contaminants, on the intestinal epithelial cell-derived innate immune response. Vet Immunol Immunopathol 2005;108(1–2):199–209.

Celi P, Cowieson A, Fru-Nji F, Steinert R, Kluenter AM, Verlhac V. Gastrointestinal functionality in animal nutrition and health: New opportunities for sustainable animal production. Animal Feed Sci Tech 2017;234:88–100.

Chao H, Ma H, Sun J, Yuan S, Dong P, Zhao A, Li L, Shen W, Zhang X. Whole-transcriptome analysis of non-coding RNA alteration in porcine alveolar macrophage exposed to aflatoxin B1. Toxins (Basel) 2022; 14(6):373.

Chatterjee A, Chatterji U. All-trans retinoic acid ameliorates arsenic-induced oxidative stress and apoptosis in the rat uterus by modulating MAPK signaling proteins. J Cell Biochem 2017;118(11):3796–3809.

Chen X, Naehrer K, Applegate TJ. Interactive effects of dietary protein concentration and aflatoxin B1 on performance, nutrient digestibility, and gut health in broiler chicks. Poult Sci 2016;95(6):1312–1325.

Duan Y, Zhang J, Dong H, Wang Y, Liu Q, Li H. Oxidative stress response of the black tiger shrimp Penaeus monodon to Vibrio parahaemolyticus challenge. Fish Shellfish Immunol 2015;46(2):354–365.

Filazi A, Sireli UT. 2013. Occurrence of aflatoxins in food. In M. Razzaghi-Abyaneh (Ed.), Aflatoxins: Recent Advances and Future Prospects, 143–170. InTech.

Ghareeb K, Awad WA, Böhm J, Zebeli Q. Impacts of the feed contaminant deoxynivalenol on the intestine of monogastric animals: Poultry and swine. J Appl Toxicol 2015;35(4):327–337.

Gregorio MCD, Bordin K, Souto PCMD, Corassin CH, Oliveira CAF. Comparative biotransformation of aflatoxin B 1 in swine, domestic fowls, and humans. Toxin Rev 2015;34(3):142–150.

Grenier B, Applegate TJ. Modulation of intestinal functions following mycotoxin ingestion: Meta-analysis of published experiments in animals. Toxins (Basel) 2013;5(2):396–430.

Grozeva N, Valchev I, Lazarov L, Hristov T, Kanakov D, Dimitrov K, Binev R. Cloacal bursa morphology in turkey broilers challenged with aflatoxin B 1 alone or co-administered with mycotoxing. Bulgarian J Vet Med 2020;23:121–129.

Guengerich FP. Forging the links between metabolism and carcinogenesis. Mutat Res 2001;488(3):195–209.

Gündüz N, Oznurlu Y. Adverse effects of aflatoxin B1 on skeletal muscle development in broiler chickens. Br Poult Sci 2014;55(5):684–692.

Hernández F, García V, Madrid J, Orengo J, Catalá P, Megías MD. Effect of formic acid on performance, digestibility, intestinal histomorphology and plasma metabolite levels of broiler chickens. Br Poult Sci 2006;47(1):50–56.

Hu LL, Chen S, Shen MY, Huang QY, Li HG, Sun SC, Wang JL, Luo XQ. Aflatoxin B1 impairs porcine oocyte quality via disturbing intracellular membrane system and ATP production. Ecotoxicol Environ Saf 2023;263:115213.

Hussain Z, Khan MZ, Khan A, Javed I, Saleemi MK, Mahmood S, Asi MR. Residues of aflatoxin B1 in broiler meat: Effect of age and dietary aflatoxin B1 levels. Food Chem Toxicol 2010;48(12):3304–3307.

Jahanian E, Mahdavi AH, Asgary S, Jahanian R. Effects of dietary inclusion of silymarin on performance, intestinal morphology and ileal bacterial count in aflatoxin-challenged broiler chicks. J Anim Physiol Anim Nutr 2017;101(5):e43–e54.

Kamdem LK, Meineke I, Gödtel-Armbrust U, Brockmöller J. Dominant contribution of P450 3A4 to the hepatic carcinogenic activation of aflatoxin B1. Chem Res Toxicol 2006;19(4):577–586.

Lewis L, Onsongo M, Njapau H, Schurz-Rogers H, Luber G, Kieszak S, Nyamongo J, Backer L, Dahiye AM, Misore A, DeCock K, Rubin C, Kenya Aflatoxicosis Investigation Group. Aflatoxin contamination of commercial maize products during an outbreak of acute aflatoxicosis in eastern and central Kenya. Environ Health Perspect 2005;113(12):1763–1767.

Li S, Liu R, Xia S, Wei G, Ishfaq M, Zhang Y, Zhang X. Protective role of curcumin on aflatoxin B1-induced TLR4/RIPK pathway mediated-necroptosis and inflammation in chicken liver. Ecotoxicol Environ Saf 2022;233:113319.

Li Y, Zhang H, Chen YP, Yang MX, Zhang LL, Lu ZX, Zhou YM, Bacillus WT. Amyloliquefaciens supplementation alleviates immunological stress in lipopolysaccharide-challenged broilers at early age. Poult Sci 2015;94(7):1504–1511.

Liu N, Wang J, Deng Q, Gu K, Wang J. Detoxification of aflatoxin B1 by lactic acid bacteria and hydrated sodium calcium aluminosilicate in broiler chickens. Livestock Sci 2018;208:28–32.

Liu X, Kumar Mishra S, Wang T, Xu Z, Zhao X, Wang Y, Yin H, Fan X, Zeng B, Yang M, Yang D, Ni Q, Li Y, Zhang M, Zhu Q, Chen F, Li D. AFB1 induced transcriptional regulation related to apoptosis and lipid metabolism in liver of chicken. Toxins (Basel) 2020;12(5):290.

Liu Y, Wang W. Aflatoxin B1 impairs mitochondrial functions, activates ROS generation, induces apoptosis and involves Nrf2 signal pathway in primary broiler hepatocytes. Anim Sci J 2016;87(12):1490–1500.

Lozano MC, Diaz GJ. Microsomal and cytosolic biotransformation of aflatoxin B1 in four poultry species. Br Poult Sci 2006;47(6):734–741.

Ma Q, Li Y, Fan Y, Zhao L, Wei H, Ji C, Zhang J. Molecular mechanisms of lipoic acid protection against aflatoxin $B_1$-induced liver oxidative damage and inflammatory responses in broilers. Toxins (Basel) 2015;7(12):5435–5447.

Monson MS, Settlage RE, McMahon KW, Mendoza KM, Rawal S, El-Nezami HS, Coulombe RA, Reed KM. Response of the hepatic transcriptome to aflatoxin B1 in domestic turkey (*Meleagris gallopavo*). PLoS One 2014;9(6):e100930.

Mughal MJ, Jing YZ. Aflatoxin B1 invokes apoptosis via death receptor pathway in hepatocytes. Oncotarget 2017;8(5):8239–8249.

Narkwa PW, Blackbourn DJ, Mutocheluh M. Aflatoxin B1 inhibits the type 1 interferon response pathway via STAT1 suggesting another mechanism of hepatocellular carcinoma. Infect Agent Cancer 2017;12:17.

Oakley A. Glutathione transferases: A structural perspective. Drug Metab Rev 2011;43(2):138–151.

Pauletto M, Tolosi R, Giantin M, Guerra G, Barbarossa A, Zaghini A, Dacasto M. Insights into aflatoxin B1 toxicity in cattle: An in vitro whole-transcriptomic approach. Toxins (Basel) 2020;12(7):429.

Ren Z, Guo C, Yu S, Zhu L, Wang Y, Hu H, Deng J. Progress in mycotoxins affecting intestinal mucosal barrier function. Int J Mol Sci 2019;20(11):2777.

Sarker MT, Wang ZY, Yang H, Wan X, Emmanuel A. Evaluation of the protective effect of lycopene on growth performance, intestinal morphology, and digestive enzyme activities of aflatoxinB1 challenged broilers. Anim Sci J 2021;92(1):e13540.

Sekine S, Ichijo H. Mitochondrial proteolysis: Its emerging roles in stress responses. Biochim Biophys Acta 2015;1850(2):274–280.

Sherratt PJ, Hayes JD. 2002. Glutathione S-Transferases. In D. Anderson, M.D. Waters, T. C. Marrs, C. Ioannides (Ed.), Enzyme Systems That Metabolise Drugs and Other Xenobiotics, 319–352. Wiley.

Souto NS, Claudia Monteiro Braga A, Lutchemeyer de Freitas M, Rechia Fighera M, Royes LFF, Schneider Oliveira M, Furian AF. Aflatoxin B1 reduces non-enzymatic antioxidant defenses and increases protein kinase C activation in the cerebral cortex of young rats. Nutr Neurosci 2018;21(4):268–275.

Tarasconi L, Dazuk V, Molosse VL, Cécere BGO, Deolindo GL, Mendes RE, Gloria EM, Ternus EM, Galli GM, Paiano D, Da Silva AS. Nursery pigs fed with feed contaminated by aflatoxin B1 (*Aspergillus flavus*) and anti-mycotoxin blend: Pathogenesis and negative impact on animal health and weight gain. Microb Pathog 2024;186:106474.

Wang F, Zuo Z, Chen K, Peng X, Fang J, Cui H, Shu G, He M, Tang L. Selenium rescues aflatoxin B1-inhibited T cell subsets and cytokine levels in cecal tonsil of chickens. Biol Trace Elem Res 2019;188(2):461–467.

Wang J, Tang L, Glenn TC, Wang JS. Aflatoxin B1 induced compositional changes in gut microbial communities of male F344 rats. Toxicol Sci 2016;150(1):54–63.

Wang XH, Li W, Wang XH, Han MY, Muhammad I, Zhang XY, Sun XQ, Cui XX. Water-soluble substances of wheat: A potential preventer of aflatoxin B1-induced liver damage in broilers. Poult Sci 2019;98(1):136–149.

Wang Z, Huang Q, Zhang F, Wu J, Wang L, Sun Y, Deng Y, Jiang J. Key role of porcine cytochrome P450 2A19 in the bioactivation of aflatoxin B1 in the liver. J Agric Food Chem 2024;72(4):2334–2346.

Wild CP, Turner PC. The toxicology of aflatoxins as a basis for public health decisions. Mutagenesis 2002;17(6):471–481.

Williams JH, Phillips TD, Jolly PE, Stiles JK, Jolly CM, Aggarwal D. Human aflatoxicosis in developing countries: A review of toxicology, exposure, potential health consequences, and interventions. Am J Clin Nutr 2004;80(5):1106–1122.

Woo LL, Egner PA, Belanger CL, Wattanawaraporn R, Trudel LJ, Croy RG, Groopman JD, Essigmann JM, Wogan GN, Bouhenguel JT. Aflatoxin B1-DNA adduct formation and mutagenicity in livers of neonatal male and female B6C3F1 mice. Toxicol Sci 2011;122(1):38–44.

Wu F, Khlangwiset P. Health economic impacts and cost-effectiveness of aflatoxin-reduction strategies in Africa: Case studies in biocontrol and post-harvest interventions. Food Addit Contam Part A Chem Anal Control Expo Risk Assess 2010;27(4):496–509.

Wu K, Jia S, Zhang J, Zhang C, Wang S, Rajput SA, Sun L, Qi D. Transcriptomics and flow cytometry reveals the cytotoxicity of aflatoxinB1 and aflatoxin M1 in bovine mammary epithelial cells. Ecotoxicol Environ Saf 2021;209:111823.

Yaacobi-Artzi S, Kalo D, Roth Z. Effect of the aflatoxins B1 and M1 on bovine oocyte developmental competence and embryo morphokinetics. Reprod Toxicol 2023;120:108437.

Yang J, Bai F, Zhang K, Bai S, Peng X, Ding X, Li Y, Zhang J, Zhao L. Effects of feeding corn naturally contaminated with aflatoxin B1 and B2 on hepatic functions of broilers. Poult Sci 2012;91(11):2792–2801.

Yilmaz S, Kaya E, Kisacam MA. 2017. The effect on oxidative stress of aflatoxin and protective effect of lycopene on aflatoxin damage. In L.B. Abdulra'uf (Ed.), Aflatoxin-Control, Analysis, Detection and Health Risks, 30, 67–90. InTech.

Yunus AW, Ghareeb K, Abd-El-Fattah AA, Twaruzek M, Böhm J. Gross intestinal adaptations in relation to broiler performance during chronic aflatoxin exposure. Poult Sci 2011;90(8):1683–1689.

Zhang FL, Ma HH, Dong PY, Yuan ZN, Zhang SE, Zhao AH, Liu HQ, De Felici M, Shen W, Zhang XF. Aflatoxin B1 disrupts testicular development via the cell cycle-related Ras/PI3K/Akt signaling in mice and pig. Environ Pollut 2023;329:121729.

Zhang LY, Zhan DL, Chen YY, Wang WH, He CY, Lin Y, Lin YC, Lin ZN. Aflatoxin B1 enhances pyroptosis of hepatocytes and activation of Kupffer cells to promote liver inflammatory injury via dephosphorylation of cyclooxygenase-2: An in vitro, ex vivo and in vivo study. Arch Toxicol 2019;93(11):3305–3320.

Zhang M, Li Q, Wang J, Sun J, Xiang Y, Jin X. Aflatoxin B1 disrupts the intestinal barrier integrity by reducing junction protein and promoting apoptosis in pigs and mice. Ecotoxicol Environ Saf 2022;247:114250.

# 2.16 Aflatoxigenic Fungi in Fish and Fish Feed

*Anandhi Anandaraj, Blessy Cleatus, Vaishnavi Thirukumaran, Maheswaran Solayappan, and Rajasekar Thirunavukkarasu*

## INTRODUCTION

Fish has become an important diet for humans and animals, which has led to the fastest growth of aquaculture and increase in fish production in the last decade. Therefore, fish feeding in farms increased the demand for aquafeeds, leading to artificial preparation of aquafeeds using fishmeal and plant-derived dietary feeds. Due to the high cost of fishmeal, which is a crude flour made by grinding and drying of fish flesh, farmers slowly switched to cost-effective plant-derived dietary foods that are rich in lipids and protein. Fish feeds derived from regionally available plant wastes such as brans of rice, maize and wheat, cottonseed cake, sunflower seedcakes and cassava are hand mixed and made into pellets using a machine [1].

Plant-derived fish feed ingredients are mostly from brans of cereals and grains, which are sites of fungal infection that is not removed during the dry milling process. Thus, brans have a higher potential to contain aflatoxin than those manufactured from milled flours or grits. Therefore, use of aflatoxin-contaminated bran in fish feed and feed ingredients paved the way for the aflatoxin to enter the fish feed product, posing hazards to fish health [2].

The disease aflatoxicosis was reported in May 1960 and involved rainbow trout hatcheries that were infected by severe liver cell carcinoma that spread among the fishes. It was found to be due to the intake by the fishes of cottonseed meal contaminated with aflatoxin. According to Ashley [3], autopsy of the infected fish showed enlarged and multinodular liver with primary liver cell carcinoma, which clarified that aflatoxins are carcinogenic. A later survey found that similar fish aflatoxicosis outbreaks were reported in other farms in Germany [4], Mexico [5], Denmark [6] and the Philippines [7], due to intake of moldy infected feed.

Contamination of fish feeds by aflatoxin and the possible transfer of toxins to humans through the food chain remains a serious concern [8] (Figure 2.16.1). Human exposure to aflatoxins can cause nausea, vomiting, abdominal pain, acute infections and its prolonged exposure can lead to complications like immunotoxicity, hepatotoxicity, and teratogenicity in chronic conditions resulting in cancer [9].

Detection of aflatoxin in food and feed is mandatory, even when present in very low concentration. Therefore, several countries have implemented strict regulations on levels of aflatoxin intake, considering individual health of human and animals. Several detection methods are being applied for detection and quantification of aflatoxins in food and feed ranging from conventional chromatographic methods to novel sensor-based techniques coupled with good agricultural practices (GAPs), good manufacturing practices (GMPs) and good storage practices (GSPs) to mitigate aflatoxin contamination.

## OCCURRENCE OF AFLATOXIN IN FISH FEED AND FEED INGREDIENTS

Some fungi produce mycotoxins, toxigenic secondary metabolites, among which are aflatoxins, the best-known and most widely researched mycotoxin. The common aflatoxigenic fungi are *Aspergillus flavus*, *A. parasiticus* and *A. nomius*. There are also some other species of *Aspergillus* that are found to produce


DOI: 10.1201/9781003487043-22

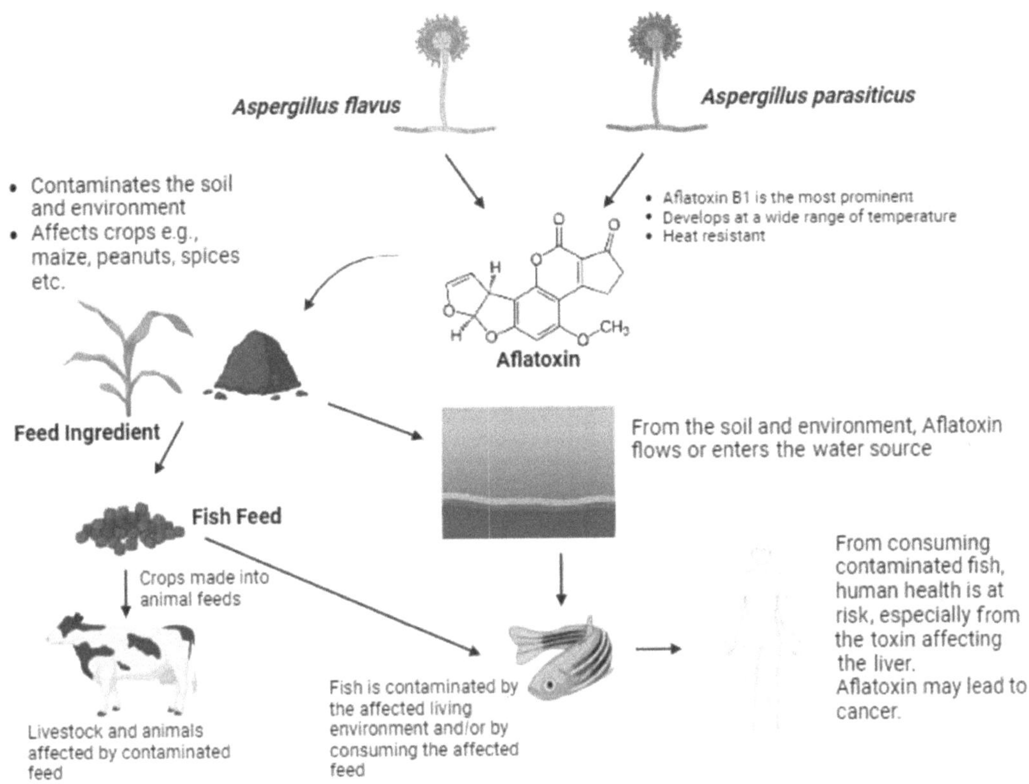


**FIGURE 2.16.1** Overview of aflatoxin contamination.

aflatoxin, for example, *A. pseudocaelatus* and *A. pseudonomius* [10]. These aflatoxigenic fungi naturally grow in soil, hay and decaying vegetation. They also colonise in various cereals and food grains such as corn, peanuts, pistachio, Brazil nuts, and copra, and contaminate them producing aflatoxins.

The occurrence of fungi and production of aflatoxins in fish feed and feed ingredients depends on various factors [11], including the following:

- Temperatures 25–28°C
- Humidity greater than 62%
- Moisture above 14%
- Species of fungi contaminating the crops
- Harvesting, processing methods and storage practices
- Seasonal variations and geographic location

This does not mean that merely the presence of fungi denotes the presence of toxin in the substrate. The aflatoxin can be present even in the absence of the fungi, though its vegetative structures are absent.

It has been reported that fish feeds are frequently contaminated by *Aspergillus* species. It is reported that in tropical countries, the major contaminants are *A. flavus* and *A. parasiticus* [12]. Studies in Egypt, Iran, Brazil and East Africa revealed the dominant species that contaminate fish feeds are *A. flavus* at 47.5%, 70%, 35% and 54.5%, respectively [13, 14]. Other *Aspergillus* species isolated from fish feeds are *A. tamarii*, *A. niger*, *A. versicolor*, *A. fumigatus*, *A. candidus*, *A. glaucus*, and *A. oryzae* [13]. Not all *Aspergillus* can produce aflatoxin, but strains that have the aflatoxin biosynthesis genes can produce it. Climate change also plays an important role in production of aflatoxin in *Aspergillus* species by altering the temperature and water activity, which greatly influences the gene expression to produce aflatoxin.

*A. parasiticus* produces all four major toxins, aflatoxin B1 (AFB1), AFB2, AFG1 and AFG2, whereas *A. flavus* produces only AFB1 and AFB2 [15]. Among 20 known aflatoxins, the four main aflatoxins are B1, B2, G1 and G2, while aflatoxin M1 and M2 are the hydroxylated metabolites of B1 and B2 [16]. The level of aflatoxin toxicity varies with the types; the order of toxicity is $AFB_1 > AFG_1 > AFB_2 > AFG_2$ [17] Aflatoxin B1 was found to be more prevalent in maize bran, other cereal products and finished fish feed with concentrations of 1.1 $\mu g \cdot kg^{-1}$, 7.4 $\mu g \cdot kg^{-1}$ and 3.8 $\mu g \cdot kg^{-1}$, respectively [18].

Studies revealed that aflatoxin contamination in terms of incidence rate was higher in farm processed feed (64.3%), feed ingredients (50%), and commercial feed (35.7%). The problem of contamination of fish feed and feed stuff with aflatoxin varies with the countries across the world. The permissible limit of aflatoxin in fish feeds is 20 $\mu g \cdot kg^{-1}$, but the samples from most tropical countries are above this limit. Table 2.16.1 shows the presence of aflatoxin contamination in various tropical and temperate countries, where it is found to be very high in East Africa at 806.9 $\mu g \cdot kg^{-1}$ [19, 20].

Survey done in Middle and East African countries revealed that sunflower seed cake has the highest level of aflatoxin contamination (806.9 $\mu g \cdot kg^{-1}$) followed by cottonseed cake with (377.9 $\mu g \cdot kg^{-1}$) [21], whereas in other countries such as the United States and Europe, cottonseed cake has the maximum contamination of 200 $mg \cdot kg^{-1}$ and 300 $mg \cdot kg^{-1}$, respectively [22].

## EFFECT OF AFLATOXIN IN FISH

The percentage of damage caused by aflatoxin in aquaculture depends on the concentration of aflatoxin and the time of exposure, age, sex, diet, weight and species susceptibility. Some species of fish were found to be sensitive to aflatoxin, although some are resistant. For example, a family of ray-finned fish called salmonids and rainbow trout are extremely sensitive to AFB1, which has LD50 (a dose that causes death in 50% of the population), whereas coho salmon appear fairly resistant (Bailey et al.) [23]. Juveniles are more susceptible than adult fishes.

Pathological effects in fish occur in three stages:

- Acute
- Subacute
- Chronic

i. **Acute stage**: This stage occurs when a moderate level of aflatoxin is ingested. Rainbow trout showed signs including anemia, pale gills, frequent hemorrhages, alteration in

**TABLE 2.16.1**
**Aflatoxin Levels in Fish Feed Samples from Developing Countries**

| Samples | Country | Concentration of Aflatoxin | Reference |
|---|---|---|---|
| Fish feed | Iran | 0.46–68.5 $\mu g \cdot kg^{-1}$ | [12] |
| | Brazil | 1.83–67.35 $\mu g \cdot kg^{-1}$, | [13] |
| | Egypt | 52.5–150 $\mu g \cdot kg^{-1}$ | [14] |
| | Kenya | <2–806.9 $\mu g \cdot kg^{-1}$ | [19] |
| | Tanzania | <2–377.9 $\mu g \cdot kg^{-1}$ | |
| | Uganda | <2–28.0 $\mu g \cdot kg^{-1}$ | |
| | Rwanda | <2–4.8 $\mu g \cdot kg^{-1}$ | |
| | East Africa | 806.9 $\mu g \cdot kg^{-1}$ | |
| | Turkey | 21 $\mu g \cdot kg^{-1}$ | [20] |

nutrient metabolism and liver damage [24], whereas Nile tilapia showed ocular opacities resulting in cataracts and blindness, lesions in fin and tail rot, yellowing of the body surface, anorexia and abnormal swimming patterns [7].

ii. **Subacute stage:** This stage includes moderate to severe liver damage, yellow eyes, yellowed mucous skin, and abnormalities in blood clotting, reduced feed conversion ratios, renal damage, reproductive failure, impaired immune responses, and premature death [25].

iii. **Chronic Stage:** This stage includes reduced feed efficiency, impaired liver function, increased susceptibility to secondary infectious diseases, necrosis and tumor development in liver and other organs, and increased mortality. Carcinogenic and genotoxic effects are more common, followed by teratogenic, hormonal, neurotoxic, or hematological changes [26]. In rainbow trout, prolonged feeding with low doses of AFB1 resulted in liver tumors [27].

After invading the liver, lipids infiltrate hepatocytes leading to necrosis causing cell death. Aflatoxins bind to DNA forming AFB1exo-8-9-epoxide, which is involved in high-level genotoxicity and carcinogenicity in fish, other animals and humans [28, 29]. Table 2.16.2 shows the various toxic effects of aflatoxin in different fish species.

## METABOLISM OF AFLATOXIN

The metabolic pathway of aflatoxin in liver involves the following:

- Hydroxylation to Aflatoxin M1, Aflatoxin Ql (AFM1, AFQl)
- Demethylation to AFP1
- Reduction to Aflatoxicol (AFL)
- Epoxidation to AFB1-8,9-epoxide (AFBO)

**TABLE 2.16.2**
**Toxic Effects of Aflatoxins in Different Fish Species**

| S. No | Fish Species | Toxicity Effect | Reference |
|---|---|---|---|
| 1 | Nile tilapia (*Oreochromis niloticus*) | Reduced growth<br>Survival rate reduced to 67%<br>Yellowing of the body surface decreased in total erythrocyte count, total leucocyte count, and haemoglobin count<br>Feed efficiency rate decreased | [7, 30–32] |
| 2 | Rainbow trout | Edema, pale gills, alteration in nutrition metabolism, liver tumors. | [24] |
| 3 | Juvenile Rainbow trout | Mortality | |
| 4 | Channel catfish (*Ictalurus punctatus*) | Decreased leukocyte count increased haematopoietic activity of blood-forming tissues | [27] |
| 5 | Rohu (*Labeo rohita*) | Reduction in production of oxygen radicals by neutrophils<br>Reduction in total protein and globulin levels<br>Decreased total erythrocyte count, total leucocyte count, haemoglobin count. | [33] |
| 6 | Sea bass (*Dicentrarchus labrax L.*) | Loss of equilibrium, rapid opercular movement haemorrhages of the dorsal skin surface | [34] |

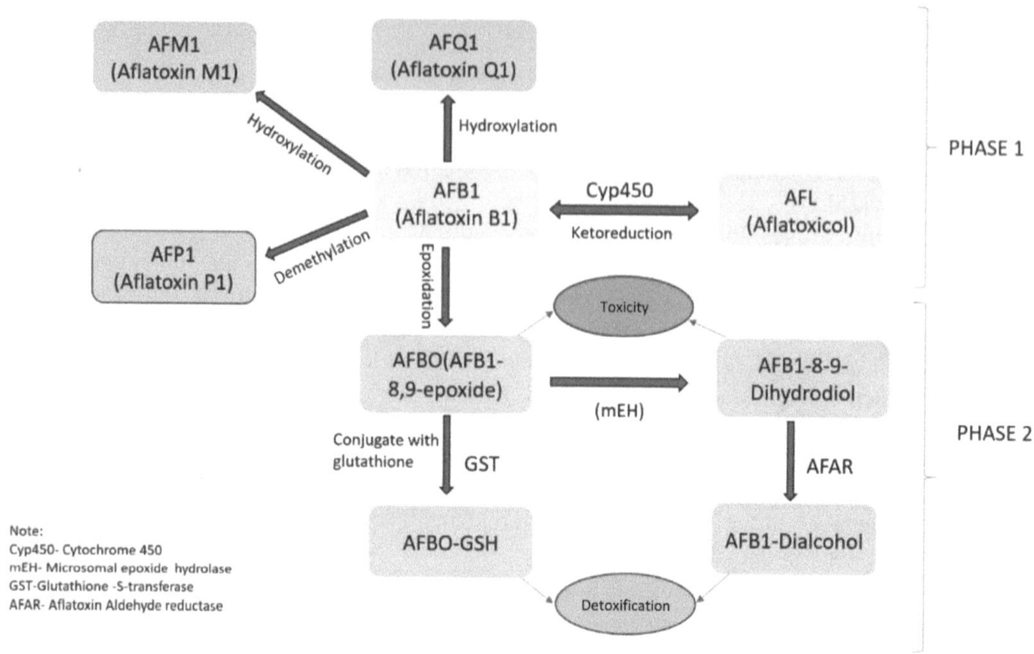


**FIGURE 2.16.2** Steps involved in the aflatoxin metabolic pathway.

In Figure 2.16.2, Phase 1 aflatoxin metabolism involves conversion of molecules into more hydrophilic substances by oxidation or reduction reactions, whereas Phase 2 involves the original molecules or metabolites conjugating with glutathione or sulfonides or glucuronides forming glutathione conjugates by glutathione S-transferase, which is considered as the main route of detoxification [35].

Therefore, the susceptibility and resistivity of an organism to aflatoxin depends on its ability to biotransform aflatoxin B1 to its less toxic stage of compounds. It is inferred that species producing more aflatoxicol are more sensitive to aflatoxin because it can be reverted to aflatoxin B1. The difference in susceptibility of a species also depends on its ability to convert AFB1 to AFBO, whereas the detoxification depends on the ability of the liver to promote a glutathione-S-transferase system forming aflatoxin-glutathione conjugates, enhancing detoxification.

In fishes, the survival rate is low due to the damage to the immune system caused by protein synthesis inhibition as a result of adduct formation between AFBO and DNA/RNA [36]. On the other hand, the glutathione-S-transferase detoxification pathway seems to be less important in fish. Therefore, the rate of conversion of AFBO and its conjugation with glutathione are the key parameters that decide the susceptibility and resistivity in fishes. It is found that trout species do not seem to have glutathione conjugate in Phase 2 metabolism; instead, they have glucuronidation as their main metabolic pathway [37].

## DETECTION AND QUANTIFICATION OF AFLATOXIN

To identify and quantify aflatoxins present in food and feed is a major challenge. Therefore, developing a sensitive and sturdy analytical method for the identification and quantification of aflatoxins that are present in low concentrations in food and feed is inevitable. There are various chromatographic and sensor-based methods used for the detection of aflatoxins, and in addition, there are several novel processing methods that have also been developed to detect the presence of aflatoxins. Some are discussed briefly below.

## THIN-LAYER CHROMATOGRAPHY (TLC)

TLC is currently the most often used separation method. Its affordability and ease of use can be attributed to its minimal instrument needs at the fundamental level. When measuring the amount of aflatoxin present on TLC plates, 0.5 μg/kg was the detection limit for red pepper, fish, maize, and wheat when combined with fluorescence densitometry [38]. In an interlaboratory investigation, corn samples spiked at 5 and 50 ng/g levels were analyzed using TLC separation and densitometric detection. According to [39], the relative repeatability standard deviation (RSDr) of the AFB1 ranged from 56.6% to 41.7%. TLC is still a recognized reference method.

## HIGH- AND ULTRAHIGH-PERFORMANCE LIQUID CHROMATOGRAPHY (HPLC/UHPLC)

Aflatoxin detection reference techniques are based on chromatography, specifically high-performance liquid chromatography (HPLC) and ultrahigh-performance liquid chromatography (UPLC). When determining aflatoxins, HPLC-fluorescence detection (FLD) is used, and most of the time, HPLC-mass spectrometry (MS)/MS systems can be used. Aflatoxin concentrations as low as micrograms per kilogram can be determined independently using HPLC equipped with an in-line photochemical reactor and FLD. One benefit of the approach is that materials for derivatization and reagents for the sensitive measurement are not required. The latter is predicated on the observation that the fluorescence characteristics of the AFB1 and AFG1 components increase in a manner analogous to electrochemical derivatization with exposure to UV light with a wavelength of 254 nm [40].

Other options exist for fluorescence-based aflatoxin detection, such as HPLC-laser-induced fluorescence (LIF). Laser-induced examination of fluorescent light released during laser irradiation forms the basis of fluorescence (LIF). According to [41] and [10], the method's sensitivity is 0.1 μg/kg for AFB1 and AFG1, and 1.2 μg/kg for AFB2 and AFG2.

## ENZYME-LINKED IMMUNOSORBENT ASSAY (ELISA)

The objective of the enzyme-linked immunosorbent assay (ELISA) technique is the qualitative or quantitative identification of mycotoxins present in the analyzed sample using particular antibodies. The technique relies on a color reaction coupled with an enzyme. Typically, competitive-type ELISA assays are employed to identify mycotoxins. As a result, the relationship between the observed color intensity and the measured compound concentration is inverse [42]. These ELISA analytical systems are great screening tools that deliver quantitative data quickly. As mentioned previously, they can frequently be employed at the sample location. Nevertheless, results could be impacted by matrix effects discovered during the examination of various products and cross-reactions with molecules that are very similar to the drug under study. Of course, the ELISA method can also be used to quantitatively determine AFB1, AFT, and AFM1 [43].

## RADIOIMMUNOASSAY (RIA)

Radioactively labelled molecules are applied during the progressive synthesis of immunocomplexes in radioimmunoassay. Radioimmunoassay (RIA) is an extremely sensitive and specialized technique. The limit of detection for agricultural samples (rice, wheat, soybean and corn) was 0.2/0.5 μg/kg for AFB1. The percentage of recovery ranged from 92% to 107% [44]. To reduce the negative effects of gamma rays, RIA requires the use of pricey, specialized equipment for different kinds of marker chemicals, which may be more advantageous for the analysis of aflatoxins to minimize health hazards [45].

## FLUORESCENCE IMMUNOASSAY (FIA)

There is already widespread usage of immunoreagents with probes based on fluorescence labelling. An easy and quick analytical process can be accomplished by combining the very sensitive fluorescence approach with the measuring instrument's sensitivity, allowing for the direct measurement of the analyte's concentration in the reaction mixture. Time-resolved fluorescence immunoassay (TRFIA) showed excellent accuracy in determining AFB1 in feed sample analysis. The range of the average recovery was 93.71%–97.80%, with a 1.25%–3.73% coefficient of variance. During the AFB1 determination of feeds, an excellent correlation was discovered between the TRFIA and HPLC methods, confirming the new method's trustworthiness [46, 47].

Apart from the methods discussed previously, the following can also be used for separation and detection of aflatoxins: over-pressured layer chromatography, capillary electrophoresis, liquid chromatography/ultra-performance liquid chromatography-mass spectrometry (LC/UPLC-MS), tandem mass spectrometry (MS/MS), supercritical fluid chromatography mass spectrometry (SFC-MS), chemiluminescence immunoassay (CLIA), microplate-based immunoassays, lateral flow immunoassay (LFIA), and biosensors.

## PREVENTION AND MANAGEMENT OF AFLATOXIN

- **Prevention in silages:** Prevention of mycotoxins in silages includes accepted practices that ensile and are aimed at preventing deterioration primarily through removal of oxygen. Silage additives can also be used to prevent mold growth [48].
- **Prevention of feed contamination:** Mold growth in feeds can be prevented by employing mold inhibitors, maintaining low moisture levels, storing feed fresh, and maintaining clean equipment. The water content should be less than 14% to prevent growth of mold [49].
- **Moisture in feed ingredients:** The most crucial element in deciding whether and how quickly molds will grow in feeds is moisture. Feed components, feed manufacturing procedures, and the storage or handling environment are the three main sources of moisture in feeds. Moisture from all three sources needs to be managed to properly regulate the moisture content of feeds. The first crucial step in managing feed moisture is managing it in the grains that are used to create the feed, as maize and other grains are the main sources of moisture and molds in feed [50].
- **Moisture in feed manufacturing processes:** Hammer mills are frequently used to grind grains to facilitate handling, increase digestibility and enhance the pelleting process. Systems for hammer mills with air assistance lessen heat accumulation in the final product, which in turn lessens moisture issues. Pelletizing is a process that combines feed and steam, presses the combination through a die, and cools the pellets to eliminate moisture and heat. Mold formation is encouraged when pellets are cooled if the extra moisture is not eliminated. When warm or hot pellets are stored in a chilly bin, moisture will condense on the interior of the bin because wet feeds are warmer than usual. If the correct circumstances are met, the residual spores may proliferate after pelleting. Thus, mold results from the pelleting process [51].
- **Moisture and the feed storage environment:** Eliminating visible sources of moisture in the feed processing and storage equipment is necessary to control the formation of mold. Leaks in feed truck compartments, augers, roofs (at the feed mill or barn), and feed storage tanks are a few examples of these sources. The fact that feed moisture varies depending on the feed's surroundings is a reality that is frequently disregarded. The air in confinement buildings can get extremely humid because the animals' breathing and feces add moisture to the surrounding air. When feed is kept in a humid atmosphere, its initial very low moisture content will increase. Therefore, proper ventilation is necessary to control the humidity in confinement housing [52].

- **Keeping feeds fresh:** Feed delivery is crucial for mold and mycotoxin production, ensuring freshness and consumption within 10 days. Feed stores should be managed using first in, first out, with feed next to the wall staying longest and susceptible to moisture migration and mold growth. Maintaining two feed stores allows for easily emptied and cleaned storage before refilling with new feed [53].
- **Equipment cleanliness:** Old, moldy feed in feed storage and delivery systems can cause mold growth and mycotoxin formation. To prevent this, remove caked, moldy feed from all feed manufacturing and handling equipment. Farms with hygienic stocking practices show lower toxin levels in their diet [54].
- **Use of mold inhibitors:** Chemical mold inhibitors are widely used in the feed industry to control mold growth. They consist of organic acids, salts of organic acids, and copper sulfate. They work equally well in solid or liquid forms, with the acid form being more active [55].
- **Dispersion:** Mold inhibitor effectiveness depends on its distribution throughout the feed. It should penetrate the entire surface of each particle and penetrate interior molds. Smaller carriers for mold-inhibiting chemicals ensure maximum contact with feed particles, enhancing their effectiveness [34].
- **Effect of feed ingredients:** Certain feed ingredients, such as protein or mineral supplements, can affect mold inhibitor performance. These supplements can reduce the effectiveness of propionic acid, while dietary fat enhances organic acid activity. Unknown factors in corn also affect these inhibitors [56].
- **Time dependence:** Mold inhibitors provide a temporary period of mold-free activity, but a longer period requires higher concentrations. The inhibitor's concentration decreases due to chemical binding, mold activity, or both. When reduced, mold uses the inhibitor as a food source, and heavily contaminated feeds require additional inhibitors. Additionally, to attain the appropriate degree of protection, feeds that are highly contaminated with molds will need to have greater levels of inhibitor added [57].
- **Influence of pelleting:** Pelletized feeds in the feed industry enhance mold inhibitor effectiveness due to the heat during pelleting. Higher pelleting temperatures are more effective. Pelletization makes feed easier for animals and molds to digest.
- **Biological methods:** Biotechnology breakthroughs have created a new approach to treating mycotoxicosis. The cell wall derivative of *Saccharomyces cerevisiae*, a glucomannan-containing yeast product, has demonstrated a significant ability to bind with commonly occurring mycotoxins [58]. It has also been found helpful in reducing the harmful effects of mycotoxins in livestock and poultry [59].

## REGULATIONS AND LIMITATION OF AFLATOXINS IN FEED

There really is no safe level. Concerning mycotoxins, the danger is strongly correlated to the concentration of the main mycotoxins and the quantity and presence of other mycotoxins in feeds. Due to variations in disease frequency and management practices, a mycotoxin level deemed safe on one farm might not be on another. The degree to which mycotoxicosis manifests itself is also influenced by variables like the way mycotoxins interact with infections, genetic variability, environmental influences, and nutritional status, among others.

Several nations control the amount of certain mycotoxins in foods and feeds to lessen their harmful and financial effects. Global laws pertaining to food and feed protect both the financial interests of animal merchants and producers as well as the public's health. Except for a few African nations, almost all nations with a fully established market economy have rules. However, the Food Safety and Standards Authority of India suggests a reasonable limit of 0.5 ppb for human foods, but the Bureau of Indian Standards has specified a maximum permissible level of 20 ppb (aflatoxin-B1) in all animal feeds [60]. The maximum level of total aflatoxins (AFB1+AFB2+AFG1+AFG2) is 20 μg/kg in feed ingredients offered to dairy animals and foods intended for human consumption (FDA, 2019 guidelines) [61].

## REFERENCES


1. Fregene, Bernadette Tosan, Rodrigue Yossa, Harrison C. Karisa, Ajibola Olaniyi, and Paul Bolorunduro. "Extension manual on quality low-cost fish feed formulation and production." *Gates Open Res* 8, no. 69 (2024): 69.
2. Kana, Jean Raphaël, Benoit Gbemenou Joselin Gnonlonfin, Jagger Harvey, James Wainaina, Immaculate Wanjuki, Robert A. Skilton, and Alexis Teguia. "Assessment of aflatoxin contamination of maize, peanut meal and poultry feed mixtures from different agroecological zones in Cameroon." *Toxins* 5, no. 5 (2013): 884–894.
3. Ashley, Laurence Marvin. "Pathology of fish fed aflatoxins and other antimetabolites." In *A symposium on diseases of fish and shellfishes*, pp. 366–379. American Fisheries Society, 1970.
4. Wunder, W., and H. Korn. "Aflatoxin cancer (hepatoma) in the liver of the rainbow trout (Salmo irideus)." *Zoologische Beitraege* 28 (1982): 99–109.
5. Perez, Ruiz. "Hepatic neoplasia in the rainbow trout (Salmo gairdneri) bred in El Zarco Fish Hatchery, Federal District." *Veterinaria Mexico* 15, no. 4 (1984): 255–261.
6. Rasmussen, H. B., K. Larsen, B. Hald, B. Moeller, and F. Elling. "Outbreak of liver-cell carcinoma among saltwater-reared rainbow trout Salmo gairdneri in Denmark." *Diseases of Aquatic Organisms* 1 (1985): 191–196.
7. Cagauan, Arsenia G., Roselyn H. Tayaban, Joselito R. Somga, and Rodora M. Bartolome. "Effect of aflatoxin-contaminated feeds in Nile tilapia (Oreochromis niloticus L.)." (2004): 172–178.
8. Panel, Efsa Contam, Dieter Schrenk, Margherita Bignami, Laurent Bodin, James Kevin Chipman, Jesús Del Mazo, Bettina Grasl-Kraupp et al. "Risk assessment of aflatoxins in food." (2020).
9. Kew, Michael C. "Hepatocellular carcinoma in developing countries: Prevention, diagnosis and treatment." *World journal of hepatology* 4, no. 3 (2012): 99.
10. Gilbert, John, and Eugenia A. Vargas. "Advances in sampling and analysis for aflatoxins in food and animal feed." *Journal of Toxicology: Toxin reviews* 22, no. 2-3 (2003): 381–422.
11. Almeida, Inês Filipa Martins, Hermínia Marina Lourdes Martins, Sara Maria Oliveira Santos, Maria Suzana Freitas, José Manuel Gaspar Nunes da Costa, and Fernando Manuel d´ Almeida Bernardo. "Mycobiota and aflatoxin B1 in feed for farmed sea bass (Dicentrarchus labrax)." *Toxins* 3, no. 3 (2011): 163–171.
12. Barbosa, Tatiana S., Carina M. Pereyra, Carla A. Soleiro, Erica O. Dias, Aguida A. Oliveira, Kelly M. Keller, Pedro PO Silva, Lilia R. Cavaglieri, and Carlos AR Rosa. "Mycobiota and mycotoxins present in finished fish feeds from farms in the Rio de Janeiro State, Brazil." *International Aquatic Research* 5 (2013): 1-9.
13. Fallah, Aziz A., Ebrahim Pirali-Kheirabadi, Mohammad Rahnama, Saeid S. Saei-Dehkordi, and Kambiz Pirali-Kheirabadi. "Mycoflora, aflatoxigenic strains of Aspergillus section Flavi and aflatoxins in fish feed." *Quality Assurance and Safety of Crops & Foods* 6, no. 4 (2014): 419–424
14. Hassan, Atef A., Manal A. Hassan, H. M. El Shafei, M. H. S. R. El Ahl, and R. H. Abd El-Dayem. "Detection of aflatoxigenic moulds isolated from fish and their products and its public health significance." *Nature and Science* 9, no. 9 (2011): 106–114.
15. D'mello Julius. P. F., and Anne M. C. Macdonald. "Mycotoxins." *Animal feed science and Technology* 69 (1997): 155–166.
16. Giray, Belma, Gozde Girgin, A. Basak Engin, Sevtap Aydın, and Gonul Sahin. "Aflatoxin levels in wheat samples consumed in some regions of Turkey." *Food control* 18, no. 1 (2007): 23–29.
17. Jaimez Ordaz, J., Cristina. A. Fente, Beatriz I. Vazquez, Carmen M. Franco, Alberto Cepeda, Gerard Mahuzier, and Pierre Prognon. "Application of the assay of aflatoxins by liquid chromatography with fluorescence detection in food analysis." *Journal of Chromatography A* 882, no. 1–2 (2000): 1–10
18. Gonçalves-Nunes, Etelvina María Carvalho, Maria M. Gomes-Pereira, Amilton P. Raposo-Costa, Carlos A. da Rocha-Rosa, Carina M. Pereyra, Rodrigo M. Calvet, Ana L. Alves-Marques, Francisco de Chagas Cardoso-Filho, and Maria C. Sanches-Muratori. "Screening of aflatoxin B1 and mycobiotarelated to raw materials and finished feed destined for fish." *Latin American Journal of Aquatic Research* 43, no. 3 (2015): 595–600.
19. Marijani, Esther, James M. Wainaina, Harrison Charo-Karisa, Louise Nzayisenga, Jonathan Munguti, Gbemenou Joselin Benoit Gnonlonfin, Emmanuel Kigadye, and Sheila Okoth. "Mycoflora and mycotoxins in finished fish feed and feed ingredients from smallholder farms in East Africa." *Egyptian Journal of Aquatic Research* 43, no. 2 (2017): 169–176.
20. Altuğ, Gülşen, and Gülsün Özyurt. "Level of aflatoxin in some fish feeds from fish farming processes, feed factories and imported feeds." *Turkish Journal of Veterinary & Animal Sciences* 27, no. 6 (2003): 1247–1252.

21. Rodrigues, Ines., Josy Handl, and Elisabeth M. Binder. "Mycotoxin occurrence in commodities, feeds and feed ingredients sourced in the Middle East and Africa." *Food Additives and Contaminants: Part B* 4, no. 3 (2011): 168–179.
22. Anthony, Makun Hussaini, Dutton Michael Francis, Njobeh Patrick Berka, Gbodi Timothy Ayinla, and Ogbadu Godwin Haruna. "Aflatoxin contamination in foods and feeds: A special focus on Africa." *Trends Vital Food Control Eng* 10 (2012): 24919.
23. Bailey, George S., Patricia M. Loveland, Donald Pierce, Jerry D. Hendricks, and John D. Groopman. "Quantitative carcinogenesis and dosimetry in rainbow trout for aflatoxin B1 and aflatoxicol, two aflatoxins that form the same DNA adduct." *Mutation Research/Environmental Mutagenesis and Related Subjects* 313, no. 1 (1994): 25–38.
24. Stewart, D., and E. Larson. "Aflatoxicosis in wildlife. Information Sheet 1582." *Mississippi State Univ. Extension Service., Cooperating with US Dept. of Agriculture* (2002).
25. Sahoo, Pramoda Kumar, Sarat Chandra Mukherjee, Sarat Kumar Nayak, and Sujit Dey. "Acute and Subchronic Toxicity of Aflatoxin $B_1$ to Rohu, *Labeo rohita* (Hamilton)." *Indian Journal of Experimental Biology* 39, no. 5 (2001): 453–458.
26. Albert C. Pier., Mary E. McLoughlin, John L. Richard, Alan L. Baetz, and Robert R. Dahlgren. 1985. "In Utero Transfer of Aflatoxin and Selected Effects in Neonatal Pigs." In *Trichothecenes and Other Mycotoxins*, edited by J. Lacey, 495–507. Chichester: John Wiley & Sons Ltd.
27. Jantrarotai, Wimol, and Richard T. Lovell. "Subchronic toxicity of dietary aflatoxin B1 to channel catfish." *Journal of Aquatic Animal Health* 2, no. 4 (1990): 248–254.
28. Gallagher, Evan P., and David L. Eaton. "In vitro biotransformation of aflatoxin B1 (AFB1) in channel catfish liver." *Toxicology and Applied Pharmacology* 132, no. 1 (1995): 82–90.
29. Gallagher, Evan P., Kent L. Kunze, Patricia L. Stapleton, and David L. Eaton. "The kinetics of aflatoxin B1oxidation by human cDNA-expressed and human liver microsomal cytochromes P450 1A2 and 3A4." *Toxicology and Applied Pharmacology* 141, no. 2 (1996): 595–606
30. El-Banna, Ragaa., Hassan M. Teleb, Maiia Hadi, and Fatma Fakhry. "Performance and tissue residue of tilapias fed dietary aflatoxin." *Veterinary Medical Journal (Giza)* 40, no. 3 (1992): 17–23.
31. Selim, Khaled M., Hana El-Hofy, and Riad H. Khalil. "The efficacy of three mycotoxin adsorbents to alleviate aflatoxin B 1-induced toxicity in *Oreochromis niloticus*." *Aquaculture International* 22 (2014): 523–540
32. Deng, Shi-Xi, Li-Xia Tian, Fu-Jia Liu, Sheng-Jie Jin, Gui-Ying Liang, Hui-Jun Yang, Zhen-Yu Du, and Yong-Jian Liu. "Toxic effects and residue of aflatoxin B1 in tilapia (*Oreochromis niloticus* × *O. aureus*) during long-term dietary exposure." *Aquaculture* 307, no. 3–4 (2010): 233–240.
33. Sahoo, Pramoda Kumar., and Sarat Chandra Mukherjee. "Immunosuppressive effects of aflatoxin B1 in Indian major carp (*Labeo rohita*)." *Comparative Immunology, Microbiology and Infectious Diseases* 24, no. 3 (2001): 143–149.
34. El-Sayed, Yasser Said, and Riad Hassan Khalil. "Toxicity, biochemical effects and residue of aflatoxin B1 in marine water-reared sea bass (*Dicentrarchus labrax* L.)." *Food and Chemical Toxicology* 47, no. 7 (2009): 1606–1609.
35. Dohnal, Vlastimil., Qinghua. Wu, and Kamil Kuča. Metabolism of aflatoxins: Key enzymes and interindividual as well as interspecies differences. *Archives of Toxicology* 88 (2014): 1635–1644.
36. Manning, B. Mycotoxins in aquaculture. In: Diaz DE, editor; *The Mycotoxin Blue Book*. United Kingdom: Context, p. 139–56, 2005.
37. Santacroce, Maria Pia., Maria Concetta Conversano, Ester Casalino, Ornella Lai, Carmela Zizzadoro, Giuseppe Centoducati. Aflatoxins in aquatic species: Metabolism, toxicity and perspectives. *Reviews in Fish Biology and Fisheries* 18 (2008): 99–130.
38. Shephard, Gordon S. "Aflatoxin analysis at the beginning of the twenty-first century." *Analytical and Bioanalytical Chemistry* 395 (2009): 1215–1224.
39. Park, Donald L., Mary W. Trucksess, Stanely Nesheim, Michael Stack, and RNewell. "Solvent-efficient thin-layer chromatographic method for the determination of aflatoxin-b1, aflatoxin-b2, aflatoxin-g1, and aflatoxin-g2 in corn and peanut products-collaborative study." *Journal Of Aoac International* 77, no. 3 (1994): 637–646.
40. Papadopoulou-Bouraoui, Andri, Joerg Stroka, and Elke Anklam. "Comparison of two post-column derivatization systems, ultraviolet irradiation and electrochemical determination, for the liquid chromatographic determination of aflatoxins in food." *Journal of AOAC International* 85, no. 2 (2002): 411–416.
41. Gan, Liang-Shang, Michael S. Otteson, Mark M. Doxtader, Paul L. Skipper, Ramachandra R. Dasari, and Steven R. Tannenbaum. "Quantitation of carcinogen bound protein adducts by fluorescence measurements." *Spectrochimica Acta Part A: Molecular Spectroscopy* 45, no. 1 (1989): 81–86.

42. Waliyar, Farid, Srinivas V. Reddy, and P. Lava-Kumar. "Review of immunological methods for the quantification of aflatoxins in peanut and other foods." *Peanut Science* 36, no. 1 (2009): 54–59.
43. Ketney, Otto, Antonello Santini, and Simona Oancea. "Recent aflatoxin survey data in milk and milk products: A review." *International Journal of Dairy Technology* 70, no. 3 (2017): 320–331.
44. Korde, Aruna, Usha Pandey, Sharmila Banerjee, H. D. Sarma, S. Hajare, Meera Venkatesh, A. K. Sharma, and M. R. A. Pillai. "Development of a radioimmunoassay procedure for aflatoxin B1 measurement." *Journal of Agricultural and Food Chemistry* 51, no. 4 (2003): 843–846.
45. Mahfuz, Mustafa, Md. Amran Gazi, Muttaquina Hossain, Md. Rezaul Islam, Shah Mohammad Fahim, and Tahmeed Ahmed. "General and advanced methods for the detection and measurement of aflatoxins and aflatoxin metabolites: A review." *Toxin Reviews* 39, no. 2 (2020): 123–137.
46. Hu, Xiaofei, Jingjing Yao, Fangyu Wang, Mengqi Yin, Yaning Sun, Mei Hu, Qiaoqiao Shi, and Gaiping Zhang. "Eu3+-labeled IgG-based time-resolved fluoroimmunoassay for highly sensitive detection of aflatoxin B1 in feed." *Journal of the Science of Food and Agriculture* 98, no. 2 (2018): 674–680.
47. Wang, Du, Zhaowei Zhang, Peiwu Li, Qi Zhang, and Wen Zhang. "Time-resolved fluorescent immunochromatography of aflatoxin b1 in soybean sauce: A rapid and sensitive quantitative analysis." *Sensors* 16, no. 7 (2016): 1094.
48. Asis, Ramon, Romina D. Di Paola, and Mario A. J. Aldao. "Determination of aflatoxin B 1 in highly contaminated peanut samples using HPLC and ELISA." *Food and Agricultural Immunology* 14, no. 3 (2002): 201–208.
49. Alp., Duygu, and Özcan. Bulantekin. "The microbiological quality of various foods dried by applying different drying methods: A review." *European Food Research and Technology* 247 (2021): 1333–1343.
50. Fumagalli, Francesca, Matteo Ottoboni, Luciano Pinotti, and Federica Cheli. "Integrated mycotoxin management system in the feed supply chain: Innovative approaches." *Toxins* 13, no. 8 (2021): 572.
51. Jarvis, Basil. "Factors affecting the production of mycotoxins." *Journal of Applied Bacteriology* 34, no. 1 (1971): 199–213.
52. Sotolu, Akeem Oladipupo, Shakiru Okanlawon Sule, Adegboyega J. Oshinowo, and Isaac. M. Ogara. "Implication of aflatoxin in fish feeds and management strategies for sustainable aquaculture." *PAT June* 10 (2014): 38–52.
53. Makinya, Kobia J., John M. Wagacha, Judith A. Odhiambo, Paddy Likhayo, Kukom Edoh-Ognakossan, Tadele Tefera, Adebayo Abass, and Christopher M. Mutungi. "The importance of store hygiene for reducing post-harvest losses in smallholder farmers' stores: Evidence from a maize-based farming system in Kenya." *Journal of Stored Products Research* 90 (2021): 101757.
54. Motalebi, Ali.Asghar, Kambiz Ardalani, and Sayed Jamili. "Effect of temperature on the produced aflatoxins in the rainbow trout feed in West Azerbaijan province." *Journal of Fisheries and Aquatic Science* 3 (2008): 392–397.
55. Nalle, Catootjie Lusje, Max A. J. Supit, Angga M. Akbar, Emiliana Langodai, and So'o Angriana. "Physical and chemical qualities of corn with different moisture levels supplemented with mold inhibitor." *Biotropia* 29, no. 3 (2022): 234 –243.
56. Jannathulla, Rajabdeen, and Jagabattula Syama Dayal. "Beneficial effects, challenges and opportunities of the filamentous fungus, *Aspergillus niger* with special reference to the shrimp feed industry—A review." *Reviews in Aquaculture* 15, no. 4 (2023): 1311–1334.
57. Wu, Haoxiang, and Jonathan Woon Chung Wong. "Mechanisms of indoor mold survival under moisture dynamics, a special water treatment approach within the indoor context." *Chemosphere* 302 (2022): 134748.
58. Devegowda, G and T.N.K. Murthy, 2005. Mycotoxins: Their adverse effects in poultry and some practical solutions. In: *Mycotoxin blue book*, Edited by Duarte Diaz. Nottingham University press, Nottingham, UK. 25–56.
59. Raju, Muppala Venkata Lakshmi Narasimha, and Gopal Devegowda. "Esterified-glucomannan in broiler chicken diets-contaminated with aflatoxin, ochratoxin and T-2 toxin: Evaluation of its binding ability (in vitro) and efficacy as immunomodulator." *Asian-Australasian Journal of Animal Sciences* 15, no. 7 (2002): 1051–1056.
60. Thakur, Sanjay., Ramesh Kumar Singh, Pradip S. De, and Avijit Dey. "Aflatoxins in feeds: Issues and concerns with safe food production." *Indian Journal of Animal Health* 61, no. 1 (2022): 01–13.
61. U.S. Food and Drug Administration (FDA). *CPG Sec. 683.100 Action Levels for Aflatoxins in Animal Feeds*. Center for Veterinary Medicine, Office of Regulatory Affairs, 2019.

# 2.17 Aflatoxigenic Fungi

## *A Challenge for Global Food Security*

*S. Umayaparvathi, S. Meenakshi, R. Saravanan, and S. Kumaran*

## INTRODUCTION

In the current environment of growing population, food security and safety rank among the most pressing issues. These are primarily influenced by three factors: (i) the availability of enough food; (ii) access to safe food; and (iii) the use of food for cultural, nutritional, and quality objectives to lead a healthy life [1]. The breakdown of any one of these facets results in hunger and food insecurity, which further affects human health in addition to the socioeconomic sphere of society. Moreover, one of the main causes of food insecurity is mycotoxin contamination of food and feed [2].

The Food and Agriculture Organization (FAO) states that mycotoxins impact 25% of the world's crop [3]. *Aspergillus, Fusarium* and *Penicillium* are the three principal genera of fungi that produce mycotoxins [4]. As one of the many types of mycotoxins, aflatoxins are extremely poisonous and have been found to infect a broad range of foods, including meat, dairy products, dried fruits, groundnuts, and maize [5]. According to Payne and Brown [6], *Aspergillus* species, specifically *A. flavus, A. nomius* and *A. parasiticus*, are the producers of aflatoxins. Tropical and subtropical climates provide warm, humid conditions that are ideal for the growth of these fungi [7]. Due to the resistance of aflatoxins to heat, food processing methods are insufficient to remove them from contaminated food and feed [8].

Both humans and animals have experienced severe health problems as a result of consuming aflatoxins from tainted food and feed [9]. To protect people's health, several nations have enacted stringent laws regarding aflatoxins in food and feed [10]. The acceptable upper limit of aflatoxins in food for humans is between 4 and 30 mg/kg. The European Union has the highest standard threshold, requiring that any product intended for direct consumption include no more than 2 mg/kg of AFB1 and 4 mg/kg of total aflatoxins [11]. In a similar vein, 20 mg/kg is the maximum allowable level for aflatoxins in the United States [12]. In addition, to improve sustainable agricultural output, a variety of cutting-edge technologies and control tactics are used for aflatoxin pre- and post-harvest management [13]. The novelty and strength of this chapter lay in the enumeration of the new methods established for aflatoxin detection in food and feed, with special reference to masked aflatoxins, despite the abundance of publications on aflatoxins in food and feed. The chapter also discusses the contamination of food and agricultural plants with aflatoxins, as well as the methods for controlling aflatoxins in food and feed to maintain food security and safety for socioeconomic growth and healthy living.

## OUTBREAKS DUE TO AFLATOXINS

A significant aflatoxin-related hepatitis outbreak was documented in the Indian states of Gujarat and Rajasthan, with an estimated mortality toll of 106 [14]. Aflatoxin was eventually shown to be present in maize, the primary staple food of the tribal people affected by the two-month-long outbreak. The initial analysis verified that *A. flavus* had been consumed [15]. In 1974, there was another aflatoxin

DOI: 10.1201/9781003487043-23

outbreak in northwest India that affected both humans and dogs [16]. There have been numerous documented aflatoxicosis outbreaks globally since 2004, resulting in 500 cases of severe sickness and 200 fatalities [17]. The ingestion of home-grown maize contaminated with moulds was the cause of most outbreaks that were documented in 2004 from rural parts of Kenya's East Province. Aflatoxin was found in food from impacted areas based on preliminary tests, as reported in 1981 [18].

## SOURCES OF AFLATOXIN

The main sources of various aflatoxins are *A. flavus, A. parasiticus* and *A. nomius* [19]. These organisms are commonly found in soil, especially in oil-rich seeds, grains, living plants [20], decomposing vegetation, crop residue like hay, deterioration of microbiological activity, and favourable environmental conditions such as more than 7% moisture and 20 °C temperature. These kinds of fungi infect a wide range of agricultural commodities and crops during incorrect or delayed harvesting, post-harvest storage and processing. *A. flavus* primarily contaminates corn, whereas *A. parasiticus* primarily contaminates peanuts. Although there are more than 20 identified aflatoxins, aflatoxin B1 (AFB1), aflatoxin B2 (AFB2), aflatoxin G1 (AFG1) and Aflatoxin G2 (AFG2) [21] are the four principal types.

## AFLATOXINS IN FOOD

Nuts, oil seeds, spices, and various cereals contain aflatoxins [22]. Aflatoxin-producing *Aspergillus* bacteria spread among themselves and contaminate grain and cereals at different stages of harvesting and storage (Figure 2.17.1). In the field, during harvest, transportation and storage, or both, fungus infection might happen [23]. An improper storage environment frequently results in aflatoxin

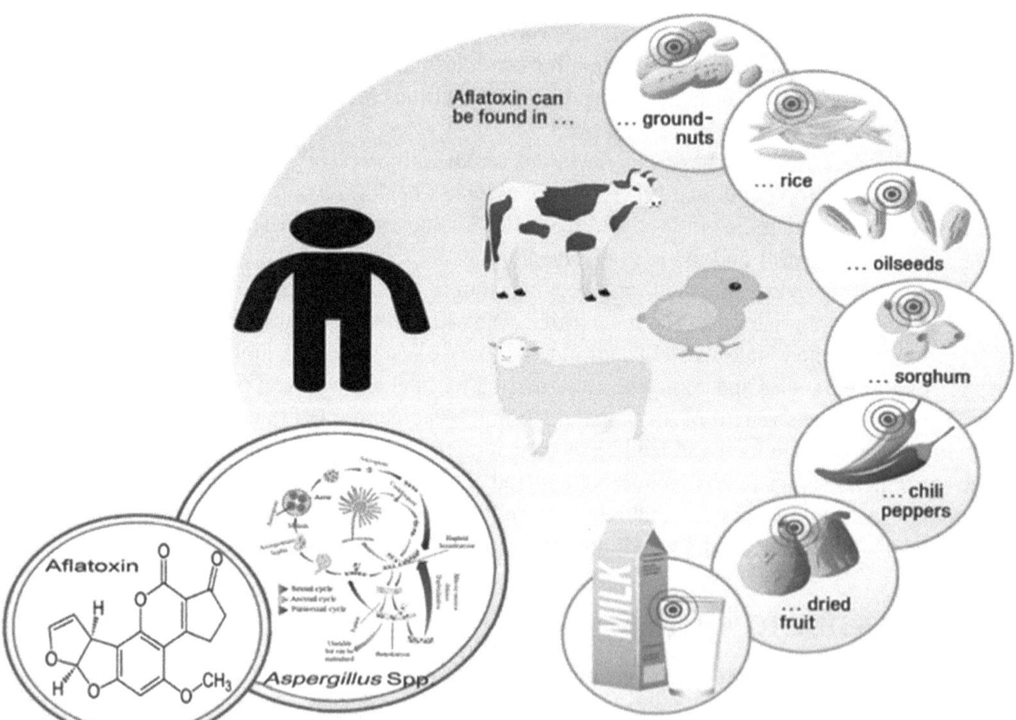


**FIGURE 2.17.1** Aflatoxins in food and feed.

contamination of wheat or barley. Aflatoxin levels in milk typically range from 1% to 6% of the feed material's total composition [24]. When humans consume foods contaminated with aflatoxins, such as eggs, meat and meat products, milk and milk products, or dairy products, they become infected with aflatoxins [25].

## AFLATOXINS CONTAMINATION IN CROP PLANTS

Globally, cereals and cereal-based products are the most common foods consumed by humans [26]. In natural settings, aflatoxins primarily contaminate rice and corn among other cereals because of modifications in agricultural techniques. Both pre- and post-harvest circumstances are used to produce the aflatoxins [27]. According to Filazi and Sireli [28], rice is more likely than other cereals to become contaminated with aflatoxins. When rice grains are improperly dried and maintain a greater moisture content (>14%), fungus grows. Consequently, the quality of the grains is degraded and the grain or husk become discoloured by this fungus. Conversely, many African diets typically include beans and groundnuts as a complement to cereal diets [29]. But under field and storage settings, these are extremely vulnerable to aflatoxin contamination [30]. Temperature, moisture, soil type, and storage conditions all affect how much fungal growth and aflatoxin generation occurs in cereals [31]. Furthermore, storage and processing conditions have a major impact on spices and they are prone to aflatoxin contamination.

In the Sultanate of Oman, Elshafie et al. [32] documented the presence of aflatoxin contamination in a broad range of spices, including ginger, cardamom, cinnamon, clove, cumin, and coriander. Moreover, aflatoxin was found in raw cow's milk and eggs from a poultry farm in Cameroon [33]. As a result, the impacted crops enabled aflatoxins to enter the food chain, which is heavily impacted by weather patterns.

## METHODS TO DETECT AFLATOXIN CONTAMINATION

The absorption and emission spectra of aflatoxins can be used to identify and detect their presence; the peak absorbance of aflatoxins is found at 360 nm wavelength. Aflatoxins G and B glow green at the 540-nm wave length and blue at 425 nm, respectively, when exposed to UV light. The detection of aflatoxins is commonly accomplished by using the florescence phenomenon, which is supported by high-performance liquid chromatography (HPLC), liquid chromatography mass spectroscopy (LCMS), and enzyme-linked immunosorbent assay (ELISA) techniques [34]. However, these conventional techniques have limitations: they take a lot of time, effort, and expert labour to perform, making them inappropriate for quick and real-time applications in food and feed samples. Therefore, for the quick and simple detection of aflatoxins, rapid and reliable techniques including polymerase chain reaction (PCR) and non-destructive techniques based on fluorescence/near-infrared spectroscopy (FS/NIRS) and hyperspectral imaging (HSI) have evolved [35]. Hussain et al. [36] used the PCR method to detect *A. flavus*, which produces aflatoxin from peanuts molecularly. Furthermore, Kim et al. [37] detected aflatoxins from *A. oryzae* isolated from several Korean foods using PCR, ELISA and HPLC. Nanoparticles (NPs) based on Au/Ag, carbon (CBNs), magnetic nanoparticles (MNPs), quantum dots (QDs), up-conversion nanoparticles (UCNPs), metal-organic frameworks (MOFs), and hybrid nanostructures have been used to further improve the sensitivity and detection of aflatoxins in food and feed [38]. To guarantee the safety of food, a number of biosensors and immunoassays have been developed to detect ultra-traces of aflatoxins.

## METHODS TO CONTROL AFLATOXIN CONTAMINATION

Aflatoxin contamination can be reduced by implementing advanced agricultural technology, good manufacturing procedures (GMPs), good storage practices (GSPs), and good agricultural practices (GAPs) [39]. According to Jalili et al. [40] and Pankaj et al. [3], there is a chance that the new processing methods utilizing a microwave, UV, pulsed light, electrolyzed water, cold plasma, ozone, electron beam, and gamma (g) irradiation treatment will help control aflatoxins and preserve and maintain the quality of agricultural and food products. Ozone was discovered to be the most efficient gas for optimizing aflatoxin degradation on grains and legumes through an electrophilic attack on the carbon bonds in the furan ring [41]. Ozone treatment, however, is less effective for post-harvest crop treatment because of its high cost. Consequently, it was discovered that exposure to gamma radiation also decreased the level of aflatoxin [42]. When paired with other technologies, gamma irradiation has a higher deterioration efficiency.

Several artificial and natural food additives have also been researched for their potential to reduce aflatoxins in feed and food. For instance, Méndez-Albores et al. [43] found that decomposing aflatoxins in extruded sorghum was possible when citric acid and moisture were combined at high temperatures (200°C) and pressures (8N). However, increasing pressure improved the effectiveness of sodium hydrosulphite ($Na2S2O_4$) for the reduction of aflatoxins in black pepper [44]. Moreover, Anjaiah et al. [45] showed that inoculating antagonistic strains of *Pseudomonas*, *Bacillus*, and *Trichoderma* spp. resulted in a considerable reduction of *A. flavus* in pre-harvest crops as part of biological control strategies. An important biological control agent against aflatoxin contamination are the non-aflatoxin-producing strains of *A. flavus* and other nontoxigenic moulds [2]. Every strategy has benefits and drawbacks when applied. To ensure food safety and security, biocontrol strategies should be used in tandem with other physical and chemical techniques and better packaging materials.

## CONCLUSION

Animal and human health are now seriously threatened by aflatoxin contamination in food crops and products made from them. In addition to posing a health risk, the toxin-contaminated items are losing money on the international food market, which makes the situation worse in developing nations. Diverse techniques involving physical, chemical, biological, and nanoparticle methods are employed to reduce and control aflatoxin in food crops. Though breeding and genetic engineering techniques are helping researchers create fungal-resistant types, the results remain a significant source of concern. Therefore, farmers and the food industry should use a combined strategy of employing resistant cultivars in conjunction with advised pre- and post-harvest techniques to decrease and degrade the aflatoxin concentration in food crops and the products generated from them.

## REFERENCES

1. FAO. 1996 13–17 November. Rome declaration on world food security and world food summit plan of action: World Food Summit. Rome: FAO.
2. Udomkun, P., A. N. Wiredu, M. Nagle, J. Müller, B. Vanlauwe, and R. Bandyopadhyay. 2017. Innovative technologies to manage aflatoxins in foods and feeds and the profitability of application–A review. *Food Control*. 76: 127–138.
3. Pankaj, S. K., H. Shi, and K. M. Keener. 2018. A review of novel physical and chemical decontamination technologies for aflatoxin in food. *Trends Food Sci. Technol.* 71: 73–83.
4. Reddy, K. R. N., B. Salleh, B. Saad, H. K. Abbas, C. A. Abel, and W. T. Shier. 2010. An overview of mycotoxin contamination in foods and its implications for human health. *Toxin Rev.* 29: 3–26.
5. Iqbal, S. Z., S. Jinap, A. A. Pirouz, and A. R. A. Faizal. 2015. Aflatoxin M1 in milk and dairy products, occurrence and recent challenges: A review. *Trends Food Sci. Technol.* 46: 110–119. doi: 10.1016/j.tifs.2015.08.005.
6. Payne, G. A., and M. P. Brown. 1998. Genetics and physiology of aflatoxin biosynthesis. *Annu. Rev. Phytopathol.* 36: 329–362. doi: 10.1146/annurev.phyto.36.1.329.
7. Battilani, P., S. Formenti, C. Ramponi, and V. Rossi. 2011. Dynamic of water activity in maize hybrids is crucial for fumonisin contamination in kernels. *J. Cereal Sci.* 54: 467–472. doi: 10.1016/j.jcs.2011.08.014.

8. Medina, A., M. K. Gilbert, B. M. Mack, G. R. OBrian, A. Rodriguez, D. Bhatnagar, et al. 2017b. Interactions between water activity and temperature on the *Aspergillus flavus* transcriptome and aflatoxin B1 production. *Int. J. Food Microbiol.* 256: 36–44. doi: 10.1016/j.ijfoodmicro.2017.05.020.
9. Sherif, S. O., E. E. Salama, and M. A. Abdel-Wahhab. 2009. Mycotoxins and child health: The need for health risk assessment. *Int. J. Hyg. Environ. Health.* 212: 347–368. doi: 10.1016/j.ijheh.2008.08.002.
10. Juan, C., A. Ritieni, and J. Mañes. 2012. Determination of trichothecenes and zearalenones in grain cereal, flour and bread by liquid chromatography tandem mass spectrometry. *Food Chem.* 134: 2389–2397. doi: 10.1016/j.foodchem.2012.04.051.
11. EC. 2010. Commission regulation (EC) no 165/2010 of 26 February 2010 amending regulation (EC) no 1881/2006 setting maximum levels for certain contaminants in foodstuffs as regards aflatoxins. *Off. J. Eur. Union L.* 50: 8–12.
12. Wu, F. 2006. Mycotoxin reduction in Bt corn: Potential economic, health, and regulatory impacts. *Transgenic Res.* 15: 277–289. doi: 10.1007/s11248-005-5237-1.
13. Prietto, L., P. S. Moraes, R. B. Kraus, V. Meneghetti, C. A. A. Fagundes, and E. B. Furlong. 2015. Post-harvest operations and aflatoxin levels in rice (*Oryza sativa*). *Crop Protection.* 78: 172–177. doi: 10.1016/j.cropro.2015.09.011.
14. Krishnamachari, K. A. V. R., R. V. Bhat, V. Nagarajan, and T. B. G. Tilak. 1975. Hepatitis due to aflatoxicosis - An outbreak in western India. *Lancet.* 1: 1061–1063. doi: 10.1016/S0140-6736(75)91829-2.
15. Bhatt, R. V., and K. A. V. R. Krishnamachari. 1978. Food toxins and disease outbreaks in India. *Arogya J. Health Sci.* 4: 92–100.
16. Reddy, B. N., and C. N. Raghavender. 2007. Outbreaks of aflatoxicoses in India. *Afr. J. Food Agric. Nutr. Dev.* 7: 1–15.
17. Azziz-Baumgartner, E., K. Lindblade, K. Gieseker, H. S. Rogers, S. Kieszak, H. Njapau, et al. 2005. Case–control study of an acute aflatoxicosis outbreak, Kenya, 2004. *Environ. Health Perspect.* 113: 1779–1783.
18. Ngindu, A., P. R. Kenya, and D. M. Ocheng. 1982. Out break of acute hepatitis by aflatoxin poisoning in Kenya. *Lancet.* 319: 1346–1348. doi: 10.1016/S0140-6736(82)92411-4.
19. Kurtzman, C. P., B. W. Horn, and C. W. Hesseltine. 1987. *Aspergillus nomius*, a new aflatoxin-producing species related to *Aspergillus flavus* and *Aspergillus parasiticus*. *Antonie Van Leeu Wen Hoek.* 53: 147–158. doi: 10.1016/S0168-6496(03)00154-5.
20. Geiser, D. M., J. W. Dorner, B. W. Horn, and J. Taylor. 2000. The phylogenetics of mycotoxin and sclerotium production in *Aspergillus flavus* and *Aspergillus oryzae*. *Fungal Genet Biol.* 31, no. 3: 169–179.
21. Inan, F., M. Pala, and I. Doymaz. 2007. Use of ozone in detoxification of aflatoxin B1 in red pepper. *J. Stored Prod. Res.* 43: 425–429. doi: 10.1016/j.jspr.2006.11.004.
22. Iqbal, S. Z., H. G. Mustafa, M. R. Asi, and S. Jinap. 2014. Variation in vitamin E level and aflatoxins contamination in different rice varieties. *J. Cereal Sci.* 60: 352–355. doi: 10.1016/j.jcs.2014.05.012.
23. Kader, A. A., and A. M. Hussein. 2009. Harvesting and post-harvest handling of dates. Aleppo: ICARDA.
24. Jacobsen, B. J. 2008. "Predicting the Incidence of Mycotoxins," in Proceeding of 26th Annual International Symposium on Man and His Environment in Health and Disease: Molds & Mycotoxins, Hidden Connections for Chronic Diseases, (Dallas, TX: American Environmental Health Foundation), 13–17.
25. Piemarini, S., L. Micheli, N. H. S. Ammida, G. Palleschi, and D. Moscone. 2007. Electrical immune sensor array using a 96-well screen-printed micro plate for aflatoxin B1 detection. *Biosens. Bioelectron.* 22: 1434–1440. doi: 10.1016/j.bios.2006.06.029.
26. Temba, M. C., P. B. Njobeh, and E. Kayitesi. 2017. Storage stability of maize groundnut composite flours and an assessment of aflatoxin B1 and ochratoxin A contamination in flours and porridges. *Food Control.* 71: 178–186. doi: 10.1016/j.foodcont.2016.06.033.
27. Hesseltine, C. W. 1974. Natural occurrence of mycotoxins in cereals. *Mycopathol. Mycol. Appl.* 53: 141–153. doi: 10.1007/bf02127204.
28. Filazi, A., and U. T. Sireli. 2013. Occurrence of aflatoxins in food. In *Aflatoxins- Recent Advances and Future Prospects*, ed. M. Razzaghi-Abyaneh, (Ankara: BoD–Books on Demand).
29. Soro-Yao, A. A., K. Brou, G. Amani, P. Thonart, and K. M. Djè. 2014. The use of lactic acid bacteria starter cultures during the processing of fermented cereal-based foods in West Africa: A review. *Trop. Life Sci. Res.* 25: 81–100.
30. Lombard, M. J. 2014. Mycotoxin exposure and infant and young child growth in Africa: What do we know? *Ann. Nutr. Metabol.* 64, no. Suppl. 2: 42–52. doi: 10.1159/000365126.

31. Achaglinkame, M. A., N. Opoku, and F. K. Amagloh. 2017. Aflatoxin contamination in cereals and legumes to reconsider usage as complementary food ingredients for Ghanaian infants: A review. *J. Nutr. Intermed. Metab.* 10: 1–7. doi: 10.1016/j.jnim.2017.09.001.
32. Elshafie, A. E., T. A. Al-Rashdi, S. N. Al-Bahry, and C. S. Bakheit. 2002. Fungi and aflatoxins associated with spices in the Sultanate of Oman. *Mycopathologia.* 155: 155–160.
33. Tchana, A., P. Moundipa, and F. Tchouanguep. 2010. Aflatoxin contamination in food and body fluids in relation to malnutrition and cancer status in Cameroon. *Int. J. Environ. Res. Public Health.* 7: 178–188. doi: 10.3390/ijerph7010178.
34. Kumar, P., D. K. Mahato, M. Kamle, T. K. Mohanta, and S. G. Kang. 2017. Aflatoxins: A global concern for food safety, human health and their management. *Front. Microbiol.* 7: 2170. doi: 10.3389/fmicb.2016.02170.
35. Tao, F., H. Yao, Z. Hruska, L. W. Burger, K. Rajasekaran, and D. Bhatnagar. 2018. Recent development of optical methods in rapid and non-destructive detection of aflatoxin and fungal contamination in agricultural products. *Trends Anal. Chem.* 100: 65–81. doi: 10.1016/j.trac.2017.12.017.
36. Hussain, A., A. Afzal, M. Irfan, and K. A. Malik. 2015. Molecular detection of aflatoxin producing strains of *Aspergillus flavus* from peanut (Arachis hypogaea). *Turk. J. Agric. Food Sci. Technol.* 3: 335–341.
37. Kim, N. Y., J. H. Lee, I. Lee, and G. E. Ji. 2014. An evaluation of aflatoxin and cyclopiazonic acid production in *Aspergillus oryzae. J. Food Prot.* 77: 1010–1016. doi: 10.4315/0362-028X.JFP-13-448.
38. Xue, Z., Y. Zhang, W. Yu, J. Zhang, J. Wang, F. Wan, et al. 2019. Recent advances in aflatoxin B1 detection based on nanotechnology and nanomaterials - A review. *Anal. Chim. Acta.* 1069: 1–27. doi: 10.1016/j.aca.2019.04.032.
39. Kamle, M., D. K. Mahato, S. Devi, K. E. Lee, S. G. Kang, and P. Kumar. 2019. Fumonisins: Impact on agriculture. Food, and human health and their management strategies. *Toxins.* 11: 328. doi: 10.3390/toxins11060328.
40. Jalili, M., S. Jinap, and A. Noranizan. 2010. Effect of gamma radiation on reduction of mycotoxins in black pepper. *Food Control.* 21: 1388–1393. doi: 10.1016/j.foodcont.2010.04.012.
41. Jalili, M. 2016. A review on aflatoxins reduction in food. *Iran. J. Health Saf. Environ.* 3: 445–459.
42. Aquino, S. 2011. Gamma radiation against toxigenic fungi in food, medicinal, and aromatic herbs. In *Science Against Microbial Pathogens: Communicating Current Research and Technological Advances*, ed. A. Mendez-Vilas, (pp. 272–281). FORMATEX.
43. Méndez-Albores, A., J. Veles-Medina, E. Urbina-Álvarez, F. Martínez-Bustos, and E. Moreno Martínez. 2009. Effect of citric acid on aflatoxin degradation and on functional and textural properties of extruded sorghum. *Anim. Feed Sci. Technol.* 150, no. 3–4: 316–329.
44. Jalili, M., and S. Jinap. 2012. Role of sodium hydrosulphite and pressure on the reduction of aflatoxins and ochratoxin A in black pepper. *Food Control.* 27: 11–15. doi: 10.1080/19440049.2010.551300.
45. Anjaiah, V., R. P. Thakur, and N. Koedam. 2006. Evaluation of bacteria and Trichoderma for biocontrol of pre-harvest seed infection by *Aspergillus flavus* in groundnut. *Biocontrol. Sci. Technol.* 16: 431–436. doi: 10.1080/09583150500532337.

# 2.18 Aflatoxigenic Fungi
## *A Global Concern for Food Safety and Human Health*

*Kajal Gaba, Samridhi Pushkarna, Shruti Kharod, Anil Kumar, Pooja Suneja, and Amita Suneja Dang*

## INTRODUCTION

Mycotoxins are secondary metabolites synthesized by diverse fungal genera, and represent a significant global concern due to their potent toxicity. Their occurrence in crops, processed foods, and animal feeds poses serious health risks. Among the mycotoxins, aflatoxins are recognized as highly hazardous, and capable of causing extensive contamination and adverse health effects, as recognized by the World Health Organization (1). Aflatoxigenic fungi are a group of filamentous fungi that have the ability to produce aflatoxins, mostly present in soil, decaying vegetation, and agricultural commodities (2). These fungi grow rapidly in warm and humid environments, with optimal growth occurring at temperature range between 25°C and 37°C and relative humidity above 80% (3).

These aflatoxins are mainly produced by two *Aspergillus* species: *A. flavus* and *A. parasiticus*. To date, more than 20 distinct forms of aflatoxins have been identified; the most common and deadly are aflatoxins B1, B2, G1, and G2 (4). Among these, aflatoxin B1 (AFB1) is the most toxic and well known for its carcinogenic properties in animals and humans, whereas AFG1, AFG2, and AFB2 exhibit varying degrees of toxicity (5). AFM1 is a hydroxyl derivative of AFB1 found in milk of animals consuming aflatoxin-contaminated feed, and represents a potential risk to human health. These aflatoxins can contaminate an ample variety of food and feed products, posing significant health hazards and requiring meticulous monitoring and control measures. Aflatoxin production is influenced by various factors, including environmental conditions, host susceptibility, and fungal genetics. High temperatures and drought stress during crop growth can increase aflatoxin production in infected crops. Additionally, poor storage practices and insect damage can enhance fungal growth and aflatoxin production in post-harvest crops.

## BIOSYNTHESIS OF AFLATOXINS IN AFLATOXIGENIC FUNGI

The biosynthesis of aflatoxins in aflatoxigenic fungi involves a complex pathway of enzymatic reactions. This process occurs in response to specific environmental conditions and involves the expression of a cluster of genes responsible for aflatoxin production (6). The biosynthesis of aflatoxins begins with the conversion of malonyl-CoA and acetyl-CoA into the polyketide norsolorinic acid anthrone (NAA) through the action of polyketide synthase enzymes. The *hypC* (hypB1) gene has played a role in the enzymatic transformation of NAA into norsolorinic acid (NOR) (7). The next step in the transformation of norsolorinic acid into averantin (AVN) is controlled by the *aflD* gene. *aflD* encodes a NOR ketoreductase required to convert the 1′-keto group of NOR to the 1′-hydroxyl group of averantin (AVN). The transformation of AVN into averufin (AVF) involves a P-450 monooxygenase and an alcohol dehydrogenase, encoded by the *aflG* and *aflH* genes, respectively (8). The transformation of versiconal hemiacetal acetate (VHA) from AVF is mediated by the *aflI* (*avfA*) gene. The gene *aflJ* encodes an esterase that catalyzes the conversion of VHA to versiconal (VAL),

DOI: 10.1201/9781003487043-24

and gene *aflK* produces a cyclase that is associated with the translation of VAL into versicolorin B (VerB) (9). VerB (versicolorin B desaturase) catalyzes the conversion of VerB to versicolorin A (VA), which has a 2,3 double bond in the dihydrobisfuran ring within the aflatoxin B1 branch (10). This double bond can undergo oxidation within the host organism, leading to the formation of a highly reactive epoxide. This epoxide is responsible for the carcinogenic, mutagenic, and cytotoxic effects associated with aflatoxin B1. Multiple enzymatic reactions participate in converting VA to demethylsterigmatocystin, and subsequently to ST and O-methylsterigmatocystin (11). In the final step in this process, catalysis by the *cytochrome P-450* monooxygenase *OrdA* yields aflatoxin B1 from O-methylsterigmatocystin (12).

## GLOBAL CONCERN FOR FOOD SAFETY

The wide contamination range of aflatoxigenic fungi refers to their ability to contaminate a diverse array of food and feed commodities. This broad range of susceptibility is due to the widespread presence of aflatoxigenic fungi in the environment and their ability to colonize a variety of substrates (13). It is typically found in soil and decaying vegetation, and it can infect a wide range of crops such as peanuts, cottonseed, corn, and tree nuts. Aflatoxin contamination poses a threat to food security, particularly in regions where food safety infrastructure is limited. Contaminated food and feed can lead to food shortages, malnutrition, and poverty, exacerbating existing socioeconomic challenges (14). Vulnerable populations, such as pregnant women and children, are at particularly high risk.

The harmful effects of aflatoxigenic fungi and aflatoxins are exacerbated by several of their characteristics:

- These fungi are commonly found in tropical and subtropical regions, where environmental conditions are conducive to their growth. However, they can also survive and grow in temperate regions under certain conditions (15).
- Because they are ubiquitous, these fungi can be found in soil, decaying vegetation, and in various agricultural commodities. They can easily spread to crops through air, water, and contaminated equipment, contributing to their wide contamination range (14).
- These fungi can infect and contaminate a wide range of host plants such as maize, peanuts, tree nuts, cottonseed, spices, and dried fruits. These fungi can colonize crops in the field and during storage, leading to aflatoxin contamination at different stages of the supply chain (16).
- Different strains of aflatoxigenic fungi can vary in their ability to produce aflatoxins. Some strains may produce high levels of aflatoxins, whereas others may produce lower levels or none at all. This variability can result in unpredictable contamination levels in different commodities and batches (4).
- Aflatoxins are extremely stable compounds capable of enduring in the environment as well as in food and feed commodities for long periods. This persistence can result in the accumulation of aflatoxins in the food chain, presenting a continued risk to human and animal health (17).

These aforementioned factors contribute to the exacerbation of aflatoxin contamination effects. Furthermore, climate change represents another notable factor in this regard. Climate change can enhance aflatoxin contamination by altering environmental conditions favorable for aflatoxigenic fungi growth (16). Changes in temperature and precipitation patterns can increase the risk of aflatoxin contamination in crops by promoting fungal growth and increasing moisture levels in certain regions, providing an environment conducive to aflatoxin production. Increased humidity, especially in regions where crops are grown, can also promote fungal proliferation (18). Moreover, extreme weather events caused by climate change, such as drought and floods, can also impact crop health and susceptibility to fungal infection, posing additional challenges for food safety and security (19).

## AFLATOXIGENIC FUNGI ON HUMAN HEALTH

Aflatoxins can induce human intoxication through contact, ingestion, and inhalation, affecting various internal organs such as the skin, colon, salivary glands, liver, kidney, stomach, and lungs. Daily aflatoxin poisoning over a period of weeks typically results in a subacute to chronic illness (13). Sudden death, numerous tissue hemorrhages, and icterus are observed in the more severe cases. The liver may become reduced in size, fibrotic, and exhibit ascites in cases of subacute to chronic hepatotoxicosis (20). It has been reported that aflatoxin exposure of 1–6 mg/day for roughly one month following ingestion of mold-damaged maize resulted in toxification characterized by jaundice, portal hypertension, ascites, and other indications of liver failure in humans (21). According to studies, children are most exposed to aflatoxin-contaminated food types, followed by adolescents and adults. This occurs due to children's immature metabolic pathway, lower capacity for detoxification, and higher intake-to-body weight ratio (22). The major effect on human health due to aflatoxins can be of the following types (Figure 2.18.1).

## LIVER DAMAGE

Aflatoxins are metabolized in the liver by enzymes such as *cytochrome P450* to form reactive intermediates. These intermediates can bind to DNA, proteins, and other cellular components, leading to cellular damage and dysfunction. Chronic exposure to aflatoxins can cause liver damage, including hepatocellular carcinoma (HCC), which is the most prevalent type of liver cancer globally. Aflatoxin exposure is particularly concerning in regions where hepatitis B virus (HBV) infection is prevalent because aflatoxins and HBV act synergistically to increase the risk of HCC (23).

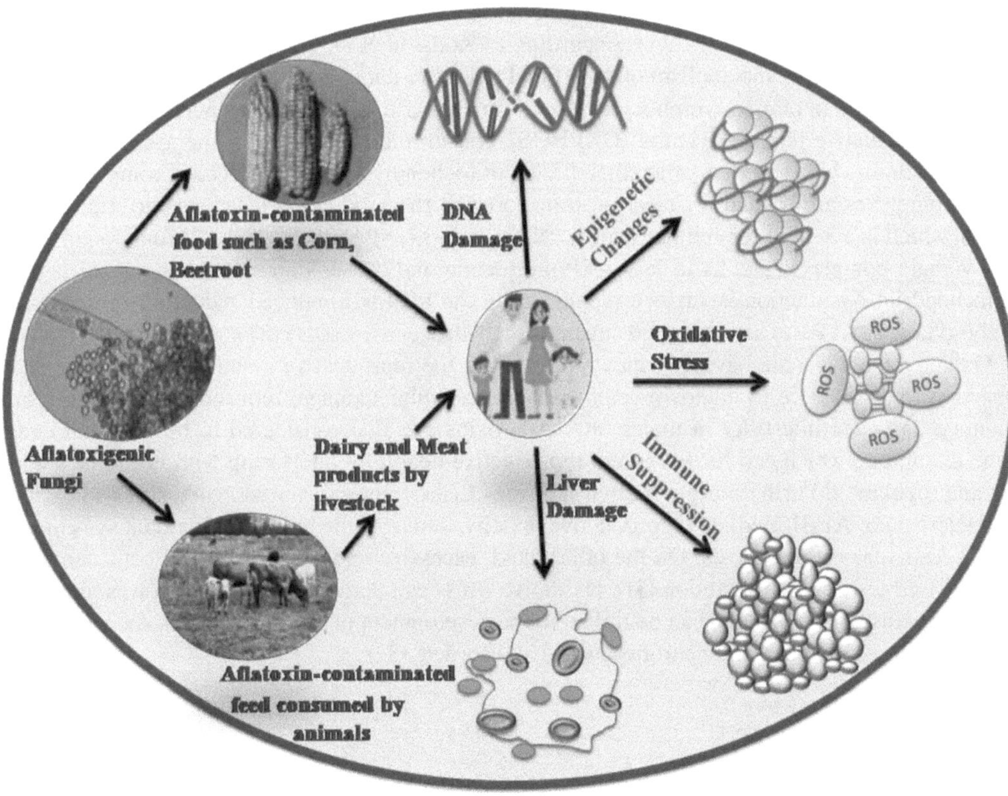


**FIGURE 2.18.1** Aflatoxigenic fungi impacts on human health.

Prolonged exposure to aflatoxins can lead to the accumulation of liver scar tissue, a condition known as liver fibrosis. This can impair liver function and lead to complications such as portal hypertension. Liver damage caused by aflatoxins can lead to various secondary complications, including ascites (accumulation of fluid in the abdominal cavity), hepatic encephalopathy (brain dysfunction due to liver failure), and coagulopathy (bleeding disorders) (24).

## IMMUNE SUPPRESSION

Aflatoxins can suppress the immune system by affecting the immune system and cytokine production. This immune suppression can make individuals more susceptible to infections and other diseases. In populations with high aflatoxin exposure, such as those relying on contaminated food staples, immune suppression can contribute to increased morbidity and mortality from infectious diseases (25). In response to AFB1, pro-inflammatory cytokines are also released, such as tumor necrosis factor-α (TNF-α), interleukin-6 (IL-6), and interleukin-1b (IL-1b) (Alameri et al., 2023). A study showed that cells treated with 50 μM AFB1 for 24 hours significantly increased the expression of pro-inflammatory factors TNF-α, inducible nitric oxide synthase (NOS2), and chemokine (C-X-C motif) ligand 2 (CXCL2), while showing no significant impact on IL-6, ARG1, TGF-β, and IL-10 expression. The study also revealed that AFB1 treatment decreased CD86 expression, but did not affect CD206 levels in macrophage cells (26).

## OXIDATIVE STRESS

Aflatoxins can trigger oxidative stress in cells by producing reactive oxygen species (ROS) and impairing antioxidant defenses. AFB1 mainly causes a decrease in enzymatic antioxidants like catalase (CAT), glutathione (GSH), and superoxide dismutase (SOD) and a noticeable increase in the level of malondialdehyde (MDA). According to Kodama et al., the generation of free radicals has been noticed during metabolism of AFB1 and oxidative damage represents one type of damage inflicted by AFB1 in human lymphocytes (27). Mitochondria in hepatocytes perform the essential process of oxidative phosphorylation (OXPHOS), consuming about 90% of the available oxygen. After undergoing bioactivation, aflatoxins disrupt mitochondrial respiratory chain complexes I–IV. These complexes are crucial for electron transport and the production of adenosine triphosphate (ATP), which is essential for cellular energy requirements (28). Disruption in the function of respiratory chain complexes results in less ATP production and causes damage to mitochondria (29). Mitochondrial dysfunction is further supported by the aflatoxin-induced reduction in adenosine triphosphatase (ATPase) activities and succinate dehydrogenase (SDH) (27).

Oxidative damage can have detrimental effects on the reproductive system. Studies have demonstrated that exposure to aflatoxins can result in testicular damage, reduced sperm quality and quantity, and even infertility in males (30). Aflatoxins are also considered to be potential endocrine disruptors, which can further affect reproductive health by interfering with hormone signaling and regulation. Furthermore, research indicates that decreased steroid synthesis is a common characteristic of AFB1-mediated reproductive toxicity, damages the body's antioxidant system, and causes testicular cell apoptosis. On the other hand, excessive apoptosis in Leydig cells can result in decreased testosterone secretion (31). Oxidative stress can damage cellular structures, including lipids, proteins, and DNA, and can be linked to the development of various diseases such as cardiovascular diseases, cancer, and neurodegenerative disorders (32).

## DNA DAMAGE

The liver can break down aflatoxin into its reactive form, aflatoxin B1-8,9-epoxide (AFBO). AFBO is a highly reactive compound that can bind covalently to DNA. It can react specifically with the N7 position of guanine bases in DNA to form AFB1-N7-guanine adducts. This covalent binding is irreversible and alters the structure of the DNA molecule. AFB1-N7-guanine adducts are particularly mutagenic because they can lead to G-to-T transversions (33). Additionally, AFB1 can directly interact with the DNA molecule, leading to DNA breakages. AFBO can also disrupt the cell cycle's miotic phases (M) as well as G1, G2, and S phase (34). Cells have several DNA repair mechanisms to correct DNA damage caused by genotoxic agents like aflatoxins. One such mechanism is nucleotide excision repair, which can recognize and remove AFB1-DNA adducts. However, if the damage is extensive or the repair mechanisms are overwhelmed, mutations may persist. This type of mutation can lead to disruption of regular cellular processes, including DNA repair, and cell cycle regulation, and can contribute to the development of cancer (35).

## EPIGENETIC CHANGES

Aflatoxins can induce epigenetic changes, such as changes in chromatin remodeling DNA methylation, non-coding RNAs (microRNAs), and histone modifications, which can affect gene expression patterns. DNA methylation involves the addition of a methyl group to the DNA molecule, typically occurring at cytosine residues within CpG dinucleotides, and forms the basis of chromatin structure (36). AFB1 can bind to DNA directly and create adducts that damage DNA and change the methylation state of nearby sites. Exposure to AFB1 may potentially alter the activity of methylation-related enzymes, including DNA methyltransferases (DNMTs), which may result in modifications to gene-specific DNA methylation patterns (37). AFB1 exposure can impact histone modifications, which are chemical changes to histone proteins that regulate chromatin structure and expression of genes. Some of the several histone modifications are methylation, acetylation, ubiquitination, and phosphorylation. AFB1 can influence the function of enzymes such as histone deacetylases (HDACs) and histone acetyltransferases (HATs); these enzymes play roles in modifications of histone protein, which in turn change the expression of genes (38). These alterations in chromatin structure can impact the accessibility of DNA to transcription factors and other regulatory proteins, thereby exerting an influence on gene expression (20). Another epigenetic change includes non-coding RNAs. AFB1 exposure can induce alteration in the expression pattern of non-coding RNAs, including long non-coding RNAs (lncRNAs) and microRNAs (miRNAs). MiRNAs are small RNAs that can modulate gene expression post-transcriptionally. They can achieve this by interacting with target mRNAs, which can result in the degradation of the mRNA or the inhibition of its translation. Alteration in the expression of these non-coding RNAs can significantly affect cellular processes and gene expression patterns. One study also revealed that aflatoxins elevated a large number of miRNAs, including miR-34a-5p, miR-33a-5p, and miR-122–5p. The overexpression of miR-34a-5p inhibited the expression of genes linked to the cell cycle, including cyclin D1 (CCDN1) and cyclin E2 (CCNE2) (39). There is growing evidence that suggests epigenetic changes induced by AFB1 exposure and can be inherited across generations. These epigenetic changes can persist over time and may be involved in the development of diseases, including cancer, by changing the gene expression that is involved in cell growth, differentiation, and apoptosis (40).

## MANAGEMENT STRATEGIES

Aflatoxin management is a multifaceted approach that encompasses a range of strategies aimed at mitigating the risks associated with aflatoxins contamination. These strategies are essential for protecting both human and animal health, ensuring the safety of food and feed supplies, complying with regulatory standards, maintaining consumer confidence, promoting sustainable agriculture, and enhancing the economic viability of the food industry. Several preventive methods can be useful and are discussed in the following section.

## AGRICULTURAL PRACTICES

Agricultural practices play a crucial role in reducing aflatoxins contamination in crops. Good agricultural practices (GAPs), such as proper irrigation, fertilization, and pest control, can help minimize fungal growth and aflatoxin production (41). Crop rotation and the use of resistant crop varieties can also reduce the risk of aflatoxin contamination (14). Additionally, educating farmers about the importance of proper crop management practices can improve overall crop quality and reduce aflatoxin levels (42).

## POST-HARVEST INTERVENTIONS

Post-harvest interventions are essential for minimizing aflatoxin contamination in stored crops. Rapid drying of harvested crops to less than 14% moisture content can inhibit fungal growth and reduce aflatoxin production (43). Proper storage conditions, including adequate ventilation and temperature control, can also help prevent aflatoxin contamination (44). Timely processing of crops into finished products can further reduce aflatoxin levels because processing can eliminate or reduce aflatoxin content in food and feed (45).

## BIOLOGICAL CONTROL

Biological control using non-aflatoxigenic strains of *Aspergillus* spp. has shown promise in reducing aflatoxin contamination in crops. These non-toxigenic strains compete with aflatoxigenic fungi for nutrients and space, reducing their ability to produce aflatoxins (46). Biological control agents can be applied to crops either as seed treatments or as biocontrol products during crop growth and storage (47). However, the effectiveness of biological control can vary depending on environmental conditions and the competitive ability of the non-toxigenic strains.

Regulatory thresholds have been set by many countries for aflatoxin concentrations in food and animal feed to safeguard public health. These limits are based on risk assessments and vary depending on the type of commodity and intended use (4). Regulatory authorities monitor compliance with these limits through sampling and testing programs. However, enforcing these regulations can be challenging, especially in developing countries with limited resources and infrastructure (48).

## CONCLUSION

Aflatoxigenic fungi represent a significant global concern for food safety and human health, given their propensity to produce carcinogenic aflatoxins that contaminate food crops. The prevalence of these fungi is influenced by environmental factors such as temperature, precipitation patterns, and humidity levels, which create conducive conditions for fungal growth and aflatoxin production. The consequences of aflatoxin exposure on human health are severe, ranging from acute toxicity to long-term carcinogenic effects. Thus, effective management strategies are imperative to mitigate the risks associated with aflatoxigenic fungi. These strategies encompass a holistic approach, including pre-harvest interventions such as crop monitoring and genetic resistance, post-harvest measures such as proper storage and processing techniques, and regulatory frameworks to enforce aflatoxin limits in food products. By integrating scientific knowledge with practical interventions, stakeholders can collectively work toward ensuring food safety and safeguarding human health against the threats posed by aflatoxigenic fungi.

## ACKNOWLEDGMENT

The authors are grateful to the Centre for Medical Biotechnology and Department of Microbiology, Maharshi Dayanand University, Rohtak, for providing infrastructure to carry out the work. KG and SK acknowledge the University Research Scholarship, Maharshi Dayanand University for providing financial support and SP is thankful to Department of Science & Technology, Ministry of Science and Technology, Government of India for providing INSPIRE fellowship.

## REFERENCES


1. World Health Organization. 2011. FAO/WHO guide for application of risk analysis principles and procedures during food safety emergencies. Food and Agriculture Organization of the United Nations.
2. Ghaemmaghami, S. S., H. Rouhanipour, & S. D. Sharifi. 2024. Aflatoxin levels in poultry feed: A comparison of mash and pellet forms. *Poultry Science 103*(1): 103254.
3. Roze, L. V., S. Y. Hong, & J. E. Linz. 2013. Aflatoxin biosynthesis: Current frontiers. *Annual Review of Food Science and Technology 4*: 293–311.
4. Shabeer, S., S. Asad, A. Jamal, & A. Ali. 2022. Aflatoxin contamination, its impact and management strategies: An updated review. *Toxins 14*(5): 307.
5. International Agency for Research on Cancer (IARC). 2002. Aflatoxins. *IARC Monographs on the Evaluation of Carcinogenic Risks to Human 82*: 171–300.
6. Ehrlich, K. C., P. Li, L. Scharfenstein, & P. K. Chang. 2010. HypC, the anthrone oxidase involved in aflatoxin biosynthesis. *Applied and Environmental Microbiology 76*: 3374–3377.
7. Caceres, I., A. Al Khoury, R. El Khoury et al. 2020. Aflatoxin biosynthesis and genetic regulation: A review. *Toxins 12*(3):150.
8. Yu, J., D. Bhatnagar, & T. E. Cleveland. 2004. Completed sequence of aflatoxin pathway gene cluster in *Aspergillus parasiticus. FEBS Letters 564*(1–2): 126–130.
9. Lin, B. K., & J. A. Anderson. 1992. Purification and properties of versiconal cyclase from *Aspergillus parasiticus. Archives of Biochemistry and Biophysics 193*: 67–70.
10. Kelkar, H. S., T. W. Skloss, J. F. Haw, N. P. Keller, & T. H. Adams. 1997. *Aspergillus nidulans* stcL encodes a putative cytochrome P-450 monooxygenase required for bisfuran desaturation during aflatoxin/sterigmatocystin biosynthesis. *Journal of Biological Chemistry 272*(3): 1589–1594.
11. Hong, S. Y., L. V. Roze, J. Wee, & J. E. Linz. 2013. Evidence that a transcription factor regulatory network coordinates oxidative stress response and secondary metabolism in *Aspergilli. Microbiology Open 2*: 144–160.
12. Cary, J. W., K. C. Ehrlich, J. M. Bland, & B. G. Montalbano. 2006. The aflatoxin biosynthesis cluster gene, aflX, encodes an oxidoreductase involved in conversion of versicolorin A to demethylsterigmatocystin. *Applied and Environmental Microbiology 72*(2): 1096–1101.
13. Kumar, P., D. K. Mahato, M. Kamle, T. K. Mohanta, & S. G. Kang. 2017. Aflatoxins: A global concern for food safety, human health and their management. *Frontiers in Microbiology 7*: 235–289.
14. Kumar, A., H. Pathak, S. Bhadauria, & J. Sudan. 2021. Aflatoxin contamination in food crops: Causes, detection, and management: A review. *Food Production, Processing and Nutrition 3*: 1–9.
15. Chandra, P. 2021. Aflatoxins: Food safety, human health hazards and their prevention. In *Aflatoxins-Occurrence, Detoxification, Determination and Health Risks*. IntechOpen.
16. Mahato, D. K., K. E. Lee, M. Kamle et al. 2019. Aflatoxins in food and feed: An overview on prevalence, detection and control strategies. *Frontiers in Microbiology 10*:483502.
17. Negash, D. 2018. A review of aflatoxin: Occurrence, prevention, and gaps in both food and feed safety. *Journal of Applied Microbiological Research 1*(1): 35–43.
18. Ajmal, M., W. Bedale, A. Akram, & J. H. Yu. 2022. Comprehensive review of aflatoxin contamination, impact on health and food security, and management strategies in Pakistan. *Toxins 14*(12): 845.
19. Daou, R., J. C. Assaf, & A. El Khoury 2022. Aflatoxins in the era of climate change: The mediterranean experience. In *Aflatoxins-Occurrence, Detection and Novel Detoxification Strategies*. IntechOpen.
20. Williams, J. H., T. D. Phillips, P. E. Jolly, J. K. Stiles, C. M. Jolly, & D. Aggarwal. 2004. Human aflatoxicosis in developing countries: A review of toxicology, exposure, potential health consequences, and interventions. *The American Journal of Clinical Nutrition 80*(5): 1106–1122.
21. Enespa, & P. Chandra. 2022. *Aflatoxins: Food Safety, Human Health Hazards and Their Prevention*. IntechOpen.
22. Matchado, A., J. W. Smith, K. J. Schulze et al. 2023. Child aflatoxin exposure is associated with poor child growth outcomes: A prospective cohort study in rural Malawi. *Current Developments in Nutrition 7*(7):101962
23. Liu, Y., & F. Wu. 2010. Global burden of aflatoxin-induced hepatocellular carcinoma: A risk assessment. *Environmental Health Perspectives 118*(6): 818–824.
24. Wu, F. 2013. Aflatoxin exposure and chronic human diseases: Estimates of burden of disease, 2020 vision briefs 20(3), International Food Policy Research Institute (IFPRI)..
25. Wu, F. 2015. Global impacts of aflatoxin in maize: Trade and human health. *World Mycotoxin Journal 8*(2): 137–142.

26. Zhu, Q., Y. Ma, J. Liang et al. 2021. AHR mediates the aflatoxin B1 toxicity associated with hepatocellular carcinoma. *Signal Transduction and Targeted Therapy 6*(1):299.
27. Kodama, M., F. Inoue, & M. Akao. 1990. Enzymatic and non-enzymatic formation of free radicals from aflatoxin B1. *Free Radical Research Communications 10*(3): 137–142.
28. Marin, D. E., & I. Taranu. 2012. Overview on aflatoxins and oxidative stress. *Toxin Reviews 31*(3–4): 32–43.
29. Hua, Z., R. Liu, Y. Chen et al. 2021. Contamination of aflatoxins induces severe hepatotoxicity through multiple mechanisms. *Frontiers in Pharmacology 11*: 605823.
30. Shuaib, F. M., J. Ehiri, A. Abdullahi, J. H. Williams, & P. E. Jolly. 2010. Reproductive health effects of aflatoxins: A review of the literature. *Reproductive Toxicology 29*(3): 262–270.
31. Jalili, C., B. Amiri, S. Kakebaraie, F. Jalili, & T. Z. Nasta. 2024. Genotoxic and cytotoxic effects of aflatoxin on the reproductive system: Focus on cell cycle dynamics and apoptosis in testicular tissue. *Toxicology 504*: 153773.
32. Hussein, H. S., & J. M. Brasel. 2001. Toxicity, metabolism, and impact of mycotoxins on humans and animals. *Toxicology 167*(2): 101–134.
33. Li, C., X. Liu, J. Wu, X. Ji, & Q. Xu. 2022. Research progress in toxicological effects and mechanism of aflatoxin B1 toxin. *PeerJ 10*: e13850.
34. Feng, W. H., K. S. Xue, L. Tang, P. L. Williams, & J. S. Wang. 2016. Aflatoxin B1-induced developmental and DNA damage in Caenorhabditis elegans. *Toxins 9*(1): 9.
35. Guengerich, F. P., W. W. Johnson, &Y. F. Ueng. 2006. Roles of cytochrome P450 enzymes in chemical carcinogenesis and cancer chemotherapy. *Cancer Chemother Pharmacol 58*: 13–20.
36. Ferreira, R. G., M. V. Cardoso, K. M. De Souza Furtado, K. M. M. Espíndola, R. P. Amorim, & M. C. Monteiro. 2019. Epigenetic alterations caused by aflatoxin b1: A public health risk in the induction of hepatocellular carcinoma. *Translational Research 204*: 51–71.
37. Mungamuri, S. K., & V. A. Mavuduru. 2020. Role of epigenetic alterations in aflatoxin-induced hepatocellular carcinoma. *Liver Cancer International 1*(2): 41–50.
38. Dai, Y., K. Huang, B. Zhang, L. Zhu, & W. Xu. 2017. Aflatoxin B1-induced epigenetic alterations: An overview. *Food and Chemical Toxicology 109*: 683–689.
39. Zhu, L., J. Gao, K. Huang, Y. Luo, B. Zhang, & W. Xu. 2015. miR-34a screened by miRNA profiling negatively regulates Wnt/β-catenin signaling pathway in aflatoxin B1 induced hepatotoxicity. *Scientific Reports 5*(1): 16732.
40. Pogribny, I. P., & I. Rusyn. 2013. Environmental toxicants, epigenetic changes, and hepatic carcinogenesis. *Journal of Hepatology 58*(5): 1103–1111.
41. Bennett, J. W., & M. A. Klich. 2003. Mycotoxins. *Clinical Microbiology Reviews 16*(3): 497–516.
42. Klich, M. A. 2007. Aspergillus flavus: The major producer of aflatoxin. *Molecular Plant Pathology 8*(6): 713–722.
43. Doster, M. A., & T. J. Michailides.1994. Aspergillus molds and aflatoxins in pistachio nuts in California. *Phytopathology 84*(6): 583–90.
44. Pitt, J. I., & A. D. Hocking. 2009. *Fungi and Food Spoilage* (3rd ed.). New York: Springer.
45. Cotty, P. J., & R. Jaime-Garcia. 2007. Influences of climate on aflatoxin producing fungi and aflatoxin contamination. *International Journal of Food Microbiology 119*(1–2): 109–15.
46. Bandyopadhyay, R., J. Atehnkeng, A. Ortega-Beltran, A. Akande, T. D. Falade, & P. J. Cotty. 2019. "Ground-truthing" efficacy of biological control for aflatoxin mitigation in farmers' fields in Nigeria: From field trials to commercial usage, a 10-year study. *Frontiers in Microbiology 10*: 2528.
47. Dorner, J. W. 2009. Development and commercialization of afla-guard, an aflatoxin biocontrol agent. *Toxin Reviews 28*(2–3): 111–33.
48. Joutsjoki, V. V., & H. J. Korhonen. 2021. Management strategies for aflatoxin risk mitigation in maize, dairy feeds and milk value chains—Case study Kenya. *Food Quality and Safety 5*: fyab005.

# 2.19 Aflatoxigenic Fungi
## *Human Health Impacts and Control Methods*

*Samridhi Pushkarna, Kajal Gaba, Shruti Kharod, Anil Kumar, S. Uma Devi, Pooja Suneja, and Amita Suneja Dang*

## INTRODUCTION

Fungi are major plant pathogens capable of producing toxic metabolites that are harmful to humans through ingestion, skin contact, or inhalation. Exposure leads to mycotoxicoses, which can be either acute or chronic. Mycotoxicoses can have irreversible detrimental effects in cases of chronic toxicity, and hence mycotoxin accumulation in food and feed poses a serious threat to human health [1, 2]. Mycotoxins are a group of low-weight secondary metabolites synthesized and secreted by filamentous fungi thriving on organic substrates, contaminating a wide range of foods, feeds, and raw materials used to produce them. *Aspergillus, Penicillium,* and *Fusarium* represent the main genera of mycotoxigenic fungi [3]. Out of thousands of mycotoxins, only a few pose significant food safety challenges and can induce severe side effects in livestock and humans [4]. Moisture content, high temperature, and water activity are some of the predisposing factors facilitating mycotoxin production in food [5]. Aflatoxins are among the most harmful mycotoxins produced primarily by the common molds *A. flavus* and *A. parasiticus.* Structurally, they belong to a group of difuranocoumarins and are synthesized as secondary metabolites resulting from the polyketide pathway [6]. Aflatoxin-producing fungal species infect various crops, including cereals (maize, wheat, sorghum, and rice), oilseeds (peanut, soybean, cotton seeds, and sunflower), spices (black pepper, chili peppers, ginger, turmeric, and coriander), and tree nuts (almonds, pistachio, coconut, walnut, Brazil nut) [7]. Currently, more than 20 different aflatoxins have been identified, but aflatoxins B1 (AFB1), AFB2, AFG1, and AFG2 are the four main and highly prevalent types [8]. Aflatoxin poisoning usually occurs through ingestion, but AFB1 can also penetrate the skin, making it the most toxic naturally occurring aflatoxin [9]. Also, the double bond at the 8,9 position of the furofuran ring and the lactone ring in the coumarin moiety are two vital toxicological sites that make AFB1 highly potent in causing deleterious effects at the molecular and cellular level [10, 11]. Aflatoxin toxicity can range from acute illness or death to chronic diseases in the long term. It is estimated that 4.5 billion people in developing countries worldwide are exposed to an uncontrolled amount of aflatoxin, thus affecting the health and economic balance globally [12]. This demands the exploration and practice of various management and control strategies to prevent and eliminate aflatoxin contamination. Hence, this chapter provides an overview of the human health impacts caused by aflatoxins and different control methods that are employed to combat the serious problem of aflatoxin contamination.

## HEALTH IMPACTS

Around 4.5 billion people across the world are affected by exposure to aflatoxins, and most of these people are in developing countries [13]. Although it impacts people of all age groups and genders, its effect is more commonly seen in pregnant women and children [14]. The primary route of exposure

DOI: 10.1201/9781003487043-25

to AFB1 is through direct ingestion of contaminated food items. Other routes include inhalation (occupational exposure) and skin contact [15]. Exposure to aflatoxins through any route can lead to the development of various toxicological effects, which may lead to acute or chronic aflatoxicosis depending on the dose and duration of exposure [16]. Moderate to large doses for a comparatively short duration can cause acute aflatoxicosis. Symptoms may include lethargy, nausea, vomiting, yellowing of the skin, itching, edema, abdominal pain, hemorrhage, acute liver damage, coma, and even death [17, 18]. There have been reports of acute aflatoxicosis outbreaks in different parts of the world. In India, a two-month-long outbreak occurred, affecting the tribal population belonging to 200 villages in the districts of Rajasthan and Gujarat, causing 106 deaths in total. The tribal population mainly survived on maize. Hence, when the maize samples were analyzed, it was found that they contained 6.25–15.6 ppm of *A. flavus* contamination, which might have caused the outbreak [19]. Apart from this, a few more incidences of aflatoxin outbreaks have been observed in India, involving both humans and animals [20]. Kenya also witnessed a severe outbreak that reported 317 cases of acute hepatic failure, with 125 deaths. After the serum samples of a few patients were analyzed, it was found the sickness was due to huge aflatoxin contamination (220 times greater than the allowed limit) in maize and not because of any virus [21]. Another serious outbreak due to high levels of aflatoxin-contaminated homegrown maize occurred in two regions of central Tanzania. It resulted in a 30% case fatality rate. Symptoms like abdominal pain, diarrhea, jaundice, vomiting, and ascites were presented by the affected cases. In addition to these clinical manifestations, the presence of aflatoxin contamination in food, along with higher levels of the aflatoxin–albumin adduct in the sera of cases, further confirmed the role of aflatoxin in causing the outbreak [22]. In all these outbreaks, acute aflatoxicosis has mainly resulted in acute hepatic toxicity, but the exact dosage of AFB1 that would cause the illness has not yet been determined.

Chronic aflatoxicosis results from exposure to low to moderate doses of aflatoxin for a prolonged duration. Although the effects of chronic aflatoxicosis are not sudden, they can result in long-developing consequences such as neurotoxicity, severe hepatotoxicity, and reproductive failure [16]. Out of all aflatoxins, it is usually AFB1 that causes serious health concerns. The fact that it can cross the blood-brain barrier and accumulate in the brain has drawn a lot of attention to its adverse effects on the central nervous system (CNS) [23, 24]. As previously studied, multiple mechanisms are involved by which AFB1 can cause neurotoxicity. These include increased production of reactive oxygen species, upregulation of cell-cycle regulatory genes, and proapoptotic genes leading to S-phase arrest and apoptosis, adenosine triphosphate (ATP) depletion, alteration of the cholinergic system, neurotransmission injury, and neuroinflammation [25]. Aflatoxins also act as genotoxins, causing DNA or chromosomal damage, thus leading to harmful mutations and the possible development of cancer. Several studies reveal that AFB1 also leads to genotoxicity. Inside the body, the active electrophilic form of AFB1 may attack nucleophilic oxygen, nitrogen, and sulfur heteroatoms present in cellular components. This results in the formation of highly reactive species, which may combine with DNA bases to generate variations in the DNA, leading to the formation of adducts and loss of control over various cellular functions (cell growth and division). In addition to DNA, it can also form adducts with RNA and proteins, thus inhibiting RNA synthesis and the activity of DNA-dependent RNA polymerase [26]. Aflatoxin-induced hepatotoxicity is the most common chronic condition that is manifested in people exposed to aflatoxin for long periods. This is because the liver is the main target organ for aflatoxin metabolism, and its conversion into extremely toxic forms (AFB1-8,9-epoxide [AFBO]) causes hepatotoxicity. These poisonous metabolites attack DNA by combining with guanine residues and forming stable aflatoxin adducts exerting acute and chronic cytotoxicities. Aflatoxins have been observed to negatively impact mitochondrial respiratory functions and can also impair the normal structure of mitochondria, leading to the development of oxidative stress. Aflatoxins further upregulate the NF-κB signaling pathway and encourage inflammasome assembly in the presence of liver immunocytes. This results in generating a local inflammatory microenvironment, thus exacerbating hepatotoxicity [27]. Furthermore, several studies have reported that the reproductive health of people consuming contaminated food items may become highly compromised and may even result in severe reproductive failure [28]. These toxins have been found to contaminate many body fluids, including maternal breast milk, maternal blood, and cord blood. The rates of contamination have been observed to be higher in samples obtained from developing countries. Although an infant needs to feed on breast milk, the contaminated milk can lead to severe health problems, including immune suppression, malnutrition, and chronic liver disease. The literature focusing on the relationship between aflatoxins and birth weight does not provide a conclusive statement, but a few studies indicate the possibility of stillbirths due to the fatal effects of higher intoxication levels [29]. In the case of studies conducted on the effects of aflatoxins on fertility, the semen of 40% of infertile men contained aflatoxins as compared to 8% of semen from fertile men. Spermatozoa were also abnormal in the contaminated semen samples, suggesting a spermatotoxic effect of aflatoxins, thereby harming fertility [30].

## CONTROL METHODS

To combat the escalating aflatoxin contamination and mitigate its detrimental impact on global health, there is an urgent need to employ methods efficient in preventing its exposure. To address this, various techniques and management strategies have been and continue to be explored. These include preventing aflatoxin contamination at the pre-harvest stage and aflatoxin detoxification using several biological, physical, and chemical methods (Figure 2.19.1), which are discussed below.

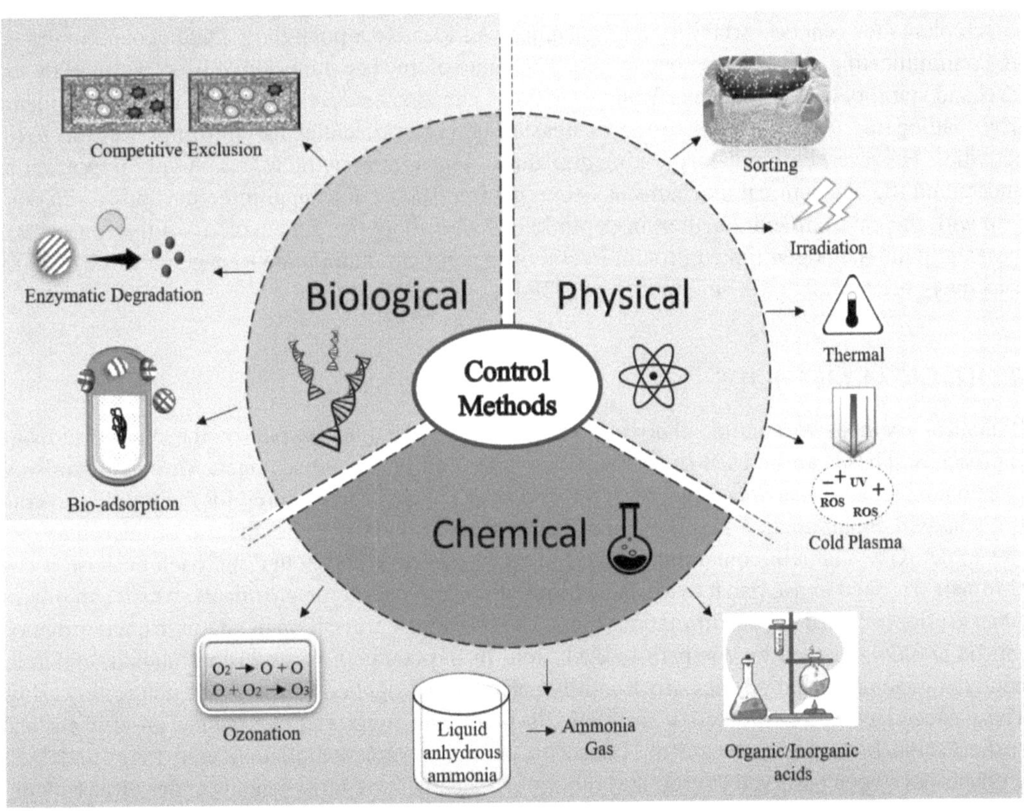


**FIGURE 2.19.1** A schematic representation of different control methods for aflatoxin contamination.

## PREVENTION OF AFLATOXIN AT THE PRE-HARVEST STAGE

### Regulated Agricultural Management

In cases of improper agronomical practices, abiotic stresses like drought, high temperature, soil moisture content, and mineral deficiencies tend to favor increased growth of aflatoxigenic fungi and aflatoxin production [31, 32]. Hence, ideal conventional steps, such as the selection of healthy seeds, proper irrigation, and crop rotation, can serve as major points of intervention at the pre-harvest stage, which can limit aflatoxin contamination if implemented properly [33]. Frequencies of insect infestation and mechanical damage to plants must also be checked and minimized [34]. The sowing time and region selected for cultivating the crop also determine the aflatoxin content and its chances of polluting under field conditions. A study showed that a maize crop cultivated early was less contaminated than one cultivated later [35, 36]. It is highly crucial for farmers and food industries to strictly follow the recommendations and guidelines provided by the Food and Agricultural Organization because primary prevention efforts at this stage may prove to be highly effective in minimizing aflatoxin contamination in food crops and goods produced from them [37, 38].

## USE OF RESISTANT VARIETIES AND GENETIC ENGINEERING

For many years, geneticists have been employing traditional techniques to develop plant cultivars that would show resistance to aflatoxin. With some advancement seen in the case of maize, but not much, plant breeders are now adopting the technique of genetic engineering to produce aflatoxin-resistant varieties [39, 40]. Various breeding programs have been deployed on maize using aflatoxin-resistant germplasm, proteins, and genes associated with host resistance, thus imparting an overall protection against aflatoxin infection [41]. One of the most important steps to achieve this goal is to screen plants for genetic variability for resistance and identify a potentially ideal genetic make-up for communicating the desirable phenotype. The value of any resistant source is determined by the level and stability of its resistance. Transgenic crops can also be developed using genome segments from pathogenic fungi to regulate the expression of genes essential for antifungal or anti-toxin activities. However, before incorporating antifungal genes into any plant, it is highly important to understand the biochemical mechanisms occurring in response to fungal infection and its relationship with the environment. Further, in-depth knowledge about the field ecology of the fungus and dynamic plant–pathogen linked protein interactions can help manipulate transgenic strategies and effectively prevent aflatoxin contamination [42–44].

## BIOLOGICAL METHODS

Biological methods to control aflatoxin contamination appear to be one of the most promising approaches. These can be used on crops at both pre- and post-harvest stages. Various organisms, like nontoxigenic *Aspergillus* fungi, bacteria, and yeasts, have been explored for their ability to control aflatoxin contamination [45]. Recent research provides evidence that the use of microbial species to detoxify aflatoxin-contaminated food is a cost-effective and better approach because it can eliminate the need to use harsh oxidizing, chlorinating, and hydrolytic chemicals, which can impart adverse effects if used beyond limitations [46, 47]. Pre-harvest prevention of aflatoxin contamination can be mainly achieved by competitive exclusion. In the competitive exclusion method, endemic, non-aflatoxigenic fungal species are carefully selected and applied to the crops before flowering. These species compete to outgrow the aflatoxin-producing fungi and can exhibit an efficient and cost-effective biocontrol mechanism. Consortia of multi-strain non-aflatoxigenic fungi and their products have been observed to produce both immediate and long-term benefits under different conditions as compared to single-strain interventions [48–51]. To maximize their efficacy, several factors need to be optimized; for example, consortia formulation, inoculum rate, herbicide application, and abiotic factors including soil temperature, substrate, and water availability [52, 53]. It is of utmost importance to ensure that the potential biocontrol strains are truly atoxigenic in nature and do not result in the production of any toxic molecules under field conditions [54]. Post-harvest biological detoxification of mycotoxins works mainly via two major processes, bio-adsorption and enzymatic degradation, both of which can be achieved by biological systems [55]. Bio-adsorption technology involves the use of inactivated microorganism cell surfaces to bind with the mycotoxins present in feed and food. In this process, the amount of toxin removed depends on various factors like the concentration of both the toxin and the adsorbent/microorganism, temperature conditions, and pH values [56]. Microbial transformation is another method of aflatoxin removal. It is an efficient and green biotechnological method in which the toxin's structure is transformed into other non-toxic derivatives by exploiting different enzyme systems and metabolic pathways present in the microorganisms. The biological enzyme degradation method works in a manner similar to biological transformation. In this method, specific enzymes are purified from biological sources. The isolated enzymes then act directly on mycotoxins, degrading them into harmless components. Based on the nature of enzymes, the procedure needs to be optimized properly for efficient degradation. Actinomycetia, Flavobacteria, Proteobacteria, and Bacillus are the classes of bacteria that have been described as potential degraders of aflatoxins present in food and feed. Studies have reported that most aflatoxin-producing fungi are also metabolically evolved to carry out aflatoxin conversion into less toxic metabolites [57, 58]. Other fungal isolates like *Trametes versicolor, Bierkandera adusta,* and the yeast *Saccharomyces cervisiae* have all been found to degrade AFB1 into less toxic compounds. In addition to these, many other examples of bacterial and fungal degradation of aflatoxins exist that can provide a more target-specific decontamination than physical and chemical control methods [59].

## PHYSICAL METHODS

Several physical methods can be used to reduce aflatoxin contamination in different food commodities. These methods include proper washing, mechanical sorting, electronic eye sorting, density segregation, sieving and dehulling, gamma, UV, and visible light irradiations, and ultrasonic and thermal treatments [60, 61]. First, the removal of fragmented and damaged grains is quite an effective initial step in reducing aflatoxin contamination because such grains usually contain a higher concentration of mycotoxins [62]. This can be achieved by hand sorting, which is certainly the most effective method of lowering the contamination, but it is not possible on a large scale. Hence, improved technologies are now being utilized for better speed and accuracy. In a study, a high-speed dual-wavelength method was applied to sort yellow corn employing absorbances at 750 and 1200 nm. As a result of this, more than 99% of aflatoxin-contaminated kernels were identified correctly, and an overall reduction of aflatoxin was found to be about 81% [63]. Thermal treatments have also proven to be beneficial in reducing aflatoxin contamination. With no standard protocol for all kinds of goods, the extent of reduction largely depends on the procedure followed and the commodity that is being detoxified. Aflatoxins are resistant to heat, so common cooking procedures like boiling, frying, or pasteurization are incapable of providing significant aflatoxin reduction. Fruits, when autoclaved at 120°C for half an hour, can remove 9%–39% of aflatoxin concentration, whereas peanuts achieve a 100% reduction when autoclaved at 1.5 atm for 1.5 hours [64]. In addition to autoclaving, extrusion is also a widely used physical detoxification technique. It is a less time-consuming method that involves high temperature, humidity, and mechanical force and has been shown to reduce aflatoxin levels by 50%–80%. This practice is largely carried out in the case of decontaminating ready-to-eat snacks, biscuits, pasta, and so on [65]. Baking is another heat treatment that was found to lower the amount of aflatoxin present in bread by 32% to 64% [66]. To reduce the microbial load but also keep the quality of food intact, non-thermal procedures like cold plasma come into play. This is a quasi-neutral ionized gas, and its application at high voltage under different humidity and treatment periods has resulted in positive outcomes in terms of inhibiting the growth of *A. flavus* and degradation of mycotoxin content to about 95% in the case of different grains and nuts [41, 67]. Other common methods include UV irradiation, in which the food sample is penetrated with ionizing radiation, which leads to fungal cell wall degradation, thus hindering the growth of fungi and increasing the shelf-life of treated food items [68]. Brown, black, and red rice grains showed 22%–79% fewer fungal colonies and lower aflatoxin content after 1 to 3 hours of UV irradiation followed by 6 months of storage [69]. The use of gamma and visible light radiation also exhibits successful management of aflatoxins. The efficiency of treatment varies with the dosage in the feed sample. In an experiment conducted on artificially contaminated maize, 10 kGy and 5 kGy doses of gamma irradiation resulted in about 90% and 70% AFB1 reduction, respectively [70]. It has been observed that the outcome of aflatoxin content reduction by irradiation improves even more when applied along with detoxifying enzymes. Additionally, various adsorbents have also been reported to eliminate aflatoxin contamination. The total aflatoxin content (TAF) of hazelnut milk was reduced by 43.7% and 71.1% with AC and Na-B treatments, respectively, as compared to the control. Similarly, activated charcoal (AC) and bentonite (Na-B) treatments led to a reduction of 42.8% and 85.9% in the AFB1 content of the samples, respectively. When exposed to the same dose and time conditions, Na-B adsorbent was found to be more effective than AC in reducing the content of AFB1, AFB2, AFG1, AFG2, and TAF in hazelnut milk. Hence, the use of binding agents offers a promising potential for reducing aflatoxin levels in hazelnut milk and preventing economic losses in the hazelnut industry [71].

## CHEMICAL METHODS

Aflatoxins tend to easily diffuse away from mycelia. Hence, products with no visible mold damage can harbor these toxins in significant amounts, which may render physical methods ineffective in imparting complete detoxification. This generates the need for adopting alternative decontamination approaches, and most researchers are now practicing chemical inactivation of aflatoxins [72]. Chemical methods for detoxifying aflatoxin contamination include the use of various chemical agents, organic and inorganic acids, ozone treatment, ammoniation, and redox-active enzymes [67]. Some acids, for example citric, tartaric, lactic, hydrochloric, and propionic, have been reported to be comparatively more effective in reducing aflatoxin content than others such as succinic, ascorbic, acetic, and formic acid [72, 73]. Chemical agents that are oxidizing in nature have also been reported as useful in bioassays carried out for aflatoxin detoxification procedures. As observed in one study, osmium tetroxide and benzoyl peroxide were able to react with AFB1 and AFG1, while NaOCl, $NaBO_3$, $KMnO_4$, 3% $H_2O_2 + NaBO_2$, and Ce $(NH_4)_2(SO_4)_3$ reacted with all four aflatoxins: AFB1, AFB2, AFG1, and AFG2, and performed detoxification [74]. Chemical agents can also be combined with physical methods to enhance the reduction of aflatoxin content in feeds and foods. For example, a 24-hour treatment with a combination of 5% w/v NaOH solution and heat (50°C) could successfully reduce 60.35% of the original AFB1 amount present in *A. flavus* contaminated low-grade maize [75]. Different factors like the antifungal activity of chemicals and duration of exposure along temperature conditions, must be optimized to further minimize aflatoxin contamination. Another chemical procedure for aflatoxin detoxification is ammoniation, which is one of the most practical approaches to decontaminate agricultural goods [72]. In developing countries, maize staples serve as a major source of dietary aflatoxin intake. For destroying these toxins, ammonia reagent is widely used because the products formed after the ammoniation reaction are stable and have negligible toxicity [76, 77]. In a study exploring the effect of ammoniation on contaminated maize, it was found that there was a proportional increase in the aflatoxin reduction with an increase in ammonia concentration. Also, treatment with ammonia under high pressure resulted in greater aflatoxin degradation than under low pressure. Another study involving the application of gaseous ammonia to cottonseed and peanut meal resulted in almost undetectable levels of aflatoxins after optimizing the moisture content, temperature, reaction time, and ammonia pressure. Hence, this procedure can be quite helpful in controlling aflatoxin contamination in food items that are specifically stored in warm, moist places on a large scale, thus making it economically practicable for commercial application too [78, 79]. Apart from acids, oxidizing agents, and ammoniation, ozonation is another method that is being widely used for reducing aflatoxin toxicity. US Food and Drug Administration (FDA) approval for the use of ozone as a potent antimicrobial agent has put the process of ozonation at the forefront of the processing of food items [80]. Ozone, a high-energy allotrope of oxygen, is a strong oxidizing and sterilizing agent that can be used for degrading chemical toxicants, including aflatoxins. It is more stable if used in gaseous form than dissolved in water [81]. The mechanism of AFB1 and AFG1 degradation by ozone is initiated with an electrophilic attack on the C8=C9 double bond present in the furan ring, resulting in the production of primary ozonides. This is followed by conversion into monozonide derivatives such as ketones, aldehydes, and organic acids [52]. The short half-life of ozone (15 to 30 minutes) also minimizes the probability of producing secondary pollution by rapid decomposition into oxygen [82], thus making detoxification of aflatoxins in foods feasible. Since the 1960s, when aflatoxins were discovered, ozone has been used to degrade aflatoxins in cottonseed meal and peanut meal [83, 84]. With time, it has been successfully used for decontaminating other crops like corn [85], pistachios [86], figs [87], and red peppers [88]. In a study on corn flour detoxification, the extent of aflatoxin depletion became more obvious with increasing ozone concentration along with exposure time. It was recorded that the application of 75 $mgL^{-1}$ ozone for one hour could decrease significant amounts of AFB1, AFG1, and AFB2. After the treatment, the moisture content of the corn flour was also reduced, making long-term storage conditions favorable [89]. To obtain good detoxification results from ozonation, various factors need to be addressed. First, the sample form must be understood. A higher level of aflatoxin degradation was observed in the case of peanut kernels than in peanut flour, possibly due to clumping of the flour. Similar consequences were noted in the case of pistachios. Hence, optimization of the ozonation apparatus based on the sample type is highly important to avoid conditions like clumping and maximize aflatoxin degradation efficiency. Second, the selection of methods to contaminate samples for studies is important. The efficiency of ozonation is seen to be higher in artificially spiked test samples, which are contaminated directly using ready-to-use aflatoxin. This is because such samples contain the toxin only on the outer surface. However, in the case of the natural method, the test samples are inoculated with aflatoxigenic fungi, which then produce aflatoxins. In these samples, the aflatoxin contamination is not limited to a specific region and may even be present inside the cotyledon. This may be due to the diffusion of aflatoxin into the core of samples and/or penetration of fungi leading to subsequent production of the toxin [90]. The level of aflatoxin contamination and moisture content in samples also influence the process of ozonation. Increased levels of aflatoxin in products require higher concentrations of ozone or a longer duration of exposure for proper intoxication. However, it is preferable to expose the samples to low ozone concentrations for extended durations to minimize the chances of any adverse effects on the quality of treated goods. Practices like reducing the moisture content of freshly harvested crops, providing optimized storage conditions, and using natural antifungal compounds can further help inhibit the levels of contamination [91]. Further, the method of ozonation, that is, dry (only ozone gas) or wet (water-vaporized ozone gas), can also affect the decontamination procedure. In some cases, like figs, gaseous ozone was found to be more effective than ozonated water, whereas in the case of aflatoxin-spiked grain samples, wet ozone could lead to better detoxification. This is because, in water, ozone forms free radicals like OH, which has a stronger oxidizing capacity than ozone gas. However, more than 95% of the studies use the dry method to carry out ozonation of various agricultural products [92]. Additionally, the concentration, exposure time, and flow rate of ozone in various temperature conditions need to be standardized accordingly.

## CONCLUSION

Aflatoxin contamination in food and feed is a serious health and economic issue, especially in developing countries, but its impact can be seen worldwide. It can lead to long-term health concerns, most commonly severe hepatotoxicity. Hence, it is crucial to apply a multifaceted approach to mitigate these risks. Some strategies at the pre-harvest stage that can help prevent aflatoxin contamination include severe monitoring of food and feed supplies, good farming practices, proper storage conditions (maintenance of temperature and humidity levels), and application of aflatoxin-detoxifying agents. Further detoxification can be achieved by appropriate biological, physical, and chemical methods. These methods can be used alone or in combination in a way that ensures an increased shelf life with minimal damage to the food samples concerning quality, taste, texture, and so on. Additionally, constant efforts to increase awareness and develop collaborations between governments, food industries, and research institutions are necessary to address global food safety and protect public health.

## ACKNOWLEDGMENT

The authors are grateful to the Centre for Medical Biotechnology, and the Department of Microbiology, Maharshi Dayanand University, Rohtak, for the infrastructure to carry out this work. SP acknowledges the Department of Science & Technology, Ministry of Science and Technology, Government of India, New Delhi for providing the INSPIRE ("Innovation in Science Pursuit for Inspired Research") Fellowship. KG and SK are thankful to the University Research Scholarship, Maharshi Dayanand University for financial support.

## REFERENCES

1. Bennett, J. W., & M. Klich. 2003. Mycotoxins. *Clinical Microbiology Reviews 16*(3): 497–516.
2. James, R. C. 1985. General principles of toxicology. In P. L. Williams and J. L. Burson (ed.), *Industrial Toxicology* p. 7–26. Lifetime Learning Publications.
3. Greeff-Laubscher, M. R., I. Beukes, G. J. Marais, & K. Jacobs. 2019. Mycotoxin production by three different toxigenic fungi genera on formulated abalone feed and the effect of an aquatic environment on fumonisins. *Mycology 11*(2): 105–117.
4. Murphy, P. A., S. Hendrich, C. Landgren, & C. M. Bryant. 2006. Food mycotoxins: An update. *Journal of Food Science 71*(5): R51.
5. Wageh, D., I. Mayumi, M. Shoata, & I. Yoshinor. 2014. Cellular interactions and metabolism of aflatoxin. *Journal of Pharmacology 65*(2): 163–92.
6. Balina, A., A. Kebede, & Y. Tamiru. 2018. Review on aflatoxin and its impacts on livestock. *Journal of Dairy and Veterinary Sciences 6*(2): e555685.
7. Mycotoxins. 2023. World Health Organization (WHO). https://www.who.int/news-room/factsheets/detail/mycotoxins#:~:text=Aflatoxins%20are%20amongst%20the%20most,frequently%20affected%20by%20Aspergillus%20spp.
8. Kumar, P., D. K. Mahato, M. Kamle, T. K. Mohanta, & S. G. Kang. 2017. Aflatoxins: A global concern for food safety, human health and their management. *Frontiers in Microbiology 7*: 235289.
9. Boonen, J., S. V. Malysheva, L. Taevernier, J. D. Di Mavungu, S. De Saeger, & B. De Spiegeleer. 2012. Human skin penetration of selected model mycotoxins. *Toxicology 301*(1–3): 21–32.
10. Heathcote, J. G., & J. R. Hibbert. 1978. Biochemical effects structure activity relationship. *Aflatoxin: Chemical and Biological Aspects 1:* 112–130.
11. Lee, L. S., J. J. Dunn, A. J. DeLucca, & A. Ciegler. 1981. Role of lactone ring of aflatoxin B1 in toxicity and mutagenicity. *Experientia 37*: 16–17.
12. Williams, J. H., T. D. Phillips, P. E. Jolly, J. K. Stiles, C. M. Jolly, & D. Aggarwal. 2004. Human aflatoxicosis in developing countries: A review of toxicology, exposure, potential health consequences, and interventions. *The American Journal of Clinical Nutrition 80*(5): 1106–1122.
13. Dhakal, A., & E. Sbar. 2021. *Aflatoxin Toxicity*. StatPearls Publishing.
14. Magoha, H., M. Kimanya, B. De Meulenaer, D. Roberfroid, C. Lachat, & P. Kolsteren. 2016. Risk of dietary exposure to aflatoxins and fumonisins in infants less than 6 months of age in Rombo, Northern Tanzania. *Maternal and Child Nutrition 12*(3): 516–527.
15. Degen, G. 2011. Tools for investigating workplace-related risks from mycotoxin exposure. *World Mycotoxin Journal 4*(3): 315–327.
16. Balan, B., A. S. Dhaulaniya, M. Kumar, M. Kumar, & P. Kumar. 2023. Aflatoxins in food: Prevalence, health effects, and emerging trends in its mitigation—An updated review. *Food Safety and Health 2*(1): 39–71.

17. Liu, Y., C. C. H. Chang, G. M. Marsh, & F. Wu. 2012. Population attributable risk of aflatoxin-related liver cancer: Systematic review and meta-analysis. *European Journal of Cancer 48*(14): 2125–2136.
18. Sarma, U. P., P. J. Bhetaria, P. Devi, & A. Varma. 2017. Aflatoxins: Implications on health. *Indian Journal of Clinical Biochemistry 32*(2): 124–133.
19. Krishnamachari, K. A., V. Nagarajan, R. V. Bhat, & T. B. Tilak. 1975. Hepatitis due to aflatoxicosis: An outbreak in Western India. *The Lancet 305*(7915): 1061–1063.
20. Reddy, B. N., & C. R. Raghavender. 2007. Outbreaks of aflatoxicoses in India. *African Journal of Food, Agriculture, Nutrition and Development 7*(5): 1–15.
21. Azziz-Baumgartner, E., K. Lindblade, K. Gieseker, et al. 2005. Case–control study of an acute aflatoxicosis outbreak, Kenya, 2004. *Environmental Health Perspectives 113*(12):1779–1783.
22. Kamala, A., C. Shirima, B. Jani, et al. 2018. Outbreak of an acute aflatoxicosis in Tanzania during 2016. *World Mycotoxin Journal 11*(3):311–320.
23. Qureshi, H., S. S. Hamid, S. S. Ali, J. Anwar, A. A. Siddiqui, & N. A. Khan. 2015. Cytotoxic effects of aflatoxin B1 on human brain microvascular endothelial cells of the blood-brain barrier. *Medical Mycology 53*(4): 409–416.
24. Huang, B., Q. Chen, L. Wang, X. Gao, W. Zhu, P. Mu, & Y. Deng. 2020. Aflatoxin B1 induces neurotoxicity through reactive oxygen species generation, DNA damage, apoptosis, and S-phase cell cycle arrest. *International Journal of Molecular Sciences 21*(18): 6517.
25. Adedara, I. A., O. E. Atanda, C. S. A. Monteiro, et al. 2023. Cellular and molecular mechanisms of aflatoxin B1-mediated neurotoxicity: The therapeutic role of natural bioactive compounds. *Environmental Research 237*(1): 116869.
26. Verma, R. J. 2004. Aflatoxin cause DNA damage. *International Journal of Human Genetics 4*(4): 231–236.
27. Hua, Z., R. Liu, Y. Chen, et al. 2021. Contamination of aflatoxins induces severe hepatotoxicity through multiple mechanisms. *Frontiers in Pharmacology 11*:605823.
28. Shuaib, F. M., J. Ehiri, A. Abdullahi, J. H. Williams, & P. E. Jolly. 2010. Reproductive health effects of aflatoxins: A review of the literature. *Reproductive Toxicology 29*(3): 262–270.
29. Lamplugh, S. M., R. G. Hendrickse, F. Apeagyei, & D. D. Mwanmut. 1988. Aflatoxins in breast milk, neonatal cord blood, and serum of pregnant women. *British Medical Journal (Clinical Research Ed.) 296*(6627): 968.
30. Ibeh, I. N., N. Uraih, & J. I. Ogonar. 1994. Dietary exposure to aflatoxin in human male infertility in Benin City, Nigeria. *International Journal of Fertility and Menopausal Studies 39*(4): 208–214.
31. Negash, D. 2018. A review of aflatoxin: Occurrence, prevention, and gaps in both food and feed safety. *Journal of Nutritional Health & Food Engineering 8*: 190–197.
32. Murphy, P., S. Hendrich, C. Landgren, & C. Bryant. 2006. Food mycotoxins: An update. *Journal of Food Science 71*: 51–65.
33. Santha, T. 1993. Effectiveness of control methods of aflatoxin. Proceedings of 3rd international food convention (IFCON), CFTRI, Mysore, India. *Trends in Food Science and Technology 4*: 1031–1039.
34. Alimentarius, C. 2003. Code of practice for the prevention and reduction of mycotoxins contamination in cereals, including annexes on ochratoxin A, zearalenone, fumonisins and tricothecenes. *CAC/RCP 51*: 1–8.
35. Warnatzsch, E. A., D. S. Reay, M. Camardo Leggieri, & P. Battilani. 2020. Climate change impact on aflatoxin contamination risk in Malawi's maize crops. *Frontiers in Sustainable Food Systems 4*: 591792.
36. Zafar, S. 2019. Grain Yield and Nutritional Quality of Commercial Maize (*Zea mays* L.) Genotypes: *Aspergillus* Exposure and Aflatoxin Contamination. Ph.D. Thesis, Government College University, Faisalabad, Pakistan.
37. Kumar, A., H. Pathak, S. Bhadauria, et al. 2021. Aflatoxin contamination in food crops: Causes, detection, and management: A review. *Food Production, Processing and Nutrition 3*: 17.
38. Mishra, H. N., & C. Das. 2003. A review on biological control and metabolism of aflatoxin. *Critical Reviews in Food Science and Nutrition 43*(3): 245–264.
39. Cardwell, K. F., & S. H. Henry. 2004. Risk of exposure to and mitigation of effect of aflatoxin on human health: A West African example. *Journal of Toxicology: Toxin Reviews 23*(2–3): 217–247.
40. Brown, R. L., Z. Y. Chen, A. Menkir, T. E. Cleveland, K. Cardwell, J. Kling, & D. G. White. 2001. Resistance to aflatoxin accumulation in kernels of maize inbreds selected for ear rot resistance in West and Central Africa. *Journal of Food Protection 64*(3): 396–400.
41. Kumar, P., A. Gupta, D. K. Mahato, et al. 2022. Aflatoxins in cereals and cereal-based products: Occurrence, toxicity, impact on human health, and their detoxification and management strategies. *Toxins 14*(10):687.

42. Umesha, S., H. M. G. Manukumar, B. Chandrasekhar, et al. 2017. Aflatoxins and food pathogens: Impact of biologically active aflatoxins and their control strategies. *Journal of the Science of Food and Agriculture 97*(6):1698–1707.
43. Amalu, U. C. 2004. Plant biotechnology and food crop development in Sub-Saharan Africa. *Technology in Society 26*(4): 537–550.
44. Wani, S. H. 2010. Inducing fungus-resistance into plants through biotechnology, *Notulae Scientia Biologicae 2*(2): 14–21.
45. Dorner, J. W. 2004. Biological control of aflatoxin contamination of crops. *Journal of Toxicology: Toxin Reviews 23*(2–3): 425–450.
46. Adebo, O. A., P. B. Njobeh, S. Gbashi, O. C. Nwinyi, & V. Mavumengwana. 2017. Review on microbial degradation of aflatoxins. *Critical Reviews in Food Science and Nutrition 57*(15): 3208–3217.
47. Pankaj, S. K., H. Shi, & K. M. Keener. 2018. A review of novel physical and chemical decontamination technologies for aflatoxin in food. *Trends in Food Science & Technology 71*: 73–83.
48. Atehnkeng, J., P. S. Ojiambo, P. J. Cotty, & R. Bandyopadhyay. 2014. Field efficacy of a mixture of atoxigenic *Aspergillus flavus* link: Fr vegetative compatibility groups in preventing aflatoxin contamination in maize (*Zea mays* L.). *Biological Control 72*: 62–70.
49. Agbetiameh, D., A. Ortega-Beltran, R. T. Awuah, et al. 2019. Potential of atoxigenic *Aspergillus flavus* vegetative compatibility groups associated with maize and groundnut in Ghana as biocontrol agents for aflatoxin management. *Frontiers in Microbiology 10*:2069.
50. Probst, C., R. Bandyopadhyay, L. E. Price, & P. J. Cotty. 2011. Identification of atoxigenic *Aspergillus flavus* isolates to reduce aflatoxin contamination of maize in Kenya. *Plant Disease 95*(2): 212–218.
51. Unnevehr, L., & D. Grace. 2013. Tackling aflatoxins: An overview of challenges and solutions. In Aflatoxins: Finding solutions for improved food safety, ed. L. J. Unnevehr & D. Grace. 2020 vision briefs 20(1), International Food Policy Research Institute (IFPRI).
52. Jalili, M. 2016. A review on aflatoxins reduction in food. *Iranian Journal of Health, Safety and Environment 3*(1): 445–459.
53. Mohale, S., A. Medina, & N. Magan. 2013. Effect of environmental factors on in vitro and in situ interactions between atoxigenic and toxigenic *A. flavus* strains and control of aflatoxin contamination of maize. *Biocontrol Science and Technology 23*(7): 776–793.
54. Abbas, H. K., R. M. Zablotowicz, B. W. Horn, N. A. Phillips, B. J. Johnson, X. Jin, & C. A. Abel. 2011. Comparison of major biocontrol strains of non-aflatoxigenic *Aspergillus flavus* for the reduction of aflatoxins and cyclopiazonic acid in maize. *Food Additives and Contaminants 28*(2): 198–208.
55. Aliabadi, M. A., F. E. Alikhani, M. Mohammadi, & R. K. Darsanaki. 2013. Biological control of aflatoxins. *European Journal of Experimental Biology 3*(2): 162–166.
56. Luo, Y., X. Liu, L. Yuan, & J. Li. 2020. Complicated interactions between bio-adsorbents and mycotoxins during mycotoxin adsorption: Current research and future prospects. *Trends in Food Science & Technology 96*: 127–134.
57. Xing, F., L. Wang, X. Liu, J. N. Selvaraj, Y. Wang, Y. Zhao, & Y. Liu. 2017. Aflatoxin B1 inhibition in *Aspergillus flavus* by *Aspergillus niger* through down-regulating expression of major biosynthetic genes and AFB1 degradation by atoxigenic *A. flavus*. *International Journal of Food Microbiology 256*: 1–10.
58. Nakazato, M., S. Morozumi, K. Saito, K. Fujinuma, T. Nishima, & N. Kasai. 1990. Interconversion of aflatoxin B1 and aflatoxicol by several fungi. *Applied and Environmental Microbiology 56*(5): 1465–1470.
59. Arimboor, R. 2024. Metabolites and degradation pathways of microbial detoxification of aflatoxins: A review. *Mycotoxin Research 40*(1): 71–83.
60. Zhang, H., M. Dong, Q. Yang, M. T. Apaliya, J. Li, & X. Zhang. 2016. Biodegradation of zearalenone by Saccharomyces cerevisiae: Possible involvement of ZEN responsive proteins of the yeast. *Journal of Proteomics 143*: 416–423.
61. Ajmal, M., W. Bedale, A. Akram, & J. H. Yu. 2022. Comprehensive review of aflatoxin contamination, impact on health and food security, and management strategies in Pakistan. *Toxins 14*(12): 845.
62. Xu, Y. A., A. Doel, S. Watson, et al. 2017. Study of an educational hand sorting intervention for reducing aflatoxin B1 in groundnuts in rural Gambia. *Journal of Food Protection 80*(1):44–49.
63. Pearson, T. C., D. T. Wicklow, & M. C. Pasikatan. 2004. Reduction of aflatoxin and fumonisin contamination in yellow corn by high-speed dual-wavelength sorting. *Cereal Chemistry 81*(4): 490–498.
64. Samarajeewa, U., A. C. Sen, M. D. Cohen, & C. I. Wei. 1990. Detoxification of aflatoxins in foods and feeds by physical and chemical methods. *Journal of Food Protection 53*(6): 489–501.

65. Milani, J., & G. Maleki. 2014. Effects of processing on mycotoxin stability in cereals. *Journal of the Science of Food and Agriculture 94*(12): 2372–2375.
66. Milani, J., S. S. Seyed Nazari, E. Bamyar, & G. Maleki. 2018. Effect of bread making process on aflatoxin level changes. *Journal of Chemical Health Risks 4*(4): 1–7.
67. Sipos, P., F. Peles, D. L. Brassó, et al. 2021. Physical and chemical methods for reduction in aflatoxin content of feed and food. *Toxins 13*(3):204.
68. Shabeer, S., S. Asad, A. Jamal, & A. Ali.2022. Aflatoxin contamination, its impact And management strategies: An updated review. *Toxins 14*(5): 307.
69. Ferreira, C. D., G. H. Lang, I. da Silva Lindemann, et al. 2021. Postharvest UV-C irradiation for fungal control and reduction of mycotoxins in brown, black, and red rice during long-term storage. *Food Chemistry* 339:127810.
70. Markov, K., B. Mihaljević, A. M. Domijan, J. Pleadin, F. Delaš, & J. Frece. 2015. Inactivation of aflatoxigenic fungi and the reduction of aflatoxin B1 in vitro and in situ using gamma irradiation. *Food Control 54*: 79–85.
71. Çakir, C., E. Turan, & A. Şimşek. 2023. The effects of bentonite and activated charcoal treatments on aflatoxin content (AFB1, AFB2, AFG1, and AFG2) and physicochemical characteristics of hazelnut milk. *Journal of Food Measurement and Characterization 17*(5): 5256–5267.
72. ParK, D. L., L. S. Lee, R. L. Price, & A. E. Pohland.1988. Review of the decontamination of aflatoxins by ammoniation: Current status and regulation. *Journal of the Association of Official Analytical Chemists 71*(4): 685–703.
73. Jubeen, F., F. Sher, A. Hazafa, F. Zafar, M. Ameen, & T. Rasheed. 2020. Evaluation and detoxification of aflatoxins in ground and tree nuts using food grade organic acids. *Biocatalysis and Agricultural Biotechnology 29*: 101749.
74. Trager, W., & L. Stoloff. 1967. Possible reactions for aflatoxin detoxification. *Journal of Agricultural and Food Chemistry 15*(4): 679–681.
75. Thakaew, R., & S. Chaiklangmuang. 2023. Aflatoxin B1 elimination in low-grade maize by co-influence of heat and chemical treatment. *Quality Assurance and Safety of Crops & Foods 15*(3): 55–67.
76. Temcharoen, P., T. Anukarahanonta, & N. Bhamarapravati. 1978. Influence of dietary protein and vitamin B12 on the toxicity and carcinogenicity of aflatoxins in rat liver. *Cancer Research 38*(7): 2185–2190.
77. Phillips, T. D., B. A. Clement, & D. L. Park. 1994. Approaches to reduction. *The Toxicology of Aflatoxins: Human Health, Veterinary, and Agricultural Significance* 383–399.
78. Nyandieka, H. S., J. O. Maina, & C. Nyamwange. 2009. Destruction of aflatoxins in contaminated maize samples using ammoniation procedures. *East and Central African Journal of Pharmaceutical Sciences 12*(3): 47–51.
79. Gardner, H. K. Jr, S. P. Koltun, F. G. Dollear, & E. T. Rayner. 1971. Inactivation of aflatoxins in peanut and cottonseed meals by ammoniation. *Journal of the American Oil Chemists' Society 48*(2): 70–73.
80. Misra, N. N., S. U. Kadam, & S. K. Pankaj. 2011. An overview of nonthermal technologies in food processing. *Indian Food Industry 30*(5–6): 45–52.
81. Karaca, H., & Y. S. Velioglu. 2007. Ozone applications in fruit and vegetable processing. *Food Reviews International 23*(1): 91–106.
82. Silva, O. F., & A. Venâncio. 2010. Ozone applications to prevent and degrade mycotoxins: A review. *Drug Metabolism Reviews 42*(4): 612–620.
83. Dwarakanath, C. T., E. T. Rayner, G. E. Mann, & F. G. Dollear. 1968. Reduction of aflatoxin levels in cottonseed and peanut meals by ozonization. *Journal of the American Oil Chemists' Society 45*(2): 93–95.
84. Dollear, F. G., G. E. Mann, L. P. Codifer Jr, H. K. Gardner Jr, S. P. Koltun, & H. L. E. Vix. 1968. Elimination of aflatoxins from peanut meal. *Journal of the American Oil Chemists' Society 45*(12): 862–865.
85. Jr, A. P., & J. M. King. 2002. Efficacy and safety evaluation of ozonation to degrade aflatoxin in corn. *Journal of Food Science 67*(8): 2866–2872.
86. Akbas, M. Y., & M. Ozdemir. 2006. Effect of different ozone treatments on aflatoxin degradation and physicochemical properties of pistachios. *Journal of the Science of Food and Agriculture 86*(13): 2099–2104.
87. Zorlugenç, B., F. K. Zorlugenç, S. Öztekin, & I. B. Evliya. 2008. The influence of gaseous ozone and ozonated water on microbial flora and degradation of aflatoxin B1 in dried figs, *Food and Chemical Toxicology 46*(12): 3593–3597.
88. Inan, F., M. Pala, & I. Doymaz. 2007. Use of ozone in detoxification of aflatoxin B1 in red pepper. *Journal of Stored Products Research 43*(4): 425–429.

89. Luo, X., R. Wang, L. Wang, Y. Li, Y. Wang, & Z. Chen. 2014. Detoxification of aflatoxin in corn flour by ozone. *Journal of the Science of Food and Agriculture 94*(11): 2253–2258.
90. Proctor, A. D., M. Ahmedna, J. V. Kumar, & I. Goktepe. 2004. Degradation of aflatoxins in peanut kernels/flour by gaseous ozonation and mild heat treatment. *Food Additives and Contaminants 21*(8): 786–793.
91. Diao, E., H. Hou, & H. Dong. 2013. Ozonolysis mechanism and influencing factors of aflatoxin B1: A review. *Trends in Food Science & Technology 33*(1): 21–26.
92. Wang, S., H. Liu, J. Lin, & Y. Cao. 2010. Can ozone fumigation effectively reduce aflatoxin B1 and other mycotoxins contamination on stored grain? *Julius-Kühn-Archiv*, (425):582.

# 2.20 Aflatoxigenic Fungi
## *Health Impacts in Infants and Children*

*G.R. Shree Kumari, V. Mohanasrinivasan, and Lokesh Ravi*

## INTRODUCTION

Aflatoxigenic fungi, a subset of fungi known for their production of aflatoxins, pose significant health risks to infants and children worldwide [1]. Aflatoxins are potent toxins produced primarily by *Aspergillus* species, notably *A. flavus* and *A. parasiticus*, and are known for their carcinogenic, mutagenic, teratogenic, and immunosuppressive properties [1, 2]. The contamination of food commodities, such as grains, nuts, and legumes, with aflatoxins presents a persistent threat to human health, particularly in regions with inadequate food storage and agricultural practices. Infants and children are particularly vulnerable to the adverse health effects of aflatoxins due to their still-developing immune systems, higher food intake relative to body weight, and reliance on breast milk or nutrition formula [3, 4]. Exposure to aflatoxins during critical periods of growth and development can lead to severe health complications, including stunted growth, malnutrition, liver damage, and an increased susceptibility to infectious diseases [4].

As fungal species (molds) develop on their substrates, they produce a class of toxins known as mycotoxins, which include aflatoxins [3, 5]. Fungi *A. flavus* and *A. parasiticus* are the primary producers of aflatoxins. Aflatoxin production occurs in other fungal species as well, although in lesser quantities. The fungus naturally exists in tropical soils, where it infects crops during preharvest (on the farm) and postharvest (during storage and processing). The most commonly affected crops are oilseeds, beans, nuts, dried fruits, dried fish, cotton, ground nuts, cassava, and spices. There are eighteen recognised varieties of aflatoxin [6, 7]. Aflatoxins B1, B2, G1, and G2 are the main ones; their names derive from how fluorescent they are in blue and green light. Mammals exposed to aflatoxin B1 (AFB1), B2 (AFB2), G1 (AFG1), and G2 (AFG2) excrete AFM1 and AFM2, which are breakdown products of AFB1, AFB2, or AFG1 and AFG2 generated in the liver. Aflatoxin affects all species of animals [2, 6]. Cattle and hens are relatively resistant; dogs, sheep, and calves are moderately vulnerable; and pigs are severely susceptible [8]. Bees are typically resistant; however, fish range from extremely susceptible to resistant [9].

It is restricted to the underprivileged and food-insecure population. Its presence in food items and animal feed poses a serious risk to the safety of food and feed. There are notable financial losses linked to their effects on the health of newborns and children [10]. The danger of food commodity contamination complies with the guidelines for dosage response and correlation with dangers to human and animal health [11, 12]. They are also the subject of more attention than any other mycotoxin due to their strong acute toxicological and carcinogenic effects in vulnerable infants and children, as well as the direct and indirect economic effects on crops, livestock, and their products, as well as the expense of regulatory initiatives aimed at lowering risks to the health of humans and animals [4]. It is inevitable for mold to grow on food products that are typically contaminated with aflatoxins during manufacturing and cooking procedures, so it is important to take sanitary precautions [3].

DOI: 10.1201/9781003487043-26

## HEALTH IMPACTS OF AFLATOXIGENIC FUNGI ON INFANTS AND CHILDREN BY EXAMINING THE MECHANISMS OF AFLATOXIN TOXICITY

The health impacts of aflatoxigenic fungi on infants and children are profound, primarily attributed to the potent toxicity of aflatoxins, which are secondary metabolites produced by certain fungi, notably *Aspergillus* species like *A. flavus* and *A. parasiticus*. Aflatoxins, including AFB1, AFB2, AFG1, and AFG2, exhibit various toxic effects, ranging from carcinogenicity to immunosuppression [1, 13]. Understanding the mechanisms of aflatoxin toxicity is crucial for comprehending its impact on infant and child health [1, 2, 13].

Aflatoxins exert their toxic effects primarily through their metabolites, which can bind to DNA and form DNA adducts, leading to mutations and potentially carcinogenesis. AFB1, in particular, is highly carcinogenic and is associated with an increased risk of hepatocellular carcinoma (HCC), especially in regions with high aflatoxin exposure levels [2, 14, 15]. In infants and children, whose bodies are rapidly developing and are more susceptible to genetic damage, exposure to aflatoxins can significantly increase the risk of childhood cancers and other long-term health complications [15].

Moreover, aflatoxins can impair immune function, making infants and children more susceptible to infections and other diseases. AFB1 has been shown to suppress immune responses by inhibiting the proliferation and function of immune cells, such as T lymphocytes and natural killer cells [3]. This immunosuppressive effect can exacerbate the severity of infectious diseases, leading to higher morbidity and mortality rates among affected children [4, 5, 15].

Additionally, aflatoxin exposure during critical periods of growth and development can lead to malnutrition and stunted growth. AFB1 can interfere with protein synthesis and metabolism, impairing nutrient absorption and utilization [5]. Chronic exposure to aflatoxins has been linked to growth faltering and malnutrition in children, contributing to long-term health consequences such as cognitive impairment and reduced physical fitness [16].

## ROUTES OF EXPOSURE OF AFLATOXIGENIC FUNGI ON INFANTS AND CHILDREN

The routes of exposure to aflatoxigenic fungi and their toxins in infants and children encompass various pathways, including dietary intake, breastfeeding, and environmental contamination. Understanding these routes is crucial for mitigating the risk of aflatoxin exposure in vulnerable populations.

1. Dietary intake: Infants and children can be exposed to aflatoxins through the consumption of contaminated food and beverages. Aflatoxin contamination commonly occurs in staple foods such as grains (e.g., maize, rice, wheat), nuts (e.g., peanuts, almonds), and legumes (e.g., soybeans, lentils). In regions with inadequate food storage facilities and poor agricultural practices, these food items are particularly susceptible to fungal contamination and subsequent aflatoxin production. Moreover, processed foods derived from contaminated raw materials, such as infant formula and cereal-based products, can also pose a risk of aflatoxin exposure to young children [2, 3, 15].
2. Breastfeeding: Breast milk, considered the optimal source of nutrition for infants, can also serve as a route of aflatoxin exposure if the mother's diet includes contaminated food. Aflatoxins ingested by breastfeeding mothers can be transferred to their infants through breast milk, albeit in lower concentrations compared to formula feeding [10]. Although the benefits of breastfeeding outweigh the potential risks of aflatoxin exposure, efforts to minimize maternal exposure to aflatoxins through dietary diversification and food safety measures are essential [6, 10, 12].
3. Environmental contamination: In addition to food and breast milk, infants and children may encounter aflatoxin contamination in their immediate environment. Aflatoxigenic fungi can proliferate in damp and warm conditions, leading to mold growth on indoor

surfaces, such as walls, floors, and furniture. Inhalation of airborne aflatoxin spores or dust particles contaminated with aflatoxins can contribute to exposure, particularly in poorly ventilated or overcrowded living spaces [10]. Furthermore, contaminated soil and water sources used for drinking or household activities can also serve as reservoirs for aflatoxigenic fungi, posing a potential route of exposure for young children [1].

## EPIDEMIOLOGICAL STUDIES HIGHLIGHTING THE PREVALENCE OF AFLATOXIN EXPOSURE IN VULNERABLE POPULATIONS

Epidemiological studies have extensively documented the prevalence of aflatoxin exposure in vulnerable populations, particularly in regions with high levels of food insecurity, inadequate food storage facilities, and limited access to healthcare. These studies provide valuable insights into the extent of aflatoxin contamination and its impact on human health, especially among infants and children. Here are some key findings from epidemiological research highlighting the prevalence of aflatoxin exposure in vulnerable populations:

1. Sub-Saharan Africa: Numerous studies conducted in Sub-Saharan Africa, where aflatoxin contamination is endemic, have consistently reported high levels of exposure among vulnerable populations, including children. For example, research in countries such as Nigeria, Kenya, and Ghana has documented elevated levels of aflatoxin biomarkers in serum, urine, and breast milk samples collected from pregnant women, lactating mothers, and young children. These studies underscore the widespread nature of aflatoxin exposure in communities heavily reliant on maize, groundnuts, and other staple crops susceptible to fungal contamination [17].
2. Southeast Asia: Aflatoxin exposure is also prevalent in Southeast Asia, where agricultural practices and climate conditions favor the growth of aflatoxigenic fungi. Epidemiological studies in countries like Bangladesh, India, and Vietnam have revealed significant levels of aflatoxin contamination in food commodities and biological samples obtained from vulnerable populations, including infants and children. High levels of aflatoxin biomarkers in maternal and cord blood samples indicate the transplacental transfer of aflatoxins, highlighting the potential risk to fetal health [3].
3. Low- and middle-income countries: Beyond specific regions, studies conducted in low- and middle-income countries (LMICs) have demonstrated the widespread occurrence of aflatoxin exposure and its adverse health effects on vulnerable populations. These studies often focus on marginalized communities facing socioeconomic challenges, such as poverty, malnutrition, and limited access to clean water and sanitation. Vulnerable groups, including children under five years of age, are particularly susceptible to aflatoxin-related health problems, including stunting, underweight, and impaired immune function [1, 17].
4. Impact on child health: Epidemiological evidence consistently links aflatoxin exposure to poor child health outcomes, including growth faltering, malnutrition, and increased susceptibility to infectious diseases. Longitudinal studies tracking aflatoxin exposure biomarkers and health outcomes in children have provided valuable insights into the cumulative effects of chronic exposure on growth and development. Additionally, research examining aflatoxin levels in complementary foods and breastfeeding practices underscores the importance of dietary diversity and food safety interventions in mitigating aflatoxin exposure among young children [3, 5, 16].

## PREVENTION

Prevention and control of aflatoxin contamination are paramount for safeguarding the long-term health and well-being of the most vulnerable members of society, particularly infants and children. Implementing comprehensive measures to mitigate aflatoxin exposure is essential for reducing the burden of aflatoxin-related health problems, including growth impairment, malnutrition, and increased susceptibility to diseases. Interventions such as improving agricultural practices, enhancing food storage and handling techniques, implementing regulatory standards for aflatoxin levels in food and feed, and promoting dietary diversification can significantly reduce the risk of aflatoxin contamination in vulnerable populations. Furthermore, raising awareness among caregivers, healthcare providers, and policymakers about the importance of aflatoxin control and its impact on child health is critical for fostering sustainable solutions to this global health challenge [14]. By prioritizing prevention and control efforts, we can create healthier environments and ensure a brighter future for children worldwide.

## CONCLUSION

In conclusion, the mechanisms of aflatoxin toxicity elucidate the significant health impacts of aflatoxigenic fungi on infants and children. From carcinogenicity to immunosuppression and malnutrition, aflatoxins pose multifaceted threats to the health and well-being of young individuals. Efforts to mitigate aflatoxin exposure in vulnerable populations, such as improving food safety measures and implementing regulatory standards, are essential for safeguarding infant and child health. Infants and children are exposed to aflatoxigenic fungi and their toxins through various routes, including dietary intake, breastfeeding, and environmental contamination. Addressing aflatoxin exposure in this vulnerable population requires a multifaceted approach, encompassing food safety measures, maternal education, environmental hygiene, and regulatory interventions aimed at minimizing the risk of exposure and safeguarding child health. Epidemiological studies have highlighted the significant prevalence of aflatoxin exposure in vulnerable populations, emphasizing the urgent need for comprehensive interventions to address this global health challenge. By understanding the extent of aflatoxin contamination and its impact on infant and child health, policymakers, healthcare professionals, and community stakeholders can work together to implement evidence-based strategies for prevention and mitigation.

## REFERENCES


1. Wild CP, Turner PC. The toxicology of aflatoxins as a basis for public health decisions. Mutagenesis 2002;17:471–81.
2. David L, Eaton JDG. The toxicology of aflatoxins: Human health, veterinary, and agricultural significance summary: Aflatoxins, natural fungal toxins found in foods and animal feeds, have public health significance. Academic Press, San Diego, 1994.
3. Soriano JM, Rubini A, Morales-Suarez-Varela M, Merino-Torres JF, Silvestre D. Aflatoxins in organs and biological samples from children affected by kwashiorkor, marasmus and marasmic-kwashiorkor: A scoping review. Toxicon 2020;185:174–83. https://doi.org/10.1016/j.toxicon.2020.07.010
4. Gebre A.T. Aflatoxin effect on livestock health: A review. Journal of Life Sciences Research and Reviews 2024; Volume 2(1): 1–5. https://doi.org/10.47363/JLSRR/2024(2)111
5. Hendrickse RG, Coulter JBS, Lamplugh SM, Macfarlane SBJ, Williams TE, Omer IA, et al. Aflatoxins and kwashiorkor: A study in Sudanese children. British Medical Journal (Clinical Research Edition) 1982;285(6345):843–6. https://doi.org/10.1136/bmj.285.6345.843.
6. Shuaib FMB, Jolly PE, Ehiri JE, Yatich N, Jiang Y, Funkhouser E, et al. Association between birth outcomes and aflatoxin B1 biomarker blood levels in pregnant women in Kumasi, Ghana. Tropical Medicine and International Health 2010;15:160–7. https://doi.org/10.1111/j.1365-3156.2009.02435.x
7. Variane ACF, Santos FC, Castro FF, Barbosa-Tessmann IP, Santos GT, Pozza MSS. The occurrence of aflatoxigenic *Aspergillus spp.* in dairy cattle feed in Southern Brazil. Brazilian Journal of Microbiology 2018;49:919–28. https://doi.org/10.1016/j.bjm.2018.05.005
8. Slater TF, Sawyer B, Strauli U. Studies on succinate-tetrazolium reductase systems. III. Points of coupling of four different tetrazolium salts. Biochimica et Biophysica Acta 1963;77:383–93.
9. Kim U, Park, Y.-S. (2006). The scientific foundation of indigenous and cultural psychology: The transactional approach. In U. Kim, K.-S. Yang, & K.-K. Hwang (Eds.), *Indigenous and cultural psychology: Understanding people in context* (pp. 27–48). Springer Science. https://doi.org/10.1007/0-387-28662-4_2

10. Tesfamariam K, Gebreyesus SH, Lachat C, Hanley-Cook GT, Roro M, Mengistu YG, et al. Chronic aflatoxin exposure during pregnancy is associated with lower fetal growth trajectories: A prospective cohort from the Butajira Nutrition, Mental Health, and Pregnancy (BUNMAP) study in rural Ethiopia. American Journal of Clinical Nutrition 2022;116:1634–41. https://doi.org/10.1093/ajcn/nqac280
11. Turner PC, Collinson AC, Cheung YB, Gong Y, Hall AJ, Prentice AM, et al. Aflatoxin exposure in utero causes growth faltering in Gambian infants. International Journal of Epidemiology 2007;36:1119–25. https://doi.org/10.1093/ije/dym122
12. Lauer JM, Duggan CP, Ausman LM, Griffiths JK, Webb P, Wang JS, et al. Maternal aflatoxin exposure during pregnancy and adverse birth outcomes in Uganda. Maternal & Child Nutrition 2019;15. https://doi.org/10.1111/mcn.12701
13. Turner PC, Moore SE, Hall AJ, Prentice AM, Wild CP. Modification of immune function through exposure to dietary aflatoxin in Gambian children. Environmental Health Perspectives 2003;11:217–20.
14. Miller JD, Baan R. Improving public health through mycotoxin control. Edited by John I. Pitt. Lyon, France: International Agency for Research on Cancer, 2012. No. 158.
15. Gourama H, Bullerman LB. *Aspergillus flavus* and *Aspergillus parasiticus*: Aflatoxigenic fungi of concern in foods and feeds: A review. Journal of Food Protection 1995;58(12):1395–1404.
16. Matchado A, Smith JW, Schulze KJ, Groopman JD, Kortekangas E, Chaima D, et al. Child aflatoxin exposure is associated with poor child growth outcomes: A prospective cohort study in rural Malawi. Current Developments in Nutrition 2023;7. https://doi.org/10.1016/j.cdnut.2023.101962
17. Hendrickse RG, Coulter JB, Lamplugh SM, MacFarlane SB, Williams TE, Omer MI, Suliman GI, and El-Zorgani GA. Aflatoxins and kwashiorkor. Epidemiology and clinical studies in Sudanese children and findings in autopsy liver samples from Nigeria and South Africa. Bulletin de La Société de Pathologie Exotique et de Ses Filiales 1983;76:559–66.

# 2.21 Aflatoxigenic Fungi
## *A Potential Human Carcinogen*

*Stanila Stoeva and Marieta Georgieva*

## INTRODUCTION

Fungi represent both a boon and a bane to humanity. Although they have been invaluable for centuries in producing foods and various other goods, they are also associated with some of the most troublesome illnesses mankind has ever encountered. The lethality risk linked to their exposure is so significant that certain species are literally regarded as the origin of a curse [1, 2]. Probably one of the famous victims of the *Aspergillus* fungi is Lord Carnarvon, a renowned Egyptian archaeology ktitor. He passed away shortly after participating in the opening of King Tutankhamun's tomb in 1922. The widely publicized "King Tut Curse" or "Mummy's Curse" could possibly be attributed to *A. flavus* and *A. niger* in certain ancient mummies, and more specifically to their potential to induce allergic bronchopulmonary aspergillosis and pulmonary hemorrhage [3–5]. Historical sources provide information on many more similar cases of sudden or delayed death after opening archaeological tombs. With the current information on aflatoxins today, it is well known that many other pathogenetic mechanisms may be the basis of the "aspergillosis curses," for instance, acute hepatitis, hepatocellular carcinoma, immunosuppression, and neutropenia.

From a global perspective, *A. flavus* is the most important fungal species producing aflatoxins in food. It releases aflatoxin B1 (AFB1) and aflatoxin B2 (AFB2), with human exposure coming primarily from contaminated corn, peanuts, and rice [6]. Another significant source of aflatoxins is *A. parasiticus*. In addition to AFB1 and AFB2, it produces aflatoxin G1 and aflatoxin G2, which can also be found in foodstuffs (maize, some nuts, and figs). Of the above, AFB1 has the most significant presence in contaminated samples, with its hydroxylated metabolite aflatoxin M1 (AFM1) often found in products of animal origin (e.g., milk and milk products) [7]. The International Agency for Research on Cancer (IARC) categorized AFB1 and AFM1 as carcinogens in humans, placing them in Group 1 and Group 2B, respectively, due to their ability to form DNA adducts [8].

Aflatoxins are commonly present in food, usually cultivated and manufactured in Africa and Asia, primarily due to favorable climatic conditions and other factors such as elevated temperatures, humidity, inadequate storage ventilation, and the proliferation of fungal spores. The ideal humidity for fungicidal growth ranges from 97% to 99%, with optimal temperatures between 25°C and 30°C. In the context of contemporary climate fluctuations, the epidemiological manifestation of aflatoxin contamination is also experiencing specific alterations. The European Commission's report on climate change indicates that Europe's temperature has increased more rapidly over the past century compared to the global average, by nearly 1°C [9]. According to Tery Vrabcheva (2018) the elevation of temperature to 30°C, attributed to global warming, causes aflatoxins, previously considered an "import" problem in Europe, to emerge as a concern even in temperate climates [10].

As per IARC, the absence of early detection and treatment options, combined with the exacerbation of the issue in low- to middle-income nations, positions liver cancer as the second leading cause of cancer-related deaths globally [11]. Hepatocellular carcinoma represents the predominant form of liver tumor, constituting 80% of all liver tumors [12]. Kucukcakan and


DOI: 10.1201/9781003487043-27

Hayrulai-Musliu (2015) stated that AFB1 exposure contributes to 4.6% to 28.2% of worldwide cases of this particular malignancy [13]. Moreover, AFB1 alone has been estimated to be responsible for 25,000 to 155,000 newly diagnosed cancer cases annually [14]. Other neoplasms that are linked to AFB1 exposure include lung, colorectal, stomach, gallbladder, esophageal, and breast cancer [7, 15].

Epidemiological research has demonstrated a connection between exposure to aflatoxins, hepatitis B virus, and the incidence of liver carcinoma. As it transpires, the coexistence of aflatoxicosis and such viral infections can elevate the risk of malignancy by up to 30 times [14, 16]. Similar findings apply to the hepatitis C virus [17]. These infectious diseases are prevalent in the populations of Sub-Saharan Africa and Asia, where climatic conditions favor the growth of *Aspergillus* spp. Unfortunately, socioeconomic factors and limited access to diverse, fresh foods further compound the risk of developing hepatocellular carcinoma associated with aflatoxin exposure. Predictably, this makes these territories the most affected by this public health emergency. At the same time, the incidence of morbidity is less common in regions such as North America, Latin America, and Europe (Figure 2.21.1) [14].

## AFB1 BIOTRANSFORMATION: THE OPENING OF PANDORA'S BOX (MECHANISM OF CARCINOGENESIS)

The mechanisms underlying the mutagenic and carcinogenic effects of AFB1 in animals and humans have been well studied over the past half century. Human mixed microsomal functioning oxidase (CYP450 system) has a critical role in metabolism and mediates the subsequent excretion of a number of endogenous and exogenous substances. In the context of aflatoxins, however, it is a collaborator in the formation of DNA-reactive epoxides at the activated double bond between carbons $C^8$ and $C^9$ in the furan ring of AFB1, which accounts for the manifestation of its mutagenic and carcinogenic effects. In addition to AFB1-8,9-epoxides, major metabolites produced during phase I of xenobiotic metabolism in humans are the less toxic AFM1, aflatoxin Q1, and aflatoxin P1. Cytochrome P450 (CYP) isoenzymes involved in oxidation reactions are CYP1A2, CYP2B6, CYP3A4, CYP3A5, and CYP3A7 [7].

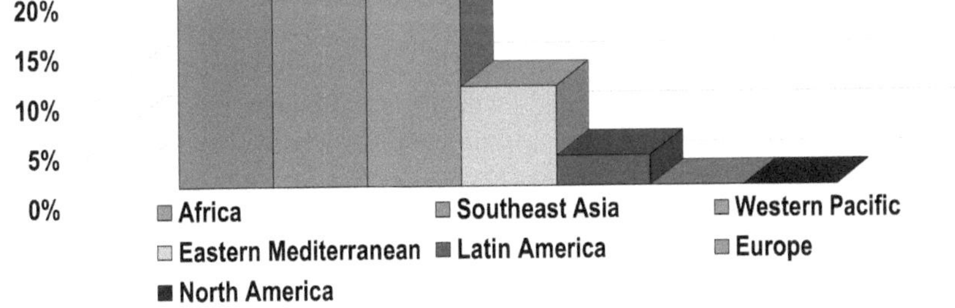


**FIGURE 2.21.1** Distribution of hepatocellular carcinoma cases linked to AFB1 in different geographical areas worldwide.

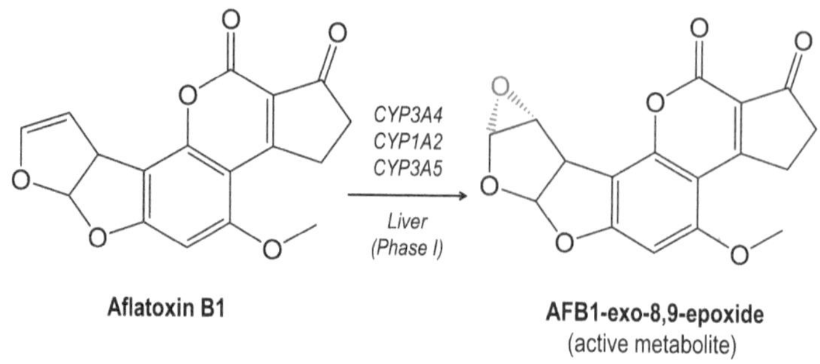


**FIGURE 2.21.2** Biotransformation of AFB1 to its active metabolite AFB1-exo-8,9-epoxide.

The carcinogenic properties of AFB1 result mostly from the formation of the exo-epoxide stereoisomer AFB1-exo-8,9-epoxide (mainly involving CYP3A4 and CYP1A2, and to a lesser extent, CYP3A5) (Figure 2.21.2).

This metabolite is particularly reactive, having been found to bind to DNA guanine at the $N^7$ position, resulting in the formation of the 8,9-dihydro-8-($N^7$-guanyl)-9-hydroxy-AFB1 (AFB1-$N^7$-Gua) adduct, consequently causing DNA mutations. Up to 80% of this process occurs at the third guanine residue of the codon 249 (5′-AG*G-3′) on the p53/PT53 tumor suppressor gene [18]. The AFB1-$N^7$-Gua adduct is highly unstable due to its positive charge, so it is released from the complex, leaving the DNA molecule apurinic [19]. The active epoxides formed can also bind to various proteins or RNA, causing dysregulation of normal cellular functions and inhibition of protein, DNA, and RNA synthesis. In addition, AFB1 induces reactive oxygen species (ROS) and oxidative stress, which activate mitochondrial ROS-dependent signaling pathways that induce apoptosis and DNA damage. The endo-epoxide metabolite reacts to a lesser extent with nucleic acids (< 1000 folds), but both stereoisomers can bind to different protein structures and thereby damage the cellular apparatus (Figure 2.21.3 provides a synthesized overview of the mechanisms of AFB1 toxicity) [20–22].

The exo-epoxide metabolite also reacts with glutathione to form conjugates, a process catalyzed by glutathione-S-transferase (GST). This enables the production of non-toxic AFB-mercapturates, which can then be eliminated from the body. Hence, this pathway plays a crucial role in detoxifying AFB1, thus reducing its potential to cause cancer. Decreased levels of this enzyme are linked to elevated levels of ROS, resulting in oxidative damage [23]. In addition, epoxides can undergo enzymatic or non-enzymatic conversion to AFB1-8,9-dihydroxydiol. The latter can be further biotransformed into dialdehyde, which is excreted in the urine as a dialcohol (with the help of aldehyde reductase) or can bind proteins, such as albumin. AFM1 is considered another detoxification product of AFB1. It undergoes identical biotransformation pathways but is a poorer substrate for epoxidation, which explains its tenfold lower mutagenicity [24].

To be excreted, phase I metabolites undergo not only glutathione conjugation but also glucuronidation and sulfation reactions. However, the enzyme GST has been found to be most important in the detoxification of aflatoxins. High levels of this enzyme in mice make them resistant to aflatoxicosis, unlike humans, whose enzyme levels are limited [25]. The products obtained during phase II of metabolism are excreted through urine and feces. It should be emphasized that the pathways of biological activation and detoxification occur simultaneously. Therefore, the risk of carcinogenesis in humans should be assessed based on the ratio between activation and inactivation processes. This includes extensive monitoring of the following:

- The degree of obtaining active exo- or endo-epoxide metabolites
- The AFB1 fraction converted to the less toxic metabolites
- The conjugating capacity of the organism during phase II of metabolism, which produces non-toxic metabolites suitable for excretion [6]

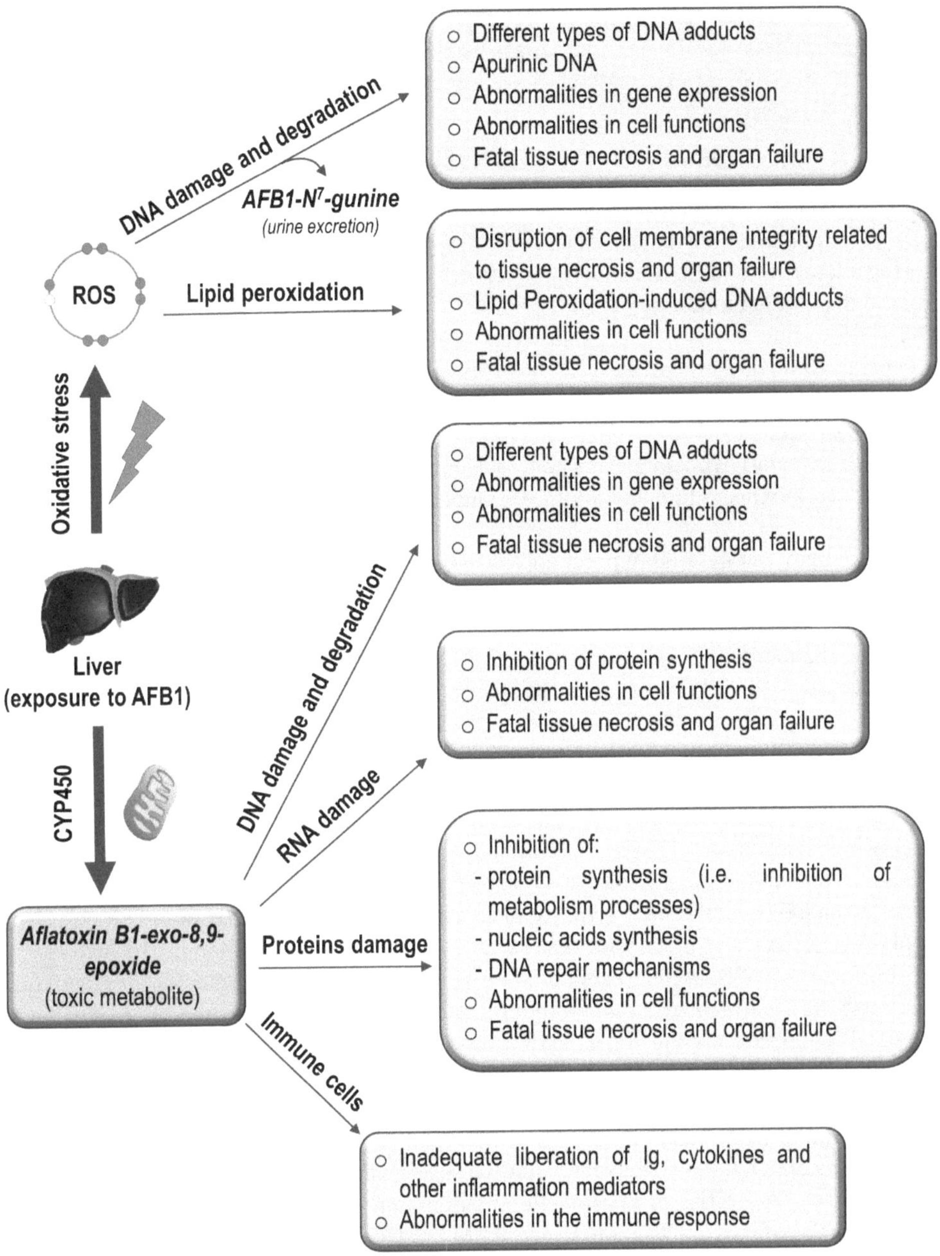


**FIGURE 2.21.3** Mechanisms of AFB1 toxicity.

## INFLUENCE OF GENETIC POLYMORPHISMS ON THE MANIFESTATION OF AFLATOXIN-ASSOCIATED LIVER CANCER

Identifying specific biomarkers for aflatoxin enables the investigation of genetic variations as potential contributors to aflatoxin-related cancers. For instance, the GST M1 null genotype and the XRCC1–399 AG DNA repair genotype heterozygote were found to be significantly linked to the development of hepatocellular carcinoma [26]. Considering the role of GST in metabolizing AFB1-exo-8,9-epoxide, it is noteworthy that infection with hepatitis B and hepatitis C can substantially decrease the activity of this enzyme. Consequently, there is a risk of heightened exposure to the toxic metabolite.

Kamdem et al. (2006) suggested that interspecies variations in hepatic CYP3A5 expression might also contribute to carcinogenesis. This is because the enzyme is involved in the conversion of AFB1 to AFB1-endo-8,9-epoxide, albeit to a lesser extent compared to CYP3A4 [27].

## BIOMONITORING OF AFLATOXIN EXPOSURE IN HUMANS

Hepatotoxicity and carcinogenesis are among the observed toxic effects associated with prolonged exposure to mycotoxins, posing a serious risk of mortality. Thus, implementing strategies for biomonitoring aflatoxins in humans is crucial. Currently, two primary approaches are utilized to assess exposure to these xenobiotics:

- **Type 1: External exposure:** This involves analyzing mycotoxins in food products, followed by determining the level of consumption of potentially contaminated foods (***e.g.,*** corn, nuts, etc.);
- **Type 2: Internal exposure:** This approach entails directly analyzing predefined biomarkers for mycotoxins in human biological samples [28].

However, employing the first indirect approach carries a high risk of yielding results with insufficient precision and notable standard deviation limits. To address the practical analytical and measurement challenges, standard operating procedures have been established. These procedures strictly define the method for selecting food specimens, the sampling protocol, sample preparation measures, and the potential chemical analysis tools. Despite the imposed regulations, this approach is considered complementary to the second one, which is viewed as a more efficient and reliable method for directly determining human exposure [14]. For the analysis, chromatographic techniques, primarily high-performance liquid chromatography coupled with tandem mass spectrometry (MS/MS) detection, as well as immunoanalytical methods such as enzyme-linked immunosorbent assay (ELISA), are predominantly utilized. Additionally, in practice, analytical procedures often concurrently monitor the contamination load of mycotoxins of various origins [29].

The examination of *Aspergillus* toxins to evaluate internal exposure has mainly focused on analyzing plasma, serum, and urine samples. Recently, there have been attempts to investigate hair analysis protocols [30–33]. After years of extensive efforts, reliable biomarkers for aflatoxins have been successfully identified. Aflatoxin–albumin adducts, which serve as the biomarker for AFB1 in peripheral blood, are crucial for assessing moderate to long-term exposure, spanning several months. While AFB1 itself can be detected in human urine, significant correlations with exposure through the consumption of contaminated food have not yet been established. Nevertheless, due to the high stability of the analytes during the storage of biological samples, Aflatoxin–albumin adducts have become the most frequently studied surrogate marker in investigations of mycotoxin intake and the associated risk of hepatocellular carcinoma. On the other hand, urinary metabolite biomarkers such as aflatoxin–$N^7$-guanine and AFM1 reflect shorter exposures in humans, typically spanning 24 to 48 hours. Another notable biomarker is 7,8-dihydro-8-oxo-2′-deoxyguanosine. It forms as a result of the oxidation of DNA guanine residues caused by the hydroxyl radical generated by AFB1-associated oxidative stress. Thus, this biomarker is widely recognized and studied as an indicator of oxidative DNA damage [34].

Currently, the determination of the aforementioned groups of biomarkers is essential for establishing the connection between exposure to aflatoxin and the development of liver cancer, along with other potential health issues [7, 28, 35, 36]. When comparing the two groups of biomarkers—for long-term and short-term exposure—it is important to consider that the former offer better insight, given the chronic nature of hepatocellular carcinoma. This underscores the reason that aflatoxin–albumin adducts have traditionally been employed in epidemiological research to evaluate the risk of tumors and other adverse effects. With the integration of the mentioned biomarkers and the systematic collection of analytical data, we have observed advancements in monitoring exposure, not just on an individual basis, but also in enabling comparative quantitative evaluations and analyses across entire populations or geographic areas [36–38]. This progress has paved the way for advocating the formulation of new health policies.

Despite notable advancements in operational procedures for evaluating exposure to aflatoxins, there remains a pressing need to enhance understanding of their toxicokinetics and toxicodynamics, as well as their quantitative association with observed carcinogenic effects. Variability exists in the biological response among individuals following toxin ingestion, adding complexity to the interpretation of biological monitoring outcomes. Moreover, the delayed onset of cancer poses an additional challenge. In this context, biomarkers of long-term exposure are considered a more reliable resource for assessing health implications [35].

## STRATEGIES TO REDUCE THE OCCURRENCE OF AFLATOXIN-ASSOCIATED RISK OF HEPATOCELLULAR CARCINOMA

The implementation and testing of highly sensitive methods for the reliable determination of biomarkers that correlate with the degree of development of aflatoxin-associated hepatocellular carcinoma represents both a reason and a motive for the search for approaches to deal with the emerging health risks. In this regard, some methods of obstruction in the formation of toxic metabolites have been established, mainly at the pharmacokinetic level:

- A reduction in biomarker levels of aflatoxin exposure was achieved following oral administration of NovaSil® clay. It is a relatively cheap sorbent product that binds to aflatoxins in the gastrointestinal tract of animals and humans, interfering with their resorption [39]. The advantage of NovaSil is that, despite its non-specific mechanism of action, it does not cause malabsorption syndromes of vitamins and microelements, which are also contained in food [40]. At the same time, it can be used in pediatric patients [41].
- Clinical data suggest that treatment with chlorophyllin (a water-soluble sodium-copper complex of chlorophyll) can also be used to reduce AFB1 resorption [42, 43].
- In their clinical study, Wang et al. (1999) observed that intermittent administration of high doses of Oltipraz inhibited the formation of active aflatoxin metabolites during phase I of their metabolism. Concomitantly, long-term administration of low doses of the organosulfur derivative increases glutathione conjugation during phase II, making it a suitable chemopreventive strategy in humans [44, 45].
- Other possible ways to reduce exposure to aflatoxins are the intake of phenols from green tea, as well as glucoraphanin, contained in broccoli sprouts [46, 47].
- Given the role of hepatitis B in reducing glutathione conjugation capacity, it should be noted that vaccination programs against the virus also have a major impact on the carcinogenic potential of aflatoxins [48].

These preventive strategies may be key to reducing morbidity and mortality from aflatoxin-associated hepatocellular carcinoma, particularly in affected geographic areas. However, they should not be taken as a guarantee of absolute protection against the toxic effects of mycotoxins, including liver cancer.

## CONCLUSION

Aflatoxin contamination in food presents a significant health concern worldwide, particularly in regions such as Africa and Asia. The fungi *A. flavus* and *A. parasiticus* are key contributors to the total annual cases of hepatocellular carcinoma globally, primarily due to their production of the mycotoxin AFB1. Despite efforts by health authorities to mitigate aflatoxin contamination, its prevalence is expected to rise in the future due to the exacerbation of global warming. Given this scenario, despite years of research into the underlying mechanisms of their carcinogenicity, it is evident that more comprehensive efforts are necessary to understand the intricate nature of AFB1-related diseases and the influence of individual characteristics on them. Nevertheless, the advent of new and increasingly sensitive analytical tools, along with the identification of reliable biomarkers for quantitatively assessing carcinogenic processes, offers a glimpse of hope for the development of novel preventive and therapeutic strategies to combat mycotoxicosis.

## REFERENCES


1. Marchant, J. 2013. *The Shadow King*. Boston: Da Capo Press.
2. Price, M., and G. Payne. 2005. Genomics and Gene regulation of the aflatoxin biosynthetic pathway. In *Handbook of Industrial Mycology* 1st Edition, ed. A. Zhiqiang, 387–404. New York: CRC Press.
3. Waness, A., Y. A. El-Sameed, B. Mahboub, et al. 2011. Respiratory disorders in the middle East: A review. *Respirology*. 16(5): 755–766. https://doi.org/10.1111/j.1440-1843.2011.01988.x
4. El-Tawil, S., and T. El-Tawil 2003. Lord Carnarvon's death: The curse of aspergillosis? *Lancet*. 362(9386): 836. https://doi.org/10.1016/S0140-6736(03)14268-7
5. Sterflinger, K., and F. Pinzari 2012. The revenge of time: Fungal deterioration of cultural heritage with particular reference to books, paper and parchment. *Environ. Microbiol.* 14: 559–566. https://doi.org/10.1111/j.1462-2920.2011.02584.x
6. Geneva: World Health Organization and Food and Agriculture Organization of the United Nations. 2018. Safety evaluation of certain contaminants in food: Prepared by the eighty-third meeting of the Joint FAO/WHO Expert Committee on Food Additives (JECFA). (WHO Food Additives Series, No. 74; FAO JECFA Monographs 19 bis). https://www.who.int/publications/i/item/9789240060760 (accessed March 12, 2024)
7. IARC. 2012. Chemical agents and related occupations. IARC Monographs on the Evaluation of Carcinogenic Risks to Humans Volume 100F. https://publications.iarc.fr/123 (accessed March 12, 2024)
8. IARC. 1993. Mycotoxins. IARC Monographs on the Evaluation of Carcinogenic Risks to Humans Volume 56. https://publications.iarc.fr/Book-And-Report-Series/Iarc-Monographs-On-The-Identification-Of-Carcinogenic-Hazards-To-Humans/Some-Naturally-Occurring-Substances-Food-Items-And-Constituents-Heterocyclic-Aromatic-Amines-And-Mycotoxins-1993 (accessed March 12, 2024)
9. European Commission. 2007. Adapting to Climate Change in Europe – Options for EU Action. Green Paper from the Commission to the Council, the European Parliament, the European Economic and Social Committee and the Committee of the Regions. https://eur-lex.europa.eu/legal-content/EN/TXT/?uri=COM:2007:0354:FIN (accessed March 12, 2024)
10. Vrabcheva., T. 2018. Aflatoxins and public health. *Bulg. J. Public Health*. 10(1): 34–41.
11. Sung, H., J. Ferlay, R. L. Siegel, et al. 2021. Global Cancer statistics 2020: GLOBOCAN estimates of incidence and mortality worldwide for 36 cancers in 185 countries. *CA Cancer J. Clin.* 71(3): 209–249. https://doi.org/10.3322/caac.21660
12. Rumgay, H., J. Ferlay, C. de Martel, et al. 2021. Global, regional, and national burden of primary liver cancer by subtype. *Eur. J. Cancer.* 161: 108–118. https://doi.org/10.1016/j.ejca.2021.11.023
13. Kucukcakan, B., and Z. Hayrulai-Musliu. 2015. Challenging role of dietary aflatoxin B1 exposure and hepatitis B infection on risk of hepatocellular carcinoma. *Maced. J. Med. Sci.* 3: 363–369. https://doi.org/10.3889/oamjms.2015.032
14. Liu, Y., and F. Wu 2010. Global burden of aflatoxin-induced hepatocellular carcinoma: A risk assessment. *Environ. Health Perspect.* 118: 818–824. https://doi.org/10.1289/ehp.0901388
15. Georggiett, O. C., J. C. Muiño, H. Montrull, et al. 2000. Estudio de la relación entre aflatoxina B1 y cáncer de pulmón relationship between lung cancer and aflatoxin B1]. *Rev. Fac. Cien. Med. Univ. Nac. Cordoba*. 57(1): 95–107.
16. Henry, S. H., F. X. Bosch, and J. C. Bowers 2002. Aflatoxin, hepatitis and worldwide liver cancer risks. *Adv. Exp. Med. Biol.* 504: 229–233. https://doi.org/10.1007/978-1-4615-0629-4
17. Chu, Y. J., H. I. Yang, H. C. Wu, et al. 2018. Aflatoxin B1 exposure increases the risk of hepatocellular carcinoma associated with hepatitis C virus infection or alcohol consumption. *Eur. J. Cancer.* 94: 37–46. https://doi.org/10.1016/j.ejca.2018.02.010
18. Deng, J., L. Zhao, N. Y. Zhang, et al. 2018. Aflatoxin B1 metabolism: Regulation by phase I and II metabolizing enzymes and chemoprotective agents. *Mutat. Res.* 778: 79–89. https://doi.org/10.1016/j.mrrev.2018.10.002

19. Wild, C. P., and P. C. Turner 2002. The toxicology of aflatoxins as a basis for public health decisions. *Mutagenesis.* 17: 471–481. https://doi.org/10.1093/mutage/17.6.471
20. Iyer, R. S., B. F. Coles, K. D. Raney, et al. 1994. DNA adduction by the potent carcinogen aflatoxin B1: Mechanistic studies. *J. Am. Chem. Soc.* 116: 1603–1609. https://doi.org/10.1021/ja00084a001
21. Benkerroum, N. 2020. Chronic and acute toxicities of aflatoxins: Mechanisms of action. *Int. J. Environ. Res. Public Health.* 17: 423. https://doi.org/10.3390/ijerph17020423
22. Nugraha, A., K. Khotimah, and I. M. C. M. Rietjens. 2018. Risk assessment of aflatoxin B1 exposure from maize and peanut consumption in Indonesia using the margin of exposure and liver cancer risk estimation approaches. *Food Chem. Toxicol.* 113: 134–144. https://doi.org/10.1016/j.fct.2018.01.036
23. Bbosa, G. S., D. Kitya, A. Lubega, et al. 2013. Review of the biological and health effects of aflatoxins on bodyorgans and body systems. In *Aflatoxins—Recent Advances and Future Prospects*, ed. M. Razzaghi-Abyaneh, 239–265. Rijeka: InTech.
24. Marchese, S., A. Polo, A. Ariano, et al. 2018. Aflatoxin B1 and M1: Biological properties and their involvement in cancer development. *Toxins.* 10: 214. https://doi.org/10.3390/toxins10060214
25. Croy, R. G., and G. N. Wogan 1981. Quantitative comparison of covalent aflatoxin-DNA adducts formed in rat and mouse livers and kidneys. *J. Natl. Cancer Inst.* 66: 761–768.
26. Kirk, G. D., P. C. Turner, Y. Gong, et al. 2005. Hepatocellular carcinoma and polymorphisms in carcinogen-metabolizing and DNA repair enzymes in a population with aflatoxin exposure and hepatitis B virus endemicity. *Cancer Epidemiol. Biomarkers Prev.* 14: 373–379. https://doi.org/10.1158/1055-9965.EPI-04-0161
27. Kamdem, L. K., I. Meineke, U. Gödtel-Armbrust, et al. 2006. Dominant contribution of P450 3A4 to the hepatic carcinogenic activation of aflatoxin B1. *Chem. Res. Toxicol.* 19(4): 577–586. https://doi.org/10.1021/tx050358e
28. Arce-López, B., E. Lizarraga, A. Vettorazzi, et al. 2020. Human biomonitoring of mycotoxins in blood, plasma and serum in recent years: A review. *Toxins (Basel).* 12(3): 147. https://doi.org/10.3390/toxins12030147
29. Beyene, A. M., X. Du, D. E. Schrunk, et al. 2019. High-performance liquid chromatography and enzyme-linked immunosorbent assay techniques for detection and quantification of aflatoxin B1 in feed samples: A comparative study. *BMC Res. Notes.* 12: 492. https://doi.org/10.1186/s13104-019-4538-z
30. De Santis, B., M. Raggi, G. Moretti, et al. 2017. Study on the association among mycotoxins and other variables in children with autism. *Toxins (Basel).* 9: 203. https://doi.org/10.3390/toxins9070203
31. Cao, X., X. Li, J. Li, et al. 2018. Quantitative determination of carcinogenic mycotoxins in human and animal biological matrices and animal-derived foods using multi-mycotoxin and analyte-specific high performance liquid chromatography-tandem mass spectrometric methods. *J. Chromatogr. B Analyt. Technol. Biomed. Life Sci.* 1073: 191–200. https://doi.org/10.1016/j.jchromb.2017.10.006
32. Mupunga, I., I. J. van Rensburg, N. Luthuli, et al. 2021. Analysis of aflatoxin biomarkers in the hair of experimental animals. *Toxins (Basel).* 13(8): 570. https://doi.org/10.3390/toxins13080570
33. Schwartzbord, J., L. Severe, and D. Brown 2017. Detection of trace aflatoxin M1 in human urine using a commercial ELISA followed by HPLC. *Biomarkers.* 22(1): 1–4. https://doi.org/10.1080/1354750X.2016.1203998
34. Chiorcea-Paquim, A. M. 2022. 8-oxoguanine and 8-oxodeoxyguanosine biomarkers of oxidative DNA damage: A review on HPLC-ECD determination. *Molecules.* 27(5): 1620. https://doi.org/10.3390/molecules27051620
35. Wild, C. P., J. D. Miller, and J. Groopman. D. 2015. Mycotoxin control in low- and middle-income countries. Lyon (FR): International Agency for Research on Cancer. (IARC Working Group Reports, No. 9.) Human exposure to aflatoxins and fumonisins. https://www.ncbi.nlm.nih.gov/books/NBK350555/ (accessed March 12, 2024)
36. Kensler, T. W., B. D. Roebuck, G. N. Wogan, et al. 2011. Aflatoxin: A 50-year odyssey of mechanistic and translational toxicology. *Toxicol Sci.* 120(1): 28–48. https://doi.org/10.1093/toxsci/kfq283
37. Xu, Y., Y. Y. Gong, and M. N. Routledge 2018. Aflatoxin exposure assessed by aflatoxin albumin adduct biomarker in populations from six African countries: REVIEW ARTICLE. *World Mycotoxin. J.* 11(3): 411–419. https://doi.org/10.3920/WMJ2017.2284
38. Kroker-Lobos, M. F., C. S. Alvarez, A. Rivera-Andrade, et al. 2019. Association between aflatoxin-albumin adduct levels and tortilla consumption in Guatemalan adults. *Toxicol. Rep.* 6: 465–471. https://doi.org/10.1016/j.toxrep.2019.05.009
39. Marroquín-Cardona, A., Y. Deng, J. Garcia-Mazcorro, et al. 2011. Characterization and safety of uniform particle size NovaSil clay as a potential aflatoxin enterosorbent. *Appl. Clay Sci.* 54(3–4): 248–257. https://doi.org/10.1016/j.clay.2011.09.009

40. Phillips, T. D., M. Wang, S. E. Elmore, et al. 2019. NovaSil clay for the protection of humans and animals from aflatoxins and other contaminants. *Clays Clay Miner.* 67(1): 99–110. https://doi.org/10.1007/s42860-019-0008-x
41. Kumi, J., R. Appiah-Opong, D. Rein, et al. 2022. Safety and efficacy of novasil clay (Calcium montmorillonite) in children exposed to aflatoxin in Ejura-Sekyedumase District in Ghana. *Glob Pediatr Health.* 9. https://doi.org/10.1177/2333794X221121243
42. Jubert, C., J. Mata, G. Bench, et al. 2009. Effects of chlorophyll and chlorophyllin on low-dose aflatoxin B(1) pharmacokinetics in human volunteers. *Cancer Prev Res (Phila).* 2(12): 1015–1022. https://doi.org/10.1158/1940-6207.CAPR-09-0099
43. Vázquez-Durán, A., G. Téllez-Isaías, M. Hernández-Rodríguez, et al. 2022. The ability of chlorophyll to trap carcinogen aflatoxin B1: A theoretical approach. *Int. J. Mol. Sci.* 23: 6068. https://doi.org/10.3390/ijms23116068
44. Wang, J. S., X. Shen, X. He, et al. 1999. Protective alterations in phase 1 and 2 metabolism of aflatoxin B1 by oltipraz in residents of Qidong, People's Republic of China. *J. Natl. Cancer Inst.* 91(4): 347–354. https://doi.org/10.1093/jnci/91.4.347
45. Sudakin, D. L. 2003. Dietary aflatoxin exposure and chemoprevention of cancer: A clinical review. *J. Toxicol. Clin. Toxicol.* 41(2): 195–204. https://doi.org/10.1081/clt-120019137
46. Tang, L., M. Tang, L. Xu, et al. 2008. Modulation of aflatoxin biomarkers in human blood and urine by green tea polyphenols intervention. *Carcinogenesis.* 9(2): 411–417. https://doi.org/10.1093/carcin/bgn008
47. Kensler, T. W., J. G. Chen, P. A. Egner, et al. 2005. Effects of glucosinolate-rich broccoli sprouts on urinary levels of aflatoxin-DNA adducts and phenanthrene tetraols in a randomized clinical trial in He Zuo township, Qidong, People's Republic of China. *Cancer Epidemiol Biomarkers Prev.* 14(11 Pt 1): 2605–2613. https://doi.org/10.1158/1055-9965.EPI-05-0368
48. DeVries, J. W., M. W. Trucksess, and L. S. Jackson. 2002. *Mycotoxins and Food Safety. Advances in Experimental Medicine and Biology.* Boston, MA: Springer.

# Section 3

## Control of Aflatoxigenic Fungi

# 3.1 Aflatoxigenic Fungi
## *Control by Nanomaterials*

*Nagashri Nanjundeshwara, Archana Lakshmaiah, and Ramya Raghavan*

## INTRODUCTION

Aflatoxins are highly toxic compounds that are produced by *Aspergillus* species, notably *A. flavus* and *A. parasiticus*. They are prevalent in hot and humid environments and readily contaminate a wide variety of food crops, including nuts, cereals, spices, and milk (Jafarzadeh et al. 2022). Aflatoxins are a family of toxic metabolites whose mycotoxins pose a significant threat to global food safety because of their ubiquitous presence in various agricultural commodities and their potent carcinogenic properties (Tiwari, Kumar Singh, and Dubey 2022). Among these toxins, aflatoxin B1 reigns supreme as the most common and potent, raising particular concern because of its well-established carcinogenic properties (Yadav et al. 2023).

Aflatoxins, carcinogenic compounds generated by moulds such as *A. flavus* and *A. parasiticus*, present a substantial risk to the health of both humans and animals. These toxic substances can negatively impact various biological systems (Kumar et al. 2021). These toxins are widespread in different food crops and can result in significant health problems, such as liver cancer, immune system suppression, and potentially fatal outcomes (Jaćević et al. 2023). Aflatoxin production thrives under specific conditions, including temperatures between 25°C and 35°C, and factors like pH, moisture, and relative humidity play crucial roles. To address this problem, diverse strategies encompassing physical, chemical, biological, and genetic engineering approaches have been used to mitigate aflatoxin levels (Salano, Mulwa, and Obonyo 2024).

Nanomaterials have rocketed to the forefront of the fight against fungal pathogens because of their remarkable properties. From nanocomposites to metal nanoparticles and even carbon nanotubes, these diverse warriors have displayed promising antifungal activity (Fouda et al. 2023). Nanoparticles synthesized from plant extracts are environmentally sustainable and can directly interact with the cell walls of fungal pathogens, disrupting their function. These nanoparticles can enter fungal cells through various mechanisms, including ion transport proteins and receptor-mediated absorption (Parveen et al. 2023).

### Importance of Controlling Aflatoxigenic Fungi

Aflatoxins, toxic compounds in food and feed, pose a serious health threat because of their ability to damage genes and trigger cancer, with liver cancer being a major concern (Balan et al. 2024). Consuming many aflatoxins, a condition known as aflatoxicosis, can trigger severe liver damage, potentially leading to death. In developing countries, where chronic exposure is common, aflatoxins cast a long shadow, hindering growth, weakening defences, and raising the spectre of liver cancer (Dhakal, Hashmi, and Sbar 2024).

Aflatoxins are chemical (Pisoschi et al. 2023) compounds that can induce extensive damage in DNA, leading to mutagenesis and carcinogenesis in both animals and humans. Detecting aflatoxins, toxic compounds produced by moulds like *A. parasiticus*, is crucial to ensuring food safety. Microbiological and molecular methods, like polymerase chain reaction (PCR), act as frontline tools for efficiently identifying aflatoxigenic fungi. However, for definitive confirmation of aflatoxin

DOI: 10.1201/9781003487043-29

presence in contaminated samples, we turn to advanced techniques like high-performance liquid chromatography (HPLC) and liquid chromatography tandem mass spectrometry (LC-MS). These sophisticated tools act like forensic detectives, providing precise measurements and undeniable proof of the toxin's presence (Miklós et al. 2020). Contamination of crops with aflatoxins not only undermines the health of consumers but also poses substantial economic burdens on the agriculture industry due to decreased productivity and stringent regulations on permissible aflatoxin levels in food and feed (Zhou et al. 2023). Therefore, the management and mitigation of aflatoxin contamination remain paramount for ensuring food security, safeguarding human health, and sustaining the economic viability of agricultural practices.

Standing watch over our food safety are regulatory bodies like the European Food Safety Authority (EFSA) and the World Health Organisation (WHO) (Azeez et al. 2021). Although regulatory bodies and organizations fight hard, aflatoxicosis outbreaks in developing countries remain a stark reminder of the ongoing battle. These incidents highlight the crucial need for constant vigilance and proactive intervention to safeguard vulnerable populations (Zachová et al. 2003). Nanoparticles, with their diverse sizes and shapes, offer a spectrum of possibilities for tackling aflatoxins. Notably, the use of natural plant-based chemicals (phytochemicals) in nanoparticle synthesis has emerged as a safer alternative to conventional detoxification methods, which often rely on potentially hazardous chemicals (Elazab et al. 2024).

## PROPERTIES OF NANOMATERIALS

Nanoparticles have unique properties that can be harnessed to control the growth of mycotoxigenic fungi, such as aflatoxins. The following are some of the key properties and effects of nanoparticles that contribute to their potential in controlling aflatoxin growth (Tables 3.1.1 and 3.1.2).

### Size and Chemical Composition

Nanoparticles can vary greatly in size, ranging from less than a nanometre to several hundred nanometres. The size and chemical composition of nanoparticles are critical factors that influence their interaction with fungal cells (Živančev et al. 2024). For instance, silver nanoparticles with sizes of 0.65 and 5 nm have been shown to penetrate the cell, form agglomerates in the cytoplasm, and associate with cell organelles, affecting the physiology of the fungus *Penicillium verrucosum* (Shalal 2023). For example, fullerene C60 nanoparticles at 10 ng/mL exhibited antioxidative action and reduced aflatoxin B1 production, whereas at 50 ng/mL, they showed prooxidative effects that unexpectedly enhanced aflatoxin production (Kovač et al. 2018).

### Antifungal and Anti-Aflatoxigenic Activities

Zinc oxide nanostructures have demonstrated inhibitory effects on the growth and aflatoxin production of *A. flavus*, a highly toxigenic fungus strain (Mohammed et al. 2023). Some nanomaterials can inhibit the activity of enzymes that are vital for the survival and proliferation of fungi. By blocking these enzymes, nanomaterials can effectively halt the growth of fungi and prevent aflatoxin production (Shabeer et al. 2022). These nanoparticles can disrupt the normal growth of the fungus in liquid media and reduce the contamination of aflatoxins in cattle feed (Hernández-Meléndez et al. 2018).

### Metal Oxides and Hydroxides

Metal oxide nanoparticles, including zinc oxide, have been characterized for their antifungal properties. They are effective against toxigenic fungi and can improve the physiological and sanitary quality of crops, such as the Mexican red maize (Zhang et al. 2022). The efficacy of these nanoparticles can be compared with that of commercial antifungal feed additives, offering a potential

**TABLE 3.1.1**
**Metal-Based Nanoparticles**

| Properties of Nanoparticles | Feature | Function | Example | Reference |
|---|---|---|---|---|
| Size and Chemical Composition | Variable sizes (less than a nanometre to several hundred nanometres); chemical composition | Influences interaction with fungal cells; affects physiology of fungus | Silver nanoparticles penetrate fungal cells and affect the physiology of *Penicillium verrucosum* | (Gómez et al. 2019) |
| Antifungal and Anti-Aflatoxigenic Activities | Flower-shaped zinc oxide (ZnO) nanostructures | Inhibits growth and aflatoxin production of *A. flavus*; disrupts normal growth of fungus | ZnO nanostructures reduce contamination of aflatoxins in cattle feeds | (Omran and Baek 2024) |
| Metal Oxides and Hydroxides | Metal oxide nanoparticles (e.g., ZnO) | Effective against toxigenic fungi; improves physiological and sanitary quality of crops | ZnO nanoparticles improve the quality of the Mexican landrace of red maize | (Padrilah et al. 2024) (Zhang et al. 2022) |
| Nanomaterials for Aflatoxins Detection | Metal nanomaterials, carbon-based nanomaterials (e.g., graphene, carbon nanotubes) | Enables sensitive and accurate detection of aflatoxins for monitoring and controlling their presence in food and feed products | Metal nanomaterials used for aflatoxin detection systems | (Zhou et al. 2023) (Balan et al. 2024) |
| Surface Properties and Antifungal Activity | Surface properties influence fungicidal properties | Positively charged nanoparticles more fungicidal than negatively charged ones | Surface roughness affects microbial adhesion and biofilm formation | (Matras et al. 2022) (Ing et al. 2012) |
| Metal Nanoparticles and Their Antifungal Properties | Various metal nanoparticles (e.g., silver, zinc oxide, copper, gold) | Enter fungal pathogens causing membrane damage and inhibiting growth | Silver nanoparticles are widely studied and commercialized for their antimicrobial properties | (Shalal 2023) (Elazab et al. 2024) |
| Chitosan Nanoparticles and Their Mechanism of Action | Chitosan nanoparticles formed spontaneously upon the addition of polyanions like tripolyphosphate | Binds tightly to fungal cells disrupting membrane integrity | Chitosan nanoparticles exhibit high affinity for binding to fungal cells | (Zhao et al. 2023) (Jan et al. 2023) |

alternative for controlling aflatoxin contamination (Omran and Baek 2024). Metal nanomaterials, carbon-based nanomaterials like graphene and carbon nanotubes, and other nanomaterials have been developed for the sensitive and accurate detection of aflatoxins, which is crucial for monitoring and controlling their presence in food and feed products (Yan et al. 2020).

## Surface Properties and Antifungal Activity

The antifungal activity of nanoparticles is also dependent on their surface properties. For instance, the fungicidal properties of silver nanoparticles can be influenced by stabilising agent molecules adsorbed on their surfaces (Thirugnanasambandan and Gopinath 2023). Positively charged

**TABLE 3.1.2**
**Nanomaterial Mechanism of Fungicidal Action**

| Mechanism | Description | Example | Reference |
|---|---|---|---|
| Fungicidal Effects | Kills fungal cells, preventing proliferation and toxin production | Silver silica nanoparticles inhibit the growth of *A. flavus*. | (Živančev et al. 2024) |
| Inhibition of Growth and Toxin Production | Inhibits fungal growth and aflatoxin production | Silver nanoparticles inhibit *A. parasiticus* growth and toxin production. | (Yousefi et al. 2024) |
| Disruption of Cellular Functions | Alters membrane permeability and inactivates respiratory chain dehydrogenases | Silver nanoparticles disrupt fungal cellular functions. | (Omran and Baek 2024) |
| Oxidative Stress Induction | Enhances oxidative stress, damaging biomolecules and causing cell death | Silver nanoparticles can induce oxidative stress in fungal cells like *C. albicans*. | (Radhakrishnan et al. 2018) |
| DNA Interaction and Damage | Interacts with fungal DNA, causing genetic damage and preventing replication | Graphene oxide nanoparticles have a high surface area and binding capacity for mycotoxins, making them a promising candidate for reducing mycotoxin contamination in food and feed products. | (Horky et al. 2018) |
| Size and Penetration Capability | Smaller nanoparticles penetrate fungal cell walls more effectively | Silica nanoparticles, particularly those with a size of 200 nm, can impact the growth and mycotoxin production of *P. verrucosum*. | (Horky et al. 2018) |
| Impact on Mycotoxin Biosynthesis | Modulates mycotoxin biosynthesis pathways | Silver nanoparticles (AgNPs) influence mycotoxin production in *P. verrucosum*. | (Barabadi et al. 2019) |
| Use of Antagonistic Microorganisms | Works with antagonistic microorganisms to inhibit fungal growth | Interaction between silver nanoparticles (AgNPs) and bacterial cells like *Escherichia coli* and *Staphylococcus aureus* | (Gibała et al. 2021) |
| Application Timing and Germination Inhibition | Timely application inhibits spore germination and subsequent growth | Colloidal silver nanoparticles applied before spore germination inhibit germination. | (Wang and Wu 2021) |

nanoparticles are more fungicidal than negatively charged ones, highlighting the importance of surface charge in shaping their antifungal properties (Matras et al. 2022).

## NANO-ENCAPSULATED BIOCONTROL

Nanomaterials are versatile and can be used in a wide range of applications, from medical imaging and diagnostics to tissue engineering and environmental sensing. Nano-encapsulated biocontrol agents refer to the use of nanotechnology to improve the delivery and efficacy of biological agents used in agriculture to control pests and diseases (Yousefi et al. 2024). This is achieved by entrapping active ingredients, such as phytohormones, beneficial microbes, or natural pesticides, within nanometre-sized capsules. Nanomaterials can interact with the cell membranes of fungi, causing structural damage that leads to cell lysis and death. This disruption of membrane integrity prevents the fungi from maintaining their essential cellular functions (Shabeer et al. 2022).

### Improved Plant Stress Tolerance

Nanoencapsulation allows for the controlled release of active compounds, ensuring that they are delivered at the right time and place and in the correct amounts. Encapsulated phytohormones can enhance plant stress tolerance without the need for plants to activate complex signal cascades (Saberi Riseh et al. 2022). The encapsulation process protects the active ingredients from environmental factors, such as UV radiation and degradation, thereby enhancing their stability and shelf life (Maruyama et al. 2020).

### Efficient Biocontrol

Encapsulation techniques, such as microencapsulation, have been shown to increase the bioavailability of biocontrol organisms like *Trichoderma harzianum* in the field, leading to better control of pathogens like *Sclerotinia sclerotiorum* (Maruyama et al. 2020). Encapsulation provides a protective barrier that shields the biocontrol agents from harmful UV radiation, which can otherwise reduce their effectiveness. The relationship between surface properties and antifungal activity is a crucial aspect of antimicrobial research (Marcus and Fox 2021). Studies have shown that surface roughness plays a significant role in microbial adhesion, biofilm formation, and colonization by microorganisms like *Candida spp*. Specifically, rough surfaces increase yeast counts because of the higher likelihood of microorganism adhesion, attachment, and retention in surface irregularities, protecting shear forces (Clack et al. 2024). In the context of antifungal surfaces, the addition of chitosan salts to materials like denture bases has been explored to enhance antifungal action. However, it has been observed that the surface roughness increased with the addition of chitosan salts, potentially impacting fungal cell attachment and biofilm formation (Saberi Riseh et al. 2022).

### Sustained Release and Viability

Encapsulated biocontrol agents can be designed to release their active ingredients slowly over time, which can lead to sustained biocontrol effects and better viability of the agents (Maruyama et al. 2020). The high hydrophilicity and lipophilicity of the encapsulated particles improve their ability to penetrate plant tissues, making them more effective (Yu et al. 2023). Nanoparticles can be engineered to target specific sites within an organism, which is particularly useful in medical applications for delivering drugs directly to the affected area, thereby enhancing the efficacy of the treatment and reducing side effects (Yousefi et al. 2024).

## TYPES OF ANTIFUNGAL NANOPARTICLES

Antifungal nanomaterials have emerged as a promising approach to combat fungal infections and reduce the threats posed by aflatoxigenic fungi-induced mycotoxins. These nanomaterials offer several advantages over traditional antifungal agents, including increased surface area-to-volume ratios, unique physicochemical properties, and lower toxicity profiles (Xu et al. 2020) (León-Buitimea et al. 2021) (Asghari et al. 2016) (Figure 3.1.1).

Antifungal nanomaterials are a diverse group of nanosized particles that are used to combat fungal infections. These materials can be engineered from various substances, including metals, lipids, polymers, and even biological extracts. The following is a comprehensive list of antifungal nanomaterials, their properties, and their applications.

### Carbon Nanotubes

Carbon nanotubes (CNTs), including single-walled and multiwalled varieties, have been recognized for their antimicrobial properties, which extend to antifungal applications. They can be used alone or as part of nanocomposites, which combine the properties of CNTs with those of other materials for enhanced antimicrobial effects (Sadare, Frances, and Daramola 2022).

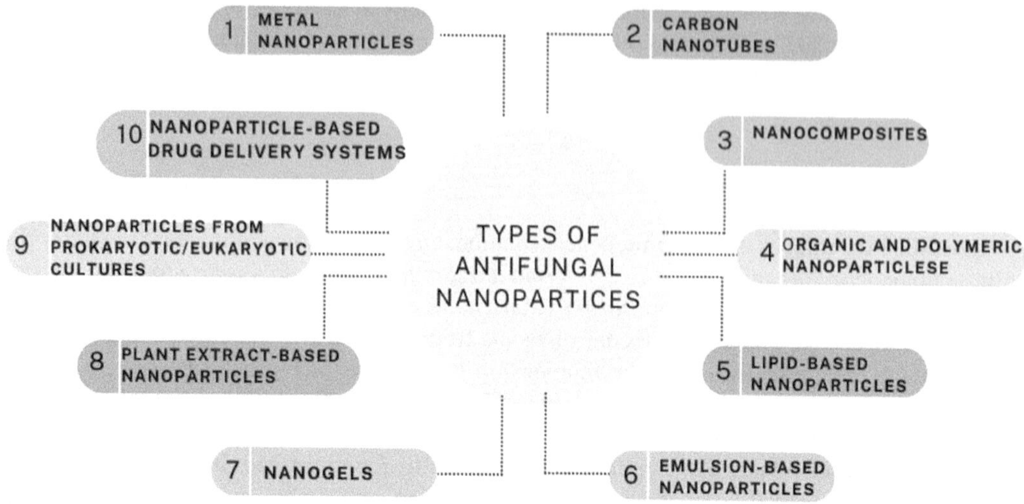


**FIGURE 3.1.1** Types of antifungal nanoparticles.

## Nano Composites

Nanocomposites, such as those combining zinc oxide with silver and multiwalled carbon nanotubes, have been synthesized and shown to possess strong antimicrobial properties. These materials are particularly effective against both gram positive and negative bacteria and have potential for use in water and air treatment applications because of their antibacterial activities (Raj et al. 2023).

## Lipid-Based Nanoparticles

Lipid-based nanoparticles, such as solid-lipid nanoparticles (Valença et al. 2023), and niosomes, carry antifungal drugs deep into the heart of the infection. By targeting the specific site of infection, these lipid warriors maximize the drug's effectiveness and minimize unwanted side effects (Mali and Javesh 2023). Nanoparticles can be used to create drug delivery systems that control the release of antifungal agents, enhancing their effectiveness and reducing side effects. These systems can be designed to target specific sites of infection (León-Buitimea et al. 2021). Emulsion-based nanoparticles allow them to dissolve antifungal drugs more effectively, making them readily available to fight infection. This translates to stronger antifungal action with minimal side effects (Valença et al. 2023).

## Nanogels

Nanogels are another form of antifungal nanomaterials that can encapsulate drugs and release them in a controlled manner. They are hydrogel particles in the nanometre range and can be responsive to environmental changes such as pH and temperature (Mali and Javesh 2023). Nanoparticles obtained from the cultures of prokaryotic or eukaryotic organisms are another type of antifungal nanomaterial. These can be engineered to have specific properties that make them effective against fungal infections (León-Buitimea et al. 2021).

## Chitosan Nanoparticles

Chitosan, a natural polymer derived from chitin, has antifungal activity in its nanoparticle form. The preparation of chitosan nanoparticles involves the spontaneous formation of particles upon the addition of polyanions like tripolyphosphate to a chitosan solution (Mahmoud 2021) (Jan et al. 2023). The antifungal activity of these nanoparticles is influenced by their molecular weight, degree of substitution, concentration, and the type of fungus targeted. Chitosan nanoparticles exhibit a high affinity for binding to fungal cells because of their nature and large surface area, which allows them to adsorb tightly onto the surface of fungal cells and disrupt membrane integrity (Yushina et al. 2023).

### Metal Nanoparticles and Their Antifungal Properties

Metal nanoparticles, such as silver, zinc oxide, copper, and gold nanoparticles, have shown significant antifungal properties (Table 3.1.1). They can enter fungal pathogens through various mechanisms, causing membrane damage and ultimately inhibiting growth (Omran and Baek 2024). Silver nanoparticles, in particular, have been widely studied and commercialized due to their broad-spectrum antimicrobial properties. They act by generating reactive oxygen species and free radicals, leading to protein denaturation, DNA damage, and cell wall disruption (Yousefi et al. 2024). The size, shape, and morphology of these nanoparticles are crucial factors that determine their antifungal efficacy (Mansoor et al. 2021). Phyto nanoparticles are used for their antimicrobial properties, including antifungal activity. Their effectiveness depends on their concentration, and they are particularly useful in agricultural applications to control fungal pathogens (Parveen et al. 2023). Silver nanoparticles, in particular, have shown remarkable effectiveness against *Candida glabrata* strains, including those resistant to traditional antifungal drugs like amphotericin B. In addition, metal nanoparticles can be incorporated into controlled drug release systems, further boosting their antifungal potency (Lotfali et al. 2021) (León-Buitimea et al. 2021). Gold nanoparticles also play a crucial role in inhibiting various fungal pathogens (Parveen et al. 2023).

## MECHANISM OF NANOMATERIALS IN CONTROLLING AFLATOXIGENIC FUNGI

Nanomaterials, particularly nanoparticles, have been extensively studied for their potential to control aflatoxigenic fungi. The mechanisms by which these nanomaterials exert their antifungal effects are varied and complex. The following are some key mechanisms identified.

### Disruption of Cellular Functions

The antifungal activity of nanoparticles can be attributed to their ability to disrupt various cellular functions (Table 3.1.2). For instance, silver nanoparticles alter the permeability of fungal cell membranes, leading to the leakage of vital cellular components such as reducing sugars and proteins (Nsengumuremyi et al. 2020). This disruption can induce a state of respiratory chain dehydrogenase inactivation, impairing the fungus's ability to produce energy (Shalal 2023) (Figure 3.1.2).

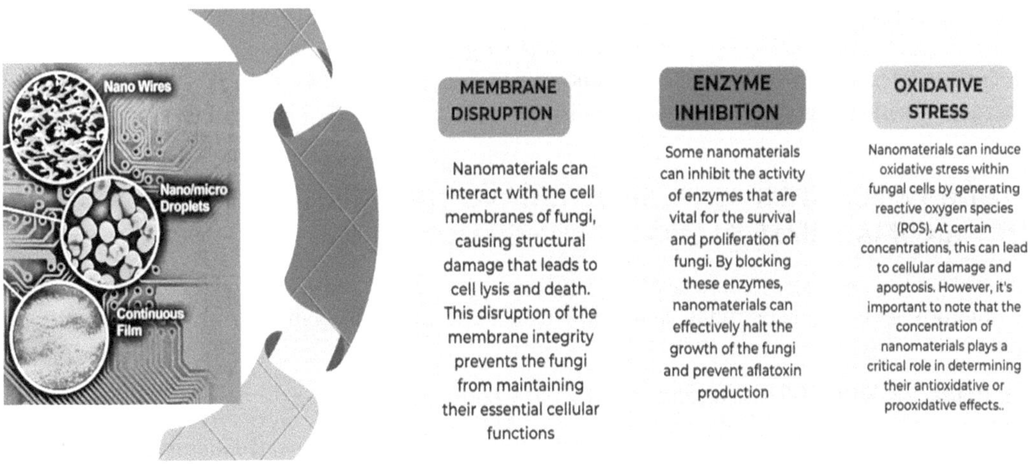


**FIGURE 3.1.2** Mechanism of fungicidal action.

### Oxidative Stress Induction

Nanoparticles can enhance cellular oxidative stress in microbes by increasing the concentration of reactive ions such as Ag(+) ions. This oxidative stress can damage critical biomolecules within fungal cells, including DNA, proteins, and lipids, leading to cell death (Mohanta et al. 2023).

### DNA Interaction and Damage

Reactive species, such as hydroxyl radicals (OH•), generated by nanoparticles can interact with DNA, causing single-strand breaks and the formation of DNA adducts like 8-hydroxyl-2′-deoxyguanosine (8-OHdG). This type of genetic damage can prevent the fungus from replicating and producing toxins (Zhao et al. 2023) (Pisoschi et al. 2023).

### Size and Penetration Capability

The size of the nanoparticles plays a significant role in their antifungal activity. Smaller nanoparticles ($\leq$ 5 nm) are more effective in inhibiting fungal growth than larger ones. This is likely due to the smaller particles' ability to penetrate the fungal cell wall and disrupt internal cellular mechanisms (Živančev et al. 2024).

### Impact on Mycotoxin Biosynthesis

Nanoparticles can influence the biosynthesis of mycotoxins in fungi. For example, certain nanoparticles increase citrinin production while decreasing ochratoxin A production in *P. verrucosum*. This indicates that nanoparticles can modulate the metabolic pathways involved in mycotoxin production (Topi et al. 2023).

### Use of Antagonistic Microorganisms

Some nanomaterials work with antagonistic microorganisms to inhibit aflatoxigenic fungi growth. These microorganisms can produce bioactive compounds that, when combined with the effects of nanoparticles, create a hostile environment for the fungi, thereby inhibiting their growth and toxin production (Ren et al. 2020).

The timing of nanoparticle application is crucial for their effectiveness. Colloidal silver nanoparticles applied before the germination of spores can strongly inhibit germination and subsequent fungal growth, whereas application after the initial growth phase has a marginal effect (Kotzybik et al. 2016). In conclusion, nanomaterials, particularly silver nanoparticles, exhibit a range of mechanisms that make them effective in controlling aflatoxigenic fungi. These mechanisms include direct fungicidal effects, disruption of cellular functions, induction of oxidative stress, DNA damage, and interaction with antagonistic microorganisms (Mohanta et al. 2023) (Figure 3.1.3).

## STUDIES AND EXPERIMENTS SHOWCASING THE EFFECTIVENESS OF NANOMATERIALS IN CONTROLLING THESE FUNGI

Several studies and experiments have been conducted to explore the effectiveness of nanomaterials in controlling aflatoxigenic fungi. The following are a few notable examples.

### Mesoporous Silica Nanoparticles with Zinc Oxide

Mesoporous silica nanoparticles incorporated with zinc oxide demonstrated significant antifungal activity against *A. flavus and Fusarium graminearum*. These nanoparticles were also able to adsorb

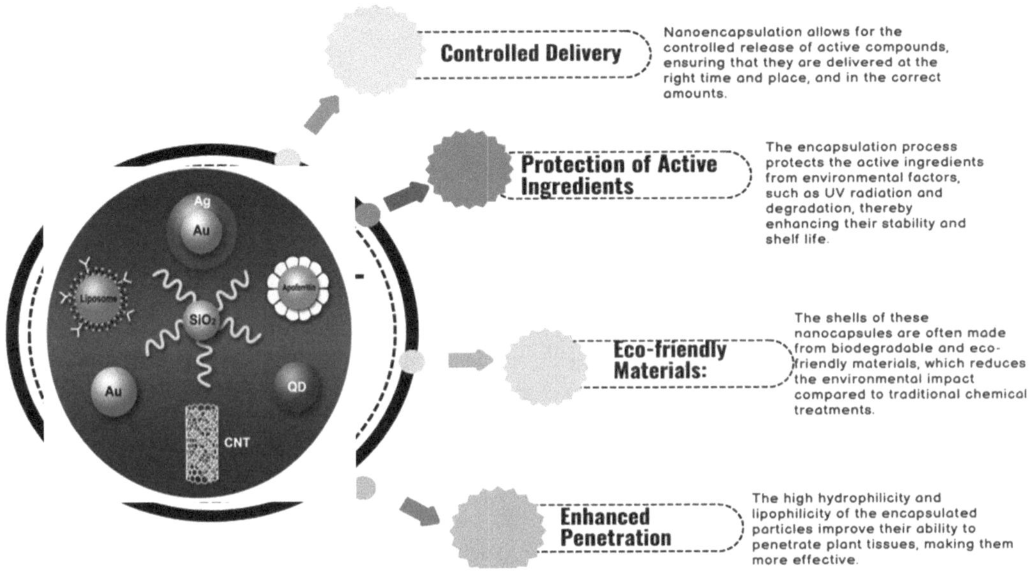


**FIGURE 3.1.3** Aflatoxigenic fungi: Control by nanomaterials.

aflatoxin B1 and deoxynivalenol, with a notable adsorption capacity for aflatoxin B1 (Zanoni 2018). An investigation into the impact of silver nanoparticles on gene expression in *A. flavus* revealed that AgNPs could reduce the production of aflatoxin B1 by affecting the transcription activity of genes involved in the aflatoxin biosynthesis pathway.

## Polyaniline-coated Nanoparticles

Polyaniline-coated nanoparticles of zinc oxide and copper oxide were synthesized and tested for their antifungal properties against *A. parasiticus*. The coated nanoparticles showed higher fungal growth inhibition compared to non-coated ones (Sohail et al. 2022).

## Silver Nanoparticles in Rice Plants

A study demonstrated that biogenic silver nanoparticles could elicit physiological, biochemical, and antioxidant modifications in rice plants to control *A. flavus*. The application of AgNPs at a concentration of 50 mg/L was effective in reducing biotic stress and controlling the proliferation of the fungal pathogen (Sultana et al. 2021).

These studies showcase the potential of various nanomaterials in controlling aflatoxigenic fungi and their toxins. The mechanisms through which these nanoparticles exert their effects include inhibiting fungal growth, adsorbing toxins, inducing genetic modifications, and enhancing the plant's defence mechanisms.

## Enhanced Diagnosis and Imaging

Nanotechnology has significantly improved the capabilities of medical imaging and diagnostics. Nanoparticles can be engineered to enhance imaging for diagnosis and treatment monitoring, providing higher resolution and specificity (Clack et al. 2024). They can also be used with in vitro assays to enable earlier diagnosis of diseases such as cancer. For example, gold nanoparticles can be functionalized with various ligands to interact with cancer cells, aiding in the understanding of molecular changes even in a small percentage of cells. (Sim and Kui Wong 2021) (Mohanta et al. 2023).

### Targeted Drug Delivery

One of the most promising applications of nanomaterials in medicine is in the development of targeted drug delivery systems. Nanoparticles can passively accumulate at tumour sites, actively target cancer cells, and traverse physiological barriers like the blood–brain barrier, enhancing the precision of drug targeting and reducing systemic toxicity (Kościelecka et al. 2023). Different types of nanoparticles, including silver, zinc oxide, gold, copper oxide, titanium dioxide, magnetic and iron oxide, and carbon nanoparticles, have been explored for their antifungal effects (Gurunathan, Reum Lee, and Kim 2022). These nanoparticles have shown potential in reducing the symptoms of infections without leaving harmful residues (Li et al. 2022).

### Nano Catalysts and Enzyme Immobilization

The valorization of agricultural wastes into biofuels is a strategy that aligns with environmental and energy security goals. It does not compromise food security and offers a solution to waste management challenges. Moreover, the process contributes to preventing environmental degradation and ensures a sustainable energy supply (Awogbemi and Vandi Von Kallon 2022).

Nanomaterials play a strategic role in the production of bioethanol and biodiesel, offering new methods for biomass processing (Khoo et al. 2020). For example, graphene-based nanomaterials have been used as cathode catalysts in microbial fuel cells, which are devices that can convert chemical energy into electrical energy using microorganisms. The use of nanomaterials in microbial fuel cells can lead to the modification of components, enhancing power generation (Khoo et al. 2020).

Fungi offer an eco-friendly and cost-effective solution for waste management through a process known as mycoremediation. This method uses fungi to degrade, restore, and heal contaminated ecosystems, proving to be a sustainable alternative to traditional waste management practices (McKenna 2023).

Nanoparticles offer interesting properties for the production of biofuels, such as second-generation ethanol and biodiesel, through transesterification of oils and fats. The small size of nanoparticles allows for quantum effects that are not present in bulk materials, which can be advantageous in catalysis (Rai et al. 2016). These nanomaterials can also improve the efficiency of microalgae harvesting, which is a critical step in biodiesel production from algae (Bin Rashid 2023). Additionally, increased human, financial, and infrastructural investments are essential to support the conversion process, promote renewable fuel utilization, and contribute to environmental sustainability (Awogbemi and Vandi Von Kallon 2022).

## CONCLUSION

This review comprehensively examined the multifaceted challenge of aflatoxin contamination, particularly aflatoxin B1, highlighting its detrimental impact on agriculture and human health. *A. flavus*, the key culprit, underscores the complexity of managing mycotoxin risks. Environmental factors significantly influence aflatoxigenic fungi growth, rendering specific crops more susceptible. The exploration of nanomaterials, including metal nanoparticles, chitosan, fullerenes, and biopolymers, offers a novel approach for effective toxin control and detection. Advanced biosensing techniques and nanotechnology-integrated food packaging hold promise for a significant leap in food safety and storage. Nevertheless, knowledge gaps regarding aflatoxin–nanomaterial interactions and the applicability of these strategies require further exploration. The potential of nanomaterials for enhanced antimicrobial testing presents a promising avenue to mitigate aflatoxin's impact on human health and agriculture. Future research should focus on closing knowledge gaps, refining detection methods, and developing cost-effective, scalable, and environmentally sustainable nanomaterial applications. By addressing these challenges, we can pave the way for a future with minimized aflatoxin threats.

## REFERENCES


Asghari, Farnoush, Zahra Jahanshiri, Mohammad Imani, Masoomeh Shams-Ghahfarokhi, and Mehdi Razzaghi-Abyaneh. 2016. Chapter 10 - Antifungal Nanomaterials: Synthesis, Properties, and Applications. In *Nanobiomaterials in Antimicrobial Therapy*, ed. Alexandru Mihai Grumezescu, 343–383. William Andrew Publishing. doi:10.1016/B978-0-323-42864-4.00010-5.

Awogbemi, Omojola, and Daramy Vandi Von Kallon. 2022. Valorization of Agricultural Wastes for Biofuel Applications. *Heliyon* 8(10):e11117. doi:10.1016/j.heliyon.2022.e11117.

Azeez, Areej, Mrwa Hindi, Maha khudiar, and Noor Khadim. 2021. Determination of Aflatoxins in Feed by HPLC and PCR. *Journal of Food and Pharmaceutical Sciences*. doi:10.22146/jfps.881.

Balan, Biji, Amit S. Dhaulaniya, Munendra Kumar, Mohit Kumar, and Prateek Kumar. 2024. Aflatoxins in Food: Prevalence, Health Effects, and Emerging Trends in Its Mitigation—An Updated Review. *Food Safety and Health* 2(1):39–71. doi:10.1002/fsh3.12030.

Barabadi, Hamed, Babak Tajani, Mona Moradi, Kaveh Damavandi Kamali, Ramovatar Meena, Soheila Honary, Mohammad Ali Mahjoub, and Muthupandian Saravanan. 2019. Penicillium Family as Emerging Nanofactory for Biosynthesis of Green Nanomaterials: A Journey into the World of Microorganisms. *Journal of Cluster Science* 30(4):843–856. doi:10.1007/s10876-019-01554-3.

Bin Rashid, Adib. 2023. Utilization of Nanotechnology and Nanomaterials in Biodiesel Production and Property Enhancement. *Journal of Nanomaterials* 2023:e7054045. Hindawi. doi:10.1155/2023/7054045.

Clack, Kimberley, Mohamed Sallam, Serge Muyldermans, Prabhakaran Sambasivam, Cong Minh Nguyen, and Nam-Trung Nguyen. 2024. Instant Candida Albicans Detection Using Ultra-Stable Aptamer Conjugated Gold Nanoparticles. *Micromachines* 15(2):216. doi:10.3390/mi15020216.

Dhakal, Aayush, Muhammad F. Hashmi, and Evelyn Sbar. 2024. Aflatoxin Toxicity. In *StatPearls*. Treasure Island (FL): StatPearls Publishing. http://www.ncbi.nlm.nih.gov/books/NBK557781/.

Elazab, Nahla T., Zakaria A.M. Baka, Hazem H. Saleh, and Mohamed M. El-Zahed. 2024. Green Synthesis of Silver Nanoparticles Using Cakile Maritima Seed Extract: Molecular, Antifungal and Physiological Studies. *Physiological and Molecular Plant Pathology* 129:102183. doi:10.1016/j.pmpp.2023.102183.

Fouda, Amr, Arnab Bhowmik, Saad El-Din Hassan, and Mohamed Hijri. 2023. Editorial: Green Nanomaterials: Prospective Biotechnological Applications. *Frontiers in Microbiology* 14. https://www.frontiersin.org/journals/microbiology/articles/10.3389/fmicb.2023.1280398.

Gibała, Agnieszka, Paulina Żeliszewska, Tomasz Gosiewski, Agnieszka Krawczyk, Dorota Duraczyńska, Joanna Szaleniec, Maciej Szaleniec, and Magdalena Oćwieja. 2021. Antibacterial and Antifungal Properties of Silver Nanoparticles—Effect of a Surface-Stabilizing Agent. *Biomolecules* 11(10):1481. doi:10.3390/biom11101481.

Gómez, José V., Andrea Tarazona, Fernando Mateo, Misericordia Jiménez, and Eva M. Mateo. 2019. Potential Impact of Engineered Silver Nanoparticles in the Control of Aflatoxins, Ochratoxin A and the Main Aflatoxigenic and Ochratoxigenic Species Affecting Foods. *Food Control* 101:58–68. doi:10.1016/j.foodcont.2019.02.019.

Gurunathan, Sangiliyandi, Ah Reum Lee, and Jin Hoi Kim. 2022. Antifungal Effect of Nanoparticles Against COVID-19 Linked Black Fungus: A Perspective on Biomedical Applications. *International Journal of Molecular Sciences* 23(20):12526. Multidisciplinary Digital Publishing Institute. doi:10.3390/ijms232012526.

Hernández-Meléndez, David, Enrique Salas-Téllez, Anai Zavala-Franco, Guillermo Téllez, Abraham Méndez-Albores, and Alma Vázquez-Durán. 2018. Inhibitory Effect of Flower-Shaped Zinc Oxide Nanostructures on the Growth and Aflatoxin Production of a Highly Toxigenic Strain of *Aspergillus Flavus* Link. *Materials* 11(8):1265. Multidisciplinary Digital Publishing Institute. doi:10.3390/ma11081265.

Horky, Pavel, Sylvie Skalickova, Daria Baholet, and Jiri Skladanka. 2018. Nanoparticles as a Solution for Eliminating the Risk of Mycotoxins. *Nanomaterials* 8(9):727. Multidisciplinary Digital Publishing Institute. doi:10.3390/nano8090727.

Ing, Ling Yien, Noraziah Mohamad Zin, Atif Sarwar, and Haliza Katas. 2012. Antifungal Activity of Chitosan Nanoparticles and Correlation With Their Physical Properties. *International Journal of Biomaterials* 2012:e632698. Hindawi. doi:10.1155/2012/632698.

Jaćević, Vesna, Jelena Dumanović, Suliman Y. Alomar, Radmila Resanović, Zoran Milovanović, Eugenie Nepovimova, Qinghua Wu, Tanos Celmar Costa Franca, Wenda Wu, and Kamil Kuča. 2023. Research Update on Aflatoxins Toxicity, Metabolism, Distribution, and Detection: A Concise Overview. *Toxicology* 492:153549. doi:10.1016/j.tox.2023.153549.

Jafarzadeh, Shima, Milad Hadidi, Mehrdad Forough, Abdorreza Mohammadi Nafchi, and Amin Mousavi Khaneghah. 2022. The Control of Fungi and Mycotoxins by Food Active Packaging: A Review. *Critical Reviews in Food Science and Nutrition* 63(23):6393–6411. doi:10.1080/10408398.2022.2031099.

Jan, Faisal, Hamza Arshad, Mehreen Ahad, Asif Jamal, and Donald L. Smith. 2023. In Vitro Assessment of *Bacillus* Subtilis FJ3 Affirms Its Biocontrol and Plant Growth Promoting Potential. *Frontiers in Plant Science* 14:1205894. doi:10.3389/fpls.2023.1205894.

Khoo, Kuan Shiong, Wen Yi Chia, Doris Ying Ying Tang, Pau Loke Show, Kit Wayne Chew, and Wei-Hsin Chen. 2020. Nanomaterials Utilization in Biomass for Biofuel and Bioenergy Production. *Energies* 13(4):892. Multidisciplinary Digital Publishing Institute. doi:10.3390/en13040892.

Kościelecka, Klaudia, Aleksandra Kuć, Daria Kubik-Machura, Tomasz Męcik-Kronenberg, Jan Włodarek, and Lidia Radko. 2023. Endocrine Effect of Some Mycotoxins on Humans: A Clinical Review of the Ways to Mitigate the Action of Mycotoxins. *Toxins* 15(9):515. doi:10.3390/toxins15090515.

Kotzybik, Kathrin, Volker Gräf, Lena Kugler, Dominic A. Stoll, Ralf Greiner, Rolf Geisen, and Markus Schmidt-Heydt. 2016. Influence of Different Nanomaterials on Growth and Mycotoxin Production of Penicillium Verrucosum. *PLoS One* 11(3):e0150855. doi:10.1371/journal.pone.0150855.

Kovač, Tihomir, Bojan Šarkanj, Tomislav Klapec, Ivana Borišev, Marija Kovač, Ante Nevistić, and Ivica Strelec. 2018. Antiaflatoxigenic Effect of Fullerene C60 Nanoparticles at Environmentally Plausible Concentrations. *AMB Express* 8(1):14. doi:10.1186/s13568-018-0544-0.

Kumar, Abhishek, Hardik Pathak, Seema Bhadauria, and Jebi Sudan. 2021. Aflatoxin Contamination in Food Crops: Causes, Detection, and Management: A Review. *Food Production, Processing and Nutrition* 3(1):17. doi:10.1186/s43014-021-00064-y.

León-Buitimea, Angel, Javier A. Garza-Cervantes, Diana Y. Gallegos-Alvarado, Macario Osorio-Concepción, and José Ruben Morones-Ramírez. 2021. Nanomaterial-Based Antifungal Therapies to Combat Fungal Diseases Aspergillosis, Coccidioidomycosis, Mucormycosis, and Candidiasis. *Pathogens* 10(10):1303. doi:10.3390/pathogens10101303.

Li, Li, Hui Pan, Lei Deng, Guoliang Qian, Zupeng Wang, Wenyi Li, and Caihong Zhong. 2022. The Antifungal Activity and Mechanism of Silver Nanoparticles Against Four Pathogens Causing Kiwifruit Post-Harvest Rot. *Frontiers in Microbiology* 13. Frontiers. doi:10.3389/fmicb.2022.988633.

Lotfali, Ensieh, Hossein Toreyhi, Kamyab Makhdoomi Sharabiani, Azam Fattahi, Amirali Soheili, Reza Ghasemi, Mahyar Keymaram, Yasaman Rezaee, and Sayna Iranpanah. 2021. Comparison of Antifungal Properties of Gold, Silver, and Selenium Nanoparticles Against Amphotericin B-Resistant Candida Glabrata Clinical Isolates. *Avicenna Journal of Medical Biotechnology* 13(1):47–50. doi:10.18502/ajmb.v13i1.4578.

Mahmoud, Mohamed A. 2021. Characterization and Antifungal Efficacy of Biogenic Silver Nanoparticles and Silver–Chitosan Nanocomposites. *Egyptian Journal of Phytopathology* 49(2):68–79. doi:10.21608/ejp.2021.97374.1043.

Mali, Ravindra, and Patil Javesh. 2023. Nanoparticles: A Novel Antifungal Drug Delivery System. *Materials Proceedings* 14(1):61. Multidisciplinary Digital Publishing Institute. doi:10.3390/IOCN2023-14513.

Mansoor, Sheikh, Irfana Zahoor, Tawseef Rehman Baba, Shahid A. Padder, Z. A. Bhat, Aabid Mustafa Koul, and Linghuo Jiang. 2021. Fabrication of Silver Nanoparticles Against Fungal Pathogens. *Frontiers in Nanotechnology* 3. Frontiers. doi:10.3389/fnano.2021.679358.

Marcus, Andrew, and Glen Fox. 2021. Fungal Biovalorization of a Brewing Industry Byproduct, Brewer's Spent Grain: A Review. *Foods* 10(9):2159. doi:10.3390/foods10092159.

Maruyama, Cintia Rodrigues, Natália Bilesky-José, Renata de Lima, and Leonardo Fernandes Fraceto. 2020. Encapsulation of Trichoderma Harzianum Preserves Enzymatic Activity and Enhances the Potential for Biological Control. *Frontiers in Bioengineering and Biotechnology* 8. Frontiers. doi:10.3389/fbioe.2020.00225.

Matras, Ewelina, Anna Gorczyca, Sebastian Wojciech Przemieniecki, and Magdalena Oćwieja. 2022. Surface Properties-Dependent Antifungal Activity of Silver Nanoparticles. *Scientific Reports* 12(1):18046. Nature Publishing Group. doi:10.1038/s41598-022-22659-2.

McKenna, Jack. 2023. Using Fungi for Waste Management. MDPI Blog. https://mdpiblog.wordpress.sciforum.net/2023/12/12/fungi-for-waste-management/.

Miklós, Gabriella, Cserne Angeli, Árpád Ambrus, Attila Nagy, Valéria Kardos, Andrea Zentai, Kata Kerekes, Zsuzsa Farkas, Ákos Jóźwiak, and Tibor Bartók. 2020. Detection of Aflatoxins in Different Matrices and Food-Chain Positions. *Frontiers in Microbiology* 11. https://www.frontiersin.org/journals/microbiology/articles/10.3389/fmicb.2020.01916.

Mohammed, Yasser Hussein Issa, Saad Alghamdi, Basit Jabbar, Dina Marghani, Saba Beigh, Amr S. Abouzied, Nasrin E. Khalifa, Weam M. A. Khojali, Bader Huwaimel, Dalal Hussien M. Alkhalifah, et al. 2023. Green Synthesis of Zinc Oxide Nanoparticles Using Cymbopogon Citratus Extract and Its Antibacterial Activity. *ACS Omega* 8(35):32027–32042. doi:10.1021/acsomega.3c03908.

Mohanta, Yugal Kishore, Awdhesh Kumar Mishra, Jibanjyoti Panda, Ishani Chakrabartty, Bhaskar Sarma, Sujogya Kumar Panda, Hitesh Chopra, Gokhan Zengin, Mark G. Moloney, and Majid Sharifi-Rad. 2023. Promising Applications of Phyto-Fabricated Silver Nanoparticles: Recent Trends in Biomedicine. *Biochemical and Biophysical Research Communications* 688:149126. doi:10.1016/j.bbrc.2023.149126.

Nsengumuremyi, Daniel, Parise Adadi, Gavers K. Oppong, Nadezhda V. Barakova, Elena F. Krivoshapkina, Daniel Nsengumuremyi, Parise Adadi, Gavers K. Oppong, Nadezhda V. Barakova, and Elena F. Krivoshapkina. 2020. The Potential Application of Nanoparticles on Grains During Storage: Part 1 – An Overview of Inhibition Against Fungi and Mycotoxin Biosynthesis. In *Mycotoxins and Food Safety*. IntechOpen. doi:10.5772/intechopen.91005.

Omran, Basma A., and Kwang-Hyun Baek. 2024. Dual Extracellular Mycofabrication of Cobalt and Zinc Nano Metal Oxides Mediated by Mycelial-Cell Free Filtrate of *Aspergillus Sojae*: Characterization and Assessment of Antibacterial Activity. *Journal of Molecular Structure* 1300:137190. doi:10.1016/j.molstruc.2023.137190.

Padrilah, Siti Nadzirah, Nik Iskandar Putra Samsudin, Mohd Yunus Abd Shukor, and Noor Azlina Masdor. 2024. Nanoemulsion Strategies in Controlling Fungal Contamination and Toxin Production on Grain Corn Using Essential Oils. *Green Chemistry Letters and Reviews* 17(1):2315138. doi:10.1080/17518253.2024.2315138.

Parveen, Javeria, Tahira Sultana, Abeer Kazmi, Khafsa Malik, Abd Ullah, Amir Ali, Bushra Qayyum, Naveed Iqbal Raja, Zia-ur-Rehman Mashwani, and Saif Ur Rehman. 2023. Phytosynthesized Nanoparticles as Novel Antifungal Agent for Sustainable Agriculture: A Mechanistic Approach, Current Advances, and Future Directions. *Journal of Nanotechnology* 2023:e8011189. Hindawi. doi:10.1155/2023/8011189.

Pisoschi, Aurelia Magdalena, Florin Iordache, Loredana Stanca, Adriana Ionescu Petcu, Letitia Purdoiu, Ovidiu Ionut Geicu, Liviu Bilteanu, and Andreea Iren Serban. 2023. Comprehensive Overview and Critical Perspective on the Analytical Techniques Applied to Aflatoxin Determination – A Review Paper. *Microchemical Journal* 191:108770. doi:10.1016/j.microc.2023.108770.

Radhakrishnan, Venkatraman Srinivasan, Surya Prakash Dwivedi, Mohammed Haris Siddiqui, and Tulika Prasad. 2018. In Vitro Studies on Oxidative Stress-Independent, Ag Nanoparticles-Induced Cell Toxicity of Candida Albicans, an Opportunistic Pathogen. *International Journal of Nanomedicine* 13(T-NANO 2014 Abstracts):91–96. doi:10.2147/IJN.S125010.

Rai, Mahendra, Júlio César dos Santos, Matheus Francisco Soler, Paulo Ricardo Franco Marcelino, Larissa Pereira Brumano, Avinash P. Ingle, Swapnil Gaikwad, Aniket Gade, and Silvio Silvério da Silva. 2016. Strategic Role of Nanotechnology for Production of Bioethanol and Biodiesel. *Nanotechnology Reviews* 5(2):231–250. De Gruyter. doi:10.1515/ntrev-2015-0069.

Raj, Riya, Monika Bhattu, Meenakshi Verma, Roberto Acevedo, Nguyen D. Duc, and Jagpreet Singh. 2023. Biogenic Silver Based Nanostructures: Synthesis, Mechanistic Approach and Biological Applications. *Environmental Research* 231(Pt 1):116045. doi:10.1016/j.envres.2023.116045.

Ren, Xianfeng, Qi Zhang, Wen Zhang, Jin Mao, and Peiwu Li. 2020. Control of Aflatoxigenic Molds by Antagonistic Microorganisms: Inhibitory Behaviors, Bioactive Compounds, Related Mechanisms, and Influencing Factors. *Toxins* 12(1):24. Multidisciplinary Digital Publishing Institute. doi:10.3390/toxins12010024.

Saberi Riseh, Roohallah, Mohadeseh Hassanisaadi, Masoumeh Vatankhah, Fatemeh Soroush, and Rajender S. Varma. 2022. Nano/Microencapsulation of Plant Biocontrol Agents by Chitosan, Alginate, and Other Important Biopolymers as a Novel Strategy for Alleviating Plant Biotic Stresses. *International Journal of Biological Macromolecules* 222(Pt A):1589–1604. doi:10.1016/j.ijbiomac.2022.09.278.

Sadare, Olawumi Oluwafolakemi, Chioma Nnaji Frances, and Michael Olawale Daramola. 2022. Antimicrobial (Antibacterial) Properties and Other Miscellaneous Applications of Carbon Nanotubes (CNTs). In *Handbook of Carbon Nanotubes*, ed. Jiji Abraham, Sabu Thomas, and Nandakumar Kalarikkal, 1875–1902. Cham: Springer International Publishing. doi:10.1007/978-3-030-91346-5_46.

Salano, Elsie Nyangweso, Richard Mwanza Mulwa, and Meshack Amos Obonyo. 2024. Peanut (Arachis Hypogea) Accessions Differentially Accumulate Aflatoxins upon Challenge by *Aspergillus Flavus*: Implications for Aflatoxin Mitigation. *Journal of Agriculture and Food Research* 15:100923. doi:10.1016/j.jafr.2023.100923.

Shabeer, Saba, Shahzad Asad, Atif Jamal, and Akhtar Ali. 2022. Aflatoxin Contamination, Its Impact and Management Strategies: An Updated Review. *Toxins* 14(5):307. Multidisciplinary Digital Publishing Institute. doi:10.3390/toxins14050307.

Shalal, Omar Sadik. 2023. Green Synthesis of Copper Nanoparticles and Their Activity against *Aspergillus* Chitin Deacetylase Gene and Cellulose Dehydrogenase Protein Expression. *Journal of Pure and Applied Microbiology* 17(1):258–265. doi:10.22207/jpam.17.1.15.

Sim, Serjay, and Nyet Kui Wong. 2021. Nanotechnology and Its Use in Imaging and Drug Delivery (Review) *Biomedical Reports* 14(5):1–9. Spandidos Publications. doi:10.3892/br.2021.1418.

Sohail, Younas, Nadeem Raza, Nasir Shakeel, Hina Raza, Suryyia Manzoor, Ghazala Yasmin, Amjad Iqbal, Shamaila Manzoor, Munirah D. Albaqami, and Saikh Mohammad Wabaidur. 2022. Polyaniline-Coated Nanoparticles of Zinc Oxide and Copper Oxide as Antifungal Agents Against *Aspergillus* Parasiticus. *Frontiers in Plant Science* 13. Frontiers. doi:10.3389/fpls.2022.925451.

Sultana, Tahira, Bilal Javed, Naveed Iqbal Raja, and Zia-ur-Rehman Mashwani. 2021. Silver Nanoparticles Elicited Physiological, Biochemical, and Antioxidant Modifications in Rice Plants to Control *Aspergillus Flavus*. *Green Processing and Synthesis* 10(1):314–324. De Gruyter. doi:10.1515/gps-2021-0034.

Thirugnanasambandan, Theivasanthi, and Subash C. B. Gopinath. 2023. Nanomaterials in Food Industry for the Protection from Mycotoxins: An Update. *3 Biotech* 13(2):64. doi:10.1007/s13205-023-03478-2.

Tiwari, Shikha, Bijendra Kumar Singh, and Nawal Kishore Dubey. 2022. Aflatoxins in Food Systems: Recent Advances in Toxicology, Biosynthesis, Regulation and Mitigation Through Green Nanoformulations. *Journal of the Science of Food and Agriculture* 103(4):1621–1630. doi:10.1002/jsfa.12265.

Topi, Dritan, Janja Babič, Breda Jakovac-Strajn, and Gabrijela Tavčar-Kalcher. 2023. Incidence of Aflatoxins and Ochratoxin A in Wheat and Corn from Albania. *Toxins* 15(9):567. doi:10.3390/toxins15090567.

Valença, Camilla A. S., Ana A. T. Barbosa, Silvio S. Dolabella, Patricia Severino, Carla Matos, Karolline Krambeck, Eliana B. Souto, and Sona Jain. 2023. Antimicrobial Bacterial Metabolites: Properties, Applications and Loading in Liposomes for Site-Specific Delivery. *Current Pharmaceutical Design* 29(28):2191–2203. doi:10.2174/1381612829666230918111014.

Wang, Pu, and Ying Wu. 2021. A Review on Colloidal Delivery Vehicles Using Carvacrol as a Model Bioactive Compound. *Food Hydrocolloids* 120:106922. doi:10.1016/j.foodhyd.2021.106922.

Xu, Jia-Le, Yu-Xuan Luo, Sheng-Hao Yuan, Li-Wen Li, and Ning-Ning Liu. 2020. Antifungal Nanomaterials: Current Progress and Future Directions. *Innovations in Digital Health, Diagnostics, and Biomarkers* 1(1):3–7. doi:10.36401/IDDB-20-03.

Yadav, Kanchan, Kalimuthu Moovendaran, Namasivayam Dhenadhayalan, Shang-Fan Lee, Man-Kit Leung, and Raman Sankar. 2023. From Food Toxins to Biomarkers: Multiplexed Detection of Aflatoxin B1 and Aflatoxin M1 in Milk and Human Serum Using PEGylated Ternary Transition Metal Sulfides. *Sensors and Actuators Reports* 5:100156. doi:10.1016/j.snr.2023.100156.

Yan, Chunlei, Qi Wang, Qingli Yang, and Wei Wu. 2020. Recent Advances in Aflatoxins Detection Based on Nanomaterials. *Nanomaterials* 10(9):1626. Multidisciplinary Digital Publishing Institute. doi:10.3390/nano10091626.

Yousefi, Shima, Weria Weisany, Seyed Ebrahim Hosseini, and Mehran Ghasemlou. 2024. Mechanisms of Nanoencapsulation to Boost the Antimicrobial Efficacy of Essential Oils: A Review. *Food Hydrocolloids* 150:109655. doi:10.1016/j.foodhyd.2023.109655.

Yu, Xuepeng, Yueyue Yang, Qing Liu, Zhengyu Jin, and Aiquan Jiao. 2023. A Hydroxypropyl Methylcellulose/Hydroxypropyl Starch Nanocomposite Film Reinforced With Chitosan Nanoparticles Encapsulating Cinnamon Essential Oil: Preparation and Characterization. *International Journal of Biological Macromolecules* 242(Pt 1):124605. doi:10.1016/j.ijbiomac.2023.124605.

Yushina, Yuliya K., Andrey V. Sybachin, Oksana A. Kuznecova, Anastasia A. Semenova, Eteri R. Tolordava, Vladislava A. Pigareva, Anastasiya V. Bolshakova, Vyacheslav M. Misin, Alexey A. Zezin, Alexander A. Yaroslavov, et al. 2023. Thin Cationic Polymer Coatings Against Foodborne Infections *Coatings* 13(8):1389. doi:10.3390/coatings13081389.

Zachová, I., J. Vytrasová, M. Pejchalová, L. Cervenka, and G. Tavcar-Kalcher. 2003. Detection of Aflatoxigenic Fungi in Feeds Using the PCR Method. *Folia Microbiologica* 48(6):817–821. doi:10.1007/BF02931519.

Zanoni, Elton. 2018. Mesoporous Silica Nanoparticles Incorporated With Zinc Oxide: Antifungal Activity on *Aspergillus Flavus* and Fusarium Graminearum and Adsorption of Their Toxins. https://www.academia.edu/37913323/MESOPOROUS_SILICA_NANOPARTICLES_INCORPORATED_WITH_ZINC_OXIDE_ANTIFUNGAL_ACTIVITY_ON_Aspergillus_flavus_and_Fusarium_graminearum_AND_ADSORPTION_OF_THEIR_TOXINS.

Zhang, Dong-Dong, Si Hu, Qiong Wu, Jin-Feng Zhao, Ke-Rui Su, Li-Qin Tan, and Xian-Qing Zhou. 2022. Construction of ZnO@mSiO2 Antibacterial Nanocomposite for Inhibition of Microorganisms During *Zea Mays* Storage and Improving the Germination. *LWT* 168:113907. doi:10.1016/j.lwt.2022.113907.

Zhao, Yihui, Yalan Zhang, Hao Dong, Weiliang Wu, Xingfen Yang, and Qi He. 2023. Functional Biopolymers for Food Packaging: Formation Mechanism and Performance Improvement of Chitosan-Based Composites. *Food Bioscience* 54:102927. doi:10.1016/j.fbio.2023.102927.

Zhou, Jiaojiao, Yuantao Liu, Xiaoping Du, Yue Gui, Jiangling He, Fang Xie, and Jie Cai 2023. Recent Advances in Design and Application of Nanomaterials-Based Colorimetric Biosensors for Agri-Food Safety Analysis. *ACS Omega* 8(49):46346–46361. doi:10.1021/acsomega.3c06409.

Živančev, Jelena, Sandra Bulut, Sunčica Kocić-Tanackov, Danica Jović, Aleksandar Fišteš, Igor Antić, and Aleksandar Djordjevic. 2024. The Impact of Fullerenol Nanoparticles on the Growth of Toxigenic *Aspergillus Flavus* and Aflatoxins Production in Vitro and in Corn Flour. *Journal of Food Science* 89(3):1814–1827. doi:10.1111/1750-3841.16952.

# 3.2 Aflatoxigenic Fungi
## *Control by Plant Bioactive Metabolites*

*M. Veerapagu, K.R. Jeya, A. Sankara Narayanan, K. Aravinthan, Pooja Suneja, and Amita Suneja Dang*

## INTRODUCTION

The growth of aflatoxins, which are exceedingly toxic mycotoxins that lead to carcinoma, makes aflatoxigenic fungi, especially *Aspergillus flavus* and *Aspergillus parasiticus*, a serious threat to public health and sanitation. Although A. *parasiticus* is frequently linked to spices and dried fruits, *A. flavus* is commonly found in crops including maize, peanuts, and tree nuts[1]. Both species thrive in warm, sticky environments that are typically found in tropical and subtropical regions. Aflatoxigenic organisms have special qualities that add to their pathogenicity and contaminate food and feed supplies. They produce strong spores that permit them to endure brutal circumstances and spread over significant distances, helping with their boundless scattering. Additionally, these parasites have particular gene clusters and metabolic pathways for aflatoxin production, including fundamental compounds like polyketide synthases (PKSs) and cytochrome P450 monooxygenases[2].

### Prevalence and Impact of Aflatoxigenic Fungi

Aflatoxigenic growths and their aflatoxins are far-reaching foreign substances in different food and feed products, introducing huge difficulties in food handling and worldwide exchange. Products like maize, peanuts, almonds, flavors, and creature feed are profoundly powerless against aflatoxin tainting[3].The worldwide commonness of aflatoxigenic parasites is impacted by factors like environment, rural practices, stockpiling conditions, and handling techniques. Worldwide exchange adds to the cross-line spread of aflatoxin tainting, underscoring the requirement for normalized guidelines and reconnaissance frameworks to guarantee sanitation[4]. Grasping the attributes, normal species, and commonness of aflatoxigenic growths is essential for executing successful control gauges and safeguarding food and feed supplies from aflatoxin pollution.

### Plant Metabolites as Antifungal Agents

Plant metabolites have arisen as possible antifungal drugs against aflatoxigenic organisms, attracting consideration because of their regular beginning and viability. These mixtures are incorporated by plants as a component of their defense system against microorganisms, including growths[5]. Among them, phenolic compounds like flavonoids, tannins, and phenolic acids have been widely investigated for their antifungal properties. These substances can obstruct contagious development by upsetting the cell membrane, impeding parasitic metabolic compounds, and influencing signal communicative systems[6]. They upset contagious layers, prompting cell spillage and interaction interruption; restrain significant parasitic biocatalysts and aflatoxin biosynthesis, hampering development and toxin creation; actuate oxidative stress, causing DNA and protein damage; and slow down parasitic signal mechanisms, upsetting development and advancement[7,8].


DOI: 10.1201/9781003487043-30

Despite their true capacity, challenges remain concerning the basic application of plant metabolites as antifungal drugs. Future examinations expect to beat these difficulties by investigating nanoencapsulation procedures for further improved stability and controlled delivery, exploring synergistic impacts of metabolite combinations, and evaluating their general effect on crop well-being, yield, and ecological supportability. All in all, plant bioactive metabolites present promising paths for controlling aflatoxigenic fungus and diminishing aflatoxin defilement in food and feed supplies. Determining their components, adequacy, and tending to related difficulties are vital stages in creating reasonable procedures for food safety and security.

## TYPES OF AFLATOXINS

Many agricultural commodities that are ingested by people and livestock are contaminated with aflatoxin, which is a serious hazard to public well-being and food sustainability worldwide. Aflatoxins are arranged into four principal types: AFB1, AFB2, AFG1, and AFG2, with AFB1 being the most strong and pervasive[9].

AFB1 is classified as a Group 1 carcinogen by the International Agency for Research on Cancer (IARC) because it likely causes solid malignant growth in people. AFB1 is notorious for its powerful cancer-causing nature, especially hepatocellular carcinoma (HCC) or malignant liver growth [10]. It goes through bioactivation in the liver through the enzyme cytochrome P450, shaping responsive intermediates like AFB1-8,9-exo-epoxide. These intermediates can bind covalently to DNA, forming DNA adducts that lead to mutation in essential oncogenes and tumor suppressor genes, contributing fundamentally to liver tumorigenesis and making AFB1 a significant worldwide general well-being concern[11]. AFB2, although less harmful than AFB1, actually presents critical well-being risks, including cancer-causing impacts. Research has demonstrated how AFB2 can experience metabolic initiation, prompting responsive metabolites equipped for DNA restriction and mutagenesis, but less significantly than AFB1[12]. Examinations somewhere in the range of AFB1 and AFB2 as far as genotoxicity and cancer-causing strength highlight the significance of thinking about them in risk evaluation and administrative structures. Aflatoxin G1 (AFG1) and aflatoxin G2 (AFG2), while less pervasive, also show cancer-causing and hepatotoxic properties. AFG1 is viewed as more poisonous than AFG2 because of its metabolic enactment to a receptive epoxide halfway like AFB1-8,9-exo-epoxide. Both AFG1 and AFG2 can cause DNA harm, oxidative pressure, and lipid peroxidation, adding to their toxicological impacts on liver cells and expected cancer-causing nature.

### MECHANISMS OF AFLATOXIN TOXICITY

Aflatoxins display their impeding impacts through different components, with an essential spotlight on liver harm. AFB1, upon ingestion, goes through metabolic actuation in the liver, creating exceptionally responsive intermediates. These intermediates can harm the DNA of liver cells, prompting mutations that start the advancement of hepatocellular carcinoma (liver malignant growth)[11]. Aside from its immediate relationship with malignant liver growth, ongoing openness to aflatoxins represents a range of health dangers beyond hepatocellular carcinoma. Some people tend to be more vulnerable to infections and have a weakened defense system as a result of aflatoxins, which may hamper the body's capacity to combat illness[13]. In summary, AFB1 remains a significant worry because of its powerful cancer-causing nature, whereas AFB2, AFG1, and AFG2 pose similar health risks, especially in communities with high aflatoxin exposure.

## SOURCES AND ROUTES OF AFLATOXIN CONTAMINATION

Aflatoxins, powerful mycotoxins principally synthesized by *Aspergillus* species, represent a significant danger to both food safety and general well-being because of their cancer-causing and hepatotoxic properties[14].

## Common Sources of Aflatoxin Contamination

Aflatoxins exhibit a surprising capacity to invade a wide scope of agrarian products such as maize (corn), peanuts (groundnuts), cottonseed, and various tree nuts like almonds, pistachios, and walnuts. These harvests are especially vulnerable to aflatoxin contamination because of a mix of elements, including positive natural circumstances set apart by warmth and humidity, deficient capacity rehearses, and improper pest management practices[15]. Maize (corn) is an essential natural site for aflatoxigenic organisms like *A. flavus* and *A. parasiticus* and faces imminent danger of aflatoxin contamination inferable from variables, for example, increased humidity levels during storage, pervasion by bugs like corn earworm and corn borers, and kernel damage during storage and maintenance processes[16]. Peanuts (groundnuts) are also inclined to aflatoxin contamination, particularly in warm and humid environments, where aflatoxigenic fungus invades peanuts during pod developments and flourishes under storage conditions. This problem is worsened by poor drying, damp storage, and insect infestation[17].

## Routes of Aflatoxin Contamination

### Pre-harvest Infection

An immense course of aflatoxin contamination comes from the pre-harvest contamination of crop plants by aflatoxigenic fungal growths, especially *A. flavus* and *A. parasiticus*. These organisms have an amazing ability to colonize crops like maize, peanuts, and cottonseed during their development stages in the field[18]. Under ideal natural circumstances of increased moisture levels and temperatures, these organisms produce aflatoxins, especially AFB1, known for its strength and pervasiveness among aflatoxins. Recent examinations have examined the multifaceted instruments of pre-harvest aflatoxin contamination, providing insight into the complicated connections between ecological variables and contagious host elements. Factors, for example, changes in environment, soil humidity levels, and farming strategies, essentially impact the degree of parasitic colonization and ensuing aflatoxin synthesis in crops[19]. Furthermore, the sensitivity of certain crop types toward fungi infection and toxin generation is largely determined by their genetic composition, which also accounts for the variation in amounts of aflatoxin seen in various agricultural produce.

### Post-harvest Contamination

After harvest, crops remain defenseless against continuous aflatoxin defilement during storage, transportation, and handling stages. Deficient storage conditions arise as an essential element in establishing conditions promoting contagious development and aflatoxin synthesis[20]. Factors, for example, deficient ventilation, raised humidity levels, and pest pervasions inside storerooms, can operate with quick contagious multiplication and elevated toxin levels in stockpiled crops. Research focused on post-reap systems has stressed the significance of carrying out sound capacity practices to successfully relieve aflatoxin contamination. Mediations like humidity-safe containers, maintaining ideal storage temperatures and moisture, and carrying out effective pest control practices can reduce the probability of aflatoxin contamination in stored produce.

# AFLATOXIN BIOSYNTHESIS

Aflatoxin biosynthesis is an unpredictable interaction represented by enzymatic responses coordinated by genes inside the aflatoxin biosynthesis gene cluster. This group includes various genes encoding biocatalysts and regulators that team up to create aflatoxins, which are exceptionally harmful secondary metabolites basically produced by specific *Aspergillus* species[21]. Polyketide synthases (PKSs) are vital in this pathway, assuming a significant part in the underlying periods of aflatoxin secretion. The excursion of biosynthesis is initiated with the actuation of the aflatoxin gene complexes, set off by natural prompts and regulator proteins[22]. The genes inside this genetic cluster encode the proteins required for packaging of the antecedents that are essential for aflatoxin blend. Remarkably, the norsolorinic acid synthase is a critical protein catalyzing the development of norsolorinic acid (NOR), a crucial intermediary in aflatoxin biosynthesis[23]. Following NOR arrangement, ensuing enzymatic responses like hydroxylation, reduction, and cyclization occur, prompting the evolution of different aflatoxin subsidiaries like AFB1 and AFG1.

The synergistic action of these biocatalysts is finely tuned and facilitated by regulators encoded inside the aflatoxin gene clusters. Transcription factors like AflR and AflS command the expression of aflatoxin biosynthesis, guaranteeing the exact guideline of aflatoxin synthesis because of external influences[24]. Furthermore, post-translational modifications and protein collaborations further regulate the properties of enzymes and substrate explicitness, adding to the intricacy of aflatoxin biosynthesis. Ongoing progressions in genomics and bioinformatics have extended how we might interpret the genetic engineering and administrative components overseeing aflatoxin biosynthesis. This includes recognizing regulatory components driving aflatoxin amalgamation, for example, PKS proteins and transcription factors like AflR and AflS, and investigating systems like catalyst hindrance and hereditary control to disturb the aflatoxin biosynthetic process[25].

## FACTORS AFFECTING AFLATOXIN SYNTHESIS

### Environmental Factors

Aflatoxin synthesis relies upon different natural factors, for example, the food source, compounds, and climatic circumstances. Be that as it may, the circumstances that support aflatoxigenic parasites may not necessarily align with ideal aflatoxin creation conditions[26]. Pre-reap conditions, including weather, pest pervasions, and rural practices, together influence the start of contagious colonization and toxin amassing in crops[27]. Post-reap factors, for example, storage temperature, moisture levels, and bug action, further add to aflatoxin improvement during maintenance and handling[28]. Ongoing advances in post-reap innovations like vacuum-sealed storage and modified gas packaging have been demonstrated as viable in diminishing aflatoxin pollution risks by establishing ominous conditions for parasitic development and toxin production. Physical factors like pH, light, dampness, temperature, and environmental gases are critical elements in aflatoxin contamination. Aflatoxigenic fungi flourish in a wide pH range (1.7–9.3), with the ideal level being 3–7[29]. Lower pH levels restrain contagious development, whereas marginally higher pH levels advance both parasitic development and aflatoxin synthesis[30]. Dim light favors aflatoxin creation, though daylight represses it[31]. High humidity content and relative moisture levels above 85% are ideal for aflatoxin synthesis[32]. Temperature likewise plays a basic role, with the ideal level for aflatoxin creation being 25–35°C[33].

### Nutritional and Biological Factors

Nutritional elements, for example, carbon sources, amino acids, nitrogen, lipids, and trace components, fundamentally impact aflatoxin synthesis. Sugar-rich substrates support more aflatoxin creation compared with oil-rich substrates because carbohydrates provide fundamental carbon for contagious development[34]. Nitrogen, lipids, nutrients, amino acids, and certain metals like zinc and magnesium additionally advance aflatoxin synthesis[35]. Biological factors include contagious fungus, weeds, bug wounds, and hereditary variety among aflatoxigenic parasites. Aflatoxin creation fluctuates based on contagious strains, with *A. flavus* being an overwhelming leader[36]. Hereditary variety among aflatoxigenic parasites adds to the varieties of toxin production limits and profiles, underscoring the requirement for strain-explicit control systems[37]. Gene editing and RNA interference are two genetic modification methods that could help decrease the production of harmful fungal toxins by focusing on key genes thar are involved in making aflatoxin. Coordinating hereditary experiences with agronomic practices and ecological administration is fundamental for developing comprehensive ways to alleviate aflatoxin contamination in agrarian frameworks.

## CURRENT STRATEGIES AND APPROACHES FOR AFLATOXIN CONTROL

### Prevention at the Farming Level

A basic part of aflatoxin control is counteraction at the farming level, zeroing in on procedures to limit parasitic defilement and aflatoxin synthesis in crops[38]. This incorporates the use of good agricultural practices (GAPs) and the improvement of safe harvest assortments[39]. To reduce aflatoxigenic fungus accumulation in the soil and disrupt the entire existence of fungi that generate aflatoxin, crop rotation is a crucial strategy[40]. Aflatoxin generation and fungal development can be encouraged by excessive moisture, which is another reason that water management is essential for preventing fungal contamination. Fungal hazards can be reduced by using irrigation systems that drip, thus preventing over-irrigation at key development periods[41]. To improve the ability to resist fungi that produce aflatoxin, plants might, for example, strengthen their defensive systems by creating more antifungal substances or activating receptors that recognize pathogens[42].

### Chemical Control Methods

Chemical control techniques plan to hinder growth of aflatoxigenic fungus or detoxify aflatoxins in yields or feed[43]. Fungicides focusing on these parasites smother contagious development and lessen aflatoxin synthesis; however, their utilization requires cautious administration to avoid ecological defilement and opposition advancement. Integrated pest management and fungicide rotation improve efficacy. Synthetic detoxification procedures adjust aflatoxins to less harmful mixtures. Ammoniation, ozonation, and hydrogen peroxide treatment are models. These strategies expect to deliver aflatoxins that are less destructive but require consideration of elements like toxin type, substrate arrangement, and principles of application[44]. Chemical control techniques are pivotal in aflatoxin control; however, they must be used judiciously to preserve the quality of food and feed.

### Biological Control Methods

Natural control specialists, including valuable microorganisms like specific bacterial and parasitic strains, assume a critical role in diminishing aflatoxin contamination in horticulture[45]. These microorganisms utilize different systems to fight aflatoxin-delivering parasites or prevent their development, eventually diminishing aflatoxin levels in crops. One powerful procedure includes utilizing hostile growths, particularly non-aflatoxigenic strains, that can outcompete aflatoxin-creating parasites in fields[46]. These organisms can colonize plant surfaces and produce AU: Do you mean "compounds"? If not, please clarify this text; "intensifies" is a verb, and it is unclear what is being intensified here. antimicrobial mixtures that directly stifle contagious development[47]. Moreover, other bacteria like *Pseudomonas fluorescens* and *Streptomyces* spp. show antifungal properties against aflatoxigenic fungus through the development of auxiliary metabolites and anti-microbials. Adversarial growths, especially non-aflatoxigenic strains, outcompete the growth of aflatoxigenic fungus for supplements and space, making ominous circumstances that diminish aflatoxin production. Also, they produce bioactive compounds like enzymes or secondary metabolites that directly target and repress aflatoxin secretion[48].

### Post-harvest Interventions

Post-harvest interventions are fundamental in controlling aflatoxin contamination and maintaining food safety. Legitimate drying and stockpiling methods are significant because they help maintain ideal moisture and temperature levels that dissuade parasitic development and aflatoxin synthesis in gathered crops[27]. Furthermore, using adsorbents like zeolites and activated charcoal as aflatoxin binders is a useful tactic in the farming of animals. To minimize the negative effects on livestock health and avoid the ingestion of aflatoxins, such binders have been included in animal diets to bind them in the gut[49]. Agricultural stakeholders may protect human and animal health and drastically lower aflatoxin levels by using these post-harvest treatments.

### Integrated Pest Management (IPM)

Integrated pest management (IPM) offers an extensive way to deal with aflatoxin control by joining systems like natural control, crop pivot, and the utilization of resistant crops. This all-encompassing system aims to economically limit aflatoxin contamination[50]. For instance, organic control uses gainful microorganisms and hostile parasites to battle aflatoxin-delivering growths, whereas crop rotation upsets contagious life cycles by rotating powerless and safe harvests. Using safe harvest assortments diminishes contagious disease and aflatoxin contamination, adding to long-haul control procedures[51]. Joint efforts among various partners are crucial for effective aflatoxin control execution. Drawing in researchers, farmers, industry stakeholders, policymakers, and customers advances development, information sharing, and adherence to administrative principles[52].

## TYPES OF PLANT BIOACTIVE METABOLITES

Plants are eminent for their different scope of bioactive metabolites, which assume essential parts in plant physiology and can fundamentally affect human well-being[53]. They are ordinarily ordered into six significant families based on their biosynthesis: phenolics, terpenes, alkaloids, saponins, glycosides, and polysaccharides[54].

### Phenolic Compounds

Phenolic compounds include a gathering of different phytochemicals containing at least one phenol group, which are responsible for the variety, flavor, and taste of various spices used in food and refreshments[55]. Vegetables, natural products, wine, espresso, and tea are rich wellsprings of phenolic compounds although their dissemination in plants is not uniform. They assume fundamental roles in plant multiplication, advancement, and development. They are striking for their cell reinforcement properties and different biological functions[56]. Phenolic acids like ferulic acid and caffeic acid display anti-inflammatory attributes and have potential in regulating defense mechanisms. Stilbenes, including resveratrol, stand out for their cardioprotective impacts and their expected role in controlling the progression of chronic illnesses. Ferulic acid, especially remarkable for its neuroprotective impacts, shows promise in battling neurodegenerative illnesses[57].

### Flavonoids

Flavonoids, have drawn extensive interest because of their extraordinary cell reinforcement qualities. For example, widely concentrated on intensifiers like quercetin and kaempferol, flavonoids are well-known for their radical-quenching capacity, which is vital for protecting plants from redox imbalance and forestalling oxidative harm in human cells[58]. These flavonoids reinforce the plant's guard systems and show promising well-being benefits in people, including modulation of inflammation and immune response[59]. A thorough comprehension of the biosynthesis pathways and accessibility of flavonoids is urgent for releasing their total helpful potential and coordinating them into useful food sources, nutraceuticals, and drugs for improving human well-being and prosperity.

### Alkaloids

Alkaloids, nitrogen-containing intensifiers tracked down in plants, have assorted anticancer, antimicrobial, and pain-relieving pharmacological properties, among others[60–62]. For example, vincristine from *Catharanthus roseus* is an essential anticancer compound used in chemotherapy, featuring the restorative capability of plant alkaloids[63]. Berberine, identified in *Berberis* species, displays antimicrobial impacts against microbes, exhibiting its worth in customary medication and present-day drug advancement[64]. These alkaloids are of interest to scientists and drug specialists because of their pharmacological functions and underlying variety. Recent exploration has uncovered new alkaloids with bioactivities ranging from anticancer and antimicrobial impacts to pain-relieving properties. The fact that the alkaloid morphine is still a mainstay of pain management treatments shows how important alkaloids have always been for treating illnesses affecting humans[65].

## Terpenoids

Terpenoids are a family of naturally occurring bioactive compounds that are derived from plants. The most common terpenoids are diterpenoids (diterpenes), triterpenoids (triterpenes), monoterpenoids (monoterpenes), and sesquiterpenoids (sesquiterpene)[66]. Isoprenoids are also known as terpenoids. More recent research has shed light on terpenoid biosynthesis pathways, revealing their various roles in both plants and humans. Monoterpenes, for example limonene and linalool, not only add to the particular fragrances of medicinal ointments but also exhibit antimicrobial and anti-inflammatory properties[67]. Sesquiterpenes like artemisinin, obtained from *Artemisia annua*, assume a significant role in treating malarial fever due to their antiparasitic action[68]. Triterpenes, including saponins and sterols, add to established protection components and have shown remedial impacts in regions like cardiovascular well-being and regulation of immune response[69].

## Glycosides

Glycosides, a gathering of plant intensifiers described by a sugar molecule connected to a non-sugar molecule, show significant organic uses and restorative properties. Among the subclasses of glycosides of great interest are the ginsenosides, which are derived from Panax ginseng, known for their adaptogenic properties, and used in traditional medicine to increase stress tolerance[70]. Ginsenosides exhibit different pharmacological effects, including modulation of hostility, weakness, and neuroprotective functions, highlighting their true capacity in stress response modulation, mental improvement, and general prosperity[71]. Cardiac glycosides, including digitoxin and ouabain from *Digitalis* species, have a longstanding history in treating heart conditions due to their constructive heart capability outcomes[72]. The saponins are another eminent subclass, which includes different natural functions like cholesterol reduction and anticancer effects. In addition, saponins found in soybeans and vegetables have shown anticancer properties by prompting cell apoptosis and upsetting cancer development [73].

## Essential Oil

Essential oils from plants have gained prominence recently because of their flexibility and potential medical advantages. Among the striking natural oils are tea tree oil (*Melaleuca alternifolia*) and lavender oil (*Lavandula angustifolia*), each having particular restorative properties. Tea tree oil is prestigious for its strong antimicrobial properties, powerful against microscopic organisms, growths, and infections. It finds normal use in skin medicines for skin diseases, skin break-outs, and minor injuries[74]. Lavender oil is valued for its quieting and relieving effects, including its famous fragrant healing for unwinding, stress mitigation, and improvement of rest quality. Research has additionally featured lavender oil's true capacity for reducing uneasiness and stress, further improving its remedial allure[75]. Citrus rejuvenating balms, acquired from natural products like oranges, lemons, and grapefruits are rich in intensifiers like limonene and citral, known for their temperament improving impacts and reviving aromas. These oils are often used in fragrance-based treatments to improve the mindset, reduce nervousness, and lift mental lucidity[76].

## ANTIFUNGAL MECHANISMS OF PLANT BIOACTIVE COMPOUNDS

The numerous ways in which plant metabolites shield plants from fungus-related diseases have long been understood. Those bioactive substances, which come from different plant components, are essential to the defensive mechanism of the plant because they prevent the growth of fungi and lower the risk of infection. Developing efficient methods to fight fungal infections in agricultural and native environments requires an understanding of the processes of the way plant metabolites exhibit their antifungal effects. The detailed explanations of the various ways in which plant metabolites protect against aflatoxigenic fungus are provided below and depicted in Figure 3.2.1.

### Disruption of Fungal Cell Membranes

Phenolic compounds, essential oils, and terpenoids from flowers were substantially studied for their capacity to destabilize fungal cellular membranes, resulting in membrane leakage, altered permeability, and eventual cellular demise[77]. The mechanisms of fungal cellular membrane disruption using plant metabolites include interactions with membrane proteins, lipids, and carbohydrates,

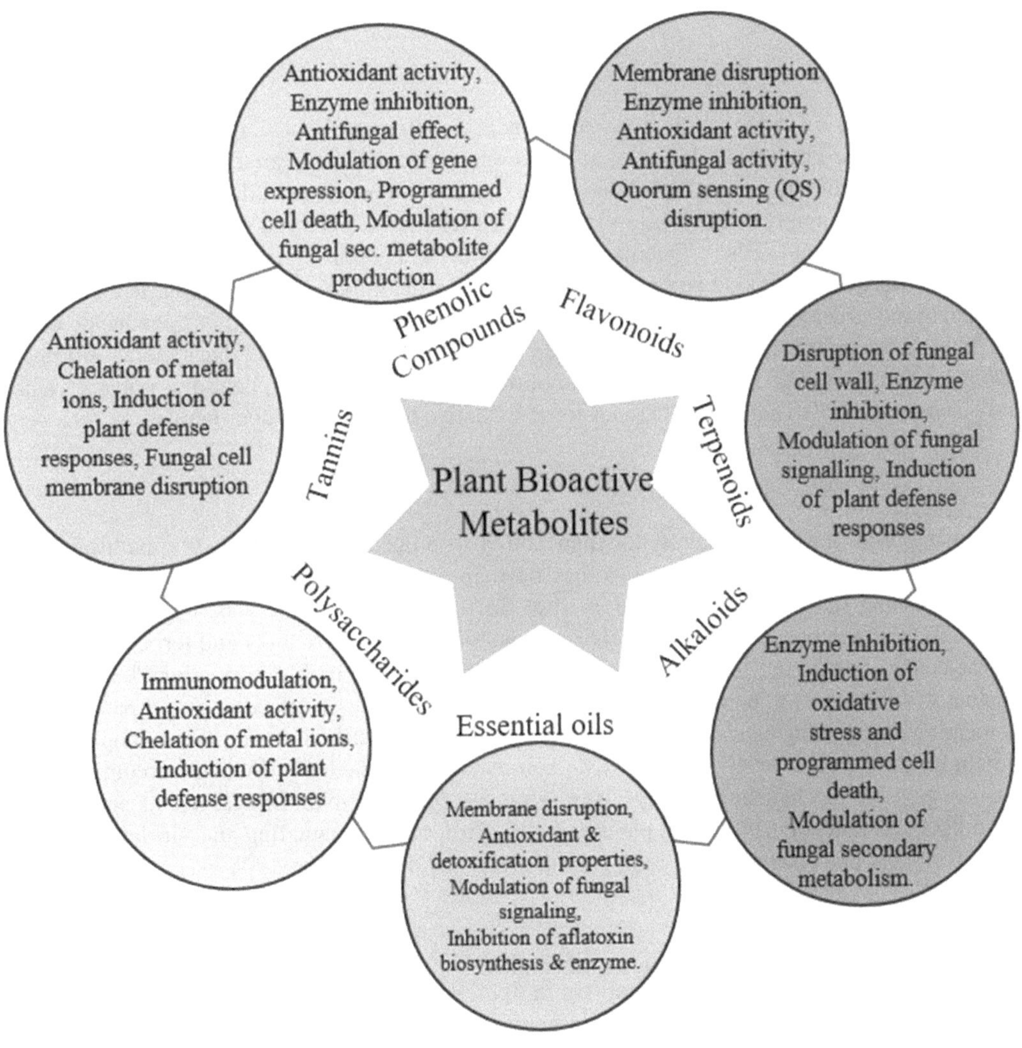


**FIGURE 3.2.1** Mechanisms of plant bioactive compounds against aflatoxigenic fungi.

causing adjustments in fluidity, permeability, and ion transport. This disruption additionally influences membrane-certain enzymes and receptors, impacting important mobile procedures and signaling pathways [78]. For instance, flavonoids together with quercetin and kaempferol can interact with the lipid bilayers of fungal cellular membranes, central to structural adjustments and permeability, eventually inflicting cellular lysis. Terpenoids like thymol and carvacrol, identified in oils like thyme and oregano, disrupt fungal membranes by changing their fluidity and integrity[79,80].

### Inhibition of Enzymes

The inhibition of cellular wall synthesis enzymes is an important mechanism that uses plant metabolites. Another important goal is ergosterol biosynthesis enzymes, vital for fungal membrane stability[81]. Plant polyphenols and alkaloids inhibit enzymes like squalene epoxidase and 14$\alpha$-demethylase, reducing ergosterol manufacturing and destabilizing fungal membranes[82]. Furthermore, positive plant metabolites interfere with nucleic acid synthesis enzymes, disrupting DNA and RNA polymerase, which can be vital for fungal replication and gene expression[83]. Furthermore, fungal metabolism and growth are impacted by plant terpenoids and alkaloids, which block cytochrome P450 and adenosine triphosphate (ATP) synthase, two enzymes that provide energy[84].

### Induction of Oxidative Stress

Oxidative stress is a crucial factor in plant secondary compounds' antifungal interest and significantly boosts their efficacy against fungal infections. However, they can also act as pro-oxidants under precise conditions, triggering oxidative stress in fungi. For instance, flavonoids like quercetin, catechins, and kaempferol were substantially studied for their capacity to generate reactive oxygen species (ROS) in fungal cells[85]. Quercetin, for example, can go through redox reactions, leading to the manufacture of superoxide anion and hydrogen peroxide, inflicting oxidative harm and disrupting membrane structures in fungi[86]. Terpenoids showcase antifungal capability by inducing oxidative damage through diverse mechanisms[87]. Berberine is a famous alkaloid that generates ROS and disrupts redox homeostasis in fungi[88]. In addition to these polyphenols, peptides, polyacetylenes, and glucosinolates also generate ROS and cause oxidative harm and cellular demise.

### Modulation of Fungal Signaling

Plant metabolites have won interest for their ability to influence fungal signaling pathways. One sizable way that plant secondary metabolites have an effect on fungal signaling is by disrupting quorum sensing (QS) mechanisms. QS is vital for fungi to coordinate gene expression primarily based on population density. Certain plant compounds, like flavonoids and terpenoids, inhibit QS molecules, hampering fungi's capacity to alter virulence and biofilm formation[89]. Flavonoids, together with kaempferol, prevent hyphal growth and morphogenesis, affecting fungal pathogenicity by focusing on signaling pathways similar to the cAMP-PKA (cyclic adenosine 3′,5′-monophosphate–protein kinase A) pathway[90]. Additionally, plant metabolites like berberine and curcumin prompt stress-responsive mechanisms in fungi, increasing susceptibility to antifungal agents and countering fungal resistance. Furthermore, plant peptides disrupt fungal signaling and virulence with the aid of receptors and enzymes, inflicting growth arrest and cellular demise[91].

### Chelation of Metal Ions

Plant secondary metabolites have a vital role in fighting fungal pathogens through diverse mechanisms, with metallic ion chelation becoming an important antifungal approach. Certain plant metabolic compounds, drastically phenolic compounds, and natural acids can chelate vital metallic ions like iron, magnesium, zinc, and copper essential for mycelial proliferation and metabolism. Organic acids like citric acid and malic acid also contribute to chelation, primarily based on antifungal interest[92]. These acids can bind to metallic ions together with iron and magnesium, disrupting vital enzymatic sports and metabolic pathways vital for fungal survival. Citric acid, for example, chelates ferrous ions, inhibiting siderophore-mediated iron acquisition in fungal pathogens and increasing their susceptibility to oxidative burden and antimycotic medications[93].

## Disruption of Fungal Cell Wall

Among the various mechanisms utilized, disruption of the mycotic cell wall using metabolites emerges as a vital antifungal approach. Certain plant metabolites, together with chitinases and glucanases, have key positions that focus on the structural integrity of the fungal cell envelope using hydrolyzing chitin and glucans, resulting in cellular wall degradation and weakening[94]. Chitinases and glucanases, enzymes found in plants, hydrolyze chitin and glucans, lead to cellular wall degradation and weakening, and hence increase fungal susceptibility to environmental stresses and antifungal agents[95]. For instance, quercetin, a flavonoid identified in diverse flowers, disrupts the fungal cell envelope by binding to chitin and interfering with its characteristic. Terpenoids and essential oils disrupt mycotic cell wall permeability and integrity, resulting in leakage of the cytoplasmic contents and impaired physiological processes[96].

## Induction of Programmed Cell Death

The activation of programmed cellular death (PCD) using plant secondary metabolites represents a captivating and complicated mechanism used to fight fungal pathogens[97]. Certain plant metabolites were observed to result in PCD, comparable to apoptosis in fungal cells. This mechanism holds unique importance in host–pathogen interactions, wherein plants make use of metabolites to eliminate invading fungi. The induction of PCD in fungal cells using plant secondary metabolites takes place through numerous pathways. One important mechanism includes the activation of apoptotic cascades, which incorporates the upregulation of pro-apoptotic genes, caspase activation, and DNA fragmentation[98]. Together, these procedures result in cellular shrinkage, chromatin condensation, membrane blebbing, and in the long run, apoptotic cellular demise.

## Modulation of Fungal Secondary Metabolite Production

Plant metabolites have the potential to encourage the synthesis of secondary metabolites in fungi, which includes mycotoxins and virulence factors[99]. Through the alteration of gene expression or enzyme interest, those metabolites can decrease the manufacturing of toxic compounds, thereby reducing fungal virulence and pathogenicity[100]. The interactions among plant metabolites and fungi include influencing fungal cells, affecting their metabolism, and changing gene expression. For instance, flavonoids and phenolics can act as signaling molecules that cause the expression of genes responsible for secondary metabolite biosynthesis in fungi. Conversely, a few plant metabolites may also prevent vital enzymes or pathways in fungal secondary metabolite biosynthesis, resulting in decreased metabolite synthesis[101]. In phases of fungal ecology, the modulation of fungal secondary metabolite biosynthesis by plant metabolites has implifications for fungal expansion, competitiveness, and virulence, influencing ecological niches, symbiotic relationships, and pathogenicity.

## Enhancement of Plant Defense Response

Plant metabolites, which exist as signals or inducers of genes linked to defense against fungal infections, may boost a plant's defense systems[102]. These metabolites cause the expression of protection-associated genes and pathways upon detecting fungal pathogens or elicitors. For instance, flavonoids together with quercetin and catechin can result in the manufacture of pathogenesis-related (PR) proteins and antioxidant enzymes[103]. Moreover, many plant secondary metabolites showcase direct antimicrobial effects that result in plant protection. For example, alkaloids such as sanguinarine and berberine have wide-spectrum antibacterial activities against fungal infections, whereas terpenoids, such as thymol and carvacrol, destroy fungal cellular membranes, increasing plant defense mechanisms[104,105]. Furthermore, those metabolites can interact with phytohormones like salicylic acid (SA), jasmonic acid (JA), and ethylene (ET), modulating plant protection responses primarily based on the pathogen's lifestyle[106].

## PLANT METABOLITES TO COMBAT AFLATOXIGENIC FUNGI

Plant-derived metabolites provide a promising opportunity because vegetation evidently produces an array of various chemical compounds as a part of their boom technique or in reaction to stress and infections[107]. Organic plant extracts were extensively used traditionally because of their antifungal traits throughout many species, including fungi, bacteria, and insects[108]. Table 3.2.1 provides an extensive list of phytobioactive compounds that can hinder the growth of aflatoxigenic fungi and the production of aflatoxin.

### PHENOLIC COMPOUNDS

Phenolic compounds, in particular the ones belonging to the subclass of flavonoids, phenolic acids, and tannins, provide a flexible approach for stopping the development of aflatoxin[109]. Strong antioxidant characteristics are exhibited through phenolic compounds, which can be crucial in disposing of free radicals and oxidative stress, which is essential in the biosynthesis of aflatoxin[110]. In particular, studies performed by Farouk et al.[111] show the effectiveness of coffee-derived chlorogenic acid in dramatically decreasing aflatoxin formation in *A. flavus* by enhancing the protecting antioxidant mechanism. Vitexin and hyperoside, along with other polyphenols found in buckwheat hulls, hindered the synthesis of AFB1 by changing the antioxidant system of fungal cells and affecting fungal growth[112]. It was similarly reported that the flower extract of *Cannabis sativa* L., which is rich in phenolic compounds, limits aflatoxin production and the growth of *A. flavus* by using its total antioxidant ability and free radical-scavenging capacity[113]. Some phenolic substances immediately block crucial enzymes that take part in the aflatoxin generation mechanisms, consisting of cytochrome P450 monooxygenases and polyketide synthases[114]. Resveratrol and quercetin, for instance, impede polyketide synthases, thereby halting the production of aflatoxin precursors[115]. Curcumin, a phenolic compound found in turmeric, displays inhibitory effects on the enzymes responsible for aflatoxin synthesis[116].

### FLAVONOIDS

Flavonoids, notably quercetin and kaempferol, exhibit multifaceted mechanisms in preventing aflatoxin manufacturing[8]. It is recognized that these two flavonoids interfere with the production of aflatoxin by blocking important enzymes and altering the expression of certain genes in fungi that produce aflatoxin. Epicatechin and catechins, discovered in vegetation like tea and cocoa, show powerful antifungal properties that grow to inhibit aflatoxin manufacturing[117]. The study by Naseer et al.[118] highlighted the impressive effectiveness of ethanolic extracts from citrus peels in hindering the growth of *A. flavus* and the production of aflatoxin in rice. Additionally, as per Tiwari and Shankar's 2018 research[119], quercetin showed notable efficacy in suppressing *A. flavus* growth and aflatoxin generation by targeting the mitogen-activated protein kinase (MARK) signaling pathway, crucial for aflatoxin synthesis. Another study in 2020 by Pok[120] demonstrated that citrus flavonoids, including naringin, neohesperidin, and quercetin, could significantly reduce aflatoxin levels in maize by up to 85%–100%.

**TABLE 3.2.1**
**Control of Aflatoxigenic Fungus and Aflatoxin Synthesis by Plant Bioactive Metabolites**

| Source / Phytochemical Group | Plant | Aflatoxigenic Activity | Phytobioactive Compounds | Influence on Aflatoxigenic Fungus/Aflatoxin | References |
|---|---|---|---|---|---|
| ***Trachyspermum ammi* essential oil (TAEO)** | *Trachyspermum ammi* | Inhibits growth and AFB1 biosynthesis | γ-terpinene | *Aspergillus flavus*, aflatoxin B1 (AFB1) | Singh et al.[160] |
| **Alkaloid** | — | Inhibits mycelial growth and decreases the total aflatoxin production | Berberine chloride and berberine sulfate | *A. flavus* (ATCC 28539) and *A. parasiticus* (ATCC 15517) | Moradi et al.[129] |
| **Plant Extracts** | *Quercus ilex* (Holm oak) | Antifungal and anti-aflatoxigenic activity | Coumaric acid and protocatechuic acid | *A. flavus*, Cq103 and Cq8, AFB1, AFB2 | Boy et al.[161] |
| **Plant extracts (Polyphenol)** | *Aloysia citrodora* | Inhibits AFB1 synthesis | Polyphenol -luteolin-7-diglucuronide | AFB1 | Cadenillas et al.[162] |
| **Polyphenol** | *Curcuma longa,* | Decreases AFB1 synthesis and suppresses *A. flavus* infection in peanut | Curcumin | *A. flavus*, AFB1 | Zhang et al.[163] |
| **Alkaloid** | *Fumaria vaillantii* and *F. asepala* | Downregulates AFB1 biosynthesis key genes (aflR, aflD, and aflM) | Isoquinoline alkaloids, protopine and sanguinarine; parfumine and allocryptopine | *A. flavus* | Mohseni et al.[164] |
| **Phenolic compound** | *Magnolia officinalis,* | Inhibits spore germination an mycelial growth | Honokiol | *A. flavus* | Zhang et al.[165] |
| **Essential oils** | Lavandins *Grosso,* lavandin *Abrial*, *Origanum virens,* and *Rosmarinus officinalis* | Inhibits mold growth, Suppresses aflatoxin biosynthesis | Linalool, linalyl acetate, carvacrol, α-pinene, 1,8-cineole, camphor | *A. parasiticus* | Loran et al.[140] |
| **Plant extracts** | *Ilex paraguariensis*, rosemary (*Rosmarinus officinalis*) and green tea (*Camellia sinensis*) | Antiradical and anti-aflatoxigenic activity | Polyphenols | *A. flavus*, AFB1 | Al Khoury et al.[166] |
| **Plant extracts** | *Ephedra sinica* | Antifungal and anti- aflatoxigenic ability | — | *A. flavus* (ATCC 28542), AFB1 | Deabes et al.[167] |
| **Plant extract** | eggplant (*Solanum melongena* L.) | Antioxidant capacity, antifungal properties, and inhibition of AFB1 | Octadecanoic acid, alpha-kaurene, p-allylphenol, linoleic acid | *A. flavus*, AFB1 | Ismail et al.[168] |
| **Plant extracts (aqueous)** | *Mimosa tenuiflora* | Inhibits growth and aflatoxin production. Downregulates *aflR*, *aflS* gene expression | Condensed tannins | *A. flavus*, AFB1 | Hernandez, et al.[124] |

*(Continued)*

**TABLE 3.2.1** *(Continued)*
**Control of Aflatoxigenic Fungus and Aflatoxin Synthesis by Plant Bioactive Metabolites**

| Source / Phytochemical Group | Plant | Aflatoxigenic Activity | Phytobioactive Compounds | Influence on Aflatoxigenic Fungus/Aflatoxin | References |
|---|---|---|---|---|---|
| **Plant extracts (Flower)** | *Cannabis sativa* L. | Inhibits growth and aflatoxin production | Caffeic and ferulic acids | *A. flavus,* AFB1 | Al Khoury et al.[113] |
| **Plant extracts** | *Glycyrrhiza glabra* L. | Inhibits growth and aflatoxin production. Downregulates aflatoxin gene expression | 3,5-dimethoxycinnamic acid, | *A. flavus,* AFB1 | Youssef et al.[169] |
| **Cumin essential oil** | *Cuminum cyminum* | Inhibits growth of *A. flavus* and AFB1 biosynthesis. Downregulates genes involved in ergosterol and AFB1 synthesis | Cuminaldehyde | *A. flavus,* AFB1 | Xu et al.[170] |
| ***Perilla frutescens* essential oil (PEO)** | *Perilla frutescens* | Suppresses the growth of *A. flavus*, destroys the normal morphology and activities of the cell wall and membrane, and causes defense dysfunction against stress response | — | *A. flavus* | Hu et al.[171] |
| **Meroterpenoids** | *Arnebia euchroma* root | Inhibits mycelial growth and aflatoxin production | Arnebinol A–D | *A. flavus* | Madarshahi et al.[126] |
| **Plant extracts** | *Heracleum persicum*, *Peganum harmala*, and *Trachyspermum ammi* | Downregulates *aflR*, *aflM*, and *aflP* gene expression | Volatile compounds and non-volatile compounds | *A. flavus,* AFB1 | Safari et al.[130] |
| **Glucosinolates** | Cruciferous vegetables | Antifungal and anti-aflatoxigenic activity | allyl isothiocyanate | *A. flavus,* AFB1 | Nazareth et al.[172] |
| **Essential oil** | *Rosmarinus officinalis* L. | Antifungal and anti-aflatoxigenic activity | cineole, camphor and α-pinene | *Aspergillus flavus*, AFB1 and AFB12 | da Silva Bomfim et al.[173] |
| **Ginger essential oil (GEO)** | *Zingiber officinale* | Antifungal and anti-aflatoxigenic properties | Zingiberene, geranial | *A. flavus,* AFB1 | Nerilo et al.[174] |
| **Thyme essential oil** | *Thymus vulgaris* | Antifungal and anti-aflatoxigenic properties, downregulation of gene (laeA) and modulation of hydrolase gene expression involved in fungal colonization and establishment. | Thymol, carvacrol, p-cymene, γ-terpinene, linalool | *A. flavus,* AFB1 | Oliveira et al.[175] |

### Tannins

Tannins are acknowledged for their astringent potential inhibitors of aflatoxins via a mechanism known as metal chelation in which tannins bind to crucial metal ions like iron (Fe) and copper (Cu) that are essential for fungal growth and metabolism[121]. The biosynthesis of aflatoxins in fungi entails enzymatic reactions that require metal ions as cofactors. By chelating those metal ions, tannins inhibit the function of those enzymes, thereby disrupting the synthesis of aflatoxins. Additionally, chelation of those metals through tannins can result in oxidative stress inside fungal cells. Condensed tannins, also known as proanthocyanidins, have shown the ability to mitigate AFB1-induced oxidative stress and apoptosis in broilers[122]. AFB1 and its metabolic precursors like norsolorinic acid and versicolorins are highly oxygenated molecules, making them susceptible to redox regulation. There is even a suggestion that the production of AFB1 might be a fungal response to oxidative stress because its biosynthesis is triggered by elevated oxidative stress levels[123]. The aqueous extract from *Mimosa tenuiflora* bark impacts both *A. flavus* growth and AFB1 production. The inhibition of AFB1 is linked to the suppression of cluster genes in the biosynthetic pathway, particularly the internal regulators aflR and aflS[124].

### Terpenoids

Terpenoids are acknowledged for their antifungal properties, which can doubtlessly manipulate aflatoxins. One outstanding mechanism by which terpenoids exert their antifungal results is through disrupting fungal membranes. This disruption compromises the structural balance of fungal cells, impairs their capacity to adjust inner strategies, and finally reduces fungal viability. The main terpene components, carvone and linalool from *Carum carvi* L. and *Coriandrum sativum* L., were found to be effective against the growth of *A. flavus* and accumulation of AFB1[125]. In a study conducted by Madarshahi[126], four meroterpenoids were isolated and identified from the root of *Arnebia euchroma*, known as arnebinol A–D, exhibiting antifungal and anti-aflatoxigenic effects against *A. flavus*. Specifically, arnebinol D demonstrated inhibition of *A. flavus* aflatoxin production and reduced ergosterol levels, inducing wrinkled and compressed conidia with phialides, damaged cell walls, and complete inhibition of mycelial growth.

### Alkaloids

Alkaloids are a category of nitrogen-containing compounds observed in diverse plant species. A few alkaloids, like berberine, exhibit antimicrobial properties against aflatoxin-generating fungi[127] and have broad-spectrum antimicrobial properties, making them powerful against *Aspergillus*. Alkaloids can, in a roundabout way, affect aflatoxin synthesis by inhibiting fungal growth and colonization, which are critical steps in the aflatoxin biosynthesis process. Berberine, a naturally occurring benzylisoquinoline alkaloid, is found in the roots, rhizomes, and stem bark of natural herbs, such as *B. aquifolium*, *B. vulgaris*, and *B. aristate*[128]. In a study, berberine salts inhibited the radial growth of *A. flavus* and *A. parasiticus* mycelium[129]. Safari et al.[130] demonstrated significant inhibition of AFB1 production by *A. flavus* when exposed to *B. vulgaris* extract (6 mg/mL). Malekinezhad et al.[131] showed that different levels of berberine added to aflatoxin-challenged chickens reduced the toxin's negative effects on feed intake. It was previously demonstrated that piperine, a significant active ingredient in long and black peppers (*Piper nigrum* L. and *Piper longum* L.), efficiently inhibits the synthesis of aflatoxin. Piperine boosts the response to stress elements like bZIP gene transcription factors and antioxidative enzymes like catalases and superoxide dismutases, and suppresses genes involved in the aflatoxin biosynthesis process, especially regulator veA[132].

### Essential Oils

Essential oils (EOs) are effective in preventing the growth of the *A. flavus* fungus and the generation of aflatoxin[133]. These plant-derived compounds are not only effective but are also considered safe for consumption, being classified as "generally recognized as safe" (GRAS) by regulatory authorities like the US FDA[114]. For instance, some studies suggest that a combination of major components from an essential oil may exhibit weaker antimicrobial properties compared to the entire EO due to the absence of certain constituents[134]. Songsamoe et al.[135] observed enhanced antifungal activity against *A. flavus* when combining linalool and caryophyllene in a 10:1 ratio using *M. alba* EOs. Furthermore, blends of peppermint and tea tree EOs demonstrated synergistic effects against *A. niger*[136]. However, predicting the antimicrobial efficacy of EO blends remains challenging due to the complex chemical composition of each EO, resulting in both synergistic and antagonistic outcomes[137]. The interactions between EO compounds and those present in the food matrix may also influence the antifungal and anti-aflatoxigenic properties of EOs, as observed in the evaluation conducted by Tian et al. on thyme EOs in brown and white rice[7].

The antifungal activity of paeonol (2-hydroxy-4-methoxyacetone), an active component commonly isolated from *Cynanchum paniculatum*, is known to modify the cell wall ultrastructure and decrease the content of both β-1,3-glucan and chitin in *A. flavus* [138]. Similarly, extracts of *Tulbaghia violacea* were found to reduce the content of β-1,3-glucan and chitin in the *A. flavus* cell wall[139]. Carvacrol, a phenolic compound in oregano EO, inhibits aflatoxin synthesis by suppressing key enzymes involved in the biosynthetic pathway[140]. Compounds like eugenol and thymol, observed in oils like clove and thyme, interfere with fungal cell signaling and gene expression related to aflatoxin synthesis[141]. Essential oils additionally show antioxidant and toxin neutralization ability, which prevents aflatoxin-triggered cell harm and complements metabolic detoxification processes[142].

### Polysaccharides

Polysaccharides decrease aflatoxin toxicity by inhibiting the genotoxicity and carcinogenicity of aflatoxins, shielding cells from harm, and decreasing the danger of aflatoxin-triggered diseases. This detoxification capacity is attributed to their ability to scavenge free radicals and adjust oxidative pressure pathways, thereby mitigating cell harm because of aflatoxins[143]. Additionally, polysaccharides inhibit aflatoxin binding to cell receptors, both via aggressive binding or interference with receptor sites[144]. Furthermore, polysaccharides show off immunomodulatory results, specifically improving immune responses in the liver, a first-rate goal organ for aflatoxin toxicity. Due to its well-established antioxidative, anti-inflammatory, and immunomodulatory qualities, Arabic gum (AG), a protein-based polysaccharide with edible qualities, is widely used in Arab traditional medicine[145]. Notable pharmacological benefits of AG include anti-inflammation[146] and antioxidation[147,148] actions. Its capacity to inhibit excessive ROS, improve oxidant scavenging, and reduce lipid peroxidation is the source of its detoxifying properties[149]. Exposure to aflatoxin can cause significant damage to the liver because of inflammatory processes, lipid peroxidation, and oxidative damage. AG supplementation can protect liver cells from damage caused by aflatoxin exposure.

## CHALLENGES AND FUTURE PERSPECTIVES

Using plant metabolites for aflatoxin control faces a few challenges. Plant metabolites, while viable, frequently face difficulties connected with formulation and stability. Many of these mixtures have unfortunate water dissolvability, making it difficult to foster viable conveyance frameworks[150]. Also, their stability under various natural circumstances, such as different temperatures and pH levels, can influence their adequacy over time[151]. Guaranteeing the viability of plant metabolites against aflatoxin-creating parasites while maintaining the specific benefits for living beings is critical. A few metabolites might show a wide range of antimicrobial action, possibly influencing non-target living beings and disturbing environments[152,153].

Progress in nanotechnology and microencapsulation may enhance the formulation and delivery of plant metabolites. Nanoformulations can improve solvency, strength, and targeted delivery, expanding the adequacy of metabolites against aflatoxin contamination[154]. Utilizing bioinformatics and high-throughput screening techniques can help disclose novel plant metabolites with explicit anti-aflatoxigenic properties[155]. Targeted screening can distinguish metabolites that hinder key compounds or pathways associated with aflatoxin biosynthesis. Investigating synergistic connections between various plant metabolites or consolidating them with other biocontrol metabolites can upgrade their adequacy against aflatoxigenic fungus[156]. Embracing accurate horticultural strategies, for example, sensor-based technology and variable rate application, can improve the use of plant metabolites in aflatoxin control. Progress in biotechnology, including genetic manipulation and metabolic engineering, can upgrade the creation of desired plant metabolites with further developed properties. Integrated pest management (IPM) provides the ecosystem with feasible and harmless methods for dealing with aflatoxin management. Interest in research and development and technology transfer is fundamental to revealing the maximum capacity of plant metabolites in battling aflatoxin defilement, guaranteeing food safety, and advancing economical farming practices[157].

## CONCLUSION

Using plant bioactive metabolites to control aflatoxigenic fungal growth provides a diverse methodology with significant potential in addition to certain difficulties. These metabolites offer the ecosystem a reasonable and harmless method instead of manufactured synthetic chemicals, adjusting great to current horticultural practices that zero in on diminishing natural effects[158]. Their different systems of activity, including enzyme hindrance, cytoplasmic membrane disturbance, and obstruction with parasitic cell signaling, exhibit their adequacy in suppressing aflatoxin synthesis and restricting contagious development[159]. Nonetheless, difficulties like stability, bioavailability, and specificity should be considered in continuous research and development. Progress in nanotechnology, bioinformatics, and synthetic biology is vital for maximizing the effective delivery of plant metabolites in aflatoxin management. Additionally, cooperation among scientists, rural partners, and administrative bodies is fundamental for advancing the protected and mindful utilization of plant bioactive metabolites. Administrative structures should be laid out to guarantee the adequacy, well-being, and accurate labeling of these metabolites. Worldwide collaboration and information sharing stages are imperative in dispersing the best information, working toward innovation, and cultivating development. As innovation advances and logical comprehension is extended, the future of plant bioactive metabolites in aflatoxin control is promising. The creation of innovative chemicals, improved formulations, and unified approaches will be fueled by ongoing research funding, strengthened capacity programs, and legislative assistance. Finally, using plant metabolites to reduce aflatoxin protects food and feed safety while encouraging sustainable farming methods for a stronger and more resilient food chain.

## REFERENCES


1. Magan, N.; Dobson, A.; Al Ghamdi, F. L.; Bokhari, F. M.; Aly, M. M. Toxigenic fungi associated with dried fruits and fruit-based products collected from Jeddah province. *IOSR J Pharm Biol Sci*, **2019**, 14(1), 10–20.
2. Khan, F. A.; Khan, S.; Khan, N. M.; Khan, H.; Khan, S.; Ahmad, S.; Rehman, N.; Aziz, R. Antimicrobial and antioxidant role of the aerial parts of *Aconitum violaceum*. *J Mex Chem Soc*, **2021**, 65(1), 84–93. https://doi.org/10.29356/jmcs.v65i1.1310.
3. Jallow, A.; Xie, H.; Tang, X.; Qi, Z.; Li, P. Worldwide aflatoxin contamination of agricultural products and foods: From occurrence to control. *Compr Rev Food Sci Food Saf*, **2021**, 20(3), 2332–2381. https://doi.org/10.1111/1541-4337.12734.
4. Ayofemi Olalekan Adeyeye, S. Aflatoxigenic fungi and mycotoxins in food: A review. *Crit Rev Food Sci Nutr*, **2020**, 60(5), 709–721. https://doi.org/10.1080/10408398.2018.1548429.
5. Chen, Y.; Xing, M.; Chen, T.; Tian, S.; Li, B. Effects and mechanisms of plant bioactive compounds in preventing fungal spoilage and mycotoxin contamination in postharvest fruits: A review. *Food Chem*, **2023**, 415, 135787. https://doi.org/10.1016/j.foodchem.2023.135787.
6. Zhou, H.; Sun, F.; Lin, H.; Fan, Y.; Wang, C.; Yu, D.; Liu, N.; Wu, A. Food bioactive compounds with prevention functionalities against fungi and mycotoxins: Developments and challenges. *Curr Opin Food Sci*, **2022**, 48, 100916. https://doi.org/10.1016/j.cofs.2022.100916.

7. Tian, F.; Lee, S. Y.; Chun, H. S. Comparison of the antifungal and antiaflatoxigenic potential of liquid and vapor phase of *Thymus vulgaris* essential oil against *Aspergillus flavus*. *J Food Prot*, **2019**, 82(12), 2044–2048. https://doi.org/10.4315/0362-028X.JFP-19-016.
8. Castano-Duque, L.; Lebar, M. D.; Carter-Wientjes, C.; Ambrogio, D.; Rajasekaran, K. Flavonoids modulate *Aspergillus flavus* proliferation and aflatoxin production. *J Fungi*, **2022**, 8(11), 1211. https://doi.org/10.3390/jof8111211.
9. Pickova, D.; Ostry, V.; Toman, J.; Malir, F. Aflatoxins: History, significant milestones, recent data on their toxicity and ways to mitigation. *Toxins*, **2021**, 13(6), 399. https://doi.org/10.3390/toxins13060399.
10. Cai, P.; Zheng, H.; She, J.; Feng, N.; Zou, H.; Gu, J.; Yuan, Y.; Liu, X.; Liu, Z.; Bian, J. Molecular mechanism of aflatoxin-induced hepatocellular carcinoma derived from a bioinformatics analysis. *Toxins (Basel)*, **2020**, 12(3), 203. https://doi.org/10.3390/toxins12030203.
11. Cao, W.; Yu, P.; Yang, K. P.; Cao, D. Aflatoxin B1: Metabolism, toxicology, and its involvement in oxidative stress and cancer development. *Toxicol Mech Methods*, **2022**, 32(6), 395–419. https://doi.org/10.1080/15376516.2021.2021339.
12. Liu, Y.; Mao, H.; Hu, C.; Tron, T.; Lin, J.; Wang, J.; Sun, B. Molecular docking studies and in vitro degradation of four aflatoxins (AFB1, AFB2, AFG1, and AFG2) by a recombinant laccase from *Saccharomyces cerevisiae*. *J Food Sci*, **2020**, 85(4), 1353–1360. https://doi.org/10.1111/1750-3841.15106.
13. Omotayo, O. P.; Omotayo, A. O.; Mwanza, M.; Babalola, O. O. Prevalence of mycotoxins and their consequences on human health. *Toxicol Res*, **2019**, 35(1), 1–7. https://doi.org/10.5487/TR.2019.35.1.001.
14. Hua, Z.; Liu, R.; Chen, Y.; Liu, G.; Li, C.; Song, Y.; Cao, Z.; Li, W.; Li, W.; Lu, C.; et al. Contamination of aflatoxins induces severe hepatotoxicity through multiple mechanisms. *Front Pharmacol*, **2021**, 605823 https://doi.org/10.3389/fphar.2020.605823.
15. Agriopoulou, S.; Stamatelopoulou, E.; Varzakas, T. Advances in occurrence, importance, and mycotoxin control strategies: Prevention and detoxification in foods. *Foods*, **2020**, 9(2), 137. https://doi.org/10.3390/foods9020137.
16. Manu, N.; Opit, G. P.; Osekre, E. A.; Arthur, F. H.; Mbata, G.; Armstrong, P.; Danso, J. K.; McNeill, S. G.; Campbell, J. F. Moisture content, insect pest infestation and mycotoxin levels of maize in markets in the Northern region of Ghana. *J Stored Prod Res*, **2019**, 80. https://doi.org/10.1016/j.jspr.2018.10.007.
17. Ayeni, K. I.; Atanda, O. O.; Krska, R.; Ezekiel, C. N. Present status and future perspectives of grain drying and storage practices as a means to reduce mycotoxin exposure in Nigeria. *Food Control*, **2021**, 126, 108074. https://doi.org/10.1016/j.foodcont.2021.108074.
18. Ching'anda, C.; Bandyopadhyay, R.; Callicott, K.; Orbach, M. J.; Cotty, P. J.; Mehl, H. L. Crop host influences on outcomes of competition between *Aspergillus* section *Flavi* species coinfecting maize and groundnuts. *PhytoFrontiers™*, **2023**, 3(2), 369–381. https://doi.org/10.1094/phytofr-07-22-0076-r.
19. Perrone, G.; Ferrara, M.; Medina, A.; Pascale, M.; Magan, N. Toxigenic fungi and mycotoxins in a climate change scenario: Ecology, genomics, distribution, prediction and prevention of the risk. *Microorganisms*, **2020**, 8(10), 1496. https://doi.org/10.3390/microorganisms8101496.
20. Cervini, C.; Verheecke-Vaessen, C.; He, T.; Mohammed, A; Magan, N.; Medina, A. Improvements within the peanut production chain to minimize aflatoxins contamination: An Ethiopian case study. *Food Control*, **2022**, 136, 108622. https://doi.org/10.1016/j.foodcont.2021.108622.
21. Liao, J.; He, Z.; Xia, Y.; Lei, Y.; Liao, B. A review on biosynthesis and genetic regulation of aflatoxin production by major *aspergillus* fungi. *Oil Crop Science*, **2020**, 5(4), 166–173. https://doi.org/10.1016/j.ocsci.2020.11.001.
22. Thakur, C.; Malhotra, A. S.; Shankar, J. Integrated bioinformatics analysis to study gallic acid-mediated inhibition of polyketide synthase a from aflatoxin biosynthesis pathway of *Aspergillus flavus*. *Chem Africa*, **2023**, 6(3), 1399–1413. https://doi.org/10.1007/s42250-023-00589-4.
23. Khan, R.; Ghazali, F. M.; Mahyudin, N. A.; Samsudin, N. I. P. Aflatoxin biosynthesis, genetic regulation, toxicity, and control strategies: A review. *Journal of Fungi*, **2021**, 7(8), 606. https://doi.org/10.3390/jof7080606.
24. Lv, C.; Wang, P.; Ma, L.; Zheng, M.; Liu, Y.; Xing, F. Large-scale comparative analysis of eugenol-Induced/Repressed genes expression in *Aspergillus flavus* using RNA-seq. *Front Microbiol*, **2018**, 9, 1116. https://doi.org/10.3389/fmicb.2018.01116.
25. Yunes, N. B. S.; Oliveira, R. C.; Reis, T. A.; Baquião, A. C.; Rocha, L.O.; Correa, B. Effect of temperature on growth, Gene expression, and aflatoxin production by *Aspergillus nomius* isolated from Brazil nuts. *Mycotoxin Res*, **2020**, 36(2), 173–180. https://doi.org/10.1007/s12550-019-00380-w.
26. Mannaa, M.; Kim, K. D. Influence of temperature and water activity on deleterious fungi and mycotoxin production during grain storage. *Mycobiology*, **2017**, 45(4), 240–254. https://doi.org/10.5941/MYCO.2017.45.4.240.

27. Pandey, M. K.; Kumar, R.; Pandey, A. K.; Soni, P.; Gangurde, S. S.; Sudini, H. K.; Fountain, J. C.; Liao, B.; Desmae, H.; Okori, P.; et al. Mitigating aflatoxin contamination in groundnut through a combination of genetic resistance and post-harvest management practices. *Toxins*, **2019**, *11*(6), 315. https://doi.org/10.3390/toxins11060315.
28. Asare Bediako, K.; Ofori, K.; Offei, S. K.; Dzidzienyo, D.; Asibuo, J. Y.; Adu Amoah, R. Aflatoxin contamination of groundnut (*Arachis hypogaea* L.): Predisposing factors and management interventions. *Food Control*, **2019**, 98, 61–67. https://doi.org/10.1016/j.foodcont.2018.11.020.
29. Nakuwa, M. S.; Mongi, R.; Ngoma, S. Food handling practices, the prevalence of aflatoxin dietary exposure and its associated factors among children aged 6–23 months in Bukombe District, Tanzania. *NFS Journal*, **2023**, 31, 162–170. https://doi.org/10.1016/j.nfs.2023.05.004.
30. Eshelli, M.; Harvey, L.; Edrada-Ebel, R.; McNeil, B. Metabolomics of the bio-degradation process of aflatoxin B1 by actinomycetes at an initial PH of 6.0. *Toxins (Basel)*, **2015**, 7(2), 439–456. https://doi.org/10.3390/toxins7020439.
31. Rushing, B. R.; Selim, M. I. Aflatoxin B1: A review on metabolism, toxicity, occurrence in food, occupational exposure, and detoxification methods. *Food Chem Toxicol*, **2019**, 124, 81–100. https://doi.org/10.1016/j.fct.2018.11.047.
32. Ding, N.; Xing, F.; Liu, X.; Selvaraj, J. N.; Wang, L.; Zhao, Y.; Wang, Y.; Guo, W.; Dai, X.; Liu, Y. Variation in fungal microbiome (Mycobiome) and aflatoxin in stored in-shell peanuts at four different areas of China. *Front Microbiol*, **2015**, 6, 1055. https://doi.org/10.3389/fmicb.2015.01055.
33. Siciliano, I.; Berta, F.; Bosio, P.; Gullino, M. L.; Garibaldi, A. Effect of different temperatures and CO2 levels on *Alternaria* toxins produced on cultivated rocket, cabbage and cauliflower. *World Mycotoxin J*, **2017**, 10(1), 63–71. https://doi.org/10.3920/WMJ2016.2108.
34. Kumar, A.; Pathak, H.; Bhadauria, S.; Sudan, J. Aflatoxin contamination in food crops: Causes, detection, and management: A review. *Food Prod Process Nutr*, **2021**, 3, 17. https://doi.org/10.1186/s43014-021-00064-y.
35. Liu, J.; Sun, L.; Zhang, N.; Zhang, J.; Guo, J.; Li, C.; Rajput, S. A.; Qi, D. Effects of nutrients in substrates of different grains on aflatoxin B1 production by *Aspergillus flavus*. *Biomed Res Int*, **2016**, 2016, 7232858. https://doi.org/10.1155/2016/7232858.
36. Godswill Awuchi, C.; Nyakundi Ondari, E.; Josiah Eseoghene, I.; Twinomuhwezi, H.; Otuosorochi Amagwula, I.; Morya, S. Fungal growth and mycotoxins production: Types, toxicities, control strategies, and detoxification. *Fungal Reprod Growth*, **2022**, 100207. https://doi.org/10.5772/intechopen.100207.
37. Ebrima, A. A. J.; Lamin, B. S.; Dumevi, R. M.; Jalil, Y. Assessment of aflatoxin awareness in the Gambia. *Afr J Agric Res*, **2022**, 18(4), 281–287. https://doi.org/10.5897/ajar2021.15915.
38. Mahuku, G.; Nzioki, H. S.; Mutegi, C.; Kanampiu, F.; Narrod, C.; Makumbi, D. Pre-harvest management is a critical practice for minimizing aflatoxin contamination of maize. *Food Control*, **2019**, 96, 219–226. https://doi.org/10.1016/j.foodcont.2018.08.032.
39. Kaale, L. D.; Kimanya, M. E.; Macha, I. J.; Mlalila, N. Aflatoxin contamination and recommendations to improve its control: A review. *World Mycotoxin J*, **2021**, 14(1), 27–40. https://doi.org/10.3920/WMJ2020.2599.
40. Singh, S. K.; Patel, M. B.; Thakker, B. N; Barad, A. J.; Gogoi, R.; Hooda, K. S. An overview of challenges and elimination of aflatoxin contamination in maize (*Zea mays*). *J Pharmacogn Phytochem*, **2019**, 8(4), 2895–2906.
41. Nada, S.; Nikola, T.; Bozidar, U.; Ilija, D.; Andreja, R. Prevention and practical strategies to control mycotoxins in the wheat and maize chain. *Food Control*, **2022**, 136, 108855. https://doi.org/10.1016/j.foodcont.2022.108855.
42. Winter, G.; Pereg, L. A review on the relation between soil and mycotoxins: Effect of aflatoxin on field, food and finance. *Eur J Soil Sci*, **2019**, 70(4), 882–897. https://doi.org/10.1111/ejss.12813.
43. Tahir, N. I.; Hussain, S.; Javed, M.; Rehman, H.; Shahzady, T. G.; Parveen, B.; Ali, K. G. Nature of aflatoxins: Their extraction, analysis, and control. *J Food Saf*, **2018**, 38(6), e12561. https://doi.org/10.1111/jfs.12561.
44. Gibellato, S. L.; Dalsóquio, L. F.; do Nascimento, I. C. A.; Alvarez, T. M. Current and promising strategies to prevent and reduce aflatoxin contamination in grains and food matrices. *World Mycotoxin J*, **2021**, 14(3), 293–304. https://doi.org/10.3920/WMJ2020.2559.
45. Mwakinyali, S. E.; Ding, X.; Ming, Z.; Tong, W.; Zhang, Q.; Li, P. Recent development of aflatoxin contamination biocontrol in agricultural products. *Biol Control*, **2019**, 128, 31–39. https://doi.org/10.1016/j.biocontrol.2018.09.012.
46. Kinyungu, S.; Isakeit, T.; Ojiambo, P. S.; Woloshuk, C. P. Spread of *Aspergillus Flavus* and aflatoxin accumulation in post harvested maize treated with biocontrol products. *J Stored Prod Res*, **2019**, 84, 101519. https://doi.org/10.1016/j.jspr.2019.101519.

47. Ren, X.; Zhang, Q.; Zhang, W.; Mao, J.; Li, P. Control of aflatoxigenic molds by antagonistic microorganisms: Inhibitory behaviours, bioactive compounds, related mechanisms, and influencing factors. *Toxins*, **2020**, 12(1), 24. https://doi.org/10.3390/toxins12010024.
48. Boukaew, S.; Petlamul, W.; Phitthayaphinant, P.; Prasertsan, P. Potential use of *Streptomyces mycarofaciens* SS-2-243 as a biofumigant to protect maize seeds against two aflatoxin producing fungi. *Eur J Plant Pathol*, **2019**, 155(2), 489–503. https://doi.org/10.1007/s10658-019-01782-3.
49. Ahn, J. Y.; Kim, J.; Cheong, D. H.; Hong, H.; Jeong, J. Y.; Kim, B. G. An in vitro study on the efficacy of mycotoxin sequestering agents for aflatoxin B1, deoxynivalenol, and zearalenone. *Animals*, **2022**, 12(3), 333. https://doi.org/10.3390/ani12030333.
50. Lagogianni, C. S.; Tsitsigiannis, D. I. Effective chemical management for prevention of aflatoxins in maize. *Phytopathol Mediterr*, **2018**, 57(1), 186–198. https://doi.org/10.14601/Phytopathol_Mediterr-22492.
51. Ojiambo, P. S.; Battilani, P.; Cary, J. W.; Blum, B. H.; Carbone, I. Cultural and genetic approaches to manage aflatoxin contamination: Recent insights provide opportunities for improved control. *Phytopathology*, **2018**, 8(9), 1024–1037. https://doi.org/10.1094/PHYTO-04-18-0134-RVW.
52. Mesterházy, A.; Oláh, J.; Popp, J. Losses in the grain supply chain: Causes and solutions. *Sustainability (Switzerland)*, **2020**, 12(6), 2342. https://doi.org/10.3390/su12062342.
53. Hashim, M.; Arif, H.; Tabassum, B.; Rehman, S.; Bajaj, P.; Sirohi, R.; Khan, M. F. A. An overview of the ameliorative efficacy of *Catharanthus roseus* extract against Cd2+ toxicity: Implications for human health and remediation strategies. *Front Public Health*, **2024**, 12, 1327611. https://doi.org/10.3389/fpubh.2024.1327611.
54. Riaz, M.; Khalid, R.; Afzal, M.; Anjum, F.; Fatima, H.; Zia, S.; Rasool, G.; Egbuna, C.; Mtewa, A. G.; Uche, C. Z.; et al. Phytobioactive compounds as therapeutic agents for human diseases: A review. *Food Sci Nutr*, **2023**, *11*(6), 2500–2529. https://doi.org/10.1002/fsn3.3308.
55. Pinto, T.; Vilela, A. Healthy drinks with lovely colours: Phenolic compounds as constituents of functional beverages. *Beverages*, **2021**, 7(1), 12. https://doi.org/10.3390/beverages7010012.
56. Sarankar, S. K.; Rajak, B.; Ojha, S.; Somkuwar, S. Exploring the utilization of phenolic compounds in pharmaceuticals and healthcare. *GSC Biol Pharm Sci*, **2023**, 24(3), 95–102. https://doi.org/10.30574/gscbps.2023.24.3.0372.
57. Carter, N. S.; Stamper, B. D.; Elbarbry, F.; Nguyen, V.; Lopez, S.; Kawasaki, Y.; Poormohamadian, R.; Roberts, S. C. Natural products that target the arginase in leishmania parasites hold therapeutic promise. *Microorganisms*, **2021**, 9(2), 267. https://doi.org/10.3390/microorganisms9020267.
58. Atala, E.; Fuentes, J.; Wehrhahn, M. J.; Speisky, H. Quercetin and related flavonoids conserve their antioxidant properties despite undergoing chemical or enzymatic oxidation. *Food Chem*, **2017**, 234, 479–485. https://doi.org/10.1016/j.foodchem.2017.05.023.
59. Jiang, Y. X.; Dai, Y. Y.; Pan, Y. F.; Wu, X. M.; Yang, Y.; Bian, K.; Zhang, D. D. Total flavonoids from *Radix glycyrrhiza* exert anti-inflammatory and antitumorigenic effects by inactivating INOS signaling pathways. *Evid Based Complementary Altern Med*, **2018**, 2018, 6714282. https://doi.org/10.1155/2018/6714282.
60. Yun, D.; Yoon, S. Y.; Park, S. J.; Park, Y. J. The anticancer effect of natural plant alkaloid isoquinolines. *Int J Mol Sci*, **2021**, 22(4), 1653. https://doi.org/10.3390/ijms22041653.
61. Bhambhani, S.; Kondhare, K. R.; Giri, A. P. Diversity in chemical structures and biological properties of plant alkaloids. *Molecules*, **2021**, 26(11), 3374. https://doi.org/10.3390/molecules26113374.
62. Watanabe, K.; Yano, S.; Horie, S.; Yamamoto, L. T.; Takayama, H.; Aimi, N.; Sarai, S.-I.; Ponglux, D.; Tongroach, P.; Shan, J.; et al. Pharmacological properties of some structurally related indole alkaloids contained in the Asian herbal medicines, hirsutine and mitragynine, with special reference to their $CA^{2+}$ antagonistic and opioid-like effects. In *Pharmacological Research on Traditional Herbal Medicines*; CRC Press, **2021**, 163–177. https://doi.org/10.1201/9781003072867-11.
63. Sharma, A.; Tiwari, P.; Arora, R.; Sankaranarayanan, A. Madagascar Periwinkle alkaloids: Biosynthesis, ethnobotanical attributes, and pharmacological functions. *S Afr J Bot*, **2022**, 151, 108–115. https://doi.org/10.1016/j.sajb.2022.09.039.
64. Nisa, S.; Shoukat, M.; Bibi, Y.; Al Ayoubi, S.; Shah, W.; Masood, S.; Sabir, M.; Asma Bano, S.; Qayyum, A. Therapeutic prospects of endophytic *Bacillus* Species from *Berberis Lycium* against oxidative stress and microbial pathogens: Therapeutic prospects of endophytic *Bacillus* Species. *Saudi J Biol Sci*, **2022**, 29(1), 287–295. https://doi.org/10.1016/j.sjbs.2021.08.099.
65. Nosek, K.; Leppert, W.; Puchała, Ł.; Łoń, K. Efficacy and safety of topical morphine: A narrative review. *Pharmaceutics*, **2022**, 14(7), 1499. https://doi.org/10.3390/pharmaceutics14071499.
66. Bose, S.; Malik, J.; Mandal, S. C. Application of phytochemicals in pharmaceuticals. In *Advances in Pharmaceutical Biotechnology: Recent Progress and Future Applications*; Springer Singapore: Singapore, **2020**, 55–68. https://doi.org/10.1007/978-981-15-2195-9_5.

67. Paulino, B. N.; Silva, G. N. S.; Araújo, F. F.; Néri-Numa, I. A.; Pastore, G. M.; Bicas, J. L.; Molina, G. Beyond natural aromas: The bioactive and technological potential of monoterpenes. *Trends Food Sci Technol*, **2022**, 128, 188–201. https://doi.org/10.1016/j.tifs.2022.08.006.
68. Mohammadi, S.; Jafari, B.; Asgharian, P.; Martorell, M.; Sharifi-Rad, J. Medicinal plants used in the treatment of malaria: A key emphasis to *Artemisia*, *Cinchona*, *Cryptolepis*, and *Tabebuia* Genera. *Phytother Res*, **2020**, 34(7), 1556–1569. https://doi.org/10.1002/ptr.6628.
69. Ahmad, M. F.; Ahmad, F. A.; Khan, M. I.; Alsayegh, A. A.; Wahab, S.; Alam, M. I.; Ahmed, F. Ganoderma lucidum: A potential source to surmount viral infections through β-glucans immunomodulatory and triterpenoids antiviral properties. *Int J Biol Macromol*, **2021**, 187, 769–779. https://doi.org/10.1016/j.ijbiomac.2021.06.122.
70. Chen, X.; Famurewa, A. C.; Tang, J.; Olatunde, O. O.; Olatunji, O. J. Hyperoside attenuates neuroinflammation, cognitive impairment and oxidative stress via suppressing TNF-α/NF-KB/Caspase-3 signaling in type 2 diabetes rats. *Nutr Neurosci*, **2022**, 25(8), 1774–1784. https://doi.org/10.1080/1028415X.2021.1901047.
71. Hyun, S. H.; Kim, S. W.; Seo, H. W.; Youn, S. H.; Kyung, J. S.; Lee, Y. Y.; In, G.; Park, C. K.; Han, C. K. Physiological and pharmacological features of the non-saponin components in Korean Red Ginseng. *J Ginseng Res*, **2020**, 44(4), 527–537. https://doi.org/10.1016/j.jgr.2020.01.005.
72. Botelho, A. F. M.; Pierezan, F.; Soto-Blanco, B.; Melo, M. M. A review of cardiac glycosides: Structure, toxicokinetics, clinical signs, diagnosis and antineoplastic potential. *Toxicon*, **2019**, 158, 63–68. https://doi.org/10.1016/j.toxicon.2018.11.429.
73. Wang, Z.; Xu, J.; Wang, Y.; Xiang, L.; He, X. Total saponins from *Tupistra chinensis* baker inhibits growth of human gastric cancer cells in vitro and in vivo. *J Ethnopharmacol*, **2021**, 278, 114323. https://doi.org/10.1016/j.jep.2021.114323.
74. Kabir Mumu, S.; Mahboob Hossain, M. Antimicrobial activity of tea tree oil against pathogenic bacteria and comparison of its effectiveness with eucalyptus oil, lemongrass oil and conventional antibiotics. *Am J Microbiol Res*, **2018**, 6(3), 73–78. https://doi.org/10.12691/ajmr-6-3-2.
75. Lin, P. C.; Lee, P. H.; Tseng, S. J.; Lin, Y. M.; Chen, S. R.; Hou, W. H. Effects of aromatherapy on sleep quality: A systematic review and meta-analysis. *Complement Ther Med*, **2019**, 45, 156–166. https://doi.org/10.1016/j.ctim.2019.06.006.
76. Agarwal, P.; Sebghatollahi, Z.; Kamal, M.; Dhyani, A.; Shrivastava, A.; Singh, K. K.; Sinha, M.; Mahato, N.; Mishra, A. K.; Baek, K. H. Citrus essential oils in aromatherapy: Therapeutic effects and mechanisms. *Antioxidants*, **2022**, 11(12), 2374. https://doi.org/10.3390/antiox11122374.
77. Ahmad, A.; Elisha, I. L.; van Vuuren, S.; Viljoen, A. Volatile phenolics: A comprehensive review of the anti-infective properties of an important class of essential oil constituents. *Phytochemistry*, **2021**, 190, 112864. https://doi.org/10.1016/j.phytochem.2021.112864.
78. Kumar, M.; Tomar, M.; Saurabh, V.; Mahajan, T.; Punia, S.; del Mar Contreras, M.; Rudra, S. G.; Kaur, C.; Kennedy, J. F. Emerging trends in pectin extraction and its anti-microbial functionalization using natural bioactives for application in food packaging. *Trends Food Sci Technol*, **2020**, 105, 223–237. https://doi.org/10.1016/j.tifs.2020.09.009.
79. Zhang, J.; Ma, S.; Du, S.; Chen, S.; Sun, H. Antifungal activity of thymol and carvacrol against postharvest pathogens *botrytis cinerea*. *J Food Sci Technol*, **2019**, 56(5), 2611–2620. https://doi.org/10.1007/s13197-019-03747-0.
80. Konuk, H. B.; Ergüden, B. Phenolic –OH group is crucial for the antifungal activity of terpenoids via disruption of cell membrane integrity. *Folia Microbiol (Praha)*, **2020**, 65(4), 775–783. https://doi.org/10.1007/s12223-020-00787-4.
81. Jamzivar, F.; Shams-Ghahfarokhi, M.; Khoramizadeh, M.; Yousefi, N.; Gholami-Shabani, M.; Razzaghi-Abyaneh, M. Unraveling the importance of molecules of natural origin in antifungal drug development through targeting ergosterol biosynthesis pathway. *Iran J Microbiol*, **2019**, 11(6), 448. https://doi.org/10.18502/ijm.v11i6.2216.
82. Jadimurthy, R.; Jagadish, S.; Nayak, S. C.; Kumar, S.; Mohan, C. D.; Rangappa, K. S. Phytochemicals as invaluable sources of potent antimicrobial agents to combat antibiotic resistance. *Life*, **2023**, 13(4), 948. https://doi.org/10.3390/life13040948.
83. Duan, W. Y.; Qin, Y. L.; Zhang, S. B.; Zhai, H. C.; Lv, Y. Y.; Wei, S.; Ma, P. A.; Hu, Y. S. Inhibitory mechanisms of plant volatile 1-octanol on the germination of *Aspergillus flavus* spores. *Food Biophys*, **2024**, 19(1), 96–108. https://doi.org/10.1007/s11483-023-09807-5.
84. Huang, W.; Wang, Y.; Tian, W.; Cui, X.; Tu, P.; Li, J.; Shi, S.; Liu, X. Biosynthesis investigations of terpenoid, alkaloid, and flavonoid antimicrobial agents derived from medicinal plants. *Antibiotics*, **2022**, 11(10), 1380. https://doi.org/10.3390/antibiotics11101380.

85. Jan, R.; Khan, M.; Asaf, S.; Lubna; Asif, S.; Kim, K. M. Bioactivity and therapeutic potential of kaempferol and quercetin: New insights for plant and human health. *Plants*, **2022**, 11(19), 2623. https://doi.org/10.3390/plants11192623.
86. Naz, R.; Roberts, T. H.; Bano, A.; Nosheen, A.; Yasmin, H.; Hassan, M. N.; Keyani, R.; Ullah, S.; Khan, W.; Anwar, Z. GC-MS analysis, antimicrobial, antioxidant, antilipoxygenase and cytotoxic activities of *Jacaranda mimosifolia* methanol leaf extracts and fractions. *PLoS One*, **2020**, 15(7), e0236319. https://doi.org/10.1371/journal.pone.0236319.
87. Elgharbawy, A. A. M.; Samsudin, N.; Benbelgacem, F. F.; Hashim, Y. Z. H. Y.; Salleh, H. M.; Santhanam, J. Phytochemicals with antifungal properties: Cure from nature. *Malays J Microbiol*, **2020**, 16(4),323–345. https://doi.org/10.21161/mjm.190551.
88. Gupta, P.; Gupta, H.; Tripathi, S.; Poluri, K. M. Biochemical and metabolomic insights into antifungal mechanism of berberine against *Candida glabrata*. *Appl Microbiol Biotechnol*, **2023**, 107(19), 6085–6102. https://doi.org/10.1007/s00253-023-12714-x.
89. Aboody, M. S. Al; Mickymaray, S. Anti-fungal efficacy and mechanisms of flavonoids. *Antibiotics*, **2020**, 9(2), 45. https://doi.org/10.3390/antibiotics9020045.
90. Sun, Z. B.; Yu, S. F.; Wang, C. L.; Wang, L. CAMP Signalling pathway in biocontrol fungi. *Curr Issues Mol Biol*, **2022**, 44(6), 2622–2634. https://doi.org/10.3390/cimb44060179.
91. Luo, X.; Wu, W.; Feng, L.; Treves, H.; Ren, M. Short peptides make a big difference: The role of botany-derived amps in disease control and protection of human health. *Int J Mol Sci*, **2021**, 22(21), 11363. https://doi.org/10.3390/ijms222111363.
92. Akhila, P. P.; Sunooj, K. V.; Navaf, M.; Aaliya, B.; Sudheesh, C.; Sasidharan, A.; Sabu, S.; Mir, S. A.; George, J.; Mousavi Khaneghah, A. Application of innovative packaging technologies to manage fungi and mycotoxin contamination in agricultural products: Current status, challenges, and perspectives. *Toxicon*, **2022**, 214, 18–29. https://doi.org/10.1016/j.toxicon.2022.04.017.
93. Bhatia, S.; Singh, S. Inhibition of Siderophores in Blocking Fungal Infection; In Fungal Siderophores: Fungal Biology. Springer, Cham; **2021**, 13–31. https://doi.org/10.1007/978-3-030-53077-8_2.
94. Preety, A. M. Purification and properties of plant chitinases: A review. *J Food Biochem*, **2019**, 43(3), e12762. https://doi.org/10.1111/jfbc.12762.
95. Vaghela, B.; Vashi, R.; Rajput, K.; Joshi, R. Plant chitinases and their role in plant defense: A comprehensive review. *Enzyme Microb Technol*, **2022**, 159, 110055. https://doi.org/10.1016/j.enzmictec.2022.110055.
96. Memar, M. Y.; Raei, P.; Alizadeh, N.; Aghdam, M. A.; Kafil, H. S. Carvacrol and thymol: Strong antimicrobial agents against resistant isolates. *Rev Res Med Microbio*, **2017**, 28(2), 63–68.
97. Oliveira, R. C.; Carvajal-Moreno, M.; Mercado-Ruaro, P.; Rojo-Callejas, F.; Correa, B. Essential oils trigger an antifungal and anti-aflatoxigenic effect on *Aspergillus Flavus* via the induction of apoptosis-like cell death and gene regulation. *Food Control*, **2020**, 110, 107038. https://doi.org/10.1016/j.foodcont.2019.107038.
98. Faridi, U. Plant secondary metabolites as apoptosis-inducing agents. *Lett Drug Des Discov*, **2022**, 20(4), 375–385. https://doi.org/10.2174/1570180819666220523092459.
99. Elhamouly, N. A.; Hewedy, O. A.; Zaitoon, A.; Miraples, A.; Elshorbagy, O. T.; Hussien, S.; El-Tahan, A.; Peng, D. The hidden power of secondary metabolites in plant-fungi interactions and sustainable phytoremediation. *Front Plant Sci*, **2022**, 13, 1044896. https://doi.org/10.3389/fpls.2022.1044896.
100. Anjali; Kumar, S.; Korra, T.; Thakur, R.; Arutselvan, R.; Kashyap, A. S.; Nehela, Y.; Chaplygin, V.; Minkina, T.; Keswani, C. Role of plant secondary metabolites in defence and transcriptional regulation in response to biotic stress. *Plant Stress*, **2023**, 100154. https://doi.org/10.1016/j.stress.2023.100154.
101. Jan, R.; Asaf, S.; Numan, M.; Lubna; Kim, K. M. Plant secondary metabolite biosynthesis and transcriptional regulation in response to biotic and abiotic stress conditions. *Agronomy*, **2021**, 11(5), 968. https://doi.org/10.3390/agronomy11050968.
102. Kumudini, B. S.; Jayamohan, N. S.; Patil, S. V.; Govardhana, M. Primary plant metabolism during plant–Pathogen interactions and its role in defense. In *Plant Metabolites and Regulation Under Environmental Stress*; Academic Press: Cambridge, MA, **2018**, 215–229. https://doi.org/10.1016/B978-0-12-812689-9.00011-X.
103. Abbey, J.; Jose, S.; Percival, D.; Jaakola, L.; Asiedu, S. K. Modulation of defense genes and phenolic compounds in wild blueberry in response to *Botrytis cinerea* under field conditions. *BMC Plant Biol*, **2023**, 23(1), 117. https://doi.org/10.1186/s12870-023-04090-5.
104. Thawabteh, A.; Juma, S.; Bader, M.; Karaman, D.; Scrano, L.; Bufo, S. A.; Karaman, R. The biological activity of natural alkaloids against herbivores, cancerous cells and pathogens. *Toxins*, **2019**, 11(11), 656. https://doi.org/10.3390/toxins11110656.

105. Redondo-Blanco, S.; Fernández, J.; López-Ibáñez, S.; Miguélez, E. M.; Villar, C. J.; Lombó, F. Plant phytochemicals in food preservation: Antifungal bioactivity: A review. *J Food Prot*, **2020**, 83(1), 163–171. https://doi.org/10.4315/0362-028X.JFP-19-163.
106. Wang, Y.; Mostafa, S.; Zeng, W.; Jin, B. Function and mechanism of Jasmonic acid in plant responses to abiotic and biotic stresses. *Int J Mol Sci*, **2021**, 22(16), 8568. https://doi.org/10.3390/ijms22168568.
107. Elshafie, H. S.; Camele, I.; Mohamed, A. A. A comprehensive review on the biological, agricultural and pharmaceutical properties of secondary metabolites based-plant origin. *Int J Mol Sci*, **2023**, 24(4), 3266. https://doi.org/10.3390/ijms24043266.
108. Vaou, N.; Stavropoulou, E.; Voidarou, C.; Tsigalou, C.; Bezirtzoglou, E. Towards advances in medicinal plant antimicrobial activity: A review study on challenges and future perspectives. *Microorganisms*, **2021**, 9(10), 2041. https://doi.org/10.3390/microorganisms9102041.
109. Ahmad, M. M.; Qamar, F.; Saifi, M.; Abdin, M. Z. Natural inhibitors: A sustainable way to combat aflatoxins. *Front Microbiol*, **2022**, 13, 993834. https://doi.org/10.3389/fmicb.2022.993834.
110. Dai, C.; Sharma, G.; Liu, G.; Shen, J.; Shao, B.; Hao, Z. Therapeutic detoxification of quercetin for aflatoxin B1-related toxicity: Roles of oxidative stress, inflammation, and metabolic enzymes. *Environ Pollut*, **2024**, 345, 123474. https://doi.org/10.1016/j.envpol.2024.123474.
111. Farouk, A.; Alsulami, T.; Ali, H. S.; Badr, A. N. In-Vitro and in-silico investigation for the spent-coffee bioactive phenolics as a promising aflatoxins production inhibitor. *Toxins (Basel)*, **2023**, 15(3), 225. https://doi.org/10.3390/toxins15030225.
112. Nobili, C.; De Acutis, A.; Reverberi, M.; Bello, C.; Leone, G. P.; Palumbo, D.; Natella, F.; Procacci, S.; Zjalic, S.; Brunori, A. Buckwheat Hull extracts inhibit *Aspergillus flavus* growth and AFB1 biosynthesis. *Front Microbiol*, **2019**, 10, 1997. https://doi.org/10.3389/fmicb.2019.01997.
113. Al Khoury, A.; Sleiman, R.; Atoui, A.; Hindieh, P.; Maroun, R. G.; Bailly, J.-D.; El Khoury, A. Antifungal and anti-aflatoxigenic properties of organs of *Cannabis sativa* L.: Relation to phenolic content and antioxidant capacities. *Arch Microbiol*, **2021**, 203(7), 4485–4492. https://doi.org/10.1007/s00203-021-02444-x.
114. Tian, F.; Woo, S. Y.; Lee, S. Y.; Park, S. B.; Zheng, Y.; Chun, H. S. Antifungal activity of essential oil and plant-derived natural compounds against *Aspergillus flavus*. *Antibiotics*, **2022**, 11(12), 1727. https://doi.org/10.3390/antibiotics11121727.
115. Tauchen, J.; Huml, L.; Rimpelova, S.; Jurášek, M. Flavonoids and related members of the aromatic polyketide group in human health and disease: Do they really work? *Molecules*, **2020**, 25(17), 3846. https://doi.org/10.3390/molecules25173846.
116. Mohajeri, M.; Behnam, B.; Cicero, A. F. G.; Sahebkar, A. Protective effects of curcumin against aflatoxicosis: A comprehensive review. *J Cell Physiol*, **2018**, 233(4), 3552–3577. https://doi.org/10.1002/jcp.26212.
117. Badr, A. N.; El-Attar, M. M.; Ali, H. S.; Elkhadragy, M. F.; Yehia, H. M.; Farouk, A. Spent coffee grounds valorization as bioactive phenolic source acquired antifungal, anti-mycotoxigenic, and anti-cytotoxic activities. *Toxins (Basel)*, **2022**, 14(2), 109. https://doi.org/10.3390/toxins14020109.
118. Naseer, R.; Sultana, B.; Khan, M. Z.; Naseer, D.; Nigam, P. Utilization of waste fruit-peels to inhibit aflatoxins synthesis by *Aspergillus flavus*: A biotreatment of rice for safer storage. *Bioresour Technol*, **2014**, 172, 423–428. https://doi.org/10.1016/j.biortech.2014.09.017.
119. Tiwari, S.; Shankar, J. Integrated proteome and HPLC analysis revealed quercetin-mediated inhibition of aflatoxin B1 biosynthesis in *Aspergillus flavus*. *3 Biotech*, **2018**, 8(1), 47. https://doi.org/10.1007/s13205-017-1067-0.
120. Pok, P. S.; García Londoño, V. A.; Vicente, S.; Romero, S. M.; Pacín, A.; Tolaba, M.; Alzamora, S. M.; Resnik, S. L. Evaluation of citrus flavonoids against *Aspergillus parasiticus* in maize: Aflatoxins reduction and ultrastructure alterations. *Food Chem*, **2020**, 318, 126414. https://doi.org/10.1016/j.foodchem.2020.126414.
121. Zhang, L.; Guan, Q.; Jiang, J.; Khan, M. S. Tannin complexation with metal ions and its implication on human health, environment and industry: An overview. *Int J Biol Macromol*, **2023**, 253, 127485. https://doi.org/10.1016/j.ijbiomac.2023.127485.
122. Rajput, S.; Zhang, C.; Feng, Y.; Wei, X.; Khalil, M.; Rajput, I.; Baloch, D.; Shaukat, A.; Rajput, N.; Qamar, H.; et al. Proanthocyanidins alleviates AflatoxinB1-induced oxidative stress and apoptosis through mitochondrial pathway in the Bursa of Fabricius of Broilers. *Toxins (Basel)*, **2019**, *11* (3), 157. https://doi.org/10.3390/toxins11030157.
123. Grintzalis, K.; Vernardis, S. I.; Klapa, M. I.; Georgiou, C. D. Role of oxidative stress in sclerotial differentiation and aflatoxin B1 biosynthesis in *Aspergillus flavus*. *Appl Environ Microbiol*, **2014**, 80(18), 5561–5571. https://doi.org/10.1128/AEM.01282-14.
124. Hernandez, C.; Cadenillas, L.; Maghubi, A. El; Caceres, I.; Durrieu, V.; Mathieu, C.; Bailly, J. D. *Mimosa tenuiflora* aqueous extract: Role of condensed tannins in anti-aflatoxin B1 activity in *Aspergillus flavus*. *Toxins (Basel)*, **2021**, 13(6), 391. https://doi.org/10.3390/toxins13060391.

125. Lasram, S.; Zemni, H.; Hamdi, Z.; Chenenaoui, S.; Houissa, H.; Saidani Tounsi, M.; Ghorbel, A. Antifungal and antiaflatoxinogenic activities of *Carum carvi* L., *Coriandrum sativum* L. seed essential oils and their major terpene component against *Aspergillus flavus*. *Ind Crops Prod*, **2019**, 134, 11–18. https://doi.org/10.1016/j.indcrop.2019.03.037.
126. Madarshahi, F. S.; Taghizadeh, S. F.; Rezaee, R.; Azizi, M.; Asili, J.; Karimi, G. Fungicidal and anti-aflatoxin activity of four meroterpenoids isolated from *Arnebia euchroma* root. *Toxicol Environ Chem*, **2022**, 104(2), 263–279. https://doi.org/10.1080/02772248.2022.2104279.
127. Thipe, V. C.; Bloebaum, P.; Khoobchandani, M.; Karikachery, A. R.; Katti, K. K.; Katti, K. V. Green nanotechnology: Nanoformulations against toxigenic fungi to limit mycotoxin production. In *Nanomycotoxicology: Treating Mycotoxins in the Nano Way*; Academic Press: London, UK / San Diego, CA, USA, **2019**,155–188. https://doi.org/10.1016/B978-0-12-817998-7.00007-0.
128. Ghavipanje, N.; Fathi Nasri, M. H.; Vargas-Bello-Pérez, E. An insight into the potential of berberine in animal nutrition: Current knowledge and future perspectives. *J Anim Physiol Anim Nutr*, **2023**, 107(3), 808–829. https://doi.org/10.1111/jpn.13769.
129. Moradi, M. S.; Kamkar, S.; Sharifzadeh, A.; Hassan, J.; Shokri, H. The in vitro effect of berberine sulfate and berberine chloride on the growth and aflatoxin production by *aspergillus flavus* and *Aspergillus parasiticus*. *Iran J Vet Med*, **2024**, 18(02), 223–232. https://doi.org/10.32598/IJVM.18.2.1005399.
130. Safari, N.; Ardakani, M. M.; Hemmati, R.; Parroni, A.; Beccaccioli, M.; Reverberi, M. The potential of plant-based bioactive compounds on inhibition of aflatoxin B1 biosynthesis and down-regulation of AflR, AflM and AflP genes. *Antibiotics*, **2020**, 9(11), 728. https://doi.org/10.3390/antibiotics9110728.
131. Malekinezhad, P.; Ellestad, L. E.; Afzali, N.; Farhangfar, S. H.; Omidi, A.; Mohammadi, A. Evaluation of berberine efficacy in reducing the effects of aflatoxin B1 and ochratoxin a added to male broiler rations. *Poult Sci*, **2021**, 100(2), 797–809. https://doi.org/10.1016/j.psj.2020.10.040.
132. Caceres, I.; El Khoury, R.; Bailly, S.; Oswald, I. P.; Puel, O.; Bailly, J. D. Piperine inhibits aflatoxin B1 production in *Aspergillus flavus* by modulating fungal oxidative stress response. *Fungal Genet Biol*, **2017**, 107, 77–85. https://doi.org/10.1016/j.fgb.2017.08.005.
133. Khan, F. A.; Khan, N. M.; Ahmad, S.; Nasruddin; Aziz, R.; Ullah, I.; Almehmadi, M.; Allahyani, M.; Alsaiari, A. A.; Aljuaid, A. Phytochemical profiling, antioxidant, antimicrobial and cholinesterase inhibitory effects of essential oils isolated from the leaves of *Artemisia scoparia* and *Artemisia absinthium*. *Pharmaceuticals*, **2022**, 15(10), 1221. https://doi.org/10.3390/ph15101221.
134. Semeniuc, C. A.; Pop, C. R.; Rotar, A. M. Antibacterial activity and interactions of plant essential oil combinations against gram-positive and gram-negative bacteria. *J Food Drug Anal*, **2017**, 25(2), 403–408. https://doi.org/10.1016/j.jfda.2016.06.002.
135. Songsamoe, S.; Matan, N.; Matan, N. Antifungal activity of Michelia alba oil in the vapor phase and the synergistic effect of major essential oil components against *Aspergillus flavus* on brown rice. *Food Control*, **2017**, 77, 150–157. https://doi.org/10.1016/j.foodcont.2017.02.010.
136. Hossain, F.; Follett, P.; Dang Vu, K.; Harich, M.; Salmieri, S.; Lacroix, M. Evidence for synergistic activity of plant-derived essential oils against fungal pathogens of food. *Food Microbiol*, **2016**, 53, 24–30. https://doi.org/10.1016/j.fm.2015.08.006.
137. Ju, J.; Xie, Y.; Yu, H.; Guo, Y.; Cheng, Y.; Qian, H.; Yao, W. Synergistic interactions of plant essential oils with antimicrobial agents: A new antimicrobial therapy. *Crit Rev Food Sci Nutr*, **2022**, 62(7), 1740–1751. https://doi.org/10.1080/10408398.2020.1846494.
138. Li, Q.; Zhao, Y.; Zhu, X.; Xie, Y. Antifungal efficacy of paeonol on *Aspergillus flavus* and its mode of action on cell walls and cell membranes. *LWT*, **2021**, 149, 111985. https://doi.org/10.1016/j.lwt.2021.111985.
139. Belewa, V.; Baijnath, H.; Frost, C.; Somai, B. M. Tulbaghia violacea Harv. Plant extract affects cell wall synthesis in *Aspergillus flavus*. *J Appl Microbiol*, **2017**, 122(4), 921–931. https://doi.org/10.1111/jam.13405.
140. Lorán, S.; Carramiñana, J. J.; Juan, T.; Ariño, A.; Herrera, M. Inhibition of *Aspergillus parasiticus* growth and aflatoxins production by natural essential oils and phenolic acids. *Toxins (Basel)*, **2022**, 14(6), 384. https://doi.org/10.3390/toxins14060384.
141. Gómez, J. V.; Tarazona, A.; Mateo-Castro, R.; Gimeno-Adelantado, J. V.; Jiménez, M.; Mateo, E. M. Selected plant essential oils and their main active components, a promising approach to inhibit aflatoxigenic fungi and aflatoxin production in food. *Food Addit Contam Part A Chem Anal Control Expo Risk Assess*, **2018**, 35(8), 1581–1595. https://doi.org/10.1080/19440049.2017.1419287.
142. Zhang, L. L.; Chen, Y.; Li, Z. J.; Li, X.; Fan, G. Bioactive properties of the aromatic molecules of spearmint (*Mentha spicata* L.) essential oil: A review. *Food Function*, **2022**, 13(6), 3110–3132. https://doi.org/10.1039/d1fo04080d.

143. Fan, T.; Xie, Y.; Ma, W. Research progress on the protection and detoxification of phytochemicals against aflatoxin B1-induced liver toxicity. *Toxicon*, **2021**, 195, 58–68. https://doi.org/10.1016/j.toxicon.2021.03.007.
144. Qu, J.; Huang, P.; Zhang, L.; Qiu, Y.; Qi, H.; Leng, A.; Shang, D. Hepatoprotective effect of plant polysaccharides from natural resources: A review of the mechanisms and structure-activity relationship. *Int J Biol Macromol*, **2020**, 161, 24–34. https://doi.org/10.1016/j.ijbiomac.2020.05.196.
145. Ahmed, N.; El-Rayes, S. M.; Khalil, W. F.; Abdeen, A.; Abdelkader, A.; Youssef, M.; Maher, Z. M.; Ibrahim, A. N.; Abdelrahman, S. M.; Ibrahim, S. F.; et al. Arabic gum could alleviate the aflatoxin B1-provoked hepatic injury in rat: The involvement of oxidative stress, inflammatory, and apoptotic pathways. *Toxins (Basel)*, **2022**, 14(9), 605. https://doi.org/10.3390/toxins14090605.
146. Kamal, E.; Kaddam, L. A.; Dahawi, M.; Osman, M.; Salih, M. A.; Alagib, A.; Saeed, A. Gum Arabic fibers decreased inflammatory markers and disease severity score among rheumatoid arthritis patients, phase II trial. *Int J Rheumatol*, **2018**, 2018. https://doi.org/10.1155/2018/4197537.
147. Hassanien, M. A. The protective and antioxidant effects of gum arabic: A review of recent evidence using the new PubMed system. *Int J Community Med Public Health*, **2019**, 7(1). https://doi.org/10.18203/2394-6040.ijcmph20195592.
148. Mirghani, M. E. S.; Elnour, A. A. M.; Kabbashi, N. A.; Alam, M. Z.; Musa, K. H.; Abdullah, A. Determination of antioxidant activity of gum Arabic: An exudation from two different locations. *Science Asia*, **2018**, 44(3), 179–186. https://doi.org/10.2306/scienceasia1513-1874.2018.44.179.
149. Avelino, A. L. N.; Silva, N. V. R. e.; Oliveira, G. B. De; Silva, A. A. D. S.; Cavalcanti, B. C.; Jamacaru, F. V. F.; Dornelas. Antioxidant and antigenotoxic actions of gum arabic on the intestinal mucosa, liver and bone marrow of Swiss mice submitted to colorectal carcinogenesis. *Nutr Cancer*, **2022**, 74(3), 956–964. https://doi.org/10.1080/01635581.2021.1931699.
150. Dewi, M. K.; Chaerunisaa, A. Y.; Muhaimin, M.; Joni, I. M. Improved activity of herbal medicines through nanotechnology. *Nanomaterials*, **2022**, 12(22), 4073. https://doi.org/10.3390/nano12224073.
151. Enaru, B.; Drețcanu, G.; Pop, T. D.; Stănilă, A.; Diaconeasa, Z. Anthocyanins: Factors affecting their stability and degradation. *Antioxidants*, **2021**, 10(12), 1967. https://doi.org/10.3390/antiox10121967.
152. Sil, A.; Pramanik, K.; Samantaray, P.; Mondal, M. F.; Yadav, V. Essential oils: A Boon towards eco-Friendly management of phytopathogenic fungi. *J Entomol Zool Stud*, **2020**, 8(4), 1884–1891.
153. Ortega-Beltran, A.; Bandyopadhyay, R. Aflatoxin biocontrol in practice requires a multidisciplinary, long-term approach. *Front Sustain Food Syst*, **2023**, 7, 1110964. https://doi.org/10.3389/fsufs.2023.1110964.
154. Das, S.; Chaudhari, A. K. A review on the efficacy of essential oils and their nanoencapsulated formulations against aflatoxins contamination of major cereals with emphasis on mode of action. *Biocatal Agric Biotechnol*, **2023**, 102861. https://doi.org/10.1016/j.bcab.2023.102861.
155. Olmo, R.; Wetzels, S. U.; Armanhi, J. S. L.; Arruda, P.; Berg, G.; Cernava, T.; Cotter, P. D.; Araujo, S. C.; de Souza, R. S. C.; Ferrocino, I.; et al. Microbiome research as an effective driver of success stories in agrifood systems – A selection of case studies. *Front Microbiol*, **2022**, 13, 834622. https://doi.org/10.3389/fmicb.2022.834622.
156. Remesh, A. V.; Raveendran, A.; Bincy, K.; Wagh, V. S.; Dastager, S. G.; Babu, C. S. V. Insights on biorational potential of *Ocimum gratissimum* essential oil and its binary combination with monoterpene phenol for control of rice weevil (*Sitophilus oryzae*) and aflatoxigenic fungi. *Food Biosci*, **2022**, 50, 102019. https://doi.org/10.1016/j.fbio.2022.102019.
157. Mishra, R.; Tripathi, P.; Kumar, P.; Rajpoot, P. K.; Verma, S.; Aman, A. S. Innovations and future trends in storage pest management. *J Exp Agric Int*, **2024**, 46(5), 155–165. https://doi.org/10.9734/jeai/2024/v46i52366.
158. Divekar, P. A.; Narayana, S.; Divekar, B. A.; Kumar, R.; Gadratagi, B. G.; Ray, A.; Singh, A. K.; Rani, V.; Singh, V.; Singh, A. K.; et al. Plant secondary metabolites as defense tools against herbivores for sustainable crop protection. *Int J Mol Sci*, **2022**, *23*(5), 2690. https://doi.org/10.3390/ijms23052690.
159. Loi, M.; Paciolla, C.; Logrieco, A. F.; Mulè, G. Plant bioactive compounds in pre- and postharvest management for aflatoxins reduction. *Front Microbiol*, **2020**, 11, 510248. https://doi.org/10.3389/fmicb.2020.00243.
160. Singh, P. P.; Jaiswal, A. K.; Singh, R.; Kumar, A.; Gupta, V.; Raghuvanshi, T. S.; Sharma, A.; Prakash, B. Assessment of *Trachyspermum ammi* Essential Oil against *Aspergillus flavus*, aflatoxin B1 contamination, and post-harvest quality of sorghum bicolor. *Food Chem*, **2024**, 443, 138502. https://doi.org/10.1016/j.foodchem.2024.138502.
161. Boy, F. R.; Casquete, R.; Gudiño, I.; Merchán, A. V.; Peromingo, B.; Benito, M. J. Antifungal effect of autochthonous aromatic plant extracts on two mycotoxigenic strains of *Aspergillus flavus*. *Foods*, **2023**, 12(9). https://doi.org/10.3390/foods12091821.

162. Cadenillas, L. F.; Hernandez, C.; Bailly, S.; Billerach, G.; Durrieu, V.; Bailly, J. D. Role of polyphenols from the aqueous extract of *Aloysia citrodora* in the inhibition of aflatoxin B1 synthesis in *Aspergillus flavus. Molecules*, **2023**, 28(13), 5123. https://doi.org/10.3390/molecules28135123.
163. Zhang, D.; Yang, Y.; Yao, B.; Hu, T.; Ma, Z.; Shi, W.; Ye, Y. Curcumin inhibits *Aspergillus flavus* infection and aflatoxin production possibly by inducing ROS burst. *Food Res Int*, **2023**, 167, 112646. https://doi.org/10.1016/j.foodres.2023.112646.
164. Mohseni, M.; Azimi, A. A.; Hashemloian, B. D.; Sepehr, M. F. Phytochemical investigation of alkaloid fractions of *Fumaria vaillantii* and *F. asepala* and their antifungal influence on gene expression pattern in *aspergillus* species. *Turk J Botany*, **2023**, 47(3), 244–255. https://doi.org/10.55730/1300-008X.2762.
165. Zhang, W.; Li, B.; Lv, Y.; Wei, S.; Zhang, S.; Hu, Y. Transcriptomic analysis shows the antifungal mechanism of honokiol against *Aspergillus flavus. Int J Food Microbiol*, **2023**, 384, 109972. https://doi.org/10.1016/j.ijfoodmicro.2022.109972.
166. Al Khoury, A.; El Khoury, A.; Rocher, O.; Hindieh, P.; Puel, O.; Maroun, R. G.; Atoui, A.; Bailly, J. D. Inhibition of aflatoxin B1 synthesis in *Aspergillus flavus* by mate (*Ilex paraguariensis*), Rosemary (*Rosmarinus officinalis*) and green tea (*Camellia sinensis*) extracts: relation with extract antioxidant capacity and fungal oxidative stress response modulation. *Molecules*, **2022**, 27(23), 8550. https://doi.org/10.3390/molecules27238550.
167. Deabes, M. M.; Khalil, W. K. B.; S.Kh, A. E. A.; Ibrahim, S. D.; Naguib, K. M. The inhibitory effect of different Ephedra plant extracts on the *Aspergillus flavus* growth and aflatoxin B1 gene expression. *Jordan J Biol Sci*, **2022**, 15(01), 59–66. https://doi.org/10.54319/jjbs/150108.
168. Ismail, I. A.; Qari, S. H.; Shawer, R.; Elshaer, M. M.; Dessoky, E. S.; Youssef, N. H.; Hamad, N. A.; Abdelkhalek, A.; Elsamra, I. A.; Behiry, S. I. The application of pomegranate, sugar apple, and eggplant Peel extracts suppresses *Aspergillus flavus* growth and aflatoxin B1 biosynthesis pathway. *Horticulturae*, **2021**, 7(12), 558. https://doi.org/10.3390/horticulturae7120558.
169. Youssef, N. H.; Qari, S. H.; Matar, S.; Hamad, N. A.; Dessoky, E. S.; Elshaer, M. M.; Sobhy, S.; Abdelkhalek, A.; Zakaria, H. M.; Heflish, A. A.; et al. Licorice, Doum, and Banana Peel extracts inhibit *Aspergillus flavus* growth and suppress metabolic pathway of aflatoxin B1 production. *Agronomy*, **2021**, *11* (8), 1587. https://doi.org/10.3390/agronomy11081587.
170. Xu, D.; Wei, M.; Peng, S.; Mo, H.; Huang, L.; Yao, L.; Hu, L. Cuminaldehyde in cumin essential oils prevents the growth and aflatoxin B1 biosynthesis of *aspergillus flavus* in peanuts. *Food Control*. **2021**, 125, 107985. https://doi.org/10.1016/j.foodcont.2021.107985.
171. Hu, Z.; Yuan, K.; Zhou, Q.; Lu, C.; Du, L.; Liu, F. Mechanism of antifungal activity of Perilla frutescens essential oil against *Aspergillus flavus* by transcriptomic analysis. *Food Control*, **2021**, 123. https://doi.org/10.1016/j.foodcont.2020.107703.
172. Nazareth, T. de M.; Alonso-Garrido, M.; Stanciu, O.; Mañes, J.; Manyes, L.; Meca, G. Effect of allyl isothiocyanate on transcriptional profile, aflatoxin synthesis, and *Aspergillus flavus* growth. *Food Res Int*, **2020**, 128, 108786. https://doi.org/10.1016/j.foodres.2019.108786.
173. da Silva Bomfim, N.; Kohiyama, C. Y.; Nakasugi, L. P.; Nerilo, S. B.; Mossini, S. A. G.; Romoli, J. C. Z.; Graton Mikcha, J. M.; Abreu Filho, B. A. de; Machinski, M. Antifungal and antiaflatoxigenic activity of rosemary essential oil (*Rosmarinus officinalis* L.) against *Aspergillus flavus. Food Addit Contam Part A*, **2020**, 37(1), 153–161. https://doi.org/10.1080/19440049.2019.1678771.
174. Nerilo, S. B.; Romoli, J. C. Z.; Nakasugi, L. P.; Zampieri, N. S.; Mossini, S.; Rocha, G.; da Gloria, E. M.; Alves de Abreu Filho, B.; Junior, M. M. Antifungal activity and inhibition of aflatoxins production by *Zingiber officinale* roscoe essential oil against *Aspergillus flavus* in stored maize grains. *Ciencia Rural*, **2020**, 50(6). https://doi.org/10.1590/0103-8478cr20190779.
175. Oliveira, R. C.; Carvajal-Moreno, M.; Correa, B.; Rojo-Callejas, F. Cellular, physiological and molecular approaches to investigate the antifungal and anti-aflatoxigenic effects of thyme essential oil on *Aspergillus flavus. Food Chem*, **2020**, 315, 126096. https://doi.org/10.1016/j.foodchem.2019.126096.

# 3.3 Aflatoxigenic Fungi
## *Biological Control – An Overview*

*I. Lavkor*

## INTRODUCTION

Aflatoxins are generated as a secondary metabolite by aflatoxigenic fungi, *Aspergillus* species, notably *A. flavus* and *A. parasiticus* [1–3]. Aflatoxin $B_1$ ($AFB_1$), aflatoxin $B_2$ ($AFB_2$), aflatoxin $G_1$ ($AFG_1$), and aflatoxin $G_2$ ($AFG_2$) are the four most recognized types of aflatoxin [4, 5]. It is important to point out that although *A. parasiticus* is capable of producing all four types, aflatoxigenic *A. flavus* is typically only capable of producing $AB_1$ and $AB_2$ [6, 7].

Aflatoxin is a well-studied toxin globally due to its potential hazards to public health. Liver cancer in humans has been caused by aflatoxins and foods [8–10]. Reportedly, 4.5 billion people in the developing world are at risk of cancer-related diseases due to exposure to high levels of aflatoxins in their food [11]. It is commonly found in oilseeds, nuts, and various spices [12, 13]. A variety of crop species, including corn, chilies, peanuts, cottonseeds, pistachio nuts, and coconuts, are highly susceptible to contamination by aflatoxins [14–16]. Aflatoxin contamination in food and feed products is a growing global concern [17]. Every year, aflatoxin contamination of crops results in significant economic losses and threatens the safety of the food and feed supply. Over the last ten years, it has been revealed that 55% of cereal grains worldwide are contaminated with aflatoxins [18].

Aflatoxins are well-known toxins that pose significant public health hazards and have been extensively studied [1]. Peanuts and beans are greatly sensitive to aflatoxin contamination during both in the field and storage [19]. Specifically, $AFB_1$, $AFB_2$, $AFG_1$, and $AFG_2$ have reached worrying levels in peanuts. This led to the introduction of contaminated peanut products into the markets of developing countries in Asia and Africa [20, 21]. It is concluded that aflatoxin contamination harms human and animal well-being, income, food safety, and trade [22, 23].

Both before and after harvest, aflatoxins can be a source of contamination for crops. Nonetheless, certain conditions can increase aflatoxigenic *Aspergillus* growth, such as dry environmental conditions when crops reach maturity, elevated moisture levels in grains, incomplete drying of commodities, and insufficient storage location [24]. Insufficiently drying rice kernels to below 14% moisture content increases the risk of aflatoxin contamination. This contamination not only affects the external appearance of the grains but also diminishes their quality [25].

Various effective methods improved control of aflatoxigenic fungi and reduced aflatoxin production during pre- or post-harvest stages, including chemical, physical, and biological approaches [26]. Physical techniques such as low-temperature food storage, rodent control in buildings, quick drying, and radiation are effective methods [27]. Various chemical compounds have been employed to inhibit fungal growth and the production of aflatoxin, including acids such as acetic, citric, propionic, and benzoic acids, oxides like hydrogen peroxide and ammonium hydroxide, as well as copper sulphate. Nevertheless, these chemicals are associated with adverse health and environmental impacts; thus they are generally regarded as products with limited benefits. Hence, improved non-toxic and environmentally friendly antifungal alternatives are urgently needed [28].

DOI: 10.1201/9781003487043-31

Microorganisms can be used as biocontrol agents and exhibit a comparatively restricted spectrum of activity compared with most chemical fungicides. The use of biocontrol agents pre- and post-harvest is a complex issue involving several biological, environmental, and economic factors. Biological control products that are newly developed should not only demonstrate effectiveness against the targeted plant pathogen but also ensure safety and cost-effectiveness [29, 30].

Several organisms are known to control aflatoxigenic fungi and aflatoxin contamination, including bacteria, yeasts, and nontoxigenic *Aspergillus* fungi [31–33]. This review examines the latest advancements in utilizing biological control methods to address aflatoxigenic fungi and aflatoxin contamination in agricultural products.

## BIOCONTROL AGENTS FOR AFLATOXIGENIC *ASPERGILLUS* SPP.

### Fungi

A new method of biological control that uses nontoxigenic *A. flavus* strains to compete with aflatoxin-producing fungi is gaining popularity due to its high success rate, cost-effectiveness, and eco-friendliness [32]. Significant progress has been achieved using nontoxigenic strains of *A. flavus* and *A. parasiticus* in biocontrol of aflatoxin contamination in pre- and post-harvest crops. This approach stems from the competition of the nontoxigenic fungus applied to the soil under field conditions with the aflatoxigenic fungus in the environment. The nontoxigenic strains of biocontrol agents must dominate the agricultural environment during the period when considering the effectiveness of competitive exclusion and the susceptibility of crops to toxigenic strains [34].

Field trials have shown that applying different nontoxigenic *A. flavus* and *A. parasiticus* strains to the soil can efficiently reduce aflatoxin contamination in peanuts, corn [35], and cottonseed [36]. Using nontoxigenic *A. flavus* strains in cotton fields was found to be highly effective in controlling aflatoxin production in cottonseed [37]. In a separate investigation, Dorner et al. [38, 39] conducted a study that applied nontoxigenic *A. flavus* and *A. parasiticus* strains to the soils in peanut fields at different levels. The results showed a significant decrease in aflatoxin contamination in peanuts, ranging from 74.3% to 99.9%. Another study determined that a non-aflatoxigenic *A. flavus* strain was found to reduce aflatoxin contamination in corn between 80% and 95% [40]. Numerous field experiments conducted with crops such as peanuts and cotton have consistently shown that using these nontoxigenic *Aspergillus* strains reduces aflatoxin contamination from 70% to 90% [41–43]. Hruska et al. [44] revealed through the exclusion of competition that the low population of toxigenic *A. flavus* under treatment was positively correlated with the decreased aflatoxin level. On the other hand, the biocontrol agent *A. flavus* strain showed increased propagation and colonization, thereby mitigating the risk of aflatoxin contamination.

Aflasafe, the biocontrol solution against aflatoxin, has been successfully manufactured in Africa as the first corn application to control aflatoxin. In Africa, Aflasafe has been effectively applied in soil to maize crops with up to 99% efficacy [45]. Building on its success with corn, it is now available for other crops, including sunflower, sesame, sorghum, and millet [46–48].

For aflatoxin reduction in the United States, AF36 and NRRL 21882 nontoxigenic *A. flavus* strains are used. The US Environmental Protection Agency has registered AF36 as a biopesticide, which is utilized to manage aflatoxin-producing fungi in cotton fields. On the other hand, *A. flavus* NRRL 21882 (Afla-guard) is applied to prevent peanut fields against aflatoxin contamination [49]. Afla-guard demonstrated significant efficacy in extensive trials, achieving a remarkable 97.5% reduction in aflatoxins in shelled groundnuts and an impressive 85.2% reduction in stored groundnuts [50]. In another similar study, Lavkor et al. [51] determined that applying Afla-guard in groundnut cultivation reduced the aflatoxin contamination rate by between 89.1 and 99.8% in 3 different treatments. Hence, this study demonstrates that implementing this biological control strategy effectively reduces aflatoxin levels. It is important that the study showed that competitive exclusion of toxigenic strains by nontoxigenic strains in the field reduced aflatoxin.

In Argentina, a nontoxigenic strain of *A. flavus* has shown biological control capabilities for peanuts. This biocontrol agent exhibited notable enhancements in both propagation and colonization, resulting in a considerable reduction in aflatoxin levels. Moreover, applying the biocontrol agent before harvest displayed a sustained effect, offering protection to peanuts throughout the storage period [52]. Dorner [53] studied the levels of aflatoxins in peanuts that were treated during storage (control-treated) and those that were not treated (control-control). The results showed that the treated peanuts had significantly lower levels of aflatoxins. On the other hand, Lavkor et al. [54] conducted a study on the biological control of aflatoxin in the corn cultivation area of Türkiye. Their findings suggest that the spray applied during the storage phase did not provide additional protection against aflatoxin contamination. Conversely, only field treatment had a significant impact on reducing the occurrence of aflatoxins during storage. These results indicate the potential of using competitive exclusion as a biocontrol method during the pre- and post-harvest period for aflatoxin.

In addition to the use of nontoxigenic *A. flavus* biocontrol strains, there is a growing interest in exploring alternative biocontrol agents to combat aflatoxin-producing molds. One noteworthy example involves the application of a strain of *Trichoderma harzianum*, which demonstrated efficacy in limiting *A. flavus* contamination. This particular strain achieved a substantial 57% and 65% reduction in aflatoxin levels in groundnut [55] and sweet corn [56], respectively. Despite these promising results, it is important to note that there are currently no commercially available products specifically targeted against *A. flavus* contamination [57]. Nonetheless, Lagogianni and Tsitsigiannis [58] assessed six biofungicides with active ingredients such as *Aureobasidium pullulans, T. harzianum, Streptomyces griseoviridis, Bacillus subtilis QST 713*, zeolite, and laminarine, and determined that most of them were effective in reducing *A. flavus* conidiospores and $AFB_1$ production with in vitro conditions.

## Bacteria

Biological control strategies using nontoxic strains and the antifungal properties of various bacteria have been investigated, with the potential to be effective against *Aspergillus* spp., and many aflatoxigenic fungi. These biocontrol agents could include enhancing different ways to fungal invasion of various bacteria species, including *B. subtilis Pseudomonas* spp., *Ralstonia* spp., *Streptomyces* spp., *Lactobacilli* spp., and *Burkholderia* spp., have demonstrated the capacity to impede the growth of fungi and suppress the production of aflatoxin biosynthesis by *Aspergillus* spp. in vitro, in vivo, and in the field.

In a study conducted by Palumbo et al. [59], it was detected that certain strains of *Bacillus, Pseudomonas, Ralstonia*, and *Burkholderia* isolated in California almond samples completely suppressed the growth of *A. flavus*. In addition, *B. subtilis AU195* showed the most potent antifungal activity when tested for antagonistic activity against *A. flavus* under in vitro conditions [60].

The extensive capacity of *Bacillus* spp. to produce diverse metabolites possessing antibacterial and antifungal properties has been heavily utilized in medical and industrial sectors. This capability serves as a crucial factor in their efficacy as biocontrol agents against aflatoxigenic [61, 62]. Biological control is based on a crucial approach for introducing a biocontrol agent into the infection site to diminish or eradicate colonization by the pathogen microorganism [63]. Nesci et al. [64] reported that several strains of *B. subtilis* and *P. solanacearum*, isolated from corn soil, were effective in inhibiting aflatoxin accumulation. The presence of *A. flavus* is significantly influenced by plant growth-promoting rhizobacteria (PGPR), which serve as essential components of the soil. Rhizosphere soils host a greater abundance and diversity of PCPR, which are essential for this interaction [65]. Rhizosphere bacteria must be highly abundant to achieve good efficacy against aflatoxigenic *Aspergillus* spp. on crops under field conditions.

In another study, *B. cereus* and *B. megaterium* were used as biocontrol agents to control kernel rot in peanuts caused by *A. flav*us [66, 67]. Reddy et al. [68] determined a 62.6% decrease in $AFB_1$ levels in sorghum grains by *Pseudomonas fluorescent* treatment.

### Yeast

Numerous studies have identified various yeast species, including *Candida, Debaryomyces, Pichia, Saccharomyces, Saccharomycopsis, Saccharomycodes, Schizosaccharomyces, Trichosporon,* and *Zygosaccharomycesi,* as biocontrol agents against aflatoxigenic *Aspergillus*, and a necessary approach to managing aflatoxin contamination [32].

Laboratory studies have shown promising results, indicating that these yeast strains can effectively inhibit the growth of *Aspergillus* species. However, it is important to note that laboratory conditions may not fully replicate the complex and dynamic environment found in the field. Therefore, further field trials are essential to assess the practical effectiveness of these yeast strains in reducing aflatoxin contamination under real-world conditions.

Field trials would take into account factors such as variations in environmental conditions, crop types and agronomic practices. These experiments are essential for understanding the feasibility of incorporating saprophytic yeast species into agricultural practices for aflatoxin management.

Potential benefits of using yeast biocontrol agents include their ability to compete with *Aspergillus* for nutrients and ecological niches and their ability to produce antifungal compounds. In addition, yeast-based biocontrol agents may offer a more environmentally friendly alternative to some chemical interventions.

In conclusion, although the laboratory results are promising, the translation of these results into practical and effective aflatoxin management strategies requires rigorous field testing. This research will provide valuable insights into the feasibility and limitations of using saprophytic yeast species as biocontrol agents in real agricultural settings.

## CONCLUSION

Biological control is a promising research approach currently being investigated for the reduction and eventual elimination of aflatoxin contamination. Using nontoxigenic isolates of *A. flavus* and *A. parasiticus* in field experiments is effective in reducing aflatoxin concentrations in peanuts, cottonseed, and corn. It is also important to note that an integrated approach, combining multiple methods and strategies, is often the most effective way to manage aflatoxin contamination in pre-harvest and post-harvest products. In the future, new and innovative approaches and technologies for more targeted and effective biological control may emerge from ongoing research and advances in biological control technology.

## REFERENCES

1. Rashid, N., M.A. Bajwa, M. Rafeeq, M.M. Tariq, F. Abbas, M.A. Awan, et al. 2013. Prevalence of aflatoxicosis in broiler chickens in Quetta, Pakistan. *Pakistan Journal of Zoology* 45: 1021–1026.
2. Lavkor, I., and I. Var. 2017. The control of aflatoxin contamination at harvest, drying, pre- storage and storage periods in peanut: The new approach. In *Aflatoxin - Control, Analysis, Detection and Health Risks*, ed. L.B. Abdulra'Uf, 45–64. Intech Open.
3. Umar, A. 2023. A call for aflatoxin control in Asia. *CABI Agriculture and Bioscience* 4: 27.
4. Saini, S.S., and A. Kaur. 2012. Aflatoxin $B_1$: Toxicity, characteristics and analysis: Mini review. *Global Advanced Research Direct Research Journal of Chemistry and Material Sciences* 1: 63–70.
5. Goldblatt, L. 2012. *Aflatoxin: Scientific Background, Control, and Implications*. Amsterdam: Elsevier.
6. Robens, J., and K.F. Cardwell. 2003. The costs of mycotoxin management to the USA: Management of aflatoxins in the United States. *Journal of Toxicology: Toxin Reviews* 22: 139–152.
7. Lavkor, I. 2019. Molecular characterization of aflatoxin biosynthesis genes of *Aspergillus flavus* from peanuts production area. *Legume Research* 508: 1–6.
8. Huwig, A., S. Freimund, O. Kappeli, and H. Dutler. 2001. Mycotoxin detoxication of animal feed by different adsorbents. *Toxicology Letters* 122: 179–188.
9. Gong, Y.Y., P.C. Turner, A.J. Hall, and C.P. Wild. 2008. Aflatoxin exposure and impaired child growth in West Africa: An unexplored international public health burden. In *Mycotoxins: Detection Methods, Management, Public Health and Agricultural Trade*, eds. J.F. Leslie, R. Bandyopadhyay, A. Visconti, 53–65. Oxfordshire, UK: CAB International.

10. Voth-Gaeddert, L.E., M. Stoker, O. Torres, and D.B. Oerther. 2018. Association of aflatoxin exposure and height-for-age among young children in Guatemala. *International Journal of Environmental Health Research* 28: 280–92.
11. Li, F.-Q., T. Yoshizawa, S. Kawamura, S.Y. Luo, and Y.W. Li. 2001. Aflatoxins and fumonisins in corn from the high-incidence area for human hepatocellular carcinoma in Guangxi, China. *Journal of Agricultural and Food Chemistry* 49: 4122–4126.
12. Lavkor, I., I. Var, S. Saglam, O. Uckun, A. Tekin, and O. Savas. 2019. Presence of some mycotoxins in peanuts from harvest to storage. *Legume Research* 42: 862–866.
13. Iqbal, H.G.M., M.R. Asi, and S. Jinap. 2014. Variation in vitamin E level and aflatoxins contamination in different rice varieties. *Journal of Cereal Science* 60: 352–355.
14. Idris, Y.M.A., A.A. Mariod, I.A. Elnour, and A.A. Mohamed. 2010. Determination of aflatoxin levels in Sudanese edible oils. *Food and Chemical Toxicology* 48: 2539–2541.
15. Cornea, C.P., M. Ciuca, C. Voaides, V. Gagiu, and A. Pop. 2011. Incidence of fungal contamination in a Romanian bakery: A molecular approach. *Romanian Biotechnological Letters* 16: 5863–5871.
16. Dolezal, A., X. Shu, G. Obrian, D. Nielsen, C.P. Woloshuk, R. Boston, G. Payne, and J.W. Cary. 2014. *Aspergillus flavus* infection induces transcriptional and physical changes in developing maize kernels. *Frontiers in Microbiology* 5: 1–10.
17. Jayabarathi, P., and R. Mohamudha Parveen. 2010. Biochemical and histopathological analysis of aflatoxicosis in growing hens fed with commercial poultry feed. *International Journal of Pharmaceutical Sciences Review and Research* 32: 127–130.
18. Lee, H.J., and D. Ryu. 2017. Worldwide occurrence of mycotoxins in cereals and cerealderived food products: Public health perspectives of their co-occurrence. *Journal of Agricultural and Food Chemistry* 65: 7034–7051.
19. Tchana, A.N., P.F. Moundipa, and F.M. Tchouanguep. 2010. Aflatoxin contamination in food and body fluids in relation to malnutrition and cancer status in Cameroon. *International Journal of Environmental Research and Public Health* 7: 178–188.
20. Jiang, Y., P.E. Jolly, W.O. Ellis, J.-S. Wang, T.D. Phillips, and J.H. Williams. 2005. Aflatoxin B1 albumin adduct levels and cellular immune status in Ghanaians. *International Immunology* 17: 807–814.
21. Waliyar, F., A. Traore, D. Fatondji, and B. Ntare. 2003. Effect of irrigation interval, planting date, and cultivar on *Aspergillus flavus* and aflatoxin contamination of peanut in a sandy soil of Niger. *Peanut Science* 30: 79–84.
22. Wu, F. 2015. Global impacts of aflatoxin in maize: Trade and human health. *World Mycotoxin Journal* 8: 137–142.
23. Bandyopadhyay, R., A. Ortega-Beltran, A. Akande, C. Mutegi, J. Atehnkeng, L. Kaptoge, et al. 2016. Biological control of aflatoxins in Africa: Current status and potential challenges in the face of climate change. *World Mycotoxin Journal* 9:771–789.
24. Stepman, F. 2018. Scaling-up the impact of aflatoxin research in Africa. The role of social sciences. *Toxins* 10: 136.
25. Oyebamiji, Y.O., K.A. Ajijolakewu, and I.A. Adebayo. 2021. Worldwide prevalence of aflatoxins in food and feed. In *Aflatoxins in Food*, ed. K.R. Hakeem, 41–45. Springer.
26. Jardon-Xicotencatl, S., R. Díaz-Torres, A. Marroquín-Cardona, T. Villarreal-Barajas, and A. Méndez-Albores. 2015. Detoxification of aflatoxin-contaminated maize by neutral electrolyzed oxidizing water. *Toxins* 7: 4294–4314.
27. Lavkor, I., and M. Bicici. 2015. Aflatoxin occurrence in peanuts grown in Osmaniye at harvest, post-harvest, drying and pre-storage periods. *Journal of Agricultural Sciences* 21: 394–405.
28. Jonathan, S.G., M.A. Adeniyi, and M.D. Asemoloye. 2016. Fungal biodeterioration, aflatoxin contamination, and nutrient value of "*Suya* Spices". *Scientifica* 2016: 1–5.
29. Nunes, C., J. Usall, N. Teixidó, M. Abadias, and I. Viñas. 2002. Improved control of postharvest decay of pears by the combination of candida sake (CPA-1) and ammonium molybdate. *Phytopathology* 92: 281–287.
30. Köhl, J., J. Postma, P. Nicot, M. Ruocco, and B. Blum. 2011. Stepwise screening of microorganisms for commercial use in biological control of plant-pathogenic fungi and bacteria. *Biological Control* 57: 1–12.
31. Pfliegler, V., I. Pócsi, Z. Gyori, and T. Pusztahelyi. 2020. The *Aspergilli* and their mycotoxins: Metabolic interactions with plants and the soil biota. *Frontiers in Microbiology* 10: 1–45.
32. Peles, F., P. Sipos, Z. Gyori, W.P. Pfliegler, F. Giacometti, A. Serraino, et al. 2019. Adverse effects, transformation, and channeling of aflatoxins into food raw materials in livestock. *Frontiers in Microbiology* 10: 2861.

33. Jalili, M. 2015. A review on aflatoxins reduction in food. *Iranian Journal of Health, Safety and Environment* 3: 445–459.
34. Rajani, P., V. Sridevi, and M.V.V. Chandana Lakshmi. 2012. A review on biological control of aflatoxin crop contamination. *International Journal of Chemical, Environmental and Pharmaceutical Research* 3: 83–86.
35. Dorner, J.W., R.J. Cole, and D.T. Wicklow. 1999. Aflatoxin reduction in corn through field application of competitive fungi. *Journal of Food Protection* 62: 650–656.
36. Cotty, P.J. 1990. Effect of atoxigenic strains of *Aspergillus flavus* on aflatoxin contamination of developing cottonseed. *Plant Disease* 74: 233–235.
37. Cotty, P.J. 1994. Influence of field application of an atoxigenic strain of *Aspergillus flavus* on the populations of *Aspergillus flavus* infecting cotton bolls and on the aflatoxin content of cottonseed. *Phytopathology* 84: 1270–1277.
38. Dorner, J.W., R.J. Cole, and P.D. Blankenship. 1998. Effect of inoculum rate of biological control agents on preharvest aflatoxin contamination of peanuts. *Biological Control* 12: 171–176.
39. Dorner, J.W., R.J. Cole, W.J. Connick, D.J. Daigle, M.R. McGuire, and B.S. Shasha. 2003. Evaluation of biological control formulations to reduce aflatoxin contamination in peanuts. *Biological Control* 26: 318–324.
40. Brown, R.L., P.J. Cotty, and T.E. Cleveland. 1991. Reduction in aflatoxin content of maize by atoxigenic strains of *Aspergillus flavus*. *Journal of Food Protection* 54: 623–626.
41. Dorner, J.W. 2004. Biological control of aflatoxin contamination of crops. *Journal of Toxicology- Toxin Reviews* 23: 425–450.
42. Pitt, J.I., and A.D. Hocking. 2006. Mycotoxins in Australia: Biocontrol of aflatoxin in peanuts. *Mycopathologia* 162: 233–243.
43. Dorner, J.W. 2008. Management and prevention of mycotoxins in peanuts. *Food Additives and Contaminants* 25: 203–208.
44. Hruska, Z., K. Rajasekaran, H. Yao, R. Kincaid, D. Darlington, R.L. Brown, D. et al. 2014. Co-inoculation of aflatoxigenic and non-aflatoxigenic strains of *Aspergillus flavus* to study fungal invasion, colonization, and competition in maize kernels. *Frontiers in Microbiology* 5: 122
45. Kinyungu, S., T. Isakeit, P.S. Ojiambo, and C.P. Woloshuk. 2019. Spread of *Aspergillus flavus* and aflatoxin accumulation in postharvested maize treated with biocontrol products. *Journal of Stored Products Research* 84: 101519.
46. Bandyopadhyay, R., J. Atehnkeng, A. Ortega-Beltran, A. Akande, T.D. Falade, and P.J. Cotty. 2019. "Ground-truthing" efficacy of biological control for aflatoxin mitigation in farmers' fields in Nigeria: From field trials to commercial usage, a 10-year study. *Frontiers in Microbiology* 10: 2528.
47. Agbetiameh, D., A. Ortega-Beltran, R.T. Awuah, J. Atehnkeng, A. Elzein, P.J. Cotty, et al. 2020. Field efficacy of two atoxigenic biocontrol products for mitigation of aflatoxin contamination in maize and groundnut in Ghana. *Biological Control* 150:104351.
48. Senghor, A., A. Ortega-Beltran, J. Atehnkeng, P. Jarju, P. Cotty, and R. Bandyopadhyay. 2021. Aflasafe SN01 is the first biocontrol product approved for aflatoxin mitigation in two nations Senegal and the Gambia. *Plant Disease* 105: 1461–1473.
49. Chang, P.-K., and S.S. Hua. 2007. Nonaflatoxigenic *Aspergillus flavus* TX9-8 competitively prevents aflatoxin accumulation by *A. flavus* isolates of large and small sclerotial morphotypes. *International Journal of Food Microbiology* 114: 275–279.
50. Tsitsigiannis, D., M. Dimakopoulou, P. Antoniou, and E. Tjamos. 2012. Biological control strategies of mycotoxigenic fungi and associated mycotoxins in Mediterranean basin crops. *Phytopathologia Mediterranea* 51: 158–174.
51. Lavkor, I., H. Arioglu, I. Var, and S. Oztemiz. 2019. Biological control of aflatoxigenic fungi on peanut: For the pre-harvest approach. *Turkish Journal of Field Crop* 24: 21–27.
52. Chulze, S.N., J.M. Palazzini, A.M. Torres, G. Barros, M.L. Ponsone, R. Geisen, et al. 2015. Biological control as a strategy to reduce the impact of mycotoxins in peanuts, grapes and cereals in Argentina. *Food Additives and Contaminants Part A* 32: 471–479.
53. Dorner, J.W. 2009. Development of biocontrol technology to manage aflatoxin contamination in peanuts. *Peanut Science* 36: 60–67.
54. Lavkor, I., T. Ay, S. Sobucovali, I. Var, H. Saghrouchni, A.M. Salamatullah, et al. 2023. Non-aflatoxigenic *Aspergillus flavus*: A promising biological control agent against aflatoxin contamination of corn. *American Chemical Society Omega* 8: 16779–16788.
55. Kifle, M.H., K.S. Yobo, and M.D. Laing. 2017. Biocontrol of *Aspergillus flavus* in groundnut using *Trichoderma harzianum* strain kd. *Journal of Plant Diseases and Protection* 124: 51–56.

56. Sivparsad, B.J., and M.D. Laing. 2016. Pre-harvest silk treatment with *Trichoderma harzianum* reduces aflatoxin contamination in sweetcorn. *Journal of Plant Diseases and Protection* 123: 285–293.
57. Ren, X., Q. Zhang, W. Zhang, J. Mao, and P. Li. 2020. Control of aflatoxigenic molds by antagonistic microorganisms: Inhibitory behaviors, bioactive compounds, related mechanisms, and influencing factors. *Toxins* 12: 24.
58. Lagogianni, C.S., and D.I. Tsitsigiannis. 2019. Effective biopesticides and biostimulants to reduce aflatoxins in maize fields. *Frontiers in Microbiology* 10: 2645.
59. Palumbo, J.D., J.L. Baker, and N.E. Mahoney. 2006. Isolation of bacterial antagonists of *Aspergillus flavus* from almonds. *Microbiology Ecology* 52: 45–52.
60. Moyne, A.-L., R. Shelby, T.E. Cleveland, and S. Tuzun. 2001. *Bacillomycin* D: An iturin with antifungal activity against *Aspergillus flavus*. *Journal of Applied Microbiology* 90: 622–629.
61. Silo-Suh, L.A., B.J. Lethbridge, S.J. Raffel, H. He, J. Clardy, and J. Handelsman. 1994. Biological activities of two fungistatic antibiotics produced by *Bacillus cereus* UW85. *Applied and Environmental Microbiology* 60: 2023–2030.
62. Leifert, C., L.H. Chidburee, S. Hampson, S. Workman, D. Sigee, H.A.S. Epton, et al. 1995. Antibiotic production and biocontrol activity by *Bacillus subtilis* CL27 and *Bacillus pumilus* CL45. *Journal of Applied Bacteriology* 78: 97–108.
63. Cook, R.J., and K.F. Baker. 1983. *The Nature and Practice of Biological Control of Plant Pathogens*. St. Paul, MN: APS Press.
64. Nesci, A.V., R.V. Bluma, and M.G. Etcheverry. 2005. *In vitro* selection of maize rhizobacteria to study potential 361 biological control of *Aspergillus* section *Flavi* and aflatoxin production. *European Journal of Plant Pathology* 113: 159–171.
65. Mickler, C.J., K.L. Bowen, and J.W. Kloepper. 1995. Evaluation of selected geocarposphere bacteria for biological control of *Aspergillus flavus* in peanut. *Plant and Soil* 175: 291–299.
66. Chen, R., F. Ma, P.W. Li, W. Zhang, X.X. Ding, Q. Zhang, et al. 2014. Effect of ozone on aflatoxins detoxification and nutritional quality of peanuts. *Food Chemistry* 146: 284–288.
67. Kumar, S.N., S.R. Sreekala, D. Chandrasekaran, B. Nambisan, and R.J. Anto. 2014. Biocontrol of *Aspergillus* species on peanut kernels by antifungal diketopiperazine producing *Bacillus cereus* associated with entomopathogenic nematode. *PLoS One* 9: e106041.
68. Reddy, K.R.N., C.R. Raghavender, B.N. Reddy, and B. Salleh. 2010. Biological control of *Aspergillus flavus* growth and subsequent aflatoxin $B_1$ production in sorghum grains. *African Journal of Biotechnology* 9: 4247–4250.

# Section 4

## Advancement in the Detection of Aflatoxigenic Fungi

# 4.1 Aflatoxigenic Fungal Communities and High-throughput Analysis

*Liqian Du, Zhongkun Zhou, Yunhao Ma, Baizhuo Zhang, and Peng Chen*

## INTRODUCTION

Traditional Chinese medicine is an important resource for developing medicine in China. Fungi in Chinese medicinal plants and mycotoxin pollution have been largely affected [1]; this pollution represents a potential threat to the health of consumers and the development of the traditional Chinese medicine industry. Fungal growth leads to decreased grain quality and contamination with mycotoxins [2, 3]. There is a significant difference in climate between the north and south of China [4]. Recently, with the globalization of the Chinese medicine trade, the contamination by fungi and mycotoxins in Chinese medicine is causing increasingly wide concern for people because it can cause a variety of toxic effects in the human body [5]. These toxic effects thus impact the effective clinical use of Chinese medicine and have become important factors affecting the quality and safety of medicine. Full and effective monitoring of fungal contamination is therefore particularly important, and there is an urgent need to find optimal detection methods and preventive measures. *Aspergillus flavus* is the main food spoilage bacteria in warmer climates [6]. High-throughput sequencing is a powerful tool for monitoring fungal contamination in Chinese medicinal plant production [7], and with the rapid evolution of sequencing technology, it offers ideas for the study of fungal diversity in complex environments. The following are examples of assistance in plant mushroom research in previous studies of fungal contamination analysis of five common Chinese medicinal plants (barbary fruit, ginseng, hawthorn, malt, and straw) using high-yield sequencing techniques [8]. In brewing and other food industries [9], it is difficult to avoid producing harmful substances, such as ethyl carbamate and histamine, in the alcoholic beverage fermentation process [10]. It is therefore necessary to target the domestication and selection of wine strains [11] and methods and techniques to improve strain selection [12], but traditional screening methods [13] still rely on manual selection and flask culture [14]. The entire screening and cultivation process is cumbersome, inefficient, and expensive [15], and the simultaneous selection of ideal target strains, and industrial serial production is also a great challenge [16]. To screen industrial probiotics from an extensive range of mutant microorganisms, we need to find a reasonable and effective method of high-throughput analysis [17, 18]. The construction of a diversified screening library to broaden the scope of screening and how identifiable target metabolites can be increased by viable detection methods and, therefore, the goal of experiential high rate screening is currently to be resolved [19]. Therefore, this chapter provides a classified review of high-throughput analysis and its application status. It also summarizes the factors influencing the accumulation of mycotoxins in cereals and hopes to provide a reference for future researchers. The development of high-throughput analysis significantly improves the rate and accuracy of strain screening and is becoming one of the emerging methods of strain screening [20].

DOI: 10.1201/9781003487043-33

## FACTORS INFLUENCING *AFLATOXIN* ACCUMULATION

### Geographical Position

Rice samples were collected from 10 provinces in Northeast China, Southeast China, and Southwest China, and the aflatoxin and other mycotoxin contents in the samples were determined. The results indicated that the fungi content in cereals in different regions had significant differences. The fungal diversity of cereals in Southeast and Southwest China was significantly greater than that in Northeast China [21]. In a study conducted in 2022, it was shown that the national nature reserve and the Tibetan Plateau are affected by their geographical location and special climatic conditions [22]; in particular, the main plants in the high cold zone above 4,000 meters, such as mosses and lichens [23]. The results showed that the change of altitude directly affected the vertical distribution of fungi, and the fungi content in vegetation at high altitude was less than that at low altitude [24]. This may be due to the harsh natural environment at high altitudes, with average annual temperatures below zero degrees Celsius, strong ultraviolet radiation, and large temperature differences between day and night [25].

### Environment variable

Environmental variables determine the distribution of fungal communities. The most important factors affecting fungal communities include winter mean temperature, annual mean precipitation, harvest mean wind speed, and grain moisture [26]. Temperature is a key parameter affecting aflatoxin pollution before and after crop harvesting. The results showed that when the ambient temperature is between 25°C and 30°C, the aflatoxin ratios will increase as the ambient temperature rises. Higher temperatures promote the growth and diffusion of aflatoxin [27].

### Baking and Paint Removal (processing)

In a study in southern Nigeria, results showed that concentrations of aflatoxin in roasted and shelled nuts were higher than in samples of raw and shelled nuts [28]. We can thus make a significant contribution to promoting human food security by adopting effective processing interventions, such as cooking and stripping, thereby reducing the aflatoxin content of seed particles intended for human consumption [29].

### Biological Control

Pre-harvest biological control, such as field inoculation with nontoxic aflatoxin *Aspergillus* strains, not only inhibited aflatoxin-producing native species at the source but effectively reduced aflatoxin contamination of corn [29]. The results showed that the reduction of aflatoxin production, continuous propagation, and rapid culture of non-aflatoxin fungi and bioregenerative strains were closely related to biological control [30].

### Testa

In cereals, fungal contamination and integument size and integrity are closely related. Small, compact grains (wheat, rice, oats, sorghum) and grains with a hard coat (beans) are less likely than larger grains, such as corn, to be contaminated by fungi and to form mycotoxins [31].

### Soil

Groundnuts grown in slightly sandy soils promote rapid mushroom growth, especially in dry conditions, whereas coated soils reduce groundnut contamination due to their high water-holding capacity, which helps plants prevent drought stress [6].

### High-throughput Analysis

High-throughput analysis is a rapid detection technique using automated equipment at the model organism, cell, pathway, or molecular level. High-throughput analysis has been used in the pharmaceutical industry for more than ten years [32]. As an early detection platform, high-speed screening involves introducing lead compounds into the pipeline, leading to the development of therapeutic drugs. High-throughput analysis has received unprecedented attention. High-throughput analysis can be an effective way to discover potential drug candidates. Further, it could become a new probe into fundamental biological processes. Ultimately, it could facilitate the development of chemical-computer technologies that control chemical–biological interactions [33]. At the genera level, *Aspergillus* was the dominant fungus, with a relative abundance of 4.37% to 82.81%. At the species level, 108 fungal taxa have been identified, including six potentially toxic fungi such as *Trichoderma, penicillium chrysogenum*, and *penicillium capsulatum* [34]. High-throughput analysis is therefore very promising for the analysis of mushroom communities in spices and herbs and provides a warning for the subsequent potential biosynthesis of mycotoxins [35]. High-yield tests generally provide the theoretical basis for the genetic modification of fungi and the improvement of the synthesis efficiency of the corresponding components [36]. The established high-throughput analysis is not only effective for the screening of mycotoxin mutant strains but can also improve the mycotoxin yield [37]. In summary, the development of high-throughput analysis combining different mutagenesis breeding methods and gene modification to improve the yield of target products and obtain good high-yield strains can provide a good reference for the selection of strains that produce other antibiotics and secondary metabolites.

## ACKNOWLEDGMENTS

This work was supported by Gansu Provincial Science and Technology Major Project (Grant No. 24ZDFA001), The Lanzhou Municipal Science and Technology Program (Grant Nos. 2024-4-2, 2025-2-100, 2025-2-103) and the College Students' Innovation and Entrepreneurship Program of Lanzhou University, China (Grant Nos. 20250260006, 20250260016 and 20250260020).

## REFERENCES

1. Xu-Dong, X.U., et al., *Stage achievements made in program of key technologies of detection and analysis for harmful mycotoxin residues in CHM.* Chinese Herbal Medicines, 2010. 2(03): p. 327–334.
2. Kononenko, G.P., and A.A. Burkin, *Mycotoxin contaminations in commercially used hay.* SEL'Skokhozyaistvennaya Biologia, 2014. 49(6): p. 116–122.
3. McMillian, W.W., et al., Mycotoxin Contamination in Grain Sorghum from Fields in Georgia and Mississippi. 1983. 60(3): p. 226–227.
4. Bjurman, J. *Fungal and microbial activity in external wooden panels as determined by finish, exposure, and construction techniques.* International Biodeterioration and Biodegradation, 1999. 44(2–4): p. 199–207.
5. Guo, T., et al., *Applications of DART-MS for food quality and safety assurance in food supply chain.* Mass Spectrometry Reviews, 2017. 36(2): p. 161–187.
6. Anthony, M.H., et al., Aflatoxin Contamination in Foods and Feeds: A Special Focus on Africa. In Trends in Vital Food and Control Engineering, 2012. IntechOpen.
7. Zhu, Q., et al., *Novel sensitive high-throughput screening strategy for nitrilase-producing strains.* Applied & Environmental Microbiology, 2007. **73**(19): p. 6053–6057.
8. Xibin, L., et al., *Analysis of microbial diversity in the 1st and 2nd fermentation cycle of Jiangxiang Baijiu based on high throughput sequencing technology.* Liquor-Making Science & Technology, 2019. 297(3): p. 52–64.
9. Delacroix, J., A. Swaminathan, and M.E. Solt, *Density dependence versus population dynamics: An ecological study of failings in the California wine industry.* American Sociological Review, 1989. **54**(2): p. 245.

10. Marshall, R.S., M. Cordano, and M. Silverman, *Exploring individual and institutional drivers of proactive environmentalism in the US wine industry.* Business Strategy & the Environment, 2010. **14**(2): p. 92–109.
11. Graça, A., *Non-invasive real-time monitoring of vineyard soils, berries and leaves with FT-NIR spectroscopy.* Oiv World Congress of Vine & Wine, 2015.
12. Mittal, A., et al., *Isolation and biochemical characterization of acido-thermophilic extracellular phytase producing bacterial strain for potential application in poultry feed.* Jundishapur Journal of Microbiology, 2011. **4**(4): p. 273–282.
13. Marston, D., P. Mirkin, and S. Deno, *Curriculum-based measurement: An alternative to traditional screening, referral, and identification.* The Journal of Special Education, 1984. **18**(2): p. 109–117.
14. Kong, J., X.B. Li, and C.F. Wu, *A molecular biological method for screening and evaluating the traditional Chinese medicine used in pi-deficiency therapy involving intestinal microflora.* Asian Journal of Traditional Medicines, 2006. 40(3): p. 52–64.
15. Huai-Liang, Z., et al., *Comparison of the consistency of screening traditional Chinese medicine on urticaria between bio-resonance technology and traditional Chinese medicine.* Chinese Journal of Dermatovenerology of Integrated Traditional and Western Medicine, 2009. 8(4): p. 12–14.
16. Jayalalitha, V., et al., *Improving the viability of probiotics in yoghurt through different methods of encapsulation.* Egyptian Journal of Dairy Science, 2011. 39(1):p. 39–44.
17. Chin, T.S., et al., *Bioprocess optimization for biomass production of probiotics yeast Saccharomyces boulardii in semi-industrial scale.* Journal of Chemical and Pharmaceutical Research, 2015. **7**(3): p. 122–132.
18. Peng, F., L. Rong-Tang, and Z. Yun-Wei, *Application of molecular marker techniques in study of space mutation breeding mechanism.* Grassland and Turf, 2008. 41(2): p. 5–9.
19. Gady, et al., *Implementation of two high through-put techniques in a novel application: Detecting point mutations in large EMS mutated plant populations.* Plant Methods, 2009. 5(1): p. 13.
20. Wang, T.Y., et al., *Systematic screening of glycosylation- and trafficking-associated gene knockouts in Saccharomyces cerevisiae identifies mutants with improved heterologous exocellulase activity and host secretion.* BMC Biotechnology, 2013. **13**(4): p. 71–71.
21. Qi, Z., et al., *Distribution of mycotoxin-producing fungi across major rice production areas of China.* Food Control, 2022. **134**: p. 108572.
22. Anwar, T., and A. Abdulla, *Quantitative classification of saxicolous lichen community in Xinjiang Tumur Peak National Natural Reserve and its correlation with environment.* Journal of Plant Resources and Environment, 2015. **24**(2): p. 80–87.
23. Lei, W., et al., Screening of high-yield strain and medium optimization for maximum production of cordycepin by Cordyceps militaris. Mycosystema, 2012. 31(3): p. 382–388.
24. Alireza, et al., *A recent scenario of mass wasting and its impact on the transportation in Alborz Mountains, Iran using geo-information technology.* Arabian Journal of Geosciences, 2011. 4(7–8): p. 1337–1349.
25. Kraft, J.R., *Closed classrooms, high mountains and strange lands: An inquiry into culture and caring.* Journal of Experiential Education, 1992. **15**(3): p. 8–15.
26. Yunhua, L., et al., *Screening of phosphorus-solubilizing strain Burkholderia cenocepacia and optimizing of phosphate-dissolving culture condition.* Journal of South China Agricultural University, 2015. 36(3): p. 78–82.
27. Rodrigues P, Venncio A., et al., *A polyphasic approach to the identification of aflatoxigenic and non-aflatoxigenic strains of Aspergillus Section Flavi isolated from Portuguese almonds.* International Journal of Food Microbiology, 2009. **129**(2): p. 187–193.
28. Clement, G., et al., *Contamination of groundnut in South-Western Nigeria by aflatoxigenic fungi and aflatoxins in relation to processing.* Journal of Phytopathology, 2015. 163(4): p. 279–286.
29. Fareed, G., et al., *Determination of aflatoxin and ochratoxin in poultry feed and feed ingredients in humid semi-tropical environment.* Journal of Advanced Veterinary and Animal Research, 2014. **1**(4): p. 201–207.
30. Zuzana, H., et al., *Co-inoculation of aflatoxigenic and non-aflatoxigenic strains of aspergillus flavus to study fungal invasion, colonization, and competition in maize kernels.* Frontiers in Microbiology, 2014. **5**(4): p. 122.
31. Moghadam, M.M., and H. Hokmabadi, *Study on the effect of pistachio Testa on the reduction of' Aspergillus flavus' growth and aflatoxin B1 production in kernels of different pistachio cultivars.* Australian Journal of Crop Science, 2010. **4**(9): p. 744–749.

32. Janzen, W.P., *High throughput screening as a discovery tool in the pharmaceutical industry.* Laboratory Robotics & Automation, 2015. **8**(5): p. 261–265.
33. An, W.F., and N. Tolliday, *Cell-based assays for high-throughput screening.* Molecular Biotechnology, 2010. 45(2): p. 180–186.
34. Iram, S., et al., *Heavy metal tolerance of fungus isolated from the soil contaminated with sewage and industrial wastewater.* Polish Journal of Environmental Studies, 2013. **22**(3): p. 691–697.
35. Jiang, W., et al., *High-throughput analysis of fungal communities in Myristicae semen.* Lwt - Food Science and Technology, 2020. **128**: p. 109499.
36. Bensch, M., et al., *High throughput screening of chromatographic phases for rapid process development.* Chemical Engineering & Technology, 2010. **28**(11): p. 1274–1284.
37. Harper, G., et al., *The reduced graph descriptor in virtual screening and data-driven clustering of high-throughput screening data.* Cheminform, 2005. 44(6): p. 2145–2156.

# 4.2 Rapid Detection of Aflatoxigenic Fungal Communities in Agricultural Products

*K. Ajith Kumar and Lokesh Ravi*

## INTRODUCTION

Aflatoxin contamination in food crops poses a critical threat to both human health and the agricultural economy. These toxic compounds are secondary metabolites produced by various fungal species, with *Aspergillus flavus* and *Aspergillus parasiticus* being the primary culprits. Under favorable conditions—such as high temperatures and moisture—these fungi contaminate a wide range of food commodities during pre- and post-harvest stages. Among the various mycotoxins, aflatoxin B1 (AFB1) stands out due to its carcinogenic, mutagenic, and teratogenic properties, as well as its ability to suppress the immune system [1, 2]. AFB1 is mainly found in cereals (maize, sorghum, pearl millet, rice, and wheat), oilseeds (peanut, soybean, sunflower, and cotton), spices (chilies, black pepper, turmeric, coriander, and ginger), nuts (almond, Brazil nut, pistachio, walnut, and coconut), and various milk products. To mitigate aflatoxin contamination, a combination of strategies—such as using resistant crop varieties and implementing recommended farming practices—is crucial for maintaining food safety and safeguarding both consumers and the agricultural sector [2]. Aflatoxins exert a profound impact on human health, manifesting in various detrimental outcomes. Chronic exposure to these toxins can result in severe liver damage, including cirrhosis and hepatocellular carcinoma (HCC), a prevalent form of liver cancer. By disrupting the liver's crucial role in detoxification, aflatoxins exacerbate these conditions. Furthermore, they weaken the immune system, rendering individuals more susceptible to infections and diseases, a particularly concerning issue for immunocompromised individuals. Children exposed to aflatoxins may suffer from stunted growth, impaired development, and malnutrition due to interference with nutrient absorption and utilization [3]. Moreover, aflatoxins are potent mutagens, causing DNA damage and elevating the risk of cancer, especially HCC, leading the International Agency for Research on Cancer (IARC) to classify them as group 1 carcinogens. High doses of aflatoxins can induce acute toxicity, presenting symptoms such as vomiting, diarrhea, abdominal pain, and, in severe cases, death [3, 4].

In the realm of animal health, aflatoxins similarly wield adverse effects, notably reducing livestock productivity. Animals consuming aflatoxin-contaminated feed often exhibit diminished growth rates, decreased milk production, and impaired reproductive performance, with poultry and dairy animals particularly vulnerable. Moreover, aflatoxins compromise animals' immune systems, heightening their susceptibility to infections and secondary diseases. Chronic exposure culminates in mycotoxicosis, encompassing various health issues in animals such as liver damage, reproductive disorders, and overall health impairment. Furthermore, consumption of meat and milk from animals fed contaminated feed exposes humans to aflatoxins, amplifying health risks [5, 6]. To counteract these perils, regulatory measures and prevention strategies are imperative. Countries worldwide have established maximum permissible limits for aflatoxins in diverse foodstuffs to mitigate health risks. Implementing good agricultural practices, including proper crop management,


DOI: 10.1201/9781003487043-34

storage, and drying techniques, serves to minimize aflatoxin contamination. Researchers explore biological control methods, including biocontrol agents and competitive exclusion, as safe means of combating aflatoxigenic fungi. Effective post-harvest interventions, such as sorting, cleaning, and utilizing aflatoxin binders, further mitigate exposure. Additionally, education and awareness initiatives targeting farmers, processors, and consumers are crucial for fostering prevention efforts [7, 8].

The urgent need for rapid detection methods for aflatoxigenic fungi arises from their severe threat to food safety and public health. Aflatoxins, potent carcinogens produced by fungi like *Aspergillus* species, pervade various agricultural products, posing grave risks to human and animal consumers alike. Exposure to aflatoxins can lead to acute toxicity, chronic liver damage, immune suppression, and heightened cancer susceptibility, highlighting the criticality of timely detection to mitigate these health hazards [3]. Beyond health concerns, aflatoxin contamination incurs substantial economic losses for stakeholders across the food supply chain. Rapid detection methods offer a preemptive solution by enabling early identification and removal of contaminated food products, averting extensive recalls and market rejections. Compliance with stringent regulatory limits on aflatoxin levels is also facilitated by rapid detection, ensuring adherence to safety standards and regulatory requirements [9]. The importance of rapid detection methods extends to the preservation of food security. Aflatoxin contamination compromises food availability and quality, threatening the stability of food supplies. By swiftly identifying contaminated products, rapid detection methods help prevent the distribution and consumption of unsafe food, thereby safeguarding food security and ensuring access to safe and nutritious food for all [10].

Moreover, these methods play a pivotal role in facilitating international trade by enhancing transparency and compliance with aflatoxin regulations. Exporters can demonstrate adherence to safety standards, thereby unlocking trade opportunities and securing market access for agricultural products. Additionally, the portability and immediacy of rapid detection technologies enable on-site testing, expediting decision-making processes and reducing delays associated with traditional laboratory analyses [11]. A list of recent studies showing the mycotoxin contamination in different food crops is presented in Table 4.2.1.

In this chapter, we emphasize the crucial role of rapid detection methods for aflatoxigenic fungi in safeguarding public health, mitigating economic losses, ensuring food security, and promoting international trade. These methods offer advantages such as timely results and on-site testing, facilitating compliance with regulatory standards and enhancing consumer confidence. By emphasizing their significance, we aim to support the adoption of effective strategies for managing aflatoxin contamination and strengthening global food systems.

## CAUSES OF AFLATOXIN CONTAMINATION

Aflatoxin production is influenced by various environmental factors, with high temperature and moisture being key contributors. Elevated temperatures, typically above 25°C (77°F), provide optimal conditions for fungal growth and aflatoxin synthesis. Additionally, high levels of moisture, typically above 70%, promote fungal colonization and toxin production, particularly during pre- and post-harvest stages. Factors such as poor storage conditions, improper drying techniques, and water damage to crops exacerbate aflatoxin contamination, emphasizing the importance of moisture control in agricultural practices [2, 13, 14].

Specific fungal species, notably *A. flavus* and *A. parasiticus*, play a pivotal role in aflatoxin synthesis. These fungi are ubiquitous in the environment and commonly colonize crops such as maize, peanuts, and tree nuts. *A. flavus* is a primary producer of aflatoxins in warm and humid regions, whereas *A. parasiticus* is more prevalent in cooler climates. These fungi infect crops during growth or post-harvest stages, with aflatoxin production occurring under favorable conditions of temperature and moisture. Understanding the ecological niche and behavior of these fungal species is crucial for implementing effective strategies to prevent and mitigate aflatoxin contamination in agricultural products [2, 15, 16].

**TABLE 4.2.1**
**Some Recent Research Findings on Mycotoxin Contamination in Different Food Crops, Including the Crop Name, Location, Fungal Species, Type of Mycotoxin, and Its Concentration Range**

| S.No | Foodcrop | Country | Fungus Species | Type of Mycotoxin | Mycotoxin Concentration (ppb) | Ref |
|---|---|---|---|---|---|---|
| 1 | Corn | Serbia | *A. flavus, A. parasiticus* | Total aflatoxins | 1.01–86.10 | [12] |
| 2 | Wheat | Saudi Arabia | *A. flavus, A. parasiticus* | AFB1, AFB2, AFG1, AFG2 | 2.3 2.6, 1.3, 0.5 | |
| 3 | Maize kernel | Hungary | *A. flavus A. parasiticus* | AFB1 | 72.6–73.3 | |
| 4 | Red chili | Turkey | *A. flavus, A. parasiticus* | AFB1, AFB2, AFG1 | 0.24–165, 0.15–11.3, 0.15–3.88 | |
| 5 | Groundnut | Ethiopia | *A. flavus, A. parasiticus* | Total aflatoxins | 15–11,900 | |
| 6 | Foniomillet | Nigeria | *A. flavus* | AFB1, AFB2 | 233.2–692.0 | |
| 7 | Corn | India | *A. flavus* | AFB1 | 48–383 | |
| 8 | Sorghum | Ethiopia | *A. flavus* | AFB1, AFB2 | 29.5, 2.56 | |
| 9 | Rice | Nigeria | *A. flavus* | AFB1 | 37.26–113.20 | |
| 10 | Cashew nuts | Brazil | *A. flavus, A. parasiticus* | Total aflatoxins | 0.60–31.50 | |
| 11 | Madidi | Nigeria | *A. flavus, A. parasiticus* | AFB1, AFB2 | 0.2–125.60, 0.95–18.4 | |
| 12 | Rice | China | *A. flavus* | AFB1 | 0.03–20 | |
| 13 | Maize | Africa | *A. flavus* | AFB1 | 0.17–5.3 | |
| | Millet | | | | 0.14–6.4 | |
| | Sorghum | | | | 0.15–210.1 | |
| 14 | Guava | Egypt | *A. parasiticus* | AFB1, AFG1 | 0.163, 0.296 | |
| 15 | Dried Fruits | Pakistan | *A. flavus* | AFB21 | 0.04–9.80 | |

## EXISTING DETECTION METHODS

### Traditional Methods

#### Thin-Layer Chromatography (TLC)

Thin-layer chromatography (TLC) is a widely used technique for the separation and identification of aflatoxins, secondary metabolites produced by certain fungi, including aflatoxigenic species. In TLC, a thin layer of stationary phase, typically silica gel or alumina, is coated onto a flat support such as a glass plate. The sample containing aflatoxins is applied near the bottom of the plate, and the plate is then placed in a developing chamber containing a solvent system. As the solvent migrates up the plate by capillary action, it carries the components of the sample with it. Aflatoxins, being differentially adsorbed to the stationary phase, separate into distinct spots along the plate. Visualization techniques such as UV light or chemical staining can be used to detect the presence of aflatoxins [17].

## HIGH-PERFORMANCE LIQUID CHROMATOGRAPHY (HPLC)

High-performance liquid chromatography (HPLC) is a powerful technique for the quantitative analysis of aflatoxins in agricultural products. HPLC separates aflatoxins based on their differential interaction with a stationary phase packed inside a column, and their elution is controlled by a mobile phase (solvent). Detection is typically accomplished using a UV detector, which measures the absorbance of aflatoxins at specific wavelengths. HPLC offers high sensitivity and selectivity, making it suitable for routine analysis of aflatoxins in agricultural commodities [18, 19].

## MASS SPECTROMETRY

Mass spectrometry (MS) is a sensitive and highly selective technique used for the identification and quantification of aflatoxins in agricultural products. In MS, aflatoxins are ionized and then separated based on their mass-to-charge ratio (m/z) in a mass analyzer. The resulting mass spectrum provides information about the molecular weight and structure of aflatoxins, allowing for their unambiguous identification. MS can be coupled with chromatographic techniques such as HPLC (LC-MS) or gas chromatography (GC-MS) to enhance sensitivity and specificity [17, 18, 20].

## ENZYME-LINKED IMMUNOSORBENT ASSAY (ELISA)

ELISA is a rapid and sensitive immunoassay used for the qualitative and quantitative detection of aflatoxins in agricultural products. In ELISA, aflatoxins are captured by specific antibodies immobilized on a solid phase, such as a microplate. After washing away unbound components, the bound aflatoxins are detected using enzyme-conjugated secondary antibodies. The enzyme catalyzes a colorimetric reaction, and the intensity of the color is proportional to the concentration of aflatoxins present in the sample [17, 21, 22].

## ELECTROCHEMICAL IMMUNOSENSORS

Electrochemical immunosensors are emerging analytical tools for the detection of aflatoxins in agricultural products. These sensors typically consist of a transducer, such as an electrode, functionalized with specific antibodies for aflatoxins. When aflatoxins bind to the antibodies on the electrode surface, they induce a change in the electrical properties of the system, which can be measured and correlated with the concentration of aflatoxins present. Electrochemical immunosensors offer advantages such as rapid detection, high sensitivity, and potential for miniaturization and automation [23, 24].

## OPTICAL TECHNIQUES

### Fluorescence Spectroscopy (FS)

Fluorescence spectroscopy (FS) is a sensitive optical technique used for the detection and quantification of aflatoxins in agricultural products. Aflatoxins exhibit intrinsic fluorescence when excited by light of a specific wavelength. By measuring the intensity and wavelength of the emitted fluorescence, aflatoxins can be detected and quantified in complex matrices such as grains or nuts. FS offers advantages such as high sensitivity, rapid analysis, and potential for miniaturization for on-site testing [17, 25, 26].

## NEAR-INFRARED SPECTROSCOPY (NIRS)

Near-infrared spectroscopy (NIRS) is a non-destructive analytical technique used for the rapid analysis of aflatoxins in agricultural products. NIRS relies on the absorption of near-infrared light by chemical bonds in the sample, which provides information about the molecular composition of the sample. By analyzing the spectral data using chemometric methods, such as principal component analysis (PCA) or partial least squares regression (PLSR), the concentration of aflatoxins can be predicted in real-time without the need for sample preparation or reagents [27, 28].

## HYPERSPECTRAL IMAGING (HSI)

Hyperspectral imaging (HSI) combines imaging and spectroscopy to provide spatially resolved spectral information about agricultural products. By acquiring images at multiple wavelengths across the electromagnetic spectrum, HSI can identify and quantify aflatoxins based on their unique spectral signatures. Advanced image processing techniques, such as spectral unmixing or machine learning algorithms, can be used to extract quantitative information about aflatoxin contamination in real-time. HSI offers advantages such as high spatial resolution, rapid analysis, and the ability to detect multiple contaminants simultaneously [17, 29].

## RECENT ADVANCES IN RAPID DETECTION

Recent advances in rapid detection of aflatoxigenic fungal communities in agricultural products have been focused on leveraging various technologies to achieve quicker and more accurate identification. Here are some notable developments:

i. **DNA-based techniques:** Polymerase chain reaction (PCR) and real-time PCR (RT-PCR) have been widely employed for the rapid detection of aflatoxigenic fungi [30, 31]. These techniques allow for the specific amplification of fungal DNA, enabling quick identification even at low fungal concentrations. Additionally, techniques like loop-mediated isothermal amplification (LAMP) offer rapid and sensitive detection without the need for sophisticated equipment [32].
ii. **Biosensors:** Biosensors are emerging as promising tools for rapid detection of aflatoxigenic fungi in agricultural products. These devices integrate biological recognition elements with transducers to detect target analytes [1]. Biosensors offer advantages such as high sensitivity, rapid response, and potential for on-site testing, making them suitable for agricultural applications [30, 33].
iii. **Nanotechnology:** Nanotechnology-based approaches, such as nanomaterial-based sensors and nanobiosensors, have shown promise for the rapid detection of aflatoxigenic fungi. These approaches offer enhanced sensitivity and selectivity, as well as the potential for miniaturization and integration into portable devices for on-site testing [17, 34].
iv. **Machine learning and bioinformatics:** Integration of machine learning algorithms and bioinformatics tools with detection techniques enables automated data analysis and interpretation, improving the speed and accuracy of aflatoxin detection. These approaches can also facilitate the prediction of aflatoxin contamination based on environmental factors and fungal genomics data [17, 35, 36].

## MANAGEMENT STRATEGIES

The management of aflatoxin contamination in agricultural products necessitates a multifaceted approach. Key strategies include early detection systems, cultivation of resistant crop varieties, adoption of recommended farming practices, integrated pest management, robust monitoring and surveillance, improved storage management, regulatory compliance, and investment in research and innovation. By integrating these measures, stakeholders can effectively mitigate the risks associated with aflatoxin contamination, safeguarding both public health and economic interests in the agricultural sector [2].

## CONCLUSION AND FUTURE DIRECTIONS

The current landscape of aflatoxin detection methods offers a variety of technologies, including PCR assays, biosensors, immunoassays, and spectroscopic techniques, each with unique strengths and limitations. Although significant progress has been made in improving speed and accuracy, challenges remain, such as the need for more cost-effective and portable solutions, especially in resource-limited settings. Advancements in data analytics and machine learning show promise for enhancing efficiency and reliability through automated analysis. Moving forward, research should focus on developing novel detection techniques integrating multiple principles, miniaturizing and integrating devices for on-site testing, and fostering interdisciplinary collaboration to address aflatoxin management comprehensively. This approach aims to safeguard public health and ensure agricultural product safety globally.

## REFERENCES


1. Kumar A, Pathak H, Bhadauria S, Sudan J. Aflatoxin contamination in food crops: Causes, detection, and management: A review. Food Prod Process Nutr 2021;3:17. https://doi.org/10.1186/s43014-021-00064-y.
2. Shabeer S, Asad S, Jamal A, Ali A. Aflatoxin contamination, its impact and management strategies: An updated review. Toxins (Basel) 2022;14:307. https://doi.org/10.3390/toxins14050307.
3. G. Lizárraga-Paulín E, Moreno-Martínez E, P. Miranda-Castro S. Aflatoxins and their impact on human and animal health: An emerging problem. In Aflatoxins - Biochemistry and Molecular Biology, InTech; 2011. https://doi.org/10.5772/26196.
4. Eskola M, Kos G, Elliott CT, Hajšlová J, Mayar S, Krska R. Worldwide contamination of food-crops with mycotoxins: Validity of the widely cited 'FAO estimate' of 25%. Crit Rev Food Sci Nutr 2020;60: 2773–89. https://doi.org/10.1080/10408398.2019.1658570.
5. Richard JL. Some major mycotoxins and their mycotoxicoses—An overview. Int J Food Microbiol 2007;119:3–10. https://doi.org/10.1016/j.ijfoodmicro.2007.07.019.
6. Gallo P, Imbimbo S, Alvino S, Castellano V, Arace O, Soprano V, et al. Contamination by aflatoxins B/G in food and commodities imported in Southern Italy from 2017 to 2020: A risk-based evaluation. Toxins (Basel) 2021;13:368. https://doi.org/10.3390/toxins13060368.
7. Mahato DK, Lee KE, Kamle M, Devi S, Dewangan KN, Kumar P, et al. Aflatoxins in food and feed: An overview on prevalence, detection and control strategies. Front Microbiol 2019;10. https://doi.org/10.3389/fmicb.2019.02266.
8. Iqbal SZ, Selamat J, Ariño A. Mycotoxins in food and food products: Current status. In Food Safety, Cham: Springer International Publishing; 2016, p. 113–23. https://doi.org/10.1007/978-3-319-39253-0_6.
9. Mitchell NJ, Bowers E, Hurburgh C, Wu F. Potential economic losses to the US corn industry from aflatoxin contamination. Food Addit Contam Part A 2016;33:540–50. https://doi.org/10.1080/19440049.2016.1138545.
10. Ellis DI, Muhamadali H, Haughey SA, Elliott CT, Goodacre R. Point-and-shoot: Rapid quantitative detection methods for on-site food fraud analysis – Moving out of the laboratory and into the food supply chain. Anal Methods 2015;7:9401–14. https://doi.org/10.1039/C5AY02048D.
11. Turner PC, Sylla A, Gong YY, Diallo MS, Sutcliffe AE, Hall AJ, Wild CP. Aflatoxin impacts and potential solutions in agriculture, trade, and health: background paper for the PACA Strategy Development-Stakeholder Consultation Workshop. Partnership for Aflatoxin Control in Africa (PACA), African Union; 2012.
12. Kumar A, Pathak H, Bhadauria S, Sudan J. Aflatoxin contamination in food crops: Causes, detection, and management: A review. Food Prod Process Nutr 2021;3. https://doi.org/10.1186/s43014-021-00064-y.
13. Pitt JI, Hocking AD. Fungi and Food Spoilage, Boston, MA: Springer US; 2009, p. 1–2. https://doi.org/10.1007/978-0-387-92207-2_1.
14. Strosnider, H., Azziz-Baumgartner, E., Banziger, M., Bhat, R. V., Breiman, R., Brune, M. N., et al. (2006). Workgroup Report: Public Health Strategies for Reducing Aflatoxin Exposure in Developing Countries. *Environmental Health Perspectives*, 114(12), 1898–1903. https://doi.org/10.1289/ehp.9302.
15. Tai B, Chang J, Liu Y, Xing F. Recent progress of the effect of environmental factors on *Aspergillus flavus* growth and aflatoxins production on foods. Food Qual Saf 2020;4:21–8. https://doi.org/10.1093/fqsafe/fyz040.
16. Cary JW, Gilbert MK, Lebar MD, Majumdar R, Calvo AM. *Aspergillus flavus* secondary metabolites: More than just aflatoxins. Food Safety 2018;6:7–32. https://doi.org/10.14252/foodsafetyfscj.2017024.
17. Liu S, Jiang S, Yao Z, Liu M. Aflatoxin detection technologies: Recent advances and future prospects. Environ Sci Pollut Res 2023;30:79627–53. https://doi.org/10.1007/s11356-023-28110-x.
18. Chen M, Liu X, Yang S, Chen Z, Di B, Liu W, et al. HPLC–MS/MS method for the simultaneous determination of aflatoxins in blood: Toxicokinetics of aflatoxin B1 and aflatoxin M1 in rats. J Anal Sci Technol 2022;13:27. https://doi.org/10.1186/s40543-022-00336-3.
19. Hamed AM, Moreno-González D, García-Campaña AM, Gámiz-Gracia L. Determination of aflatoxins in yogurt by dispersive Liquid–Liquid microextraction and HPLC with photo-induced fluorescence detection. Food Anal Methods 2017;10:516–21. https://doi.org/10.1007/s12161-016-0611-6.

20. Zhang K, Banerjee K. A review: Sample preparation and chromatographic technologies for detection of aflatoxins in foods. Toxins (Basel) 2020;12:539. https://doi.org/10.3390/toxins12090539.
21. Tudorache M, Bala C. Sensitive aflatoxin B1 determination using a magnetic particles-based enzyme-linked immunosorbent assay. Sensors 2008;8:7571–80. https://doi.org/10.3390/s8127571.
22. Espinosa-Calderón A, Miguel C-ML, Francisco M-HR, Roberto M-AJ, Gerardo GGR, Torres-Pacheco I. 7 Methods for Detection and Quantification of Aflatoxins. n.d.
23. Azri F, Selamat J, Sukor R. Electrochemical immunosensor for the detection of aflatoxin B1 in palm kernel cake and feed samples. Sensors 2017;17:2776. https://doi.org/10.3390/s17122776.
24. Paniel N, Radoi A, Marty J-L. Development of an electrochemical biosensor for the detection of aflatoxin M1 in milk. Sensors 2010;10:9439–48. https://doi.org/10.3390/s101009439.
25. Chandraleka T, Abinayaa S, Rathour A, Mahalakshmi A, Heema C. Detection of aflatoxin in food products using UV fluorescence spectroscopy. Int J Innov Technol Explor Eng 2020;9:1068–71. https://doi.org/10.35940/ijitee.E2272.039520.
26. Hruska Z, Yao H, Kincaid R, Brown R, Cleveland T, Bhatnagar D. Fluorescence excitation–Emission features of aflatoxin and related secondary metabolites and their application for rapid detection of mycotoxins. Food Bioproc Tech 2014;7:1195–201. https://doi.org/10.1007/s11947-014-1265-2.
27. Yao W, Liu R, Zhang F, Li S, Huang X, Guo H, et al. Detecting aflatoxin B1 in peanuts by Fourier transform near-infrared transmission and diffuse reflection spectroscopy. Molecules 2022;27:6294. https://doi.org/10.3390/molecules27196294.
28. Zhang J, Xu B, Wang Z, Cheng F. Application of hyperspectral imaging in the detection of aflatoxin B1 on corn seed. J Food Meas Charact 2022;16:448–60. https://doi.org/10.1007/s11694-021-01171-7.
29. Chakraborty SK, Mahanti NK, Mansuri SM, Tripathi MK, Kotwaliwale N, Jayas DS. Non-destructive classification and prediction of aflatoxin-B1 concentration in maize kernels using Vis–NIR (400–1000 nm) hyperspectral imaging. J Food Sci Technol 2021;58:437–50. https://doi.org/10.1007/s13197-020-04552-w.
30. Bintvihok A, Treebonmuang S, Srisakwattana K, Nuanchun W, Patthanachai K, Usawang S. A rapid and sensitive detection of aflatoxin-producing fungus using an optimized polymerase chain reaction (PCR). Toxicol Res 2016;32:81–7. https://doi.org/10.5487/TR.2016.32.1.081.
31. Douksouna Y, Kwallah AO, Nyerere A, Runo S, Ambang Z. Application and evaluation of the loop-mediated isothermal amplification assay for the detection of aflatoxigenic fungi contaminants of rice grains in Kenya. J Plant Pathol Microbiol 2020;11. https://doi.org/10.35248/2157-7471.20.11.491.
32. Mellikeche W, Ricelli A, Casini G, Gallo M, Baser N, Colelli G, et al. Development of loop-mediated isothermal amplification (LAMP) assays for the rapid detection of toxigenic *Aspergillus flavus* and *A. carbonarius* in nuts. Int J Mol Sci 2024;25:3809. https://doi.org/10.3390/ijms25073809.
33. Sun X, Yan L, Tang Y, Zhang Y. A rapid and specific immunosensor for the detection of aflatoxigenic *Aspergilli*. Eur Food Res Technol 2012;234:1013–21. https://doi.org/10.1007/s00217-012-1716-9.
34. Yan C, Wang Q, Yang Q, Wu W. Recent advances in aflatoxins detection based on nanomaterials. Nanomaterials 2020;10:1–19. https://doi.org/10.3390/nano10091626.
35. Camardo LM, Mazzoni M, Battilani P. Machine learning for predicting mycotoxin occurrence in maize. Front Microbiol 2021;12. https://doi.org/10.3389/fmicb.2021.661132.
36. Kim Y-K, Qin J, Baek I, Lee K-M, Kim S-Y, Kim S, et al. Detection of aflatoxins in ground maize using a compact and automated Raman spectroscopy system with machine learning. Curr Res Food Sci 2023;7:100647. https://doi.org/10.1016/j.crfs.2023.100647.

# 4.3 Detection of Aflatoxin by Surface-Enhanced Raman Spectroscopy (SERS)

*José Silvestre Mendoza-Figueroa, Juan Salvador Acevedo-Gil, and Janet Jan-Roblero*

## INTRODUCTION

Raman spectroscopy is a physical phenomenon that measures the inelastic scattering of photons and was first described in 1928 by Sir Chandrasekhara Venkata Raman [1]. Raman scattering occurs during the interaction of photons with molecules, in which the incident light can be scattered in elastic and inelastic ways. During the elastic scattering, the incident light is scattered with the same energy without inducing energy transitions in the molecular bonds of the molecule, whereas during inelastic scattering, the incident light is scattered with lower or higher energy compared to the incident radiation. This occurs because the incident photons induce energy transitions in the molecular bonds from basal states to virtual states, with consequent molecular vibrations (Figure 4.3.1). In the first case, the incident photons cause the molecular bonds to pass from a basal energy state to a virtual state. Part of the incident photon energy is absorbed in the molecular bonds and produces vibration in the bonds (vibrational energy). The remaining photon energy is then scattered at a lower frequency (longer wavelength), which is called Raman Stokes. In the second case, if the molecule is

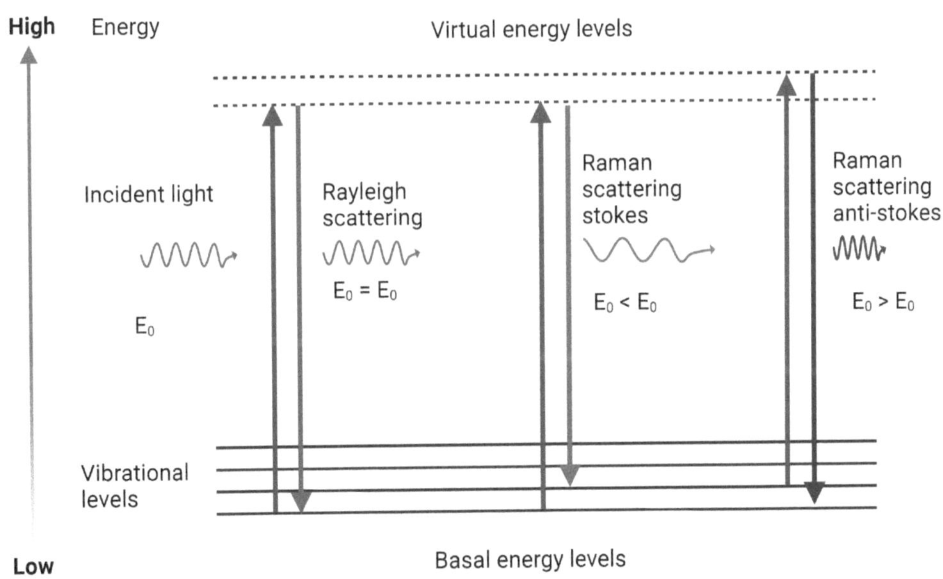


**FIGURE 4.3.1** Raman scattering process. The incident light (excitation radiation) with energy $E_0$ can be scattered in an elastic (Rayleigh) or inelastic (Raman Stokes and anti-Stokes) process. Stokes shows low frequency and longer wavelength, whereas anti-Stokes presents higher frequency and shorter wavelength.

DOI: 10.1201/9781003487043-35

already presented in an excited energy state, the energy of incident photons will be absorbed in the molecular bond, which also produces bond vibration, but the scattered energy of the photons will be higher than the original incident radiation. This indicates a higher frequency (shorter wavelength) and is called anti-Stokes [2, 3]. Standard Raman instruments measure Stokes frequencies.

Raman spectroscopy has been widely used as an analytical method for the detection of biogenic analytes; however, the traditional Raman scattering technique is not enough sensible for the detection of biogenic analytes in complex biological matrices. One of the most important variations of Raman spectroscopy is the technique called *surface-enhanced Raman scattering* (SERS), which improves the Raman signal intensity up to approximately $10^{14}$ times [4]. This method is based on the use of a nanostructured analytical substrate (rigid, flexible, or liquid matrix that contains nanostructures such as nanoparticles) or a rough surface to enhance the signal. These structures serve as a platform through which the analyte of interest (chemical compound, antibody, toxin, bacteria, etc.) is added to interact on it. SERS phenomena are achieved when the analyte is located within the three-dimensional space formed by the nanostructure architecture. These sites with high energy are called "hot spots" and should not be longer than 5 nm [5, 6]. The signal enhancement by the surface structure can be explained by two main mechanisms: (1) chemical enhancement, which is derived from charge transfer during the interaction of analytes and nanostructures in the substrate (hot spots), and (2) electromagnetic enhancement, which is the main mechanism accounting for metallic nanostructures in the substrate. This results from the resonance of the substrate plasmon and the incident radiation wavelength. Figure 4.3.2 shows a schematic representation of SERS.

Another addition to SERS is called the SERS label or SERS tag, which consists of adding molecules that contain functional groups with vibrational modes that are highly active in Raman spectroscopy. These labels result in a well-defined spectrum and are used to detect and quantify molecules whose molecular vibrations are not active in Raman spectroscopy or whose concentration is too low for detection. These label molecules bind to metal nanoparticles such as gold; examples include p-mercaptobenzoic acid (MBA) and 5′-dithiobis-2-nitrobenzoic acid (DTNB).

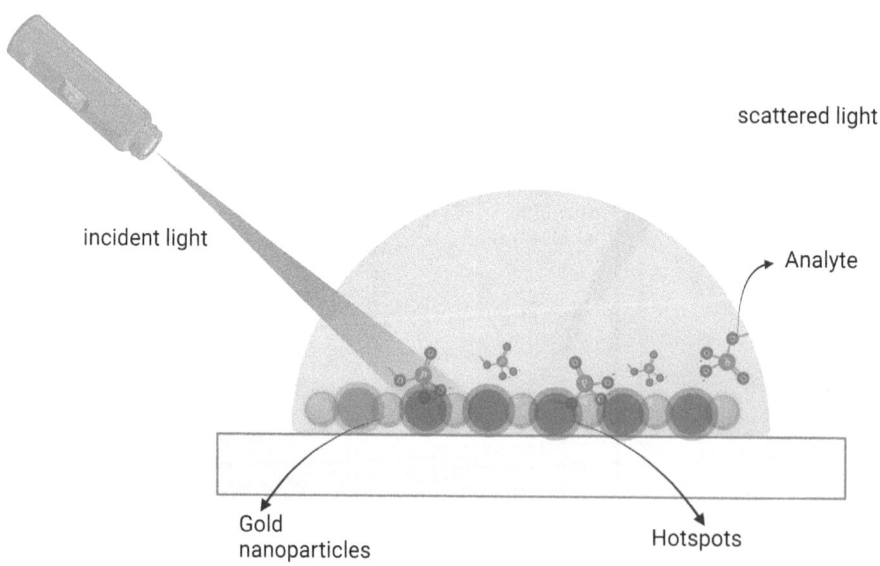


**FIGURE 4.3.2** Schematic representation of the SERS effect using a nanostructured substrate. Hot spots are shown in red and represent the concentrated energy sites between nanoparticles. When analyte is located in the hot spot regions, the Raman signal is boosted due to the charge transfer between nanoparticles and analyte or due to the electromagnetic enhancement.

## METALLIC NANOPARTICLES FOR FABRICATION OF SERS SUBSTRATES

Metallic nanoparticles such as gold nanoparticles (AuNP) and silver nanoparticles (AgNP) have been widely explored for the fabrication of SERS substrates. The main reason for this interest is the localized surface plasmon resonance (LSPR), which is an intrinsic feature of each nanoparticle (see Figure 4.3.3). LSPR is formed when electrons localized on the surface of metallic nanoparticles interact with light, inducing the oscillation of the electrons in phase with the incident electromagnetic radiation and creating a new localized electromagnetic wavelength around the nanoparticles. This electromagnetic wavelength results in a visible nanoparticle solution color, which is measurable as a visible absorbance spectrum. LSPR depends on the shape, size, and aggregation state of the nanoparticles. In the case of SERS, it is important that the analytical substrate exhibit an LSPR spectrum with absorption maxima close to or similar to the excitation laser wavelength to promote the resonance between the excitation laser and the substrate to create the electromagnetic enhancement of Raman intensity. The shape and size of nanoparticles will directly affect the LSPR curve (Figure 4.3.3); for example, isotropic nanoparticles such as spheres, show a simple LSPR curve due to the spherical shape. However, anisotropic nanoparticles such as nanorods show a complex LSPR curve because the two resonance modes (transverse and longitudinal modes) are derived from their morphology. The existence of multiple resonance modes in the anisotropic nanostructures allows better tuning of the LSPR of the SERS substrate [7].

## FABRICATION OF SERS SUBSTRATES

There are several reports showing the fabrication of SERS substrates based on nanoparticles. The most frequently used methods are covalent attachment of nanoparticles on glass surfaces, electrochemical formation of nanoparticles, spraying of nanoparticles on surface, high coverage-controlled deposition of nanoparticles on surfaces (glass, flexible polymers), and nanolithography. Although each method has its own advantages and disadvantages, we consider those methods related to the detection of mycotoxins. The main points of discussion concerning use on a specific substrate are (1) reproducibility of

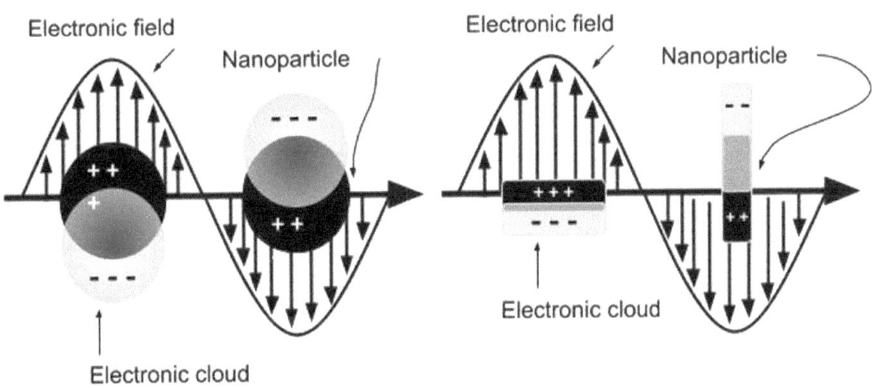


**FIGURE 4.3.3** Localized surface plasmon resonance (LSPR) of a spherical nanoparticle and a rod nanoparticle, showing the longitudinal LSPR and the transverse LSPR. LSPR is formed when electrons from the surface of gold nanoparticles are displaced along the metallic surface due to the interaction with an external electromagnetic field. Then this moving electron cloud forms a localized electromagnetic field with a wavelength depending on the shape and size of the nanoparticle and is measurable as an absorption spectrum in the visible region. In the case of spherical particles, an absorption spectrum curve with a single maximum is observed, but in the case of anisotropic nanoparticles such as nanorods, two maxima are observed in the absorption spectrum, derived from the longitudinal and transverse shape of the rod. The wavelength position of the maximum absorption values is called a plasmon.

the fabrication, (2) cost, (3) analytical sensitivity (nowadays the detection limit using SERS can reach the femtomolar units of an analyte in solution) [8, 9], and (4) the flexibility of application in different chemical environments (e.g., solvents, aqueous solutions, biological matrices).

## Detection of Mycotoxins Using SERS-based Lateral Flow Sensors

SERS can be coupled to lateral flow-based sensors. The SERS lateral flow sensor (strips) is based on the absorption of liquid by capillary flow through a paper strip. The paper strips contain two marked analytical areas. One area is for the test analyte (T), which contains a specific test (probe) for the analyte of interest, such as specific antibodies or DNA aptamers. The other area is a control area (C) to validate the assay. If the liquid containing the analyte passes through the T region, the analyte is captured and stopped for detection. If the analyte is not in the liquid, the liquid will continue to flow across the strip to the C region, and there will be no signal in the T region. Depending on the SERS setup, detection of the analyte in the T zone can be based on color change or coupled with a spectroscopic technique to enhance detection and even make it quantitative.

Zhang et al. [10] reported the design and use of SERS lateral flow paper strips for the detection of six different mycotoxins: aflatoxin B1 (AFB1), zearalenone (ZEA), deoxynivalenol (DON), fumonisin B1 (FB1), ochratoxin A (OTA), and T-2 toxin (T2). This system consisted of a SERS substrate based on core-shell Au@Ag nanoparticles. These Au@Ag core-shell nanostructures are fabricated using gold nanoparticles as the core and then coated with a silver layer as the shell (Figure 4.3.4a). The core-shell nanoparticles contain the SERS label p-mercaptobenzoic acid (pMBA) as a signal for Raman spectroscopy, adsorbed on the surface of the gold core nanoparticle for the detection of the toxins ZEA, DON, and T2 (group 1). Later, the specific antibodies for the toxins ZEA, DON, and T2 were immobilized separately on 52-nm Au@AgNPs to capture these mycotoxins present in a liquid

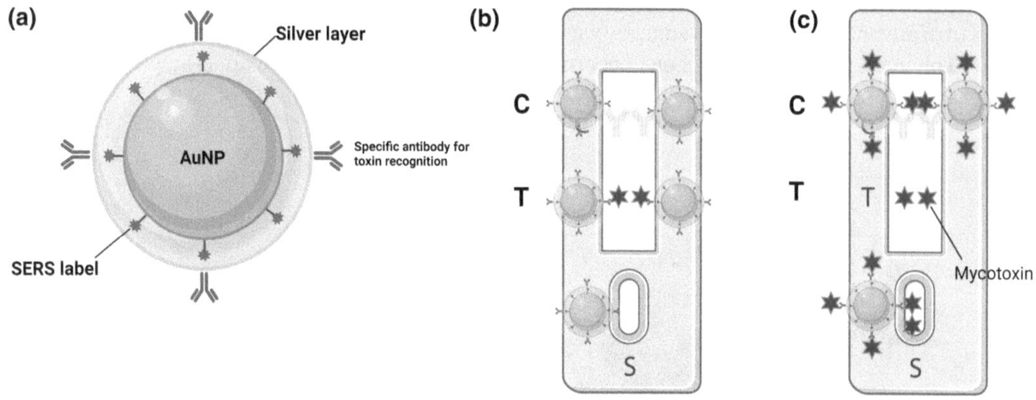


**FIGURE 4.3.4** Schematic representation of SERS based on a lateral flow system using core-shell metallic nanoparticles. (a) Representation of an Au@Ag core-shell metallic nanoparticle SERS probes. On the surface of a gold nanoparticle, a Raman label like p-mercaptobenzoic acid is adsorbed; then a thin layer of silver covers the gold nanoparticle surface. The metallic interface between the metals serves as a hot spot for enhancing the Raman signal of the label molecule. Then the surface of the silver shell is coated with specific recognition antibodies for mycotoxins. In (b) and (c) the coupling of SERS probes in a lateral flow strip is illustrated. In the test line (T), the mycotoxin is absorbed or immobilized as a capture antigen. Then, in the control line (C), an antibody is immobilized to capture the antibodies in the SERS probe to ensure lateral flow has been achieved. (b) A sample without mycotoxin (negative sample) is represented. If the sample is negative for the antibodies loaded in the SERS probe (a), then the SERS probe will be captured in the T line (interaction between antibodies and capture antigen). Once the T line is analyzed with a Raman spectrometer, it is expected that the maximum intensity of the Raman peak in the label molecule will be seen. On the other hand, in (c), if the sample contains mycotoxin, then the SERS probe will capture the mycotoxin in the sample, and it will not be able to interact with the immobilized capture antigen in the T line. Then, the Raman spectrum of the label molecule will decrease in intensity in proportion to the increase in mycotoxin concentration. This effect is called off-on.

or in a sample and thus integrate them into the SERS substrate system. Meanwhile, the specific antibodies to AFB1, FB1, and OTA (group 2) were separately immobilized on 91-nm Au@AgNPs and contain the SERS label DTNB as a detection signal for Raman spectroscopy. The test lines (T) in the strip consist of the different mycotoxins linked to bovine serum albumin (BSA). The complex BSA mycotoxin was dispersed in three test lines in the order of the six mycotoxins described here. The first test line includes BSA-AFB1 and BSA-ZEA, the second test line contains BSA-FBA and BSA-DON, and the third test line contains BSA-OTA and BSA-T2. In each test line, it is possible to detect one mycotoxin with the SERS system labeled with pMBA or DTNB (for example, in test line 1, AFB1 is detected with the Raman label DTNB and ZEA is detected with the Raman label pMBA) allowing the discrimination and quantification of the six mycotoxins. The procedure for toxin detection was as follows: A solution containing the six toxins was mixed with the SERS probes (Au@AgNPs+Ab+ Raman label). This mixture was placed in a microtiter well, and the absorbent pad of the strip was immersed in the mixture and allowed to flow through the strip. After lateral flow, Raman spectra were recorded over the three lines of the strip using a handheld Raman instrument equipped with a 785-nm laser line. The interpretation of the detection was that the Raman peak of pMBA located at 1589 $cm^{-1}$ in the first, second, and third lines detected the mycotoxins ZEA, DON, and T2, respectively. Likewise, the Raman peak of DTNB located at 1332 $cm^{-1}$ detected AFB1, FB1, and OTA in their respective lines of the strip. As the concentration of the mycotoxin in the sample increased, the signal intensity of the Raman peaks decreased because the SERS probes cannot be coupled by the antigen capture in the test line. With this assay, the six mycotoxins were detected qualitatively and quantitatively. The application of this system was also used in food, and it was observed that the response of the sensor strip regarding the concentration of the mycotoxin and the detection limit on spiked corn was recorded between 0.9 and 15 pg/mL of the toxins.

Another report by Yin et al. [11] shows the use of a SERS-based flow strip for the detection of ZEA. In a manner similar to the previously described studies, these authors use SERS probe detection based on the principle of an internal chemical standard in a core-shell Au@Ag (Figure 4.3.4a). They used two antibodies, one that specifically recognizes ZEA (anti-ZEA antibody) and another that recognizes bovine serum albumin (anti-BSA antibody), in addition to two antigens, ZEA bound to BSA (ZEA-BSA) and BSA protein, which were placed in the test line and control line of the strip, respectively. To fabricate the SERS system, a 31-nm core of AuNPs labeled with p-mercaptobenzoic acid (MBA) was first coated with a 9-nm silver shell. The AuMBA@Ag was then functionalized with the anti-ZEA antibody and finally coated with the BSA protein to block empty spaces on the nanoparticle surface, avoid interaction with other molecules, and as a target for specific antibodies to BSA present on the control line of the strip. The reactive strip consisted of two reactive lines: T and C. The test line was coated with the BSA-ZEA antigen, and the control line was coated with specific antibodies to recognize BSA to capture the SERS system substrate probe (Figures 4.3.4b and c). The operation of this strip consists of incubating the absorbent pad of the strip with a solution of ZEA or test sample together with the SERS detection system AuMBA@AgNP. Once the reaction mixture is absorbed, if the sample contains ZEA, it will be recognized by the antibody specific for ZEA attached to the AuMBA@AgNP and having the AuMBA@AgNP+AntiZEA-ZEA system. When the mixture starts to flow through the paper strip and reaches the T-line (containing the BSA-ZEA antigen), the AuMBA@AgNP+AntiZEA-ZEA SERS system will not be successfully attached to the BSA-ZEA capture system because the anti-ZEA antibodies are already occupied by the ZEA present in the sample. Therefore, the nano-system will continue to follow until the C line, where the AuMBA@AgNP+AntiZEA-ZEA SERS system will be captured by the anti-BSA antibodies that recognize the BSA protein present in the SERS system.

In the case of a negative ZEA sample, the SERS probe is completely captured in the T line through the interaction of BSA-ZEA molecules with the antiZEA antibody attached to the system AuMBA@AgNP. Once the lateral flow occurs, the Raman spectra are recorded with the 785-nm laser line directly on the T line. The maximum spectra intensity of the MBA probe in the T line indicates negative results because all the SERS systems are captured in the T line; as the amount of ZEA increases in the sample, the intensity of MBA spectra in the T line is reduced because the amount of captured SERS probes decreases. The detection limit of this system is reported to be 3.6 μg/Kg of a corn sample, which is below the tolerable limit reported for ZEA in corn (60 μg/Kg). A schematic representation of the core-shell SERS system in a lateral flow strip is illustrated in Figure 4.3.4.

Similar to the previous description, Chen et al. [12] described a lateral flow SERS system based on core-shell nanoparticles, but these authors show the use of Au@$SiO_2$ nanoparticles as SERS substrates. This system is also called SHINERS (*shell-isolated nanoparticle-enhanced Raman spectroscopy*), which refers to the use of metallic nanoparticles as the core and a thin coating film of silicon oxide as the shell. In this way, gold nanoparticles are protected against aggregation and maintain the electromagnetic enhancement from the gold or silver core [13]. The aim of this work was the detection of AFB1 and OTA mycotoxins. Briefly, the design includes two SHINER systems for the detection of both mycotoxins in the same lateral flow strips, consisting of a core of 60-nm AuNPs for Raman probe adsorption. The Raman labels MBA and DTNB were used for detection of AFB1 and OTA, respectively. A thin layer of $SiO_2$ was then coated on the AuNPs core, followed by functionalization of a silicon oxide shell with 3-triethoxysilylpropyl succinic anhydride (TEPSA) and a further reaction with the NHS/EDC system for antibody conjugation (anti-AFB1, anti-OTA, and immunoglobulin G [IgG]). Two working lines, T and C, were created for the design of the lateral flow strip, with equal proportions of BSA-AFB1 and BSA-OTA in the T line. During the detection test, the sample was mixed with SHINERS probes and placed on an absorbent pad for lateral flow, similar to the previously described work. If toxins are present in the sample, they are captured in the SHINERS by its specific antibody, and once they get to the T lines, they are not able to interact with the toxins linked in this line. If the toxins are missing in the sample, the SHINERS will capture the immobilized toxins in the T lines and will remain in the T line, allowing maximum detection of the Raman probe signals. The detection limit reported in this report is 0.24 pg/mL for AFB1 and 0.37 pg/mL for OTA and the test was able to detect mycotoxin in corn, rice, and wheat samples.

Recently, Limei et al. [14] reported the combination of magnetic nanoparticles with a lateral flow system for the detection of AFB1 and ZEA. The principle of this assay was to design a nanocomposite of iron oxide nanoparticles combined with a core-shell Au@Ag NP to concentrate the sample and improve the SERS detection in the lateral flow system, respectively. These authors fabricated iron oxide nanoparticles that were then capped with polyethyleneimine (PEI). The PEI gives a positive charge on the iron nanoparticles' surface. Separately, AuNPs were synthesized, and then MBA was adsorbed on the gold surface as a Raman label. After that, a thin layer of silver shell was deposited on the AuNPMBA system; the net charge of the surface of AuNPMBA@Ag was negative. This feature was used to attach the AuNPMBA@Ag on the modified $Fe_3O_4$@PEI particles because PEI contains multiple positive charges under certain pH. Then, in transmission electron microscopy (TEM), a structure with the appearance of a flower was observed, where the core of the nanocomposite was the iron oxide nanoparticles and around it, smaller Au@Ag nanoparticles decorated the surface. The idea of placing Au@Ag nanoparticles close to each other was to promote the formation of SERS hot spots. Then, the lateral flow assay was carried out. As a first step, the sample was concentrated using the magnetic property of iron nanoparticles. Briefly, the sample containing the mycotoxin was incubated with the magnetic nanocomposite and was then concentrated by pulling down with a magnet. The supernatant was removed, the pellet was resuspended in running buffer, and then deposited on the absorbent pad of the lateral flow strip. The principle of detection was as described in the previous report; the capture antigen was bound to the T line, and the antibodies for AFB1 and ZEA were bound on the surface of $Fe_3O_4$@AuNPMBA@Ag. If the sample contains mycotoxins, it should be captured by the specific antibodies once the SERS probes reach the T line. If the SERS probe is not able to bind to the capture antigen fixed on the T line, then the Raman signal of the MBA is decreased to increase the mycotoxin concentration in the sample. If the sample is missing toxins, then all SERS probes will be captured by the immobilized antigen in the T line, given the maximum Raman signal intensity of the MBA. The detection limit reported for this system was 1 0.1–10 μg/kg for AFB and 4–400 μg/kg for ZEA in corn samples. Table 4.3.1 summarizes the SERS-based lateral flow systems used in these three example reports.

**TABLE 4.3.1**
**Examples of SERS-based Lateral Flow Systems for the Mycotoxin Detection**

| Mycotoxin(s) | SERS System | Raman Probe | Raman Detection Mode | Detection Limit | Ref. |
|---|---|---|---|---|---|
| AFB1, ZEA, DON, FB1, OTA, and T2 | Core@shell (AuNP@Ag) | MBA for ZEA, DON and T2<br>DTNB for AFB1, FB1 and OTA | Off-On | 0.9 to 15 pg/mL | 10 |
| ZEA | Core@shell (AuNP@Ag) | MBA | Off-On | 3.6 μg/Kg | 11 |
| AFB1 and OTA | SHINERS ($Au@SiO_2$) | MBA for AFB1<br>DTNB for OTA | Off-On | AFB1 0.24 pg/mL<br>OTA 0.37 pg/mL | 12 |
| AFB1<br>ZEA | $Fe_3O_4$@ AuNP@Ag | MBA | Off-On | AFB1 0.1–10 μg/kg<br>ZEA 4–400 μg/kg | 14 |

*Notes:* Aflatoxin B1 (AFB1), zearalenone (ZEA), deoxynivalenol (DON), fumonisin B1 (FB1), ochratoxin A (OTA), and T-2 toxin (T2). MBA: p-mercaptobenzoic acid, DTNB: 5′-dithiobis-2-nitrobenzoic acid. The term off-on refers to the increase in signal of the Raman peak as the concentration of the analyte decreases.

## DETECTION OF MYCOTOXINS USING SERS-BASED MAGNETIC SUBSTRATES

In the report by Ko et al. [15], the use of magnetic-based substrates coupled with gold nanoparticles for the detection of AFB1 toxin is described. The authors design gold nano-hollows as SERS probes. The gold surface is coated with malachite green isothiocyanate as a Raman label. Then the modified gold hollows are coated with silicon oxide in a protocol similar to that referenced in [12]. This SERS probe is named *silica-encapsulated hollow gold nanoparticles* (SEHGNs). On the surface of the silica, a capture antibody for AFB1 was immobilized. Then, a secondary capture system was designed using magnetic beads. On the surface of these magnetic beads, another capture antibody for AFB1 was immobilized. To perform the SERS detection, first the sample containing AFB1 was incubated with the magnetic bead, and the AFB1 toxin was captured with the anti-AFB1 in the magnetic system. Then the SEHGNs system is added, and the capture antibodies immobilized on the SEHGNs will also recognize the mycotoxin already captured by the magnetic bead, forming a "sandwich-like structure." This magnetic complex is separated from the rest of the sample matrix using a magnet and then washed three times to remove non-reacting SEHGNs and impurities. If the sample is negative for AFB1, the SEHGNs will not interact with the magnetic beads, and it will not be able to detect the SERS signal of the Raman label. The magnetic pellet is then placed on the Raman instrument and irradiated with a 633-nm laser line. A positive result is shown by capturing the Raman spectra of malachite green isothiocyanate and is increasing in reason to increase the concentration of AFB1 in the sample; a negative sample will not record the spectra of malachite green. Interferences have not been observed with other common mycotoxins common in barley like fumonisin B ( FMB) and OTA, showing high selective detection to AFB1. The reported detection limit for this system was 0.1 ng/mL.

## SERS APTASENSORS FOR MYCOTOXIN DETECTION

Aptamers are short sequences of ssDNA or ssRNA that have been preselected for specific recognition of a molecule; they bind to a molecular target like carbohydrates, small molecules, toxins, proteins, and peptides [16]. Sensors based on aptamers are called aptasensors; the use of aptamers coupled with a nanoparticle core for SERS detection of mycotoxins is described here. Zhao et al. [17] described a protocol for a single-step dual detection of OTA and AFB1 using the specific DNA aptamers of each mycotoxin. The SERS probe is based on a core-shell of Au@Ag nanoparticles, where the AuNPs were first used as a template for adsorption of Raman labels 4-aminothiophenol (4-ATP) and 4-nitrophenol (4-NTP) for the detection of OTA and AFB1, respectively. The capture system consists of a magnetic nanobead that carries the complementary ssDNA of each aptamer. The detection process is as follows: If the sample does not contain any toxins, then the aptamer will complement the ssDNA capture sequence in the magnetic bead; therefore, the SERS probes will be bound in the magnetic bead, and the Raman signal of the 4-ATP and A-NTP will show the maximum SERS intensity spectrum of Raman labels. On the other hand, if the sample contains toxins, the toxins will be bound to the aptamer in the SERS probe system, blocking the capability of the aptamers to hybridize their complementary ssDNA on the magnetic bead. Then the Raman signal of the analyzed Raman labels will be reduced in proportion to the toxin concentration. The detection limit of this system is reported to be 0.006 ng/mL for OTA and 0.03 ng/mL for AFB1 in cornmeal.

A similar strategy was reported for the detection of ZEN and OTA by Chen et al. [18], where they used an Au@Ag core-shell system as a Raman probe. The Raman labels adsorbed on the gold core were MBA as a probe for OTA and DTNB as a probe for ZEA. The DNA aptamer specific for the mycotoxin was attached to the surface of the silver shell. In contrast to previously described work, the complementary DNA was absorbed on gold nanorods. Following the same principle of detection, if the sample lacks the mycotoxins, then the aptamers will be hybridized by the complementary DNA on the nanorods, and the Raman signal will be exhibited at maximum. On the other hand, if the concentration of mycotoxin increases, the aptamer will capture the toxin and will not be able to complement the complementary ssDNA on the nanorods; then the Raman signal of MBA or DTNB will decrease in proportion to an increased concentration of mycotoxins.

The shape of gold nanoparticles used as SERS probes directly affects the sensitivity of the SERS detection limit. In a study similar to [17], Chen et al. [19] reported the use of a core-shell Au@Ag NP, where the core Au particle is a gold nanorod, as compared with the traditional spheres used in various reports described in this chapter. In summary, this author shows that the AFB1 detection can increases 10-fold using gold nanorods, giving a limit of detection value of 0.0036 ng/mL compared with the SERS probe based on spheres for the detection of AFB1; the SERS detection system was tested in peanut oil spiked with AFB1.

Table 4.3.2 summarizes of SERS-based magnetic and aptasensor design strategies for mycotoxin detection.

**TABLE 4.3.2**
**Mycotoxin Detection Using Magnetic Beads and Aptasensors**

| Mycotoxin (s) | SERS System | Raman Labeled Probe | Detection Limit |
|---|---|---|---|
| AFB1 | SEHGNs | green isothiocyanate | 0.1 ng/mL |
| AFB1 and OTA | Aptasensor (DNA) immobilized on Au@Ag<br>Magnetic beads with complementary ssDNA | 4-ATP for OTA<br>4-NTP for AFB1 | 0.006 ng/mL for OTA and 0.03 ng/mL for AFB1 |
| ZEA and OTA | Aptasensor (DNA) immobilized on Au@Ag spheres<br>Gold nanorods for immobilization of complementary ssDNA | MBA for OTA<br>DTNB for ZEA | 0.018 ng/mL |
| AFB1 | Aptasensor (DNA) immobilized on Au@Ag nanorods<br>Gold nanorods for immobilization of complementary ssDNA | | 0.0036 ng/mL |

*Notes:* Aflatoxin B1 (AFB1), zearalenone (ZEA), ochratoxin A (OTA), MBA: p-mercaptobenzoic acid, DTNB: 5′-dithiobis-2-nitrobenzoic acid, 4-ATP 4-aminothiophenol, 4-NTP 4-nitrophenol, (SEHGNs) silica-encapsulated hollow gold nanoparticles. The term off-on refers to increase signal of Raman peak as the concentration of analyte decreases.

## CONCLUSION

SERS spectroscopy has been tested as an analytical method for the detection of mycotoxins, including aflatoxin at trace levels, and other mycotoxins can be detected in cereals, corn meal, and edible oils within minutes. The combination of lateral flow systems and specific capture molecules allows specific and reliable detection and quantification of these molecules. The application of SERS can be achieved with portable Raman equipment and as an inspection method in the food industry.

## ACKNOWLEDGMENTS

The authors would like to acknowledge the financial support provided by Grants No. SIP20241031and SIP20250269 from the Instituto Politécnico Nacional (IPN) México, as well as COFAA and EDI, IPN fellowships, and support from SNII-CONAHCyT.

## REFERENCES

1. Smith, M., W. Karen, and L. Ashton. 2016. Raman spectroscopy: An evolving technique for live cell studies. Analyst 141: 3590–600. https://doi.org/10.1039/c6an00152a
2. Orlando, A., F. Franceschini, C. Muscas, S. Pidkova, M. Bartoli, M. Rovere, and A. Tagliaferro. 2021. A comprehensive review on Raman spectroscopy applications. Chemosensors 9: 262. https://doi.org/10.3390/chemosensors9090262
3. Robin J.R, C.H. Hooper, L. Zhang, D. Wolverson, and V.K. Valev. 2019. Raman techniques: Fundamentals and frontiers. Nanoscale Research Letters 14: 231. https://doi.org/10.1186/s11671-019-3039-2
4. Kneipp, K., W. Yang, K. Harald, L.T. Perelman, I. Itzkan, R.D. Ramachandra, and M.S. Feld. 1997. Single molecule detection using surface-enhanced Raman scattering (SERS). Physical Review Letters 78: 1667–70. https://doi.org/10.1103/PhysRevLett.78.1667
5. Ding, S.Y., E.M. You, Z.Q. Tian, and M. Moskovits. 2017. Electromagnetic theories of surface-enhanced Raman spectroscopy. Chemical Society Reviews 46(13): 4042–76. https://doi.org/10.1039/C7CS00238F
6. Kleinman, S.L., R.R. Frontiera, A.I. Henry, J. Dieringer, and R.P. Van Duyne. 2013. Creating, characterizing, and controlling chemistry with SERS hot spots. Physical Chemistry Chemical Physics 15: 21–36. https://doi.org/10.1039/C2CP42598J
7. Furong, T., F. Bonnier, A. Casey, A.E. Shanahan, and H.J. Byrne. 2014. Surface enhanced Raman scattering with gold nanoparticles: Effect of particle shape. Analytical Methods 6: 9116–23. https://doi.org/10.1039/C4AY02112F
8. Luo, S.C., K. Sivashanmugan, J.D. Liao, C.K. Yao, and H.C. Peng. 2014. Nanofabricated SERS-active substrates for single-molecule to virus detection in vitro: A review. Biosensors and Bioelectronics 61: 232–40. https://doi.org/10.1016/j.bios.2014.05.013
9. Betz, J.F., W.W. Yu, Y. Cheng, I.M. White, and G.W. Rubloff. 2014. Simple SERS substrates: Powerful, portable, and full of potential. Physical Chemistry Chemical Physics 16: 2224–39. https://doi.org/10.1039/C3CP53560F
10. Zhang, W., S. Tang, Y. Jin, C. Yang, L. He, J. Wang, and Y. Chen. 2020. Multiplex SERS-based lateral flow immunosensor for the detection of major mycotoxins in maize utilizing dual Raman labels and triple test lines. Journal of Hazardous Materials 393: 122348. https://doi.org/10.1016/j.jhazmat.2020.122348
11. Yin, L., T. You, H.R. El-Seedi, I.M. El-Garawani, Z. Guo, X. Zou, and J. Cai. 2022. Rapid and sensitive detection of zearalenone in corn using SERS-based lateral flow immunosensor. Food Chemistry 396: 133707. https://doi.org/10.1016/j.foodchem.2022.133707
12. Chen, R., H. Wang, C. Sun, Y. Zhao, Y. He, M. Shemyal Nisar, and W. Wei. 2023. Au@SiO2 SERS nanotags based lateral flow immunoassay for simultaneous detection of aflatoxin B1 and achratoxin A. Talanta 258: 124401. https://doi.org/10.1016/j.talanta.2023.124401
13. Li, J.F., Y.F. Huang, Y. Ding, Z.L. Yang, S.B. Li, X.S. Zhou, and F.R. Fan. 2010. Shell-isolated nanoparticle-enhanced Raman spectroscopy. Nature 464(7287): 392–5. https://doi.org/10.1038/nature08907
14. Limei, Y., J. Cai, L. Ma, T. You, M. Arslan, H. Jayan, X. Zou, and Y. Gong. 2024. Dual function of magnetic nanocomposites-based SERS lateral flow strip for simultaneous detection of aflatoxin B1 and zearalenone. Food Chemistry 446: 138817. https://doi.org/10.1016/j.foodchem.2024.138817

15. Ko, J., C. Lee, and J. Choo. 2015. Highly sensitive SERS-based immunoassay of aflatoxin B1 using silica-encapsulated hollow gold nanoparticles. Journal of Hazardous Materials 285: 11–17. https://doi.org/10.1016/j.jhazmat.2014.11.018
16. Ni, X., M. Castanares, A. Mukherjee, and S.E. Lupold. 2011. Nucleic acid aptamers: Clinical applications and promising new horizons. Current Medicinal Chemistry 18(27): 4206–14. https://doi.org/10.2174/092986711797189600
17. Zhao, Y., Y. Yaxin, L. Yaodong, Y. Xuan, M. Li, and Q. Song. 2015. Double detection of mycotoxins based on SERS labels embedded Ag@Au core–shell nanoparticles. ACS Applied Materials and Interfaces 7(39): 21780–86. https://doi.org/10.1021/acsami.5b07804
18. Chen, R., L. Shuang, S. Yunfeng, H. Bingyang, X. Yating, Q. Yingkai, and S. Li. 2021. Surface-enhanced Raman spectroscopy aptasensor for simultaneous determination of ochratoxin A and zearalenone using Au@Ag core-shell nanoparticles and gold nanorods. Microchimica Acta 188(8): 281. https://doi.org/10.1007/s00604-021-04919-6
19. Chen, Q., Y. Mingxiu, Y. Xiaojing, L. Huanhuan, G. Zhiming, and M.H. Rahma. 2018. A large Raman scattering cross-section molecular embedded SERS aptasensor for ultrasensitive aflatoxin B1 detection using CS-Fe3O4 for signal enrichment. Spectrochimica Acta Part A: Molecular and Biomolecular Spectroscopy 189: 147–53. https://doi.org/10.1016/j.saa.2017.08.029

# Index

**Note:** Page numbers in **bold** and *italics* refer to tables and figures, respectively.

## B

## H



## I



## K



## L

## Q



## R



## S

FSC
www.fsc.org
MIX
Papier | Fördert
gute Waldnutzung
FSC® C083411

Kolibri 360 GmbH
Ferdinand-Jühlke-Straße 7
99095 Erfurt, Deutschland
produktsicherheit@kolibri360.de